I0095712

THE SCIENCE
(volume 1)

- tENTATIVELY, a cONVENIENCE

Through the eye of a needle

in a haystack press

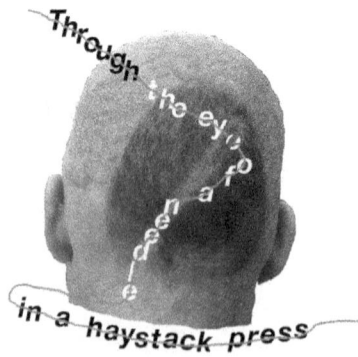

ISBN: 979-8-9871573-0-5

THE SCIENCE: Science Books Springboard-Reviewed:

Introduction
by tENTATIVELY, a cONVENIENCE - October 12, 2022

1st, *there's no such thing as* **THE SCIENCE** -

by that I don't mean that there's no such thing as SCIENCE, I mean that there's no such thing as a monolithic scientific community opinion in wch everyone's in agreement. Scientists, after all, are people 1st & foremost & have the same foibles & varieties that other people do. Even if there's an agreement about what constitutes 'scientific method' & even if an application of that method to a particular subject yields similar results from a multitude of sources it doesn't mean that there can't be underpinning problems w/ axioms &/or distortions of data based on ulterior motives, etc, etc..

During the time of the so-called COVID-19 pandemic it's been my observation that people have been going along w/ medical propaganda based on their subculturally approved 'news'feeds that purport to represent THE SCIENCE despite there being substantially adversarial opinions amongst scientists. In the liberal circles that I'm most directly exposed to, anyone who espouses an opinion that goes against the grain of what's packaged as liberal science is dismissed as a right-wing bone-head.

Since I don't think that politicizing science in that way does any good whatsoever toward helping to understand, I've tried to educate myself somewhat using a variety of medical science bks (mainly, if not entirely, for the lay reader) from sources whose philosophy I may or may not agree w/. This approach, I hope, helped me keep a more open mind & kept me aware of my biases.

In general, I just wanted to educate myself about medical science since the people who wd potentially argue w/ me asking for THE SCIENCE usually aren't 1. intellectuals, 2. scientists, 3. people capable of any significant research. This bk enables me to explain what my studies have yielded so far. Anyone who might want to argue w/ me can read this bk instead & save me the time & energy of otherwise engaging w/ what I feel is endless combatative ignorance.

My most sensationalist take-away from this reading is reinforcement for opinions that've been coming to the fore in recent yrs: viz., that medical science has replaced the Inquisition as a means of dominating & instilling fear into the 'little people' (meaning the vast majority). Just as the Inquisition robbed, kidnapped, tortured, & murdered people for 'the good of their souls' & in accordance w/ 'God's will' so does medical science bully & intimidate people 'for their own good' by inducing a constant fear of impending death. I share many of the same opinions w/ Ivan Illich (author of **Medical Nemesis**) - esp the opinion that *iatrogenesis* is responsible for more deaths than the medical establishment is ever likely to admit to. This latter opinion was further reinforced by reading Dr. Edgar March Crookshank's **History and Pathology of Vaccination - Volume 1**.

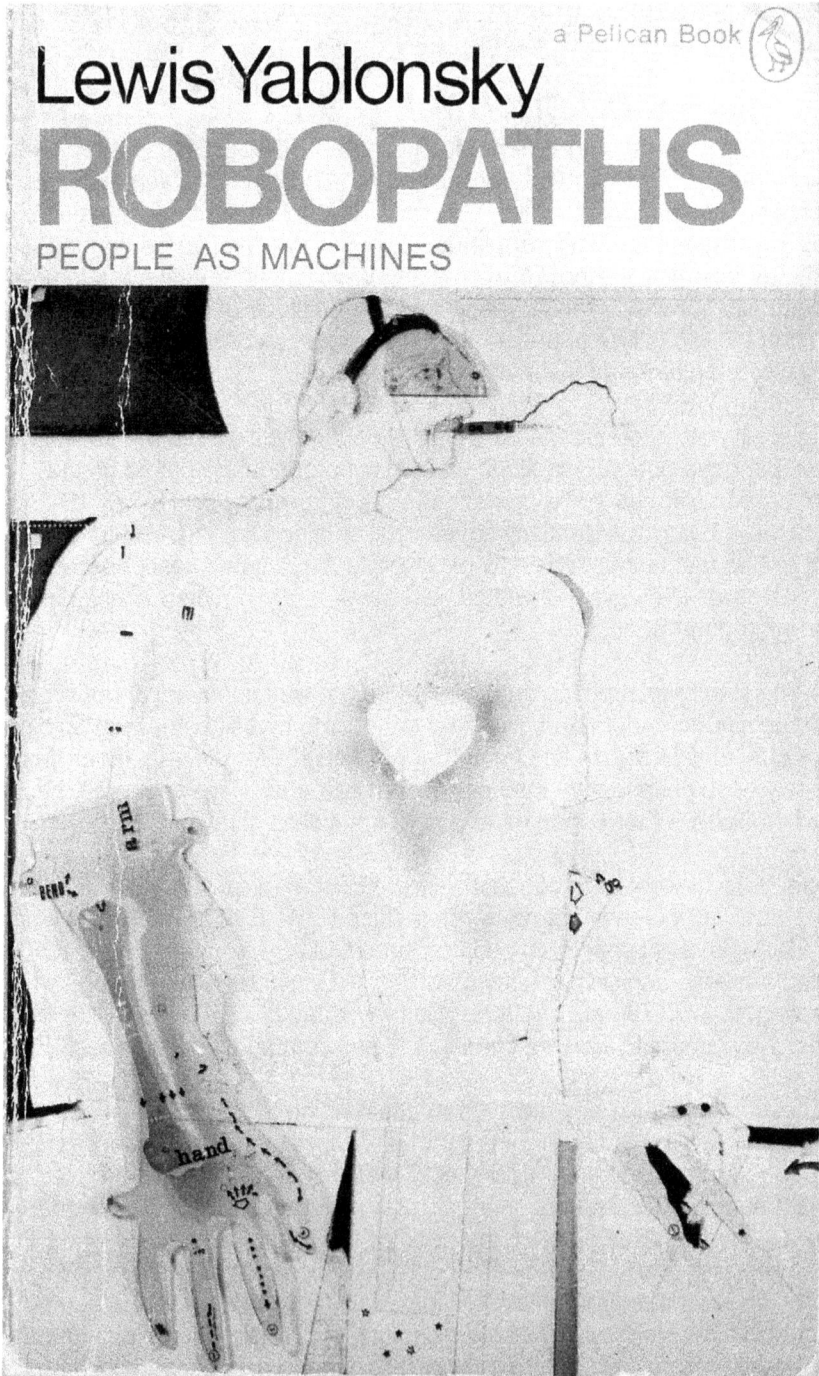

a Pelican Book

Lewis Yablonsky

ROBOPATHS

PEOPLE AS MACHINES

review of
Lewis Yablonsky's Robopaths
by tENTATIVELY, a cONVENIENCE - March 14, 2008

This isn't, necessarily, a GREAT bk. Yet, I give it a 5 star rating & recommend it to everyone. Published in 1972 when I was 18 & 19, this describes the world I grew up in as perfectly as anything I've ever read. The filmic companion to it cd be Peter Watkins' "Punishment Park". I'll be making a short movie called "Robopaths" wch excerpts text from the bk. [*May 1, 2014 interpolation: I actually made a feature-length movie (1:48:20) that I finished in May, 2012. More info can be found about that by looking at entry 369 here*: http://idioideo.pleintekst.nl/ tENTMoviesIndex.html . *Despite at least 5 tries to screen it somewhere I've been unsuccessful as of 2 yrs later.*] Below are a few of those quotes from early in the bk:

Paradoxically, although it is increasingly a distinct possibility, the final outcome of people versus their technological robots may not be the total physical annihilation of people. People may in a subtle fashion become robot-like in their interaction and become human robots or robopaths. This more insidious conclusion to the present course of action would be the silent disappearance of human interaction. In another kind of death, social death, people would be oppressively locked into robot-like interaction in human groups that had become social machines. In this context, the apocalypse would come in the form of people mouthing ahuman, regimented platitudes on a meaningless dead stage.

The relationship between potential social death and imminent megamachine wars that cause physical death is complex. A fact that can not be ignored is that it is after all the masses of people who ultimately permit their energies and financial resources to be heavily spent on ecologically suicidal technology and doomsday machines. If a majority of people in a society permit, or desire, this condition to exist they must be relatively devoid of compassion and humanistic values; or, to take a more charitable view, they have become so out of touch with reality, and have become so powerless, that they no longer exert any control over their elected acompassionate robopathic leaders.

Whatever the reasons, the people in power are actually developing the technological machinery for "a world wired for death," and a majority of people in contemporary societies are socially dead, living a day-to-day robopathic existence.

- page xiii, Robopaths - People As Machines: Preface, Lewis Yablonsky, 1972

Robopaths enact ritualistic behavior patterns in the context of precisely defined and accepted norms and rules. Robopaths have a limited ability to be spontaneous, to be creative, to change direction, or to modify their behavior in terms of new conditions. They are comfortable with the all-encompassing social machine definitions for behavior. Even the robopath's most emotional behavior is ritualistic and programmed. Sex, violence, hostility, recreation are all preplanned, pre-packaged activities, and robopaths respond on cue. The frequency, quality, and duration of most robopaths' behavior is predetermined by societal definition.

- page 7, Robopaths - People As Machines: Robopaths, Lewis Yablonsky, 1972

In a robopathic-producing social machine, conformity is a virtue. New or different behavior is viewed as strange and bizarre. "Freaks" are feared. Originality is suspect.

- page 8, Robopaths - People As Machines: Robopaths, Lewis Yablonsky, 1972

As a child a strong attempt was made to impose a completely robopathic regimen onto me: I was expected to mow the lawn regardless of whether the grass had grown to the height of the cutting blades, I wasn't allowed to sit on the furniture in the living room, there was a certain routine for putting butter on bread that was to be strictly followed, it went on & on. Naturally, I was in trouble a fair amt.

Yablonsky differentiates between sociopaths & robopaths by explaining that sociopaths commit their victimizations outside of the rules of society & that robopaths commit them w/in. B/c of this latter, no matter how heinous the effects of a robopath's behavior, it's all well & good & sanctioned by society. The robopaths can even be self-righteous about it. War? Genocide? No problem. All approved by the robopathic society, the social machine.

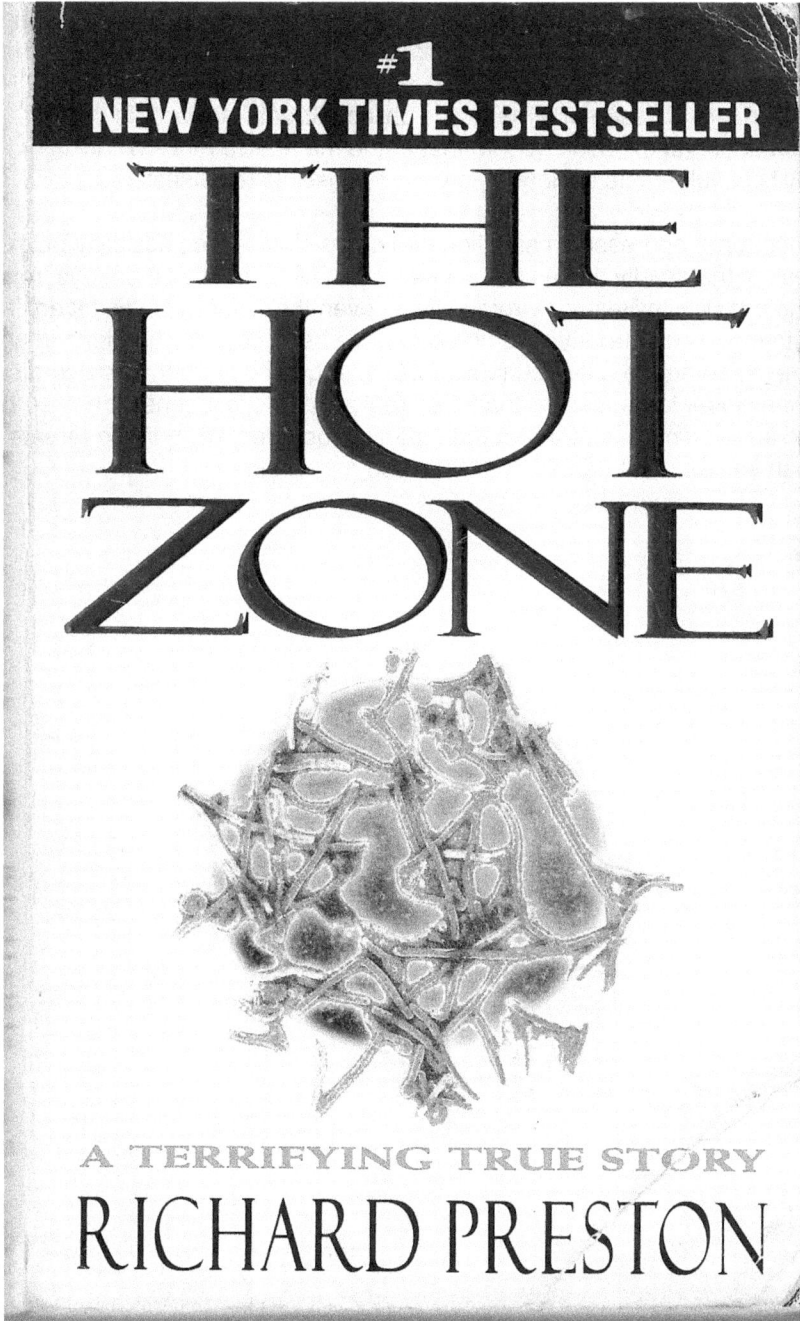

#1
NEW YORK TIMES BESTSELLER

THE HOT ZONE

A TERRIFYING TRUE STORY

RICHARD PRESTON

review of
Richard Preston's **The Hot Zone: The Terrifying True Story of the Origins of the Ebola Virus**
by tENTATIVELY, a cONVENIENCE - April 12, 2008

In 1989 I worked for a medical lab in BalTimOre in the electrical engineering department. I've written about some aspects of this in a RATicle called "Chemical & Biological Warfare Research at what was formerly known as Maryland Medical" in the 4th issue of a magazine called "Street Rat Bag". In this environment I had exposure to info that made me realize more vividly than I had already how vulnerable society-at-large is to possible lab disasters.

"The Hot Zone" addresses a situation that came close - the possibility of an outbreak of the deadly ebola virus caused by moving infected monkeys from Africa to a facility in Reston, Virginia. What even the authors of this bk don't know is that that particular facility is rumored to have been casually burglarized around that time by someone out of curiousity w/o ever getting caught. The eco-disaster was even closer to happening than they realized. This is a sensational(ist) bk but it's also an even more dire threat than many people realize. & these threats are in an urban environment nearer YOU than you may realize.

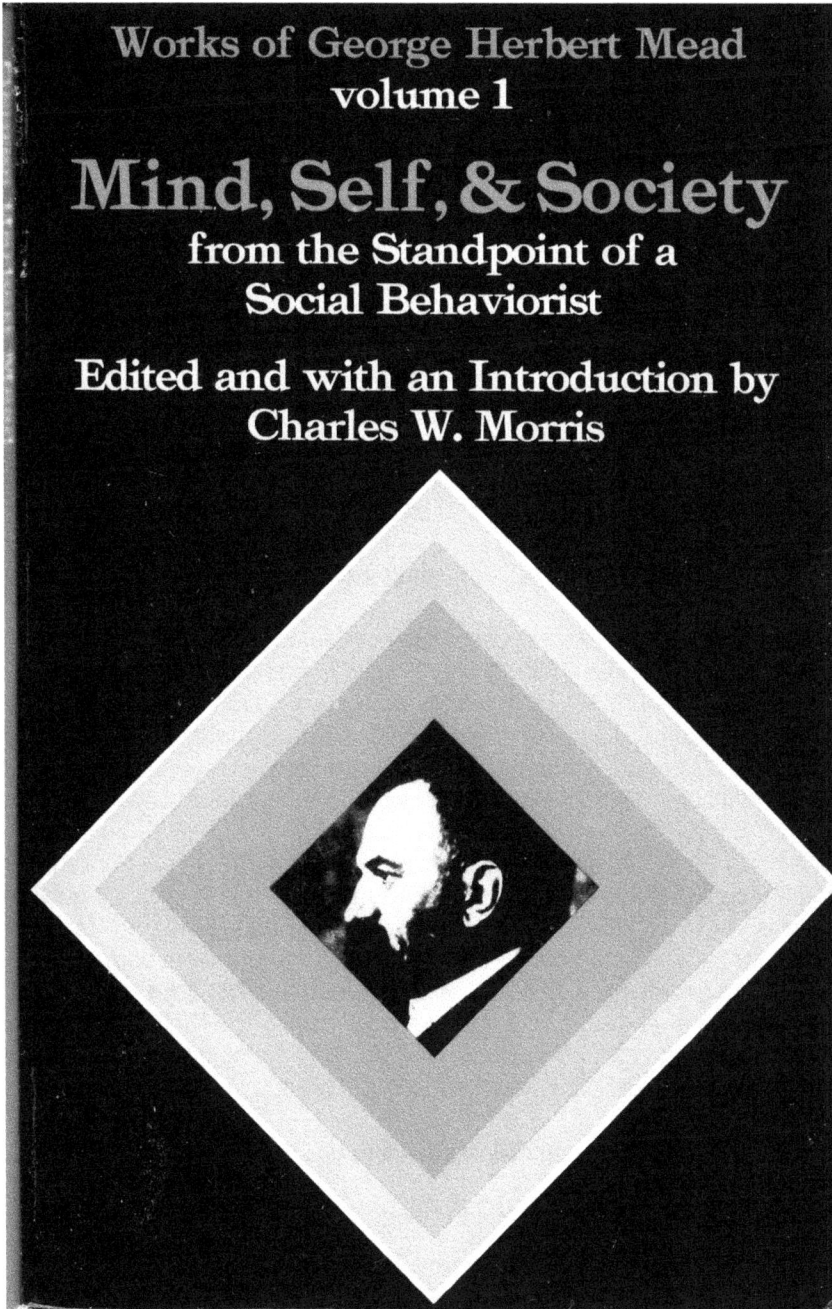

review of
Works of George Herbert Mead - volume 1
- Mind, Self, & Society - from the Standpoint of a Social
Behaviorist
- Edited and with an Introduction by Charles W. Morris
- by tENTATIVELY, a cONVENIENCE - August 16-24, 2015

"The Generalized Other don't know SHIT!"

The genesis of my reading this bk may interest some. In December of 2000 I rc'vd a letter from a man named Detlev Hjuler from Flensburg, Germany. This letter contained a want list of rare recordings of avant-garde music + a catalog of things that Hjuler published himself. My days of being a prompt replier were long since behind me & I didn't answer. Hjuler was very persistent & we finally started corresponding. He bought 23 tapes that I publish (here's the link to my tape company website: http://idioideo.pleintekst.nl/WdmUindex.html - usually somewhat out-of-date these days) & we began trading.

Somewhere along the line I saw a list of Hjuler's record collection. It was very impressive. We started trading recordings. I'm usually very open to trading, my tape company isn't really a 'business' insofar as I usually lose money on it & have no aggressive commercial intentions, but I don't always like what I receive &, therefore, don't want to continue trading w/ that particular sender. That was the case here. By December, 2003, I stipulated that I wdn't trade w/ Hjuler any more.

Hjuler goes by the name "Kommissar Hjuler", reputedly b/c he was a police detective. He was also what, for simplicity's sake, one might call an "Outsider Musician". While I found his taste in music to be very sophisticated I found his own performances to be unbearably primitive. Still, given that I'm an anarchist & that he was a policeman & that these 2 types are usually in opposition to each other I found it somewhat fascinating that we shared similar musical interests.

12 yrs later, in 2015, Hjuler got in touch w/ me again b/c he's now publishing records & wants to publish work by Franz Kamin that I had previously published. At 1st I was wary of this b/c I'd disliked Hjuler's publications from my 1st correspondence w/ him but he sent me samples & I found them somewhat interesting so I eventually agreed. One thing led to another & he put out a short piece of mine on a record w/ longer tracks by himself & the Nihilist Spasm Band. He also invited me to collaborate w/ him by doing something w/ a CD-R that he sent me that's somehow based on the ideas of Mead:

" This is my invitation to you to collaborate with us.

"10. (SHMF-019+...) - Collaboration Project:

"Kommissar Hjuler and Mama Baer run a project called (SHMF-019+...) for which any artists are allowed to create versions of the reading DIE ANTIZIPATION DES GENERALIZED OTHER. A tape, several CD-Rs and some LPs still have become released in this series on Der Schoene-Hjuler-Memorial-Fond. A list of all artists that have been committing you will find at file (SHMF), just see no. (SHMF-019) following!

"The Generalized Other refers to George Herbert Mead's psychological explanation for the origin of social self-consciousness. Within Mead's theory, is the act of 'role-taking' in which individuals react to social gestures, and adjust to common attitudes. Through 'role-taking', people adapt to social exchanges based on gesture-response action sequences. Self-consciousness is then developed through these social actions and completed upon personal reflection. This text is hard to handle for other artists, we now have given away quite a lot of free Audio-CD-Rs to other artists, but only few were able to work with our spoken text. It is a stumbling dialogue with reading parts and conversation parts and in the result we do by far not justice to the grandilocant or intellectual theme.

"Mainly artists and musicians from experimental music scene have contributed, but not at least, this project is to create a mix of most different music styles, one of the stranges contributions was by the Afro-French Urban-Rap- and Dub-musician LO daam, who normally creates music far from any experimental scene, and the crazy version by the dark metal band HELLMOUTH from Rotterdam.

"Artists and musicians, who are interested into creating their personal version for this project, could get a promotional Audio-CD-R with the spoken text, the versions sent back become released on our label, the artists will get some free copies. Especially artists from very different music scenes are invited for their contribution, also film works or collages and paintings as limited prints are possible, it need not be the medium music, anything goes.

"Several more collaboration works like the mail-collaboration between Rudolf Eb.er of Schimpfluch Group (SHMF - Eb.er), LP in limited edition, re-issued as a CD by Blossoming Noise/USA in edition of 1000 copies, or the experimental smalltalk with Juergen O. Olbrich of NO-Institue/Paper Police (SHMF - 155), CD-R in limited edition, which is also a set for other anti-live acts, are possible."

I wasn't previously familiar w/ Mead or, if I was, I'd forgotten about him. I wasn't necessarily interested in the collaboration at 1st but "The Generalized Other refers to George Herbert Mead's psychological explanation for the origin of social self-consciousness. Within Mead's theory, is the act of 'role-taking' in which individuals react to social gestures, and adjust to common attitudes." resonated w/ me b/c one of the anarchist Street Rat slogans that I use is "Evict the Ruling Elites from your Mental Real Estate!" - the idea being that mind control is largely accomplished by behavior modification mass media techniques that colonize people's thought processes & bring them in-line w/ ruling elite interests that're particularly harmful to impoverished free thinkers.

SO, I decided to read a bk by Mead that explores this idea of the "Generalized Other" & to write a review of it. The idea being to then record my reading the review & to send the txt & the recording to a German friend of mine in the Netherlands w/ the request that he either translate my English into multiple languages & then make a recording of it & send me back his translation(s) & recording &/or to do whatever else he might feel inclined to do if anything at all. In the meantime, I haven't listened to Hjuler's CD-R b/c I don't want it to bias my procedure. My plan being to then put my recording in one channel, my German friend's in the other, & to mix in the Hjuler material as the finishing touch. THEN, this is to be sent to Hjuler for possible publication, hopefully on vinyl rather than K7 or CD-R.

Mead was a "Social Behaviorist" as the title of the bk states. I've generally had a negative attitude toward Behaviorism b/c it seems to take a strictly mechanistic appraisal of human interaction w/ an eye toward being able to control behavior. For me, even if it were possible to reduce all processes to strict cause & effect sequences that *can be controlled*, wch I don't believe it is, it wdn't be a goal worth pursuing b/c the result wd be oppressively reductionist. Still, I decided to approach the bk w/ somewhat of an 'open mind' since I'm hardly an expert on Behaviorism, let alone psychology in general, & can, therefore, stand to learn much more.

1st off, I have to give credit to the compilers of this bk:

"The volume is in the main composed of two sets of excellent student notes on the course, together with excerpts from other such notes and selections from unpublished manuscripts left by Mr. Mead. A stenographic copy of the 1927 course in social psychology has been taken as basic. This set, together with a number of similar sets for other courses, owes its existence to the devotion and foresight of Mr. George Anagnos. Sensing as a student, the importance of the material of Mr. Mead's lectures (always delivered without notes), he found in Mr. Alvin Carus a sympathetic fellow-worker who was able to provide the means necessary to employ persons to take down verbatim the various courses." - p vi of Charles W. Morris's "Preface"

Having (a) student(s) pay (a) stenographer(s) to transcribe such a course is mind-boggling to me. It's very hard for me to imagine anyone doing anything nearly so *caring* or labor-intensive today. As such, I'm deeply impressed by the studiousness that went into making this bk. Then again, maybe these students were just rich enuf to hire people to take notes that they cd copy later rather than pay attn in class (or even attend?) - thusly doing the same-old-same-old thing that rich people usually do: take advantage of their privilege to give themselves the appearance of scholarliness they're actually lacking & to give themselves an unfair competitive edge. Whatever the circumstances, compiling this bk is an achievement.

On the other hand, I think the substances of Mead's ideas wd've been better served if Mead himself had organized them into carefully outlined & developed logical progressions of the type of 'I think 1. pertains & conclude that 2. follows logically' etc.. - rather than the somewhat tediously repetitive & meandering flow of the lectures - but Mead *didn't do that* so this is what the interested researcher gets.

I have no idea whether Mead really fits into the lineage suggested in the following but this is what Morris begins his "Introduction" w/: "Philosophically, Mead was a pragmatist; scientifically, he was a social psychologist. He belonged to an old tradition—the tradition of Aristotle, Descartes, Leibniz; of Russell, Whitehead, Dewey—which fails to see any sharp separation or any antagonism between the activities of science and philosophy, and whose members are themselves both scientists and philosophers." (p ix)

While I'm all in favor of ethics, I'm more relieved than convinced by the way Mead combines the 'cold' rationality of Behaviorism w/ the community-mindedness of his social values. Here's what Morris says: "The pragmatic reliance upon the experimental method, coupled with the moral and valuational relation of the movement to the democratic tradition, has resulted in a conception of philosophy as having a double concern with fact and value; and a conception of the contemporary moral problem as the redirection and reformulation of human goods in terms of the attitudes and results of the experimental method. Darwinism, the experimental method, and democracy are the headwaters of the pragmatic stream." (p x)

Morris gets me more interested in Social Psychology by posing its newness (in the early '20th century', ie): "The terms "social" and "psychologist" have not long appeared together, nor in company with biological categories, Tradition has identified psychology with the study of the individual self or mind. Even the post-Darwinian influence of biological concepts did not for a long time break up the inherited individualistic presuppositions (as is evidenced by a Huxley to find a place for moral behavior in the evolutionary process), though it did formulate the problem as to how the human mind appeared in the history of animal conduct." (p xii)

At 1st I thought "redintegration" was a typo meant to be "reintegration". Then I read it twice in the same paragraph & figured it for a term I don't know: "Mead in some places admits the facts of redintegration" & "one event leads at some organic center to the expectation of and redintegration of some other event." (p xiv) SO, I found these definitions to quote for those of you who're also unfamiliar w/ the word: "1 archaic : restoration to a former state, 2 a : revival of the whole of a previous mental state when a phase of it recurs, b : arousal of any response by a part of the complex of stimuli that originally aroused that response" (http://www.merriam-webster.com/dictionary/redintegration) "Evocation of a particular state of mind resulting from the recurrence of one of the elements that made up the original experience." (http://www.thefreedictionary.com/redintegration) What does redintegration have to do w/ the price of beans? Mead "feels that such processes do not come under the classification of "significant symbol" or "mind."" (p xiv)

Therefore, if I understand this correctly, wch I quite possibly don't, an element from a previous experience capable of stimulating some type of mental revival of sd experience is NOT a "significant symbol" & this redintegration (or reinstantiation?) is NOT a part of the "mind". Morris says: "it seems to me that he has shown that mind and the self are, without remainder, generated in a social process, and that he has for the first time isolated the mechanism of the genesis." (p xv) To wch I query: Is there, then, any process that *is not a social process* insofar as it's hypothetically 'impossible' for something to occur in a 'content vacuum'? &, given the possibility that all processes are social in the sense of non-isolated, is it then possible that a redintegration is 'inevitably' a social process that 'inevitably' generates significant symbols in a 'playing field' that can be accepted as a mind? Just sayin'. I mean I sure as shit don't 'know'.

"Mind was not to be reduced to non-mental behavior, but to be seen as a type of behavior genetically emerging out of non-mental types. Behaviorism accordingly meant for Mead not the denial of the private nor the neglect of consciousness, but the approach to all experience in terms of conduct." - p xvii

The notion of one's POV (Point-of-View) being something that prevents the possibility of objectivity or even any 'rational' basis for a belief in objectivity doesn't seem to bother Mead at all. Given the possibility that everything is interconnected &, therefore, centerless in terms of our own hypothetical subjectivity, when I was in my early 20s I posed the idea of "ogjectivity": a state that's neither objective or subjective, a state that's a hypothetically infinite flux of interpenetrating subjectivities that come as close to objectivity as we're likely to get. The idea being that solipsism is 'impossible' b/c, despite superficial appearances, we have no center, no fixed POV that can be the center of the universe (or multiverse). Whatever the case, I will most likely continue to act as if I believe there's a world outside me that there are desirable responses to - such as pleasurable engagement &/or self-protective evasion. I fully expect that no matter how expert I become at such responses my subjective center will eventually deteriorate & I will disintegrate in a very obvious way & reintegrate piecemeal into an environment wch no longer houses my POV.

"Certain of the radical behaviorists have frankly identified "I see *x*" with "my ocular muscles have contracted"; and have as frankly admitted that this identification leads into a behavioristic form of solipsism. Such a situation is simply the appearance in psychology of the logical and methodological scandal which has long harassed scientific thought: on the one hand science has prided itself upon being empirical, on bringing its most subtle theories to the test of observation; on the other hand science has tended to accept a metaphysics which regards the data of observation as subjective and mental and which denies that the objects studied have the characters which as experienced they appear to have." - p xviii

Is that really solipsism tho? It seems to me that it isn't b/c the notion that there are such things as "ocular muscles" implies a belief in physical reality outside of the POV.

"The individual must know what he is about; he himself, and not merely those who respond to him, must be able to interpret the meaning of his own gesture. Behavioristically, this is to say that the biological individual must be able to call out in himself the response his gesture calls out in the other, and then utilize the response of the other for the control of his own further conduct. Such gestures are significant symbols. Through their use the individual is "taking the role of the other" in the regulation of his own conduct." - p xxi

It's this feedback that generates mind, self, & society - making those nouns more processual than object-oriented although Mead uses words like "form" to, apparently, refer to people - returning them to object status, 'objectifying' them. What I want to know is: Are there, then, 'insignificant symbols'? Symbols that *don't* signify? I find Mead's position interesting & well-thought-out *except that* I can't really accept the notion of 'objectivity':

"Mind is the presence of behavior of significant symbols. It is the internalization within the individual of the social process of communication in which meaning emerges. It is the ability to indicate to one's self the response (and implicated objects) that one's gesture indicates to others, and to control the response in these terms. The significant gesture, itself a part of a social process, internalizes and makes available to the component biological individuals the meanings which have themselves emerged in the earlier, non-significant, stages of gestural communication. Instead of beginning with individual minds and working out to society, Mead starts with an objective social process and works inward through the importation of the social process of communication into the individual by the medium of the vocal gesture. The individual has then taken the social act into himself. Mind remains social; even in the inner forum so developed thought goes on by one's assuming the roles of others and controlling one's behavior in terms of such role-taking." - p xxii

I find Morris's summary above to be marvelously succinct & I appreciate Mead's working from the outside-in instead of the inside-out. However, I'm still not convinced that our *subjective* perceptual 'apparatus' can have *objective* data to work from no matter how we roll the die. Hence, I return to my admittedly fanciful 'ogjectivity': an infinite network of interpenetrating 'subjectivities' that are *all us* at the same time that *none of them are us exclusively*. These enable us to have multiple POVs & the more of these we have the closer we get to 'objectivity' w/o ever actually getting there.

"It is presumably the human cortex (whose place in the higher reflexes the reflexologists have made abundantly clear) and the temporal dimension of the nervous system (which allows the control of the gesture in terms of the consequences of making it) which permit the human animal alone to pass from the level of the conversation of gestures to that of the significant language symbol, and the absence of which prevent the talking birds from really talking. These two characteristics, coupled with the place of the human hand in the isolation of the physical object, are supposedly the organic bases which determine the biological differentiations of man and the animals." - p xxiii

I also always have a problem w/ 'scientific' differentiating humans from animals. Such reasoning usually smacks of speciesism, of creating a hierarchy that then gets used to justify acts of brutality. Remember that it wasn't so long ago that the notion of "subhumans" was used to justify Death Camps. Many other than me have drawn the parallel between slaughterhouses & Death Camps. I'm a meat eater & the meat I eat comes from slaughterhouses - as such, I'm not taking a more-moral-than-thou position, I, too, am culpable - but I don't want to delude myself w/ justifying ideology.

I consider myself to be an animal. If I don't succeed in communicating w/ non-humans I don't blame it on their lack of 'significant symbols' any more than I blame my own lack of communicative sensitivity. A dog recently barked at me, probably excited at the prospect of going for a walk. The barking was somewhat loud indoors so I yowled in response. The dog stopped barking & came over & licked me consolingly. To me, that was some form of communication. Mead is very good at making observations about non-humans & explaining convincingly why what THEY do is different from what WE do. But I'm still not convinced.

Morris further synopsizes Mead's position: "In so far as one can take the role of the other, he can, as it were, look back at himself from (respond to himself from) that perspective, and so become an object to himself. Thus again, it is only in a social process that selves, as distinct from biological organisms, can arise— selves as beings that have become conscious of themselves." (p xxiv) In my above story about the dog, weren't we both taking the role of the other? Or was only I taking the role of the dog while the dog was responding to an instinctually recognized cue? Perhaps that falls under this category: "In play the child simply assumes one role after another of persons and animals that have in some way or another entered into its life." (p xxiv) Note that the sexless pronoun "its" is used here instead of, eg, "his".

Given my bias against Behaviorism as a potential tool for anti-individualist control, I found this particularly interesting: "As Mead says of Dewey's views, "the individual is no thrall of society. He constitutes society as genuinely as society constitutes the individual." Indeed, every action of the individual at either the non-linguistic or linguistic levels of communication changes the social structure to some degree, slightly for the most part, greatly in the case of the genius and the leader." (p xxv)

While I find such a claim to be accurate insofar as it goes, I find it either astoundingly naive or disingenuous otherwise: some people are far more skilled in imposing their changes on the social structure than others & people w/ wealth have enormous advantages that people w/o it don't have. To act as if every person in a society is equally influential upon the social procedures that rule them is highly unrealistic. Even adding the qualification re geniuses & leaders is hardly sufficient insofar as both can exercise power over others w/ an ability that may be extremely unwelcome by those others. In the end, the implied 'fairness' is completely delusional.

This political naiveté (to give the authors the benefit of the doubt) seems further expressed by Morris here: "Under the penalty of stagnation, society cannot but be grateful for the changes which the moral act of the creative "I" introduces upon the social stage." (p xxvi) Wd that it were so. Instead I, personally, find society resistant to any new idea that doesn't just reinforce the status quo of the powers-that-be. Fashion changes are ok if it means that people feel compelled to buy, buy, BUY, but if the change were to internationally cap profits for arms dealers or pharmaceutical companies I don't think that the steerers of society wd be very "grateful for the changes".

"to the degree that the self assumes the role of the generalized other, its values are the values of the social process itself." (p xxxii) Wch social process? Wch generalized other? Isn't that going to be different for each individual? Thusly making the generalized other only 'universal' (to use a term Mead uses later) as a category rather than as a pin-pointable thing w/ specific characteristics? My own "generalized other" is far from homogeneous, there're parts to it that don't mesh, there're others whose manifestations I respect (a minority) & the much greater pile of others who're little more than a mob-disaster-waiting-for-an-excuse-to-turn-utterly-barbaric (the majority) & then still others that're a grey zone of possibilities. While there's definitely an 'other' that other is far from generalizable.

But Mead is thorough. In a footnote to Morris's intro it's stated that "Mead rather brusquely states that it is *not* the position that "the standard of morality is that which will do the most social good" (1927). Mead stresses the particular situation, not the vague and unmanageable utilitarian "society in general."" (p xxxiii) Mightn't it be just as accurate to say that any individual is going to act out of fear of reprisal from the mob unless they're truly courageous (or foolish)?

"The right act, as relative to the situation, is nevertheless objective and universal in that it demands the assent of all rational beings." (p xxxiii) Mead is a believer in rationality & democracy. While I can understand why those things wd be ideal from Mead's POV I find them highly tenuous.

"This society would not have as its goal the bare sustenance and attainment of any set of existent or authoritatively defined values" [..] "On the contrary its philosophy of history would be as experimental as the experimental method itself. It would be concerned with the technique for remaking values through the reinterpretation of the situation in terms of the best knowledge available, and that technique, it would appear, could be nothing but morality itself.

"Such a society of moral beings would seem to be Mead's version of the democratic ideal." - p xxxiii

I'm not exactly opposed to "Mead's version of the democratic ideal" as much as I'm pessimistic that "the best knowledge available" is reliably specified & that its application "could be nothing but morality itself". It seems to me that Nazism is as likely to result from a situation so described as Democracy.

"The religious attitude, based upon the pattern of helpfulness in family relations,"

Mead, at least as initially described by Morris, seems to largely prefer a more positive acceptance of what I might call "PR" versions of institutional purpose. Hence, he seems largely content to work w/ the idea of the "religious attitude" being "based upon the pattern of helpfulness in family relations". I, contrarily, see the religious attitude as largely fostering a mindset that accepts international mind control.

Furthermore, there's the notion of "the economic attitude of offering to others some surplus for what one himself needs". (Morris, p xxiv) "The economic development is one which starts off on the basis of exchange. You offer what you do not want in exchange for something which another does not want." (Mead, p 287) That's all well & good & if economics stuck to such a program life wd certainly be better than it is now. Alas, what we have instead is 'Free Trade' trying to enslave as many people as possible for the benefit of the most predatorial marketeers whose interests veer wildly from any notion of an equal surplus traded for an equal surplus. Forced inequality is the guiding 'light' of 'Free Trade'.

Mead seems to believe in a rational democracy founded on morality fostered by religion & fair trade fostered by economics - w/ Mutual Aid fostered by both. Instead we have military chaplains supposedly based on a morality of "Thou Shalt Not Kill" providing justification for mass murder & 'Free Traders' concerned w/ profit at the expense of all - their ideal business wd be one w/ minimal (or NO) outlay & maximal grist for their greed mill. Addicting sex slaves to heroin & taking all their pay is the perfect business model. That's hardly what Mead had in mind.

The religious & economic attitudes, Morris continues explaining, "are potentially universal, and language can extend as far as common activity extends. In this sense the capacity to take the role of the other in greater degree by more and more people would seem to move in the direction of the democratic ideal, provided that the selves become moral selves."* (p xxxiv)

*"In the non-moral sense of the term "social," wars and discord and disorganization are as social as their opposites. Mead's failure to stress the fact that the problem is one of getting moral selves, and not simply social selves, gives at times an impression of uncritical confidence in the future development of human society, even though at other times he is sufficiently sensitive to the socially disruptive aspects of behavior" [..]. "The pragmatist's emphasis upon education is the logical corollary of his ethical theory: education is to provide the technique by which moral selves—intelligent and socialized selves—are to be developed." - p xxxiv

SO, Mead's position is perhaps not as naive as I've been saying but he does seem to have a preference for idealizing democracy & education as incorruptible. I'm all for education but I also see it as susceptible to ulterior motive biases in the 'educators' that can entirely negate the education that Mead seems to favor.

A case in point: I was at a social gathering of educators in fairly highly placed administrative positions. They were discussing their perceived needs for keeping

out fundamentalist christians from the curriculum-determining processes. While I
was sympathetic to their concerns about having myopic special interests erode
what we all perceived as a more 'rational' education, I was still dissatisfied by *any*
elite undermining egalitarianism - even one I agreed w/.

Morris continues: "Such a democracy, as Mead clearly sees, has no undesirable
leveling tendency, and puts no premium on mediocrity. Rather it is compatible
with great differences of ability and contribution." - xxxiv

IMO, "leveling" is unlikely to happen, regardless. But that doesn't mean that what
we have is "From each according to his ability, to each according to his need" ("In
the Marxist view, such an arrangement will be made possible by the abundance
of goods and services that a developed communist system will produce; the idea
is that, with the full development of socialism and unfettered productive forces,
there will be enough to satisfy everyone's needs." - https://en.wikipedia.org/wiki/
From_each_according_to_his_ability,_to_each_according_to_his_need).
Instead what we have is privileged parasites attempting to control what the
"generalized other" is for the masses thru mass media in such a way that, as I
wrote previously, they can "occupy their mental real estate" by homogenizing
their mindset for easier manipulation. This might appear to be a form of
internationalism but it's about as internationalist as having a McDonald's
franchise w/in walking distance of every thousand people.

"The democratic society has no place for the superiority of class or possession or
power as such: it must cherish deeply the superiorities and pride in superiority
which arise in the performance of diverse social functions.

"What applies to individuals here applies to nations. Mead is an internationalist,
since the social attitude he describes can theoretically stop short of nothing less
than conscious identification with and participation in the society of man as such."
- p xxxiv

Lovely. Alas, this is theory & not praxis.

I shd probably interject at this point that I doubt that this bk represents the
currently most sophisticated thinking in psychology given that much of the
substance of it is from roughly 90 yrs ago. Added to this is my general ignorance
of psychology wch doesn't help me put Mead's ideas into the context he'd be
likely to be taught in. Therefore, my writing about this as if it's detached from a
specific time & place is bound to make any criticisms or comments of mine a bit
off the mark for any purposes other than my own idiosyncratic ones.

Finally I get to the transcription of Mead himself as I pass Morris's excellent
Preface & Introduction. "Social Psychology is especially interested in the effect
which the social group has in the determination of the experience and conduct of
the individual member." (p 1)

"There has been of late in philosophy a growing recognition of the importance of

James's insistence that a great deal has been placed in consciousness that must be returned to the so-called objective world." (p 4) That wd be *William* James, an influence on Gertrude Stein, something I understood better while reading this. Note that Mead says "so-called objective" & hearken back to my own "I'm still not convinced that our *subjective* perceptual 'apparatus' can have *objective* data to work from no matter how we roll the die."

"In social psychology we get at the social process from the inside as well as from the outside. Social psychology is behavioristic in the sense of starting off with an observable activity—the dynamic, on-going social process, and the social acts which are its component elements—to be studied and analyzed scientifically. But it is not behavioristic in the sense of ignoring the inner experience of the individual—the inner phase of that process or activity." - p 7

"The earlier conception of the central nervous system assumed that one could locate certain faculties of the mind in certain parts of the brain, but a study of the central nervous system did not reveal any such correlation." (p 21) Is that still the case 90 yrs later? It's my limited understanding that different parts of the brain may be able to provide back-up for each other in case of damage that there are still, nonetheless, areas that do the primary work for certain functions: the backbrain for autonomic functions, the visual cortex for sight, "Broca's brain" for language, etc..

A footnote on p 22 makes an interesting assertion: "We are always conscious of what we have done, never of doing it. We are always conscious only of sensory processes; hence we are conscious of motor processes only through sensory processes, which are their resultants. The contents of consciousness have, therefore, to be correlated with or fitted into a physiological system in dynamic terms, as processes going on." As a test of this I tried being conscious of reading the preceding while doing so. I succeeded. I do agree that "we are conscious of motor processes only through sensory processes" insofar as motor processes are autonomic.

"A person continues to live when he is under a general anesthetic. Consciousness leaves and consciousness returns, but the organism itself runs on. And the more completely one is able to state the psychological processes in terms of the central nervous system the less important does this consciousness become." - pp 27-28

Do processes even qualify as "psychological" when they're autonomic? Aren't they then just motor functions? reflexes? When I just now ran a Google search for the word "psychology" to see how it's commonly defined I got the following result: "Your search - psychology - did not match any documents." So, apparently, the autonomic Google search doesn't even recognize psychology. So there.

Mead emphasizes the "central nervous system" instead of 'the mind', apparently preferring the term as evocative of something physically observable. Can't they simply be synonymous? I reckon much of Mead's purpose was to explain how this cd be the case.

"There are characters which we want to separate, and it is here that the legitimacy of our parallelism lies, namely, in that distinction between the object as it can be determined, physically and physiologically, as common to all, and the experience which is peculiar to a particular organism, a particular person." (p 31) 'Til death do us part. Even tho Mead says "so-called objective" he quickly defaults to it in statements like "the object as it can be determined, physically and physiologically, as common to all". I'd be more interested in this if Mead had explored the "so-called" aspect more - but then that brings up the conundrum that Morris mentioned earlier of "the logical and methodological scandal which has long harassed scientific thought: on the one hand science has prided itself upon being empirical, on bringing its most subtle theories to the test of observation; on the other hand science has tended to accept a metaphysics which regards the data of observation as subjective". Indeed, the conundrum remains problematic, tho, regardless of whether we ignore the "so-called" possibility or not.

Behaviorism "is interested in getting such" objective "statements and correlations so it can control conduct as far as possible." "We have to lead the intelligences of infants and children into certain definite uses of media, and certain definite types of responses. How can we take the individual with his peculiarities and bring him over into a more nearly uniform type of response?" (p 35) This is where my fears & objections to Behaviorism come into play.

Most people have probably heard of the "Skinner Box", an environment used to study the application of reinforcement & punishment on test subjects. When I think of it I think of a friend's claim that he was raised 'in a Skinner Box' wch I always took to mean a crib designed to stimulate his learning. The thing about reinforcement & punishment procedures is that they're meant to enforce *what the controller deems desirable* w/o consulting the desires of the subject. Putting a lab rat in a box & then giving it food when it does one thing & punishing it when it does another is *not* in the lab rat's best interests. The lab rat's interests are treated as of secondary importance in relation to the researcher/controller's.

José M. R. Delgado, M.D., author of Physical Control of the Mind is (in)famous for experimenting "with electric stimulation of the brain, carried out in cats, monkeys, chimpanzees, and a few human patients, as well as in the somewhat romantic experimental animal, the fighting bull" (Philip Morrison as quoted on the back cover of Physical Control of the Mind). Delgado was able to stop a charging bull w/ radio-controlled electrical stimulus to its brain. This is Behaviorism. From the perspective of the bull it's not exactly fair now, is it? I'm rooting for the bull to gore Delgado, he was a bit too, ahem, *intrusive* for my tastes.

"How can we take the individual with his peculiarities and bring him over into a more nearly uniform type of response?" is something that I've been resisting my whole life. This is one lab rat who'd rather choose his own direction w/ a minimum of reinforcement/punishment intervention from Draconian controllers who fancy themselves 'Godlike', no thank you very much. This is what I mean by "evicting the ruling elites from one's mental real estate".

As Michael Kandel expresses it in his Introduction to the Harvest edition (1992) of the Stanislav Lem authored collection of short stories featuring robots, <u>Mortal Engines</u>: "Both Dostoyevski and Gastev, on either side of the ideological fence, believed that there was no essential difference between the scientific "explanation" of human beings and the literal "engineering" of human beings." (p x) Thank you, no. I prefer to grow according to my own explorations & needs rather than be engineered like a topiary plant cut into the shape of one of the bulls that Delgado tortured.

"We are trying to state the experience of the individual and situations in just as common terms as we can, and it is this which gives the importance to what we call behavioristic psychology." (p 39) Now, that's more like it. Mead's emphasis is ultimately more on understanding & ethics than it is on punishment & control. I like him for that - but control appears to be implicit in Behaviorism - maybe in all psychology - even when it does "not match any documents."

The final footnote of the 1st part entitled "The Point of View of Social Behaviorism" says this: "It is the work of the discoverer through his observations and through his hypotheses and experiments to be continually transforming what is his own private experience into a universal form. The same may be said of other fields, as in the work of the great artist who takes his own emotions and gives them a universal form so that others may enter into them." (p 41) I reckon that's what I'm doing w/ this review if we take such concepts as "universal" w/ a grain of salt.

"The vocal gesture becomes a significant symbol" [..] "when it has the same effect on the individual to whom it is addressed or who explicitly responds to it, and thus involves a reference to the self of the individual making it." [..] "The function of the gesture is to make adjustment possible among the individuals implicated in any given social act with reference to the object or objects with which the act is concerned" (p 46) I imagine that there are other "function"s also possible.

"In Wundt's doctrine, the parallelism between the gesture and the emotion or the intellectual attitude of the individual, makes it possible to set up a like parallelism in the other individual. The gesture calls out a gesture in the other form which will arouse or call out the same emotional attitude and the same idea." - p 48

Well, that's Wundt, & you know how *hhheeeeee* was! Actually, I know nothing about Wundt, I'm trying to be 'funny' here but did you get that? B/c if you didn't we may've just disproven Wundt. See how easy that was?

"For if, as Wundt does, you presuppose the existence of mind at the start, as explaining or making possible the social process of experience, then the origin of minds and the interaction among minds become mysteries. But if, on the other hand, you regard the social process of experience as prior (in a rudimentary form) to the existence of mind and explain the origin of minds in terms of interaction among individuals within that process, then not only the origin of minds, but also the interaction among minds" [..] "cease to be mysterious or miraculous." [..] "Wundt thus overlooks the important fact that communication is fundamental to the nature of what we term "mind"" - p 50

I find Mead's theory that social process creates mind very interesting. It reminds me of all the stories of children being isolated & not spoken to in an attempt to learn if they'd spontaneously speak 'the 1st language' w/o socialization (see Umberto Eco's excellent The Search for the Perfect Language for more details). While Wundt presupposes mind, at least according to Mead, Mead presupposes the possibility of objectively perceivable socialization between separate entities, at least according to tENTATIVELY, a cONVENIENCE. **They are both wrong.** Just kidding.

"The conversation of gestures does not carry with it a symbol which has a universal significance to all the different individuals." (p 55) That's more like it! Sorry if I was a bit rough on you back there, Herb. One of the things I like the most about what Gerorgie-Porgie Herbie-Poo-Dumpling does here is his wide-ranging use of observations about non-human behavior. After discussing the way birds imitate each other, Mead contrasts humans (forms): "We argued that there is no evidence of any general tendency on the part of forms to imitate each other." (p 58) My note to myself regarding this particular quote is: 'He ignores the obvious - there must be a 1st non-imitation'. Both quotes seem too out-of-context: in my experience there's plenty of tendencies for humans to imitate each other (eg: in music - wch cd be a potential corollary to the birds described) & my comment seems better placed somewhere else as a comment on there having to be a "1st" of anything in Mead's social process, a "1st" that he seems to leave out. Still on the subject of bird sounds, Mead states that "There is no evidence that the gesture generally tends to call out the same gesture in the other organism." (p 60) to wch I reply 'but it does help w/ mutual identification'.

"If the vocal gesture which the sparrow makes is identical with that which it hears when the canary makes use of the same note, then it is seen that its own response will be in that case identical with the response to the canary's note. It is this which gives such peculiar importance to the vocal gesture: it is one of those social stimuli which affect the form that makes it in the same fashion that it affects the form when made by another, That is, we can hear ourselves talking, and the import of what we say is the same to ourselves that it is to others." (p 62) But doesn't the context of its coming from another person change its meaning? If we're to take the above literally then hearing another person imitating our words cd evoke mockery, eg, & wd certainly have an opposite meaning from our own saying of it.

I don't really mean to be such a contrarian here. "In the case of the vocal gesture the form hears its own stimulus just as when this is used by other forms, so it tends to respond also to its own stimulus as it responds to the stimulus of other forms. That is, birds tend to sing to themselves, babies to talk to themselves. The sounds they make are stimuli to make other sounds. Where there is a specific sound that calls out a specific response, then if this sound is made by other forms it calls out this response in the form in question." (p 65) Perhaps, but that sounds so cut-&-dry. If a baby hears a nurturing adult imitate its sound won't the baby's reply be one of satisfaction in the knowledge that it's being pd attn to? Rather than a reply created under circumstances where that element is lacking?

According to Mead, when a parrot imitates human speech, "As far as a parrot is concerned, its "speech" means nothing, but where one significantly says something with his own vocal process he is saying it to himself as well as to everybody else within reach of his voice." (p 67) Granted, it's unlikely that when a parrot says "15 men on a dead man's chest" it's referring to the same thing as Stevenson's pirates did but that doesn't mean that the parrot doesn't take note of the reaction of humans to its sounding.

"A behaviorist, such as Watson, holds that all our thinking is vocalization." (p 69) That, then, denies the possibility of what I call Pre-Lingual Thought. In a February 25, 2010 interview between myself & Gerry Fialka available online as an audio recording on the Internet Archive (https://archive.org/details/ GerryFialkaInterviewingTentativelyAConvenience-22510) as track 71 "pre-concretized thought" I ask Gerry: "Have you ever tried to understand what goes on in your mind before you formalize a thought into language?" & then later I say: "In order for the word to come into existence, in my opinion at least, there has to be a pre-word. Now the question is: Do you accept the pre-word as being thought?"

Then, in the Vol. 21/No. 1 issue of the late lamented magazine *Rampike* (2012) there's an article by poet/essayist Alan Davies entitled "Prelinguistic Thinking" that I contributed to. As Alan says in this article: "to compose poetry is to write the composition down in the language of poetry (which is more-or-less some version of our everyday language). On the other hand / the inability to write the poem down well signals (also) an inability to compose in poetic language / in fact it is (is) that inability. So perhaps there is something like being able to write poetry (somewhere within/around the person writing) / but being (at the same time) unable to write it down. This (then) would tend to indicate a strong pre-linguistic component in the manufacture of the poem / a moment when the poem exists in-and-around the body/mind of the poet / but not yet (at all) in words." (p 39) Alan later quotes from an August 18, 2010 email from me:

"Just bringing up the idea of pre-lingual thought is challenging the common notion that thought is entirely lingual. To me that's like saying that a spilling cup exists only as itself & ignoring the hand that pushed it. In other words, I reckon I'm proposing that thought is an infinite chain of events that the lingual manifestation of is just one link of. In other other words: Is it absolutely necessary to start w/ language?

"Perhaps it's a matter of how one defines thought - is thought any process that takes place in the brain or is it only the self-aware ones? Is it possible to be self-aware w/o language? Are out memories of our early yrs so vague &/or non existent b/c we didn't have the language yet to fix them? Mainly I'm preoccupied w/ this issue b/c i sometimes write sentences like: "& then I thought _____" & I'm never satisfied w/ what I fill in the blank w/ b/c, if nothing else, it doesn't feel right - it's not an accurate quote. In my thoughts I don't seem to think of my thoughts as neatly organized phrases or whatnot - not even as stream-of-consciousness. In my attempts to accurately transcribe my thoughts, words, no matter how they're arranged, never seem right.

"Of course, the problem here is: What medium/technique does one use then to 'quote' thoughts? I don't think it's a matter of choosing, say, images over words or sounds over words or whatever. Maybe it's more a matter of creating some sort of process flow chart (& I don't mean to say that the brain's analogous to a computer). In other words, if I cd somehow convey a neuronal firing pattern to another mind I'd be more accurately quoting my thought(s). This wd be where we'd get into telepathy. Maybe I'm saying that telepathy wdn't be the transmission of words or images but, rather, a resonant pattern of energy flow." - pp 42-43

According to Mead: "It is possible that one could have the meaning of "chair" in his experience without there being a symbol, but we would not be thinking about it in that case." (p 146) Then what wd 'having the meaning of "chair"' consist of if not thinking?

Are words necessary for memory? A. R. Luria's marvelous bk The Mind of a Mnemonist - A Little Book About A Vast Memory indicates otherwise:

"It is no wonder that S.'s memories of early childhood were incomparably richer than ours. For his memory was never transformed into an apparatus for reshaping reminiscences into words" - p 76

S: "I can see my mother taking me in her arms, then she put me down again . . . I sense movement . . . a feeling of warmth, then an unpleasant sensation of cold." - p 77, 1976 Henry Regnery Company edition

Are S.'s memories of being taken in his mother's arms & the resultant warmth memories *before* he describes them w/ words? I think so. Maybe I'm making too much of an (il)logical leap here but that's potentially an example of what I call pre-lingual thought - at least if one accepts memories as thoughts & thinks of putting those thoughts into words as simply a step in a thinking process.

"With a blind person such as Helen Keller, it is a contact experience that could be given to another as it is given to herself. It is out of that sort of language that the mind of Helen Keller was built up. As she has recognized, it was not until she could get into communication with other persons through symbols which could arouse in herself the responses they arouse in other people that she could get what we term a mental content, or a self." - p 149

I find Mead's example fascinating - just as I find Keller & her teacher fascinating. Keller was, of course, both blind & deaf. My question is: is the contact experience "lingual" &, if so, then wd a contact experience w/ a sighted & hearing person also be lingual? OR, if it's not "lingual" then is it "pre-lingual" but still qualifying as "thinking" enuf to enable Keller to 'have a mind'?

Mead puts forth the idea that an object becomes food when something is capable of eating it. "There would, for example, be no food—no edible objects—if there were no organisms which would digest it. And similarly, the social process in a sense constitutes the objects to which it responds" (p 77) That seems to me to depend on definition & perspective. There was a performer named "Monsieur Mangetout" ("Mr Eats-Everything"). Mangetout is reputed to've eaten 18 bicycles, 15 shopping carts, 7 televisions, 6 chandeliers, 2 beds, 1 pair of skis, 1 Cessna aircraft, 1 coffin, & 1 computer. The question is: If Mangetout had shat out a tv w/ different programming than it played before he ate it wd that make him a social critic, a food critic, or a media critic? I'm further reminded of John Safran, Master Chef. Mangetout, like Keller, was an exceptional person &, as a pataphysician, I'm interested in what can be gleaned from the *exception* & how it then elucidates the boundaries of the 'unexceptional'.

"The logical structure of meaning, we have seen, is to be found in the threefold relationship of gesture to adjustive response and to the resultant of the given social act." [..] "for the existence of meaning depends upon the fact that the adjustive response of the second organism is directed toward the resultant of the given social act as initiated and indicated by the gesture of the first organism." - p 80

Then what initiates the 1st act? Innovation as independent of the social process?

Anticipatory Behaviorism: "If you assert a proposition and add, "but," you determine the attitude of the hearer toward it." (p 86)

"Now, if one can bring in a number of these and get a multiform reflection of all of these attitudes into harmony, he calls out an aesthetic response which we consider beautiful. It is the harmonizing of these complexities of response that constitutes the beauty of the object." - p 87

Notions of beauty & harmony go unquestioned. The royal "we" might've been fair enuf in Mead's social milieu but it doesn't fly w/ me.

"Our so-called laws of thought are the abstractions of social intercourse. Our whole process of abstract thought, technique and method is essentially social (1912)." - p 90

But what about the dumbing-down? What about the fear-imposed, punishment-threatening LCD (Lowest Common Denominator)?

"The very universality and impersonality of thought and reason is from the behavioristic standpoint the result of the given individual taking the attitudes of others toward himself, and of his finally crystallizing all these particular attitudes into a single attitude which may be called that of the "generalized other."" - p 90

"You can explain the child's fear of the white rat by conditioning its reflexes, but you cannot explain the conduct of Mr. Watson in conditioning that stated reflex by means of a set of conditioned reflexes, unless you set up a super-Watson to condition his reflexes." - p 106

It all still begs the question of what came 1st? Even in Mead's view it seems to me that if there's an attitude that's a response to the generalized other wch is created from the responses to the individual in whom the attitude forms that there still has to be the same process in the influencing individuals from here to eternity that implies a super-attitude-generator.

I don't see "thought and reason" as universal & impersonal. Also, if I were to've adapted myself "into a single attitude which may be called that of the "generalized other[.]"" I probably wd be a cowed masochistic imbecile b/c the social milieu I was surrounded by was hardly supportive. Instead, I took the 'high road', the road of swimming up-dam against the forces of dumbing-down. My purpose wasn't to succumb to the generalized other but to increase the quality of the social milieu's intellectual life until it was more compatible w/ my own desired development - *not the other way around*.

"We may respond to a musical phrase and there may be nothing in the experience beyond the response; we may not be able to say why we respond or what it is we respond to. Our attitude may simply be that we like some music and do not like other music." (p 91) IE: be critically ignorant. "It is an unusual gift which can analyze that sort of an object and pick out what the stimulus is for so complex an action." (p 92) Yep.

"Pointing out the characters which lead to the response is precisely that which distinguishes a detective office that sends out a man, from a bloodhound which runs down a man. Here are two types of intelligence, each one specialized; the detective could not do what the bloodhound does and the bloodhound could not do what the detective does. Now, the intelligence of the detective over against the intelligence of the bloodhound lies in the capacity to indicate what the particular characters are which will call out his response of taking the man." - p 93

Maybe. I'm still not convinced that we understand non-humans enuf to assert this so strongly. What if the bloodhound is *more intelligent* b/c it can understand the msg from the human trainer while the human might not be able to understand what the dog's trying to get across? Are there dogs out there who cd train humans? A human can tell another human what to look for in a hunted culprit, I'm not convinced that non-humans can't do something similar. It's between the 2 species that the communication seems to be most lacking rather than w/in the individual species.

"In the case of the vocal gesture there is a tendency to call out the response in one form that is called out in the other, so that the child plays the part of parent, of teacher, or preacher." - p 96

In other words, the child has begun to internalize the imposed mindset of the authority figure, the authority figure is *occupying the mental real estate* of the child. This isn't necessarily unhealthy given that these figures may be benevolent but what about *the mental 'grain of salt'* that's so essential to 'free thinking'?!

"The soldier is trained through a whole set of evolutions. He does not know why this particular set is given to him or the uses to which it will be put; he is just put through his drill, as an animal is trained in a circus. The child is similarly exposed to experiments without any thinking on his part." - p 105

& applied Behaviorism helps make this possible. IMO that's not healthy. It's in the vested interests of powers-that-be to send 'tools' to their slaughter w/ the 'higher' motives remaining ulterior. But I can't really blame that on Mead, he's truly preoccupied w/ ethics. Alas, if one helps develop something that can be used as a weapon someone WILL use it as a weapon.

"But what must be asked of behaviorism is whether it can state in behavioristic terms what is meant by having an idea, or getting a concept." - p 107

"Mental processes take place in this field of attitudes as expressed by the central nervous system; and this field is hence the field of ideas: the field of the control of present behavior in terms of its future consequences, or in terms of future behavior; the field of that type of intelligent conduct which is peculiarly characteristic of the higher forms of life, and especially of human beings. The various attitudes expressible through the central nervous system can be organized into different types of subsequent acts; and the delayed reactions or responses thus made possible by the central nervous system are the distinctive feature of mentally controlled or intelligent behavior." - p 118

"The young squirrel is born in the summer time, and has no directions from other forms, but it will start off hiding nuts as well as the older ones. Such action shows this experience could not direct the activity of the specific form." - p 119

Instinct goes unexplained. Later, Mead distinguishes between "impulse" wch he attributes to humans & "instinct" wch he attributes to 'lower forms of life'.

"We have to recognize that language is a part of conduct. Mind involves, however, a relationship to the character of things. Those characters are in the things, and while the stimuli call out the response which is in one sense present in the organism, the responses are to things out there. The whole process is not a mental product and you cannot put it inside of the brain. Mentality is that relationship of the organism to the situation which is mediated by sets of symbols." - pp 124-125

Aren't the "sets of symbols", then, "inside the brain" in the sense that w/o the brain the "process" wdn't take place? One can take away the "things" but not take away the brain. One can think about a cup, eg, w/o the cup's having to be physically present w/in immediate perception, but one *cannot* take away the brain & still think about the cup - & putting the brain *in a cup* just won't do.

Concepts are objects in Mead's world but he appears to question this in the footnote:

"Take the case of food. If an animal that can digest grass, such as an ox, comes into the world, then grass becomes food. That object did not exist before, that is, grass as food. The advent of the ox brings in a new object. In that sense, organisms are responsible for the appearance of whole sets of objects that did not exist before."*

*"It is objectionable to speak of the food-process in the animal as constituting the food-object. They are certainly relative to each other (MS)." - p 129

As is typical w/ language, it's possible to simplify things in a way that seems reasonable b/c it borders on self-contained. However, 'logic' like the above becomes suspect when one examines this self-containment & compares it to life-outside-of-language. In other words, The ox does not 'come into the world', there is no clear "advent of the ox". Mead is speaking in this way b/c he's trying to make an understandable point w/ accessible language. The ox & the grass exist & one can't accurately say that the ox is any more responsible for the grass being its food than one can accurately say the grass is responsible for the ox's digestive system. Hence, "They are certainly relative to each other", as Mead's manuscript says.

"And since organism and environment determine each other and are mutually dependent for their existence, it follows that the life-process, to be adequately understood, must be considered in terms of their interrelations." - p 130

& that's where I can begin to identify & agree much more than previously.

"But it is where one does respond to that which he addresses to another and where that response of his own becomes a part of his conduct, where he not only hears himself but responds to himself, talks and replies to himself as truly as the other person replies to him, that we have behavior in which the individuals become objects to themselves." - p 139

Fair enuf, this is a description of self-conscious behavior, Fine. What I then think of is Arrested Development, a type of behavior all-too-common in people I know whose childhoods were stunted by being forced prematurely into adulthood. These folks tend to act toward other people in ways they'd be unlikely to act if they cd objectify themselves & imagine the same type of behavior directed towards them.

"We realize in everyday conduct and experience that an individual does not mean a great deal of what he is doing and saying. We frequently say that such an individual is not himself. We come away from an interview with a realization that we have left out important things, that there are parts of the self that did not get into what was said. What determines the amount of the self that gets into communication is the social experience itself." [..] "It is the social process that is responsible for the appearance of the self; it is not there as a self apart from this type of experience." - p 142

I have no problem w/ agreeing that "in everyday conduct and experience that an individual does not mean a great deal of what he is doing and saying". I'm not so sure that I agree w/ "What determines the amount of the self that gets into communication is the social experience itself" insofar as I think that people are in different states of mind *when* the social experience occurs & may not be being mindful enuf to care much about whether what they're saying is adequately self-representative. Often, there are regrets if & when one realizes this later. Also, it seems to me that an extrapolation of the logic of "It is the social process that is responsible for the appearance of the self; it is not there as a self apart from this type of experience" is that individuals can be held not-responsible in general. That wd make for an interesting legal argument but I don't think I'd buy it.. or even rent it.. but I might take a free trial (pun intended).

Lately I've been thinking alot about how we live in an age of unprecedented communication opportunities but that I find communication between people to be astonishingly abysmal anyway. Take, eg, this personal instance: a friend invites me to her home, knowing that she's unreliable I try texting her, emailing her, & leaving a phone msg before going to her house b/c I want to make sure she's going to be there. She replies to none of these outreaches. I don't go & later find out that the purpose for the visit had been cancelled but that she didn't bother to tell me. As such, I was correct to not go. This instance is an extreme one b/c the person in the story is exceptionally uncommunicative but I use it as an example b/c I sometimes feel that in this day & age people take these communication methods so much for granted that they begin to delude themselves into feeling omniscient & telepathic & forget to be pragmatic. This cd be sd to be an instance of someone 'not being themselves' insofar as they're taking for granted that people 'know what they're thinking' &, therefore, don't bother to express it. I don't attribute this behavior to the "social process", tho, I attribute it to self-delusion.

"The illustration used was of a person playing baseball. Each one of his own acts is determined by his assumption of the action of the others who are playing the game. What he does is controlled by his being everyone else on the team, at least in so far as those attitudes affect his own particular response. We get then an "other" which is an organization of the attitudes of those involved in the same process." - p 154

What we *don't get* is participants in a game who *invented the game*, at least as of this time. Instead there's a body of rules & expectations that determines the "generalized other" more than the actions of the participants in the social process. A heightened self-awareness might lead to a questioning of these pre-determined rules & their generally unstated affect. EVENTUALLY, Mead gets around to addressing such questioning possibilities:

"A person may reach a point of going against the whole world about him; he may stand out by himself over against it. But to do that he has to speak with the voice of reason to himself. He has to comprehend the voices of the past and of the future. That is the only way in which the self can get a voice which is more than the voice of the community." [..] "That is the way, of course, in which society gets ahead, by just such interactions as those in which some person thinks a thing out. We are continually changing our social system in some respects, and we are able to do that intelligently because we can think." - p 168

EVENTUALLY, Mead even gets to what I might call a borderline non-Behaviorist position:

"The "me" represents a definite organization of the community there in our own attitudes, and calling for a response, but the response that takes place is something that just happens. There is no certainty in regard to it. There is a moral necessity but no mechanical necessity for the act." [..] "The separation of the "I" and the "me" is not fictitious. they are not identical, for, as I have said, the "I" is something that is never entirely calculable." - p 178

"The self is not so much a substance as a process in which the conversation of gestures has been internalized within an organic form." (p 178) Once again, this implies a pre-existing process rather than a pre-existing mind insofar as it involves the "me" under scrutiny being created by a conglomerate of other "me"s *not under scrutiny* who, in turn are implied to come out of an identical process from here to infinity. What makes that ultimately any different from presupposing a mind? If no origin can be provided than any presupposition is equally valid.

"We, of course, tend to endow our domestic animals with personality, but as we get insight into their conditions we see there is no place for this sort of importation of the social process into the conduct of the individual. They do not have the mechanism for it—language." - p 182

This, of course, has been disproven. It was even disproven as of the time of Mead's lecture. The main source of this bk is a stenographic record of Mead's 1927 lectures. But as of April of 1925 there was a bk published entitled Chimpanzee Intelligence and its Vocal Expressions by Robert M. Yerkes, a psychologist, & Blanche W. Learned, a musician. The bk has large sections of pitched & rhythmed transcriptions of chimp responses to specific described stimuli. These responses were found to be recurring. As such, eg, there is sd to be a chimp "food-word": ngak. Playful human re-enactments of this chimp vocal expression can be found in my movie entitled "mmm057: Remote Viewing / Remote Grooving": http://youtu.be/-nQ0V6-M4EI .

Since then, the famous experiments w/ Koko, a gorilla, have disproven Mead much more demonstrably:

"American Sign Language (ASL) was selected by Dr. Penny Patterson as the primary language to teach Koko because of the success that other researchers had with chimpanzees. It turned out to be a good choice, as Koko (and later Michael) learned it quickly. Within just a few weeks the gorillas were using sign combinations. Much later, observations by other researchers at zoos revealed that gorillas seem to have a natural gestural language of their own, using dozens of gestures consistently to communicate with one another. This may explain why Koko and Michael learned ASL so quickly; it's built on their intrinsic capabilities." - http://www.koko.org/sign-language

"We put personalities into the animals, but they do not belong to them; and ultimately we realize that those animals have no rights. We are at liberty to cut off their lives; there is no wrong committed when an animal's life is taken away. He has not lost anything because the future does not exist for the animal" - p 183

"There is no evidence of animals being able to recognize that one thing is a sign of something else and so make use of that sign . . . . (1912)." - p 183

"There is another procedure by which the organism selects the appropriate stimulus, where an impulse is seeking expression. This is found in the relation to imagery. It is most frequently the image which enables the individual to pick out the appropriate stimulus for the impulse which is seeking expression. This imagery is dependent on past experience. It can be studied only in man, since the image as a stimulus or a part of the stimulus can only be identified by the individual, or through his account of it given in social conduct." - p 338

I remember a Koko story that I either read or saw as part of a documentary. Alas I can't pinpoint the source at the moment so the reader might take this as apocryphal. In the story, Koko was unwelcome to enter Dr. Patterson's office in the trailer that Koko lived in. Unable to discourage Koko, Dr. Patterson either intentionally or accidentally placed a small rubber toy crocodile at the threshold & found that Koko responded to it as a threat or as a *symbol* of threat & didn't enter the rm when the toy croc was there. That might be an example where imagery can be witnessed as a stimuli w/o the individual having to identify it.

"In some lowland areas including swamps, there are issues with crocodiles and gorillas. While they usually aren't able to kill them, they can definitely cause some serious injuries that the gorilla isn't able to recover from. Sometimes these crocodiles can successfully take down a weak gorilla though." - http://

www.gorillas-world.com/gorilla-predators/

In the light of the Koko references above, Mead's sad program of justification of brutality fits right into a seemingly general Behaviorist modus operandi that finds the earlier quoted passage, "The soldier is trained through a whole set of evolutions. He does not know why this particular set is given to him or the uses to which it will be put; he is just put through his drill, as an animal is trained in a circus.", to be acceptable. The "future" may "not exist for the animal" in the sense that humans think about it but the animal will still fight for its life if it knows that it's being threatened - wch implies a future sense even if it's only an instinctual one. There're many, many salves for the conscience in such philosophy of justification - they're all worse than bullshit b/c when you step in them they do more than just make yr foot smell bad.

Mead reasons that our 'attribution of personality' to animals is similar to our feelings for inanimate objects: "A similar attribution is present in the immediate attitude we take toward inanimate physical objects about us. We take the attitude of social beings toward them. This is most elaborately true, of course, in those whom we term nature poets." (p 183) "The physical object is an abstraction which we make from the social response to nature. We talk to nature; we address the clouds, the sea, the tree, and objects about us." (p 184)

But what about hylozoism?: "in philosophy, any system that views all matter as alive, either in itself or by participation in the operation of a world soul or some similar principle. Hylozoism is logically distinct both from early forms of animism, which personify nature, and from panpsychism, which attributes some form of consciousness or sensation to all matter." - http://www.britannica.com/topic/hylozoism I tend to be a hylozoist, I tend to think of everything as being alive but not b/c of "participation in the operation of a world soul". It's more a matter of thinking of life as a process, of any process as involving change, & of all things being subjected to change.

When I die, I don't expect to continue to exist as the entity that I now perceive myself to be. To all appearances, I'll be dead meat. I don't expect some wispy translucent 'soul' to slip out of my meat envelope & fly away to Ghost World to either be fed grapes by naked maidens or be strapped into a roller coaster eternally dipping in flames. I expect to rot, I expect to *transform*. The components that I currently consist of will no longer function as one unit, instead, these components will go into the ground, will be eaten by various organisms of various sizes. When I eat something it becomes alive as a part of me even though it wd've been dead when I ate it.

"And if mind or thought has arisen in this way," [thru the internationalization of the conversation of significant gestures] "then there neither can be nor could have been any mind or thought without language; and the early stages of the development of language must have been prior to the development of mind or thought." - p 192

However, since there's no information that I know of of the *very 1st thought n'at* that's a matter of speculation & definition, eh?! What came 1st? The chicken or the egg? What came 1st? The thought to define mind in terms of language or the language to define mind in terms of language?

"Take a person's attitude toward a new fashion. It may at first be one of objection. After a while he gets to the point of thinking of himself in this changed fashion, noticing the clothes in the window and seeing himself in them. The change has taken place in him without his being aware of it. There is, then, a process by means of which the individual in interaction with others inevitably becomes like others in doing the same thing, without that process appearing in what we term consciousness." - p 193

There you have the definition of a robopath but I don't think it's as 'universal' as Mead seems to imply. When I started growing my hair long I was 14 yrs old. It was 1968. When I walked down the street, men wd shout out: "Are yooouuu a buoy or a gurrrl?" Now, most likely, if they thought I was a girl, back in those days, they might've been embarrassed to ask such a question. The upshot was that I was supposed to be shamed into conforming to the expected male appearance of the time. It didn't work.

When I was in high school, the generally unintelligent but highly aggressive sports-inclined males, the 'jocks', wd throw food at the 'hippies' in the cafeteria. The cowards were testing the waters: 1st they wd band together to torment the pioneers, then when the pioneers weathered their storm of vituperation & it was safe for the jocks to do it, they'd imitate them in a watered-down fashion. As such, the jocks had long-hair w/in a decade or less after assaulting us for having it. There was no consciousness there b/c the robopaths were incapable of consciousness. The 'hippies', however, were conscious.

"When an ant from another nest is introduced into the nest of other forms, these turn on it and tear it to pieces." (p 193) That's not that much different from my high school experience as described above. If one thinks of the 'interloper' ant as a metaphor for visionaries introducing new ideas into a novelty-resistant society then the following makes sense: "if it creates a situation which is in some sense novel, if one puts up his side of the case, asserts himself over against others and insists that they take a different attitude toward himself, then there is something important occurring that is not previously present in experience." (p 196) "Another fundamental change has taken place in the form of the world through the reaction of an individual—Einstein. Great figures in history bring about very fundamental changes." (p 202) In other words, paradigm shifts. Paradigm shifts are the great warriors against the entropy of the Idiocracy.

"Such an individual is divergent from the point of view of what we call the prejudices of the community; but in another sense he expresses the principles of the community more completely than any other. Thus arises the situation of an Athenian or a Hebrew stoning the genius who expresses the principles of his own society, one the principle of rationality and the other the principle of complete neighborliness." - pp 217-218

But since most people are robopaths only paying lip-service to the PR versions of their community's ethics out of a camouflaging self-defensive phoniness, the average person is ready at any minute to abuse anyone who actually tries to live up to the principles in case that person might cause their facade to become transparent - at least as long as they can do it safely as a member of a mob.

"We are careful, of course, not directly to plume ourselves. It would seem childish to intimate that we take satisfaction in showing that we can do something better than others. We take a great deal of pains to cover up such a situation; but actually are vastly gratified." (p 205) In other words, most people are disingenuous & have a friend or employee post their self-glorifying Wikipedia entry for them in order to safely manipulate others into believing what they want them to believe about themself. On the other hand, some of the people who most hate people who honestly speak about what they think they've accomplished are those who've never accomplished anything even remotely close but want other people to *think* they've done so but are 'just too modest to say so.' Phoniness abounds.

"This sense of superiority does not represent necessarily the disagreeable type of assertive character, and it does not mean that the person wants to lower other people in order to get himself in a higher standing. That is the form such self-realization is apt to appear to take, to say the least, and all of us recognize such a form as not simply unfortunate but as morally more or less despicable." (p 205) 'All of us' don't recognize shit.

"There is a certain enjoyableness about the misfortunes of other people, especially those gathered about their personality. It finds its expression in what we term gossip, even mischievous gossip." (p 206) Yep, but is every human uncritical of this process? I think there's a somewhat subconscious attitude that accompanies this, one in wch the person not suffering the misfortune feels temporarily exempt from such misfortune - as if the odds have now lowered: if there are x number of stabbings in an area on average per yr & Joe Blow gets stabbed instead of you then you can reassure yrself that he got it b/c he's different from you & feel satisfaction in that difference. Personally, I'd rather work toward a lessening of the misfortunes w/ an eye toward lowering the statistic. I think Mead worked in a similar direction (except in his relations to non-humans) & that statements like the following are *warnings*:

"The sense of superiority is magnified when it belongs to a self that identifies itself with the group. It is aggravated in our patriotism, where we legitimize an assertion of superiority which we would not admit in the situations to which I have been referring. It seems to be perfectly legitimate to assert the authority of the nation to which one belongs over other nations, to brand the conduct of other nationalities in black colors in order that we may bring out values in the conduct of those that make up our own nation. It is just as true in politics and religion in the putting of one sect against the others." - p 207

Thus making a pretty good argument against patriotism, nationalism, & religion. Politics strikes me as a bit trickier but I don't think I'd miss it if it were *really gone* - by wch I don't mean just replaced by dictatorship. Mead's specifically political observation helps explain my own vacillating attitude: "Without parties we could not get a fraction of the voters to come to the polls to express themselves on issues of great public importance, but we can enrol a considerable part of the community in a political party that is fighting some other party. It is the element of the fight that keeps up the interest." (p 220) I think that's wise.. stupidly sad but true. Mead was an internationalist, as am I (or I might be a patanationalist), so we're somewhat on the same page here:

"Nations ought to be able to express themselves in the functional fashion that the professional man does. There is the beginning of such an organization in the League of Nations. One nation recognizes certain things it has to do as a member of the community of nations." - the same page being p 209

But then Mead gets back to something that's unacceptable to me: "Social control is the expression of the "me" over against the expression of the "I." It sets the limits, it gives the determination that enables the "I," so to speak, to use the "me" as the means of carrying out what is the undertaking that all are interested in." (p 210) All? I hardly think so.

"We distinguish very definitely between the selfish man and the impulsive man. The man who may lose his temper and knock another down may be a very unselfish man." - pp 211-212

I find that a particularly appealing clarification.

"One can very well just carry out certain processes in which his contribution is very slight, in a more or less mechanical fashion, and find himself in a better position because of it. Such men as John Stuart Mill have been able to carry on routine occupations during a certain part of the day, and then give themselves to original work for the rest of the day." - p 212

If one is working class, one might find oneself carrying "out certain processes in which his contribution is very slight, in a more or less mechanical fashion" in order to make enuf money to survive w/ - wch might be considered putting oneself "in a better position because of it" but I think very few of the creative & imaginative ones of us prefer that to devoting ourselves "to original work". It's something we're essentially forced into by economic conditions not of our making - esp ones that deliberately devalue our more original worth - even if those devaluing it stand to make a killing off it post our death. I don't necessarily object to working for other people for a living b/c one can hone one's social skills in that way & learn a plethora of other skills as a result. But being *forced* into it by absence of economic privilege is hardly a condition to recommend it.

Mead vacillates in his attitude toward religion &, more particularly, toward the church, but I get the idea that he, at least, *tries* to pay lip service to what little he can find of value in it given its social dominance:

"The great characters have been those who, by being what they were in the community, made that community a different one. They have enlarged and enriched the community. Such figures as great religious characters in history have, through their membership, indefinitely increased the possible size of the community itself. Jesus generalized the conception of the community in terms of the family in such a statement as that of the neighbor in the parables." - p 216

In general, Mead seems to give Christianity credit for expanding the notion of brotherhood. I just see Christinanity's "brotherhood" as yet-another guise for colonialism & intolerance of anything but superficial diversity. As for Christ?: I agree w/ director Brian Flemming that Jesus is a myth concocted out of earlier myths w/ almost identical content. Flemming expressed this opinion eloquently in his 2005 documentary entitled <u>The God Who Wasn't There</u>.

Much of Mead's references to other scientist-philosophers are made to refute their positions in favor of his own. He & I are alot alike in that respect. "Cooley's social psychology, as found in his *Human Nature and the Social Order*, is hence inevitably introspective, and his psychological method carries with it the implication of complete solipsism: society really has no existence except in the individual's mind, and the concept of the self as in any sense intrinsically social is a product of imagination." (footnote, p 224) Clearly, Cooley isn't a Social Behaviorist. One of the many positive side-effects of being a critical reader (I can't think of any negative ones at the moment) is that instead of writing off Cooley b/c the main author disagrees w/ him I now might pick up a Cooley bk if I come across it for the sake of reading his take on the philosophical debate. Mead is far from thoughtless & is capable of imagining future refutation & of addressing it in advance:

"Certain of the social processes within which this communication takes place are dependent upon physiological differences, but the individual is not in the social process differentiated physiologically from other individuals. That, I am insisting, constitutes the fundamental difference between the societies of insects and human society. It is a distinction which still has to be made with reservations, because it may be that there will be some way of discovering in the future a language among ants and bees." - p 235

"The waggle dance includes information about the direction and energy required to fly to the goal. Energy expenditure (or distance) is indicated by the length of time it takes to make one circuit. For example a bee may dance 8-9 circuits in 15 seconds for a food source 200 meters away, 4-5 for a food source 1000 meters away, and 3 circuits in 15 seconds for a food source 2000 meters away.

"Direction of the food source is indicated by the direction the dancer faces during the straight portion of the dance when the bee is waggling. If she waggles while facing straight upward, than the food source may be found in the direction of the sun.

"If she waggles at an angle 60 degrees to the left of upward the food source may be found 60 degrees to the left of the sun.

"Similarly, if the dancer waggles 120 degrees to the right of upward, the food source may be found 210 degrees to the right of the sun. The dancer emits sounds during the waggle run that help the recruits determine direction in the darkness of the hive.

"Source: The information in this article was taken from "The Dance Language and Orientation of Bees" by Karl von Frisch" - http://www.extension.org/pages/26930/dance-language-of-the-honey-bee

Since humans don't use dance as a primary form of communication the way bees do it's no surprise that it wasn't the 1st place that humans wd look for language. Obviously, there's chemical communication, even if it's an automatic pilot type, & there cd also be heat information or information conveyed thru a wide variety of wave-forms that're off the human register. Mead had enuf sense to recognize this even if only in an unspecified way.

"The form which has no distant experience, such as an amoeba, or which has such distant experience involved only functionally, has not the sort of environment that other forms have. I want to bring this out to emphasize the fact that the environment is in a very real sense determined by the character of the form." - p 247

Isn't that just a matter of relativity? Mightn't there be a life form that might perceive humans as myopic to the point of insignificance b/c we only get information from stars light yrs after it's occurred? & if "the environment is in a very real sense determined by the character of the form" isn't that just another solipsistic variant?

"We speak of Darwinian evolution, of the conflict of different forms with each other, as being the essential part of the problem of development; but if we leave out some of the insects and micro-organisms, there are no living forms with which the human form is in basic conflict. We determine what wild life we will keep; we can wipe out all the form of animal or vegetable life that exist; we can sow what seed we want, and kill or breed what animals we want. There is no longer a biological environment in the Darwinian sense to set our problem." - p 250

Too much so! I'm glad there're no wolves at the door as there were in Paris during Villon's time. Then again, I think one of the greatest (of many) acts of foolishness of people is to try to control the weather. We live as part of an organism far more complex than the arrogance of even the greatest collection of human genius is likely to understand at a level that can do it justice. Even if we WERE to understand it at a sufficiently profound level I think that the wisest thing we cd do is leave well enuf alone. The last thing we need is a series of eco-disasters like the importation of hoofed animals, rabbits, & the Cane Toad into Australia or the creation of copyrighted seeds à là Monsanto.

One of the primary purposes of writing this review is to try to get an understanding of what Mead means by the "generalized other" & to take a philosophical position vis à vis the (gray) matter: "We have seen that the process or activity of thinking is a conversation carried on by the individual between himself and the generalized other; and that the general form and subject matter of this conversation is given and determined by the appearance of some sort of problem to be solved." (footnote, p 254) When he puts it like that it seems so positive!

Then again, I'm not totally convinced people are such 'problem-solving animals' that all such internal conversations are based on that. What about problem-avoiding? That might be considered to be a problem response too akin to problem-solving to be worth considering as an exception but it might open the perceptual door to other situations encountered that may be more properly classified as something other than problems. Are there non-goal oriented non-decisions made more along the line of unconscious paths-of-least-resistance? Mightn't people just *blunder*?

Whatever. I keep returning to the idea of the generalized other being more vehemently established by maestros of manipulation. What if the problem-solving conversation is more a situation of the individual being *against* the encroachment of a generalized other that's recognized as being an unwelcome invasive species rather than a product of willingly entered-into egalitarian interaction?

"Freud's conception of the psychological "censor" represents a partial recognition of this operation of social control in terms of self-criticism, a recognition, namely, of its operation with reference to sexual experience and conduct." - footnote, p 255

"That is to say, self-criticism is essentially social criticism, and behavior controlled by self-criticism is essentially behavior controlled socially. Hence social control, so far from tending to crush out the human individual or to obliterate his self-conscious individuality, is, on the contrary, actually constitutive of and inextricably associated with that individuality; for the individual is what he is, as a conscious and individual personality, just in as far as he is a member of society, involved in the social process of experience and activity, and thereby socially controlled in his conduct." - p 255

I don't completely disagree w/ his point but to requote this: "Great figures in history bring about very fundamental changes." (p 202) In other words, while self-criticism may be social criticism doesn't that then mean that the 2 sides of the equation can reverse position & that social criticism = self-criticism too? By wch I mean the greater pressure of the greater mass of beings becomes reversed in importance? By wch I also mean that the inspired person, the visionary, prefers to assert their own vision over what cd be perceived of as the bullying LCD (Lowest Common Denominator) of the Generalized Other? Most people find safety in numbers - but when you know that *their safety* relies on unity around a scapegoat & that you're likely to be that scapegoat then it's time to turn the tables. Learn from the lesson of Giordano Bruno, folks, if you're a 'heretic' (read 'free thinker') don't be tricked into being put in chains & burned at the stake! Finally, Mead goes out on a limb & starts recognizing that the PR versions of dominant social forces aren't what they purport to be:

"Oppressive, stereotyped, and ultra-conservative social institutions—like the church—which by their more or less rigid and inflexible unprogressiveness crush or blot out individuality, or discourage any distinctive or original expressions of thought and behavior in the individual selves, or personalities implicated in and subjected to them, are undesirable but not necessary outcomes of the general social process of experience and behavior. There is no necessary or inevitable reason why social institutions should be oppressive or rigidly conservative, or why they should not rather be, as many are, flexible and progressive, fostering individuality rather than discouraging it." - p 262

"We are apt to assume that our estimate of the value of the community should depend upon its size. The American worships bigness as over against the qualitative social content. A little community such as that of Athens produced some of the greatest spiritual products which the world has ever seen; contrast its achievements with those of the United States, and there is no need to ask whether the mere bigness of one has any relation to the qualitative contents of the achievements of the other." - p 266

"In the case of the universal religions we have such forms as that of the Mohammedan, which undertook the force of the sword to wipe out all other forms of society, and so found in opposition which it took either to annihilate or to subordinate to itself. On the other hand, we have the propaganda represented by Christianity and Buddhism, which merely undertook to bring the various individuals into a certain spiritual group in which they would recognize themselves as members of one society." - p 282

Why "universal religions"? Can't we just perceive them as akin to having intestinal gas? As a result of a poor choice of food for thought? Something that will *pass* eventually, leaving our bowels more settled? Passing gas in the way that the Christian souls leave their assholes in the hundreds of tedious pages in the Tales of the Arabian Nights after the Mohammedan superhero miraculously kills overwhelming odds?

As for Christinanity undertaking "to bring the various individuals into a certain spiritual group in which they would recognize themselves as members of one society"? That's pure wishful thinking. Moslems & Christians both have a long history of oppressing by force whoever they can't control. Even at the missionary end of things where violence was somewhat superseded by more subtle forms of mind control the invaded victims weren't encouraged to respect their own society 1st & foremost but to fear the imposed society represented by Jesuits & the like. While Mead takes a chance & criticizes religion, we get back to the PR version pretty quickly:

"Christianity paved the way for the social progress—political, economic, scientific —of the modern world, the social progress which is so dominantly characteristic of that world. For the Christian notion of a rational or abstract universal human society or social order, though originating as a primarily religious and ethical doctrine, gradually lost its purely religious and ethical associations, and expanded to include all the other main aspects of concrete human social life as well" - p 293

"Christianity paved the way for the social progress—political, economic, scientific"?!! Tell that to Bruno, to Galileo, to all victims of the Inquisition & the Crusades. Christinanity has paved the way for dominance by its upper echelon, the people who disguise themselves as humble vessels for divinity. As I quoted Mead as saying earlier, "We are careful, of course, not directly to plume ourselves." In this case, such carefulness is to disguise that the 'divinity' that the religious leaders are 'vessels' for is nothing but smoke & mirrors to hide that old stage magic of who's really behind the curtain pulling the strings of the puppets.

"We are rather scornful of the attitude of salesmanship which modern business emphasizes—salesmanship which seems always to carry with it hypocrisy, to advocate putting one's self in the attitude of the other so as to trick him into buying something he does not want." - p 298

Alas, people aren't scornful of this *enuf* from my perspective b/c such salesmanship has come to dominate the economy w/ its pre-planned obsolescence far more than it did in Mead's time - despite his expression of scorn. if you had a record player well that just '*had to be replaced*' by a CD player wch just '*had to be replaced*' by an iPod ad nauseum. Rent yr own life from someone & be FREE!

Again, Mead was a believer in democracy & his lectures reflect the social POV of the optimism of his time:

"Even in the society erected on the basis of castes there are some common attitudes; but they are very restricted in number, and as they are restricted they cut down the full development of the self." [..] "The development of the democratic community implies the removal of castes as essential to the personality of the individual" - p 318

I'm w/ you there, Mead, but I'd go a step further & say that even in 'democracy' there're castes based on privilege rather than on merit. I reckon that's somewhat 'inescapable' but a feller can dream can't s/he? But maybe it's not only a dream after all: "Ethical ideas, within any given human society, arise in consciousness of the individual members of that society from the fact of the common social dependence of all these individuals upon one another" (pp 319-320) Mutual Aid n'at.

"Every human individual must, to behave ethically, integrate himself with the pattern of organized social behavior which, as reflected or prehended in the structure of his self, makes him a self-conscious personality. Wrong, evil, or sinful conduct on the part of the individual runs counter to this pattern of organized social behavior which makes him" - p 320

Then again, isn't "Wrong, evil, or sinful conduct" in the mind of the beholder? Did Lieutenant Calley think he was being wrong, evil, or sinful when he committed the Mai Lai massacre? Apparently not, he was just takin' care of business like many another good ole boy. & the American public apparently was w/ him b/c he came back from Viet Nam & became a successful jewelry store owner w/o being negatively branded as the mass murderer he was - & that's certainly a part of the "organized social behavior which" made him!

"Take, for example, the labor movement. It is essential that the other members of the community shall be able to enter into the attitude of the laborer in his functions. It is the caste organization, of course, which makes it impossible; and the development of the modern labor movement not only brought the situation actually involved before the community but inevitably helped to break down the caste organization itself." - p 325

Thank you, George Herbert Mead. Alas, there're still MANY people today who *just don't get the fucking message*. Since someone else is doing the work it's taken for granted. It's Herbert George Wells's Morlocks & Eloi from <u>The Time Machine</u> all over again except that the Eloi aren't really that beautiful as a result of their ignorant leisure & the workers aren't really such guiding lights either.

"There can still be leaders, and the community can rejoice in their attitudes just in so far as these superior individuals can themselves enter into the attitudes of the community which they undertake to lead." - p 326

Yep, that's an ideal but have you ever watched the rise of a politician? At 1st, maybe they're a 'person of the people', attaining popularity thru being able to articulate common goals - but, then, as they attain more power, doesn't the situation become "salesmanship which seems always to carry with it hypocrisy, to advocate putting one's self in the attitude of the other so as to trick him into buying something he does not want"? I find myself in the peculiar position here of quoting a politician to demonstrate my point: "As I would not be a slave, so I would not be a master. This expresses my idea of democracy. Whatever differs from this, to the extent of the difference, is no democracy." --ca. August 1, 1858 Fragment on Democracy (http://www.abrahamlincolnonline.org/lincoln/speeches/ slavery.htm) Instead of "master" insert "leader", instead of "slave" insert "follower" & you'll get my drift.

The idea of the "significant symbol" recurs frequently: "It is only when the stimulus which one gives another arouses in himself the same or like response that the symbol is a significant symbol. Human communication takes place through such significant symbols, and the problem is one of organizing a community which makes this possible." (p 327) Maybe there's a deeper problem of recognizing that symbols *do not mean exactly the same thing to everyone* & learning how to work around that w/ a minimum of animosity - &, actually, Mead seems exceptionally aware of this:

"What we call the ideal of a human society is approached in some sense by the economic society on the one side and by the universal religions on the other side, but it is not by any means fully realized. Those abstractions can be put together in a single community of the democratic type. As democracy now exists, there is not this development of communication so that individuals can put themselves into the attitudes of those whom they affect." - p 328

These days, w/ the internet & other mass media, one might not even have much of an idea of who's being affected. Take, eg, this review: I can take it somewhat for granted that none of my friends will ever read it in its entirety if at all. They'd find it too intellectual, too tedious, too long. Therefore, if anyone *does* read it, it cd very well be read by someone I don't know. That makes it extremely hard for me to be aware of the affect - esp given that since it's on the internet it cd be read by someone in another country whose culture I'm woefully underinformed about.

Mead, the Behaviorist, explains things in terms of process & relativity: "If we accept those two concepts of emergence and relativity, all I want to point out is that they do answer to what we term "consciousness["]" [..] "There are certain micro-organisms that are dangerous to human beings, but they would not be dangerous unless there were individuals susceptible to the attack of these germs." [..] "There are certain objects that are beautiful but would not be beautiful if there were not individuals that have an appreciation of them." (p 330)

Of course, danger & beauty are in the relationship of the beholder - but does that mean that nothing can have any intrinsic characteristics w/o there being a beholder to name them relative to itself? I think not. Beauty, eg, may be a concept that's intrinsically based on 'subjective' standards but what one finds beautiful can be based on characteristics of the object that one can say constitute "beautiful" w/in one's system of determining such. I might say that I find an object to my liking b/c it possesses certain attributes that appeal to me - that, in turn, might be a form of "beauty" to me but I often just prefer to explain why I find the attributes appealing or interesting. Are those attributes then only in the mind of me, the beholder? I might like the ability of a plant to turn toward the sun, I might find that 'beautiful' - does that mean that the plant doesn't turn toward the sun?

"Consciousness arises from the interrelation of the form and the environment, and it involves both of them. Hunger does not create food, nor is an object a food object without relation to hunger. When there is that relation between form and environment, then objects can appear which would not have been there otherwise; but the animal did not create the food in the sense that he makes an object out of nothing. Rather, when the form is put into such relation with the environment, then there emerges such a thing as food. Wheat becomes food; just as water arises in the relation of hydrogen and oxygen." - p 333

I find Mead's use of some terms confusing, "form" & "object" particularly. It seems to me that "food" is *not* an object - at least in the sense of the word "food" representing a category. Contrarily, "wheat" is an object insofar as it's a specific plant. The relationship between wheat + the category of food & water + its components hydrogen & oxygen is not analogous. "Food" is a set that includes many things & hydrogen & oxygen are specific things.

"Consciousness as such refers to both the organism and its environment and cannot be located simply in either." (p 332) But isn't the environment a collective of organisms also? The advantage of my somewhat flippant "ogjectivity" is that the consciousness can exist in a flux in wch organisms are simultaneously environment &, therefore, *doesn't have to be located in either per se.*

"What we term "reason" arises when one of the organisms takes into its own response the attitude of the other organism involved." [..] "When it does so, it is what we term "a rational being."" (p 334) Doesn't that have to be a 'two-way street'? What if you're driving w/ a friend who's superstitious & a black cat crosses the road in front of you? One possible way of taking "into [one']s own response the attitude of the other organism involved" wd be to turn the car around, as you know yr friend wd do, rather than continue past where the cat crossed. Wd that be being a "rational being"?

"Human behavior, or conduct, like the behavior of lower animal forms, springs from impulses. An impulse is a congenital tendency to react in a specific manner to a certain sort of stimulus, under certain organic conditions. Hunger and anger are illustrations of such impulses. They are best termed "impulses," and not "instincts," because they are subject to extensive modifications in the life-history of individuals, and these modifications are so much more extensive than those to which the instincts of lower animal forms are subject that the use of the term "instinct" in describing the behavior of normal adult individuals is seriously inexact." - p 337

To make the distinction between "impulses" & "instinct" is fine w/ me. However, I think Mead's straining at the bit to make the distinction between humans & "lower animal forms". Given that non-humans are also demonstrably capable of learned behavior why not just say that both humans & non-humans have both instincts & impulses?

"The life of the child in human society subjects these and all the impulses with which human nature is endowed to a pressure which carries them beyond possible comparison with the animal instincts, even though we have discovered that the instincts in lower animals are subject to gradual changes through long-continued experience of shifting conditions." (pp 349-350) Thusly making them "impulses" by Mead's definition - bringing humans & non-humans closer together.

"The association of ideas has been superseded by associations of nerve elements." (p 341) In other words, Mead is putting the physical Behaviorist spin on things to avoid waxing metaphysical. "Imagery thus far considered no more exists in a mind than do the objects of external sense perception." (p 342) Even if it's now significant symbolism?

"As an example of simple reflection we may take the opening of a drawer that refuses to give way to repeated pulls of ever increasing energy. Instead of surrendering one's self to the effort to expend all his strength until he may have pulled off the handles themselves, the individual exercises his intelligence by locating, if possible, the resistance, identifying a little give on this side or that, and using his strength at the point where the resistance is greatest, or attending to the imagery of the contents of the drawer and removing the drawer above so that he may take out the obstacle that has defeated his efforts." - pp 355-356

This bk is full of observations such as the above that show the fruits of Mead *paying attn*.

"However complex and intricate this conduct may become, as in the life of the bee and the ant, or in building such habitats as those of the beaver, no convincing evidence has been gained by competent animal observers that one animal gives to another an indication of an object or action which is registered in what we have termed a "mind"; in other words, there is no evidence that one form is able to convey information by significant gestures to another form." - p 359

"A bonobo named Kanzi, who had learned to communicate using a keyboard with lexigrams, picked up some sign language from watching videos of Koko; Kanzi's researcher, Sue Savage-Rumbaugh, did not realize he had done so until Kanzi began signing to anthropologist Dawn Prince-Hughes, who had previously worked closely with gorillas." - Prince-Hughes, Dawn (1987). Songs of the Gorilla Nation. Harmony. p. 135. ISBN 1-4000-5058-8 - https://en.wikipedia.org/wiki/Koko_(gorilla)#cite_note-Prince-Hughes135-32

"We find among birds a curious phenomenon. The birds make an extensive use of the vocal gesture in their sexual and parental conduct. The vocal gesture has in a peculiar degree the character of possibly affecting directly the animal using it, as it does the other form." - p 359

Anyone who's watched birds in the wild has noticed birds making alarm cries when a hawk is in the area as a warning to potential prey of the hawks to be alert. I'd call that communication. Perhaps a difference between that & humans is that a bird is not likely to make that alarm sound if no threat is nearby while a human might call out "Fire" in a movie theater regardless of whether there was a fire.

"The attitude that we characterize as sympathy in the adult springs from this same capacity to take the role of the other person with whom one is socially implicated. It is not included in the direct response of help, support, and protection." [..] "Helplessness in any form reduces us to children, and arouses the parental response in the other members of the community to which we belong." [..] "The parental attitudes, like the infantile attitudes, serve first of all the purpose of the self-stimulation which we have noted in birds, and thus emphasize valuable responses, but secondarily they provide the mechanism of mind." - pp 366-267

"The more we become interested in persons the more we become interested in life." [..] "Similarly, to get an intellectual motive is one of the greatest boons which one may have, because it expands interest so widely," - p 384

YES. I've noticed that my memory was best when I was the most social & remembered thousands of people's names aided by an interest in all of their social details such as: who they were lovers w/, who they were related to, who they were friends w/, what they did for money, what their creative areas were, etc.

"A man has to keep his self-respect, and it may be that he has to fly in the face of the whole community in preserving this self-respect. But he does it from the point of view of what he considers to be a higher and better society than that which exists." - p 389

That's a perfect final quote for my purposes & it just happens to be the last paragraph of the main text of Mead's bk too. Maybe we are on the same page after all. Following that is a list of Mead's published works, there're 68 articles listed. One of them is a pamphlet entitled *The Conscientious Objector* (1917), another is entitled "National-Mindedness and International-Mindedness" in *The International Journal of Ethics* (1929). I have to give Mead credit for being a thoughtful man - regardless of whether we're really on the same page or not. I wonder who the contemporary psychologist is who's studied Mead et al & reached more sophisticated opinions based on more contemporary knowledge?

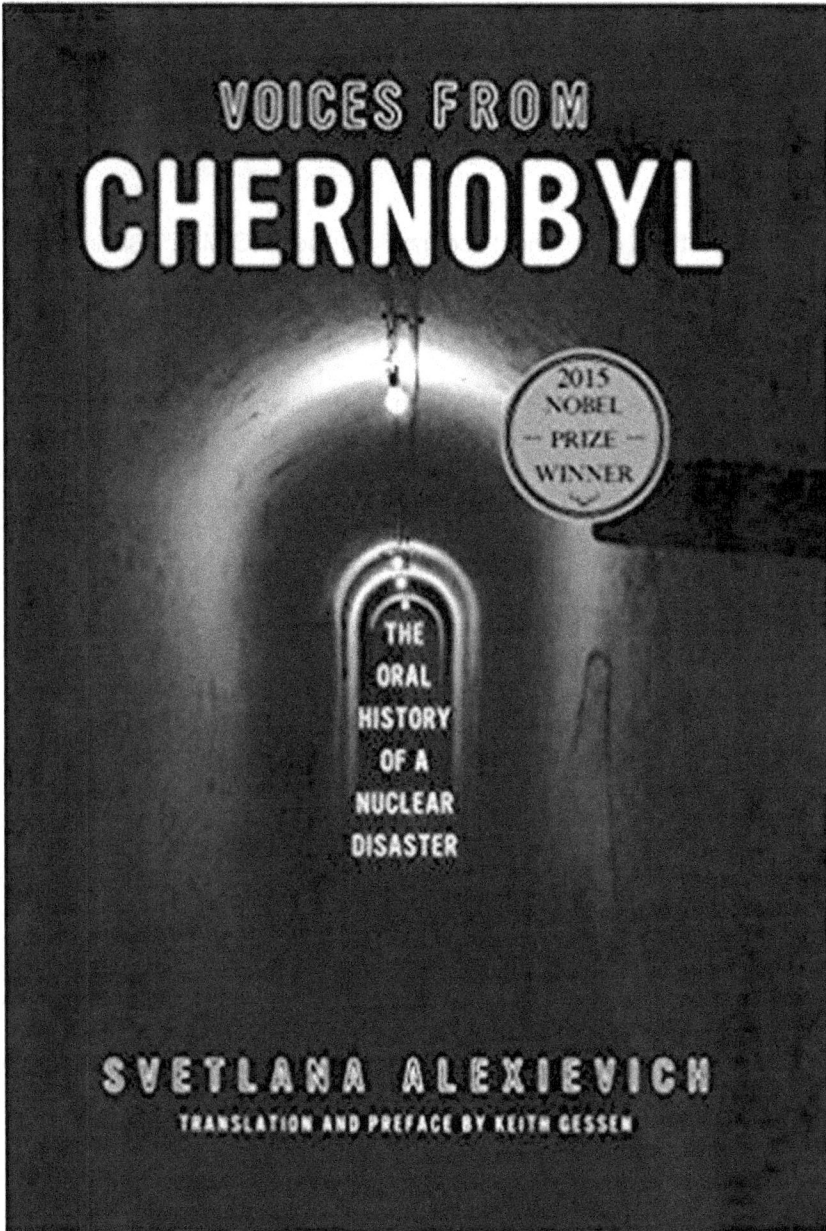

review of
Svetlana Alexievich's **Voices from Chernobyl**
by tENTATIVELY, a cONVENIENCE - May 15, 2016

"Chernobyl Hibakusha"

In 1978 I was a canvasser for Maryland Action, a consumer activist group that resisted price raising by the local gas & electric company. The canvassers were told not to bring up nuclear power, since that wasn't what we were canvassing about. Nonetheless, the people whose doors we went to often wanted to debate w/ us about nuclear power, taking it for granted that we were actually a group opposed to it.

In 1979 I was part of a group most commonly called "B.O.M.B." (Baltimore Oblivion Marching Band). B.O.M.B. was founded by Richard Ellsberry as a guerrilla performance unit. His initial vision was that we wd do things like go to shopping mall openings.

In April of 1979, the nuclear power plant on 3 Mile Island in Pennsylvania had a problem w/ its cooling towers that threatened to result in a nuclear meltdown, a very dangerous thing. B.O.M.B. had a meeting at wch we debated whether or not to go as close to it as we cd get to stage a performance action. Some people decided not to, others were all for it. I debated against it as foolish but went there anyway b/c I felt that it was of historic importance.

On April 3, 1979, 6 of us left Baltimore pre-dawn to go to 3 miles south of Middletown, PA, to the 3 Mile Island Visitor's Center that was at the edge of the Susquehanna River that the island is located in the midst of. We performed an action that parodied the scientific optimism that nuclear power plants can be kept safe & that such a powerful force can be controlled & made a movie of it. A short version of that movie can be witnessed here: https://youtu.be/WFnEj9c35fE . This action became national, if not international, news. One of the members of B.O.M.B. was shortly thereafter hired to be an assistant photographer of the inside of the plant.

I mention these 2 things to demonstrate that I was not unaware of the dangers of nuclear power plants. Decades later I read Frederik Pohl's Chernobyl, A Novel, a docudrama of sorts based on his research about Chernobyl. It was the 1st bk I read by Pohl & I was impressed by its apparent level-headedness, its carefulness of description. This bk was released a mere yr after Chernobyl's infamous April 26, 1986 disaster. In a sense while it was timely it was also premature - the more long-term negative effects weren't as clear by then as they are now.

Now, around the time of the 30th anniversary of the nuclear accident, I decided to read Voices from Chernobyl in preparation for a video-tele-conference w/ artist activists in Belarus that I was co-organizing w/ Monty Canstin [sic] in Mogilev & his friends & collaborators in Minsk [an unedited screen recording of this by Ryan Broughman can be witnessed here: https://youtu.be/DiklpJ_RX3E]. I got copies of this bk from my local library in both Russian & English so that I cd compare the 2 languages. Chernobyl was located in the Ukraine next to the Dnieper River across from Belarus. Reputedly 60 to 70 % of the fallout went into Belarus.

From the "Translator's Preface" to Voices from Chernobyl:

"In Belarus, very little has changed since these interviews were conducted. Back in 1996, Alesandr Lukashenka was the lesser-known of Europe's "last two dictators." Now Slobodan Milosevic is on trial at The Hague and Lukashenka has pride of place. He stifles any attempt at free speech and his political opponents continue to "disappear." On the Chernobyl front, Lukashenka has encouraged studies arguing that the land is increasingly safe and that more and more of it should be brought back into agricultural rotation. In 1999, the physicist Vaily Borisovich Nesterenko (interviewed on page 210), authored a report criticizing this tendency in government policy and suggesting that Belarus was knowingly exporting contaminated food. He has been in jail ever since. — *Keith Gessen, 2005*"

I wonder if Lukashenka himself wd be willing to eat food grown on such radioactive lands? - esp a regular diet of it? It's worth noting that this bk is published by the Dalkey Archive Press, a press that I esteem highly as the publisher of difficult experimental literature. It was quite a surprise for me to see that it published this.

"On April 26, 1986, at 1:23:58, a series of explosions destroyed the reactor in the building that housed Energy Block #4 of the Chernobyl Power Station. The catastrophe at Chernobyl became the largest technological disaster of the twentieth century.

"For tiny Belarus (population: 10 million), it was a national disaster. During the Second World War, the Nazis destroyed 619 Belarussian villages along with their inhabitants. As a result of Chernobyl, the country lost 485 villages and settlements. Of these, 70 have been forever buried underground. During the war, one out of every four Belarussians was killed; today, one out of every five Belarussians lives on contaminated land. This amounts to 2.1 million people, of whom 700,000 are children. Among the demographic factors responsible for the depopulation of Belarus, radiation is number one. In the Gomel and Mogilev regions, which suffered the most from Chernobyl, mortality rates exceed birth rates by 20%." - p 1

"In a year they evacuated all of us and buried the village. My father's a cab driver, he drove there and told us about it. First they'd tear a big pit in the ground, five meters deep. Then the firemen would come up and use their hoses to wash the house from its roof to its foundation, so that no radioactive dust gets kicked up. They wash the windows, the roof, the door, all of it. Then a crane drags the house from its spot and puts it down into the pit. There's dolls and books and cans all scattered around. The excavator picks them up. Then it covers everything with sand and clay, leveling it. And then instead of a village, you have an empty field." - p 223

"A while ago in the papers it said that in Belarus alone, in 1993 there were 200,000 abortions. Because of Chernobyl." - p 174

Considering that the total population of Belarus is only 10,000,000 & that, obviously, less than half of those are women of a fertile age 200,000 abortions is pretty phenomenal. Imagine being afraid to give birth, imagine being afraid that yr DNA has been hopelessly derailed, that you're the last of yr line.

Frederik Pohl's Chernobyl begins w/ an interesting quote:

"From *The Revelation of St. John the Divine:*

"And the third angel sounded, and there fell a great star from heaven, burning as it were a lamp, and it fell upon the third part of the rivers, and upon the fountains of the waters; and the name of the star is called wormwood; and the third part of the waters became wormwood; and many men dies of the waters, because they were made bitter.

"The Ukrainian word for wormwood is *chernobyl.*"

At least 2 European women friends of mine claim to've been effected by radiation from Chernobyl. It's quite possible in both cases despite their both having been a considerable distance away from it. Imagine the stigma of being possibly (or definitely) irradiated. Even if one shows no external signs of ill-health one becomes a pariah, no healthy person is likely to want to risk having children w/ such a person. reading these 1st-person accts impressed that upon me, impressed upon me that there are now MILLIONS of people in this unfortunate position. Think of how many people have emigrated from the Ukraine alone who've kept their origins veiled in order to avoid this stigma.

"The large differences between 1990 and 2000 in the numbers of Ukrainian and Russian speakers for the 1987-1990 immigrants are more puzzling. One hypothesis is that many of the Ukrainians recorded in the 2000 census were illegal migrants at the time of the 1990 census, and that by 2000 they had permanent status and/or felt more comfortable responding to the census.

"The total number of persons of Ukrainian ancestry was 893,055 in 2000. The number of all immigrants was 253,400, and 56 percent of them arrived between 1991 and 2000. If we add the 1987-1990 immigrants (12.5 percent of all immigrants), we have a total of 68.5 percent of all immigrants belonging to the Fourth Wave. In absolute numbers there were 142,000 immigrants between 1991-2000, and 31,600 arrived between 1987 and 1990." - http://www.ukrweekly.com/old/archive/2003/410319.shtml

How many of these Ukrainian immigrants, legal & illegal, were fleeing from Chernobyl? Voices from Chernobyl is a bk of transcribed interviews w/ people directly effected by the proximity of Chernobyl, often people who were married to the people who tried to do damage control but who were dead by a decade later as a result of their irradiation.

"I'm not a writer. I won't be able to describe it. My mind is not capable of understanding it. And neither is my university degree. There you are: a normal person. A little person. You're just like everyone else—you go to work, you return from work. You get an average salary. Once a year you go on vacation. You're a normal person! And then one day you're suddenly turned into a Chernobyl person. Into an animal, something that everyone's interested in, and that no one knows anything about. You want to be like everyone else, and now you can't. People look at you differently. They ask you: was it scary? How did the station burn? What did you see? And, you know, can you have children? Did your wife leave you? At first we were all turned into animals. The very word "Chernobyl" is like a signal. Everyone turns their head to look at you. He's from there!" - p 34

"The world has been split in two: there's us, the Chernobylites, and then there's you, the others. Have you noticed? No one here points out that they're Russian or Belarussian or Ukranian. We all call ourselves Chernobylites. "We're from Chernobyl." "I'm a Chernobylite." As if this is a separate people. A new nation." - p 126

Imagine the stigmatization, imagine how horrible it is to be a Chernobylite:

"I go home, I'd go dancing. I'd meet a girl I liked and say, "Let's get to know one another."

""What for? You're a Chernobylite now. I'd be scared to have your kids."" - p 79

Afraid to have kids is right:

"My little daughter is different—she's different. She's not like the others. She's going to grow up and ask me: "Why aren't I like the others?"

"When she was born, she wasn't a baby, she was a little sack, sewed up everywhere, not a single opening, just the eyes. The medical card says: "Girl, born with multiple complex pathologies: aplasia of the anus, aplasia of the vagina, aplasia of the left kidney." That's how it sounds in medical talk, but more simply: no pee-pee, no butt, one kidney. On the second day I watched her get operated on, on the second day of her life. She opened her eyes and smiled, and I thought that she was about to start crying. But, God, she smiled!

"The ones like her don't live, they die right away. But she didn't die, because I loved her.

"In four years she's had four operations. She's the only child in Belarus to have survived being born with such complex pathologies.." - p 85

"I'm afraid. I'm afraid to love. I have a fiancé, we already registered at the house of deeds. Have you ever heard of the Hibakusha of Hiroshima? The ones who survived after the bomb? They can only marry each other. No one writes about it here, no one talks about it, but we exist. The Chernobyl Hibakusha. He brought me home to his mom, she's a very nice mom. She works at a factory as an economist, and she's very active, she goes to all the anti-Communist meetings. So this very nice mom, when she found out I'm from a Chernobyl family, a refugee, asked: "But, my dear, will you be able to have children?" And we've already registered! He pleads with me: "I'll leave home. We'll rent an apartment." But all I can hear is: "My dear, for some people it's a sin to give birth." It's a sin to love." - p 108

Maybe that "very nice mom" who believes in "sin" shd go to hell.

"There was a black cloud, and hard rain. The puddles were yellow and green, like someone had poured paint into them. They said it was dust from the flowers. Grandma made us stay in the cellar. She got down on her knees and prayed. And she taught us, too. "Pray! It's the end of the world. It's God's punishment for our sins." My brother was eight and I was six. We started remembering our sins. He broke a glass can with the raspberry jam, and I didn't tell my mom that I'd got my new dress caught on a fence and it ripped. I hid it in the closet." - p 221

NO, it's *not* the "end of the world" but I'm sure that if humans can manage to *really* end the world by blowing it to smithereens it'll be considered by somebody - esp if the almighty dollar's in there somewhere. &, NO, it's *not* "God's punishment for our sins", there is NO God & NO sin - but there's an endless supply of sniveling robopaths who'll fall back on any mythology before they try to actually look at what's happening. &, NO, the little girl isn't going to go to 'HELL' for ripping her dress. If her mom were less delusional she might realize that a ripped dress is a good entry point into sewing lessons.

"I heard—the adults were talking—Grandma was crying—since the year I was born [1986], there haven't been any boys or girls born in our village. I'm the only one. The doctors said I couldn't be born. But my mom ran away from the hospital and hid at Grandma's. So I was born at Grandma's'" - p 222

Imagine that. For me, that outdoes Greek tragedy by a mile - not that it's a competition. People are afraid to continue to exist, afraid of the mutations, of the failures to adapt, of the deformities.

Literally EVERYTHING is effected. It's not like a bomb blowing up one bldg, like one field being destroyed - so that production goes on elsewhere - or, rather, some places were far more dangerously radioactive than others but almost anything cd be radioactive & it wasn't always easy to tell unless it manifested like this:

""And the chickens had black cockscombs, not red ones, because of the radiation. And you couldn't make cheese. The milk didn't go sour—it curdled into powder, white powder. Because of the radiation."" - p 40

But not all milk curdled into powder. Some, apparently, seemed 'normal':

"'I go in to see a doctor. 'Sweety,' I say, 'my legs don't move. The joints hurt.' 'You need to give up your cow, grandma. The milk's poisoned.' 'Oh, no,' I say, 'my legs hurt, my knees hurt, but I won't give up the cow. She feeds me.'"" - p 43

"I was in a taxi one time, the driver couldn't understand why the birds were all crashing into his window, like they were blind. They'd all gone crazy, or like they were committing suicide." - p 89

Typically, ignorance reigns. Maybe it's not 'reasonable' to expect people to understand the extraordinary, the things that they don't personally have direct dealings w/. How 'reasonable' is it to expect people whose lives center around having kids & farming to understand where the electricity comes from that's powering their precious tv?

"My son calls from Gomel: Are the May bugs out?"

""No bugs, there aren't even any maggots. They're hiding."

""What about worms?"

""If you'd find a worm in the rain, your chicken'd be happy. But there aren't any."

""That's the first sign. If there aren't any May bugs and no worms, that means strong radiation."

""What's radiation?"

""Mom, that's a kind of death. Tell Grandma you need to leave. You'll stay with us."" - pp 50-51

Still, one wd hope that the lack of insects, regardless of what cause it wd be attributed to, wd be recognized as a dire warning sign by people accustomed to tending the soil. An interesting aside found on the Wikipedia entry re May bugs is this:

"Both the grubs and the imagines have a voracious appetite and thus have been and sometimes continue to be a major problem in agriculture and forestry. In the pre-industrialized era, the main mechanism to control their numbers was to collect and kill the adult beetles, thereby interrupting the cycle. They were once very abundant: in 1911, more than 20 million individuals were collected in 18 km² of forest.

"Collecting adults was an only moderately successful method. In the Middle Ages, pest control was rare, and people had no effective means to protect their harvest. This gave rise to events that seem bizarre from a modern perspective. In 1320, for instance, cockchafers were brought to court in Avignon and sentenced to withdraw within three days onto a specially designated area, otherwise they would be outlawed. Subsequently since they failed to comply, they were collected and killed. (Similar animal trials also occurred for many other animals in the Middle Ages.)" - https://en.wikipedia.org/wiki/Cockchafer

Are trials of insects any more ridiculous than all the rest of it?

"What's it like, radiation? Maybe they show it in the movies? Have you seen it? Is it white, or what? What color is it? Some people say it has no color and no smell, and other people say that it's black. Like earth. But if it's colorless, then it's like God. God is everywhere, but you can't see Him. They scare us! the apples are hanging in the garden, the leaves are on the trees, the potatoes are in the fields. I don't think there was any Chernobyl, they made it up. They tricked people. My sister left with her husband. Not far from here, twenty kilometers. They lived there two months, and the neighbor comes running: "Your cow sent radiation to my cow! She's falling down." "How'd she send it?" "Through the air, that's how, like dust. It flies." Just fairy tales! Stories and more stories." - pp 51-52

Right. It's not enuf to have Holocaust deniers, now we have Chernobyl deniers. Why, I've never seen anyone die so death doesn't exist. However, for people who were refugees from the war in Tajikistan, the irradiated area around Chernobyl seemed like a nice alternative:

"They come onto the bus one day to check our passports. Just regular people, except with automatic weapons. They look through the documents and then push the men out of the bus. And then, right there, right outside the door, they shoot them. They don't even take them aside. I would never have believed it." - p 55

No doubt that was justified by somebody's idea of a Motherland or a Fatherland.

The "Chernobylites", the stigmatized victims of the backfiring of overconfident technological 'progress' partially 'adapted' by having a sense of humor. Peppered throughout the tales of misery are Chernobylite jokes. I used most of these jokes in a performance I gave 3 days after the anniversary on April 29, 2016 (https://vimeo.com/164947710):

"They asked the Armenian broadcaster: 'Maybe there are Chernobyl apples?' 'Sure, but you have to bury the core really deep.'"

"There was a Ukranian woman at the market selling big red apples. 'Come get your apples! Chernobyl apples!' Someone told her not to advertise that, no one will buy them. 'Don't worry!' she says. 'They buy anyway. Some need them for their mother-in-law, some for their boss.'"

"Guy comes home from work, says to his wife, "They told me that tomorrow I either go to Chernobyl or hand in my Party card." "But you're not in the Party." "Right, so I'm wondering how do I get a Party card by tomorrow morning?"

"After Chernobyl you can eat anything you want, but you have to bury your own shit in lead."

"This prisoner escapes from jail, and runs to the thirty-kilometer zone at Chernobyl. They catch him, bring him to the dosimeters. He's "glowing" so much, they can't possibly put him back in prison, can't take him to the hospital, can't put him around people."

"'Now all the stores have radio-products.' 'Impotents are divided into the radioactive & the radiopassive.'"

One way or another, the typical bullshit rises to the top:

""During the war they burned us, and we lived underground. In bunkers. They killed my brother and two nephews. All told, in my family we lost seventeen people. My mom was crying and crying. There was an old lady walking through the villages, scavenging. 'You're mourning?' she asked my mom. 'Don't mourn. A person who gives his life for others, that person is holy.' And I can do anything for my Motherland. Only killing I can't do. I'm a teacher, and I taught my kids to love others. That's how I taught them: 'Good will always triumph.' Kids are little, their souls are pure."" - p 45

People don't necessarily die to give their "life for others." They're more likely to die by being caught in the mayhem caused by power struggles instigated by megalomaniacs & made further possible by robopaths. When the Nazis went on their genocidal rampage thru Belarus the Belarussian victims were mostly people who were unfortunate enuf to be in the way of German expansionism, an expansionism justified by the most extreme racist arrogance - an arrogance based in appeals to the 'Fatherland'. The idea of the 'Motherland' isn't any better.

It makes sense for people to defend themselves against this genocide but to justify it by appealing to a Motherland or a Fatherland is likely to result in more genocide down the line. 'Good' will *not* 'always triumph' but stupidity sure does have a good track record.

""There was one guy, he came back here from jail. Under the amnesty. He lived in the next village. His mother died, the house was buried. He came over to us. 'Lady, give me some bread and some lard. I'll chop wood for you.' He gets by."" - p 47

What "amnesty" was that? In an April 27, 1992 Washington Post article I found this: "Seated in the living room of his apartment in Kiev, 18 months after his early release from prison as part of a general amnesty for Chernobyl officials, Dyatlov tells a quite different story from the official version. He said he and other Chernobyl operators were made scapegoats for the designers of a dangerously unstable reactor." (https://www.washingtonpost.com/archive/politics/1992/04/27/chernobyls-shameless-lies/96230408-084a-48dd-9236-e3e61cbe41da/) Is that the "amnesty" meant? An amnesty for people held responsible for the Chernobyl accident who did time for it?

Of course, some humans fuck up & some humans come in to clean up the mess - but they're not necessarily the same people, eh?:

"So they brought us in, and they took us right to the power station. They gave us white robes and white caps. And gauze surgical masks. We cleaned the territory. We spent a day cleaning down below, and then a day above, on the roof of the reactor. Everywhere we used shovels. The guys who went up, we called them the storks. The robots couldn't do it, their systems got all crazy. But we worked. And we were proud of it." - p 68

"The robots couldn't do it"! Can you imagine that?! So humans go in & do it instead.

"We came home. I took off all my clothes that I'd worn there and threw them down the trash chute. I gave my cap to my little son. He really wanted it. And he wore it all the time. Two years later they gave him a diagnosis: a tumor in his brain . . . You can write the rest of this yourself. I don't want to talk anymore." - pp 73-74

"They asked us to kill a wild boar for a wedding. It was a request. The liver melted in your hands, but they wanted it anyway, for the wedding. For the christening.

"We shoot for science too. One time we shot two rabbits, two foxes, two wild goats. They're all sick, but we still tenderize them and eat. At first we were afraid to, but now we're used to it. You have to eat something, and we can't all move to the moon, to another planet.

"Someone bought a fox-fur hat at the market, and he went bald. An Armenian bought a cheap machine gun from a guy from the Zone—the idea. People frighten one another." - p 103

The radioactive area is called "The Zone" by many of those interviewed:

"They say there used to be this guy, he'd walk around. If he got drunk, he'd start reading everyone lectures. He'd studied philosophy at the university, then he'd been in prison. You meet someone in the Zone, they'll never tell you the truth about themselves. Or very rarely. But this one was intelligent. "Chernobyl," he'd say, "happened so that philosophers could be made." He called animals "walking ashes," and people, "talking earth." The earth talked because we eat earth, that is, we are built from earth.

"The Zone pulls you in. You'll miss it, I tell you. Once you've been there, you'll miss it." - p 97

Now maybe you've noticed that everyone's grammar is 'correct', all the punctuation's in the 'right' places. That's one of those things that reporters do to what other people say to make things palatable for their editors & publishers. It's also something that diminishes accuracy.

But back to The Zone: People who've witnessed the great Russian filmmaker's movie entitled "Stalker" or read the great Russian SF writing duo the Strugatsky Brother's "Roadside Picnic" (that "Stalker" is based on) will recognize "The Zone" as a militarily-enclosed area that appears normal but has invisible hazards that the illegal "Stalker" guides people thru. The Stalker has a child who's deaf & speechless but who has telekinetic powers.

But "Roadside Picnic" was written in 1971 & the movie was made in 1979, both well before Chernobyl - &, yet, the parallels w/ Chernobyl are stunning. "The Zone" surrounding the catastrophe is closed off by the military. Nonetheless, people enter it, esp looters & people returning to their homes. Children born of people exposed to the radioactivity are likely to be born deformed or not at all or to die soon thereafter.

Is the reference to "The Zone" coincidental? A common Soviet phrase? ""The Zone is a separate world. A different world in the midst of the rest of the world. It was invented by the Strugatsky Brothers, but literature stepped back in the face of reality."" (p 132) The fact that an SF bk wd have its central premise repurposed for an obvious parallel situation astounds me enuf & it's far from the only literary reference that comes up in the interviews.

"Everywhere you went, people would say, "Ah, movie people. Hold on , we'll find you some heroes," And they'd produce an old man and his grandson who spent two days chasing cows off from right near Chernobyl. After the shoot the livestock specialist calls me over to a giant pit, where they're burying the cows with a bulldozer. But it didn't even occur to me to shoot that. I turned my back on the pit and shot the scene in the great tradition of our patriotic documentaries: the bulldozer drivers are reading *Pravda*, the headline in huge black letters: "The nation will not abandon those in trouble!" I even get lucky: I look and there's a stork landing in a field. A symbol! No matter what catastrophes befall us, we will triumph! Life goes on!" - p 110

At the same time that I was reading this bk, I was checking out movies relevant to Belarus too. One of them is called "Primo Levi's Journey" (2006) by Davide Ferrario. In it, the filmmaker reconstructs the journey that Italian writer Levi went thru after he was liberated from a Nazi concentration camp. While Ferrario was filming he was stopped by a local party bureaucrat who proceeded to try to steer the filming process. In one scene this party figure hosts the filmmaker & crew to a dinner where he begins a toast to everyone's health w/ a reference to the stork. I used the sound from this scene to sample from in a piece I made for the afore-mentioned video-tele-conference. A screen recording of this conference made by Ryan Broughman can be found here: https://youtu.be/DiklpJ_RX3E - w/ the sampling piece starting around 8:23 (unfortunately, the sound is intermittent, a better version will be online eventually).

"I started filming the apple tress in bloom. The bumble bees are buzzing, there's a white, bridal color. Again, people are working, the gardens are in bloom. I'm holding the camera in my hands, but I don't understand it. This isn't right. The exposure is normal, the picture is pretty, but something's not right. And then it hits me: I don't smell anything." - p 111

The radioactivity effects everything. Maybe there aren't any bugs, maybe the wild boar's liver melts in yr hands (preferably not in yr mouth), maybe you can't smell anymore. The world is topsy-turvy, reality has changed fundamentally.

"And suddenly I catch myself filming everything just the way I saw it filmed in the war movies. And then I notice that the people are behaving in the same way. They're all carrying themselves just like in that scene from everyone's favorite movie, *The Cranes Are Flying*—a lone tear, short words of farewell. It turned out that we were all looking for a form of behavior that was familiar to us." - p 113

But how familiar can a radioactive environment be? Does all behavior immediately become hebephrenic?

"You can't compare it to a war, not exactly, but everyone compares it anyway. I lived through the Leningrad Blockade as a kid, and you can't compare them. We lived there like it was the front, we were constantly being shot at. And there was hunger, several years of hunger, when people were reduced to their animal instincts. Whereas here, why, please, go outside to your garden and everything's blooming! These are incomparable things. But I wanted to say something else— I lost track—it slipped away. A-ah. When the shooting starts, God help everyone! You might die this very second, not some day in the future, but right now. In the winter there is hunger. In Leningrad people burned furniture, everything wooden in our apartment we burned, all the books, I think, we even used some old rags for the stove. A person is walking down the street, he sits down, and the next day you walk by and he's still sitting there, that is to say he froze, and he might sit there like that another week, or he might sit until the spring. Until it warms up. No one has the strength to break him out of the ice." - pp 123-124

The things that humans impose on each other, as the saying goes: man's inhumanity to man, become a comparative touch point for negative consequences that're almost incomprehensible. "There are ten million Belarussians, and two million of us live on poisoned land. It's a huge devil's laboratory. People come to us from everywhere, they write dissertations, from Moscow and Petersburg, from Japan and Germany and Austria. They're preparing for the future." (p 125) Right, they're preparing for the *next nuclear disaster*. Be prepared. But why not avert it altogether?!

"I remember a conversation with this scientist: "This is for thousands of years," he explained. "The decomposition of uranium: that's 238 half-lives. Translated into time: that's a billion years. And for thorium: it's fourteen billion years."" - p

127

"Although thorium (Th), with atomic number 90, has 6 naturally occurring isotopes, none of these isotopes are stable; however, one isotope, 232Th, is relatively stable, with a half-life of 14.05 billion years, considerably longer than the age of the earth, and even slightly longer than the generally accepted age of the universe. This isotope makes up nearly all natural thorium." - https://en.wikipedia.org/wiki/Thorium-232

"In radioactive decay, the half-life is the length of time after which there is a 50% chance that an atom will have undergone nuclear decay. It varies depending on the atom type and isotope, and is usually determined experimentally." - https://en.wikipedia.org/wiki/Half-life

"Radioisotopes with short half-lives are dangerous for the straightforward reason that they can dose you very heavily (and fatally) in a short time. Such isotopes have been the main causes of radiation poisoning and death after above-ground explosions of nuclear weapons."

[..]

"Long-term isotopes are more complicated. They don't dose as heavily, but there are a lot more issues than just that. Plutonium for example is comparatively long-lived, but some of its decay products can be quite nasty. Also, plutonium happens to be particularly toxic due to its chemistry, which aggravates the damage it can do.

"The biggest danger from radioisotopes with mid-to-long half lives is that they can keep an entire region of earth nastily radioactive for a very long time, e.g. hundreds or thousands or even tens of thousand of years. That's the main reason why disposing of reactor wastes, which often contain just such isotopes, is such a contentious issue." - http://physics.stackexchange.com/questions/8904/why-do-they-consider-radioactive-matter-with-long-half-lives-more-dangerous-than

Therefore, it's possible that the region around Chernobyl will be "nastily radioactive for a very long time", possibly longer than the earth will exist, certainly longer than humanity is likely to exist. &, yet, one imagines that humans will 'adapt' - or, at least, have some sort of chance of doing so. Is adapting to such a thing even possible? I certainly don't know. The Chernobylites are finding out. But keep in mind that "In the Gomel and Mogilev regions, which suffered the most from Chernobyl, mortality rates exceed birth rates by 20%." At that rate it won't take long for the indigenous people of the area to die off completely. Maybe that trend has reversed already.

""Some instructors came from the Central Committee. Their route: hotel to regional Party headquarters in a car, and back, also in a car. They study the situation by reading the headlines of the local papers. They bring whole cases of sandwiches from Minsk. They boil their tea from mineral water. They brought that too. The woman on duty at the hotel told me. People don't believe the papers, television, or radio—they look for information in the behavior of the bosses, that's more reliable."" - pp 127-128

"*From rumors:*

[..]

"The forest animals have radiation sickness. They wander around sadly, they have sad eyes. The hunters are afraid and feel too sorry for them to shoot. And the animals have stopped being afraid of humans. Foxes and wolves go into the villages and play with the children.

"The Chernobylites are giving birth to children who have an unknown yellow fluid instead of blood. There are scientists who insist that monkeys became intelligent because they lived near radiation. Children born in three or four generations will be Einsteins. It's a cosmic experiment being carried out on us . . ." - p 133

Wishful thinking. Just as in "Roadside Picnic" where the Stalker's daughter compensates the loss of hearing & speech w/ a gain in the ability to move things w/ her mind so people have the hope that Belarus's loss of "485 villages and settlements" will give rise to a better world. No pain, no gain. Maybe that'll really happen.

On the other hand, the snail's pace of improving education & decreasing brutality are a bit more likely to achieve the desired goal. I'm reminded of the 2011 movie "Limitless" in wch a pill makes a person smarter. It's so typical: people don't want to *work to be smarter*, they want to take a pill to do that - Presto, Chango! People can be Einsteins NOW, not in "three or four generations" by having their intelligence *encouraged* instead of suppressed b/c it's so damned inconvenient for the status quo.

"There was a moment when there existed the danger of a nuclear explosion, and they had to get the water out from under the reactor, so that a mixture of uranium and graphite wouldn't get into it—with the water they would have formed a critical mass. The explosion would have been between three and five megatons. This would have meant that not only Kiev and Minsk, but a large part of Europe would have been uninhabitable. Can you imagine it? A European catastrophe. So here was the task: who would dive in there and open the bolt on the safety valve? They promised them a car, an apartment, a dacha, aid for their families until the end of time. They searched for volunteers. And they found them! The boys dove, many times, and they opened that bolt, and the unit was given 7,000 rubles, They forgot about the cars and apartments they promised—but that's not why they dove! Not for the material, least of all for the material promises. [*Becomes upset.*] Those people don't exist anymore, just the documents in our museum, with their names. But what if they hadn't done it? In terms of our readiness for self-sacrifice, we have no equals." - pp 136-137

My stepfather was a sign painter for the Baltimore Gas & Electric Company. BG&E have a nuclear power plant at Calvert Cliffs. Every yr, BG&E had a lottery where male employees were picked to clean the power plant. It didn't matter what department you were in, it didn't matter how irrelevant yr skills were to such a dangerous job. As I understand it, the reason for the lottery was that *there were no professionals for the job, anyone who knew best how to do it knew it was too dangerous*. Therefore, other people had to be found to do it. According to my stepfather, refusal meant chance of losing job. However, he never had to do it b/c there were always younger guys eager to do it for the abnormally high pay.

"Since you're writing this book, you need to have a look at some unique video footage. We're gathering it little by little. It's not a chronicle of Chernobyl, no, they wouldn't let anyone film that, it was forbidden. If anyone did manage to record any of it, the authorities immediately took the film and returned it ruined. They didn't allow anyone to film the tragedy, only the heroics. There are some Chernobyl photo albums now, but how many photo and video cameras were broken! People were dragged through the bureaucracy. It required a lot of courage to tell the truth about Chernobyl. It still does." - p 138

So much for "Pravda", "Truth", right?! Cover-ups everywhere, not just in the U.S.S.R., the PR images of nations must be preserved. Having the nations actually live up to the PR is quite another story. Think of the bombing of the MOVE group in Philly. At 1st, an 'expert witness' was brought out to proclaim that MOVE had committed suicide. THEN, it was gradually proven that the authorities had dropped a bomb on the MOVE house, deliberately killing the inhabitants in as cold-blooded a mass murder as any that those very same authorities wd decry as 'evil' when perpetrated by anyone else.

Voices from Chernobyl is a bk of interviews. I found them all telling. But the one that might've been the most telling for me is the one I've been quoting from from pp 133-143. It's one of the longest ones. It's w/ Sergei Vasilyevich Sobolev, Deputy Head of the Executive Committee of the Shield of Chernobyl Association, the keeper of a Chernobyl museum. He reveals:

"Those who worked on the reactor or near it, their—they—it's a common symptom for rocketeers also, this is well-known—their urino-genital system ceases to function. But no one talks of this out loud." - p 140

In other words, they're impotent.

"It's a well-known statistic that there are 800 waste burial sites around Chernobyl. He was expecting some fantastically engineered structures, but these were ordinary ditches. They're filled with "orange forest," which was cut down in an area of 150 hectares around the reactors. [*In the days after the accident, the pines and evergreens around the reactor turned red, then orange.*] They're filled with thousands of tons of metal and steel, small pipes, special clothing, concrete constructions. He showed me a photo from an English magazine that had a panoramic view from above. You could see thousands of individual pieces of automative and aviation machinery, fire trucks and ambulances."

[..]

"it was taken apart long ago and carried off to the market, for spare parts for the kolkhoz and people's homes. Everything's been stolen" - p 141

Those things that were isolated to protect people from the radioactivity have been scavenged & dispersed. Chernobyl is the gift of death that keeps on giving.

"I remember that first radioactive rain—"black rain," people called it later. First off, you're just not ready for it, and second, we're the best, most extraordinary, most powerful country on Earth, My husband, a man with a university degree, an engineer, seriously tried to convince me that it was an act of terrorism. An enemy diversion. A lot of people at the time thought that. But I remembered how I'd once been on a train with a man who worked in construction who told me about the building of the Smolensk nuclear plant: how much cement, boards, nails, and sand was stolen from the construction site and sold to neighboring villages. In exchange for money, for a bottle of vodka." - p 149

In Frederik Pohl's aforementioned acct of Chernobyl, he partially attributes the failure of the nuclear power plant to poor concrete manufactured to try to meet quotas. Never forget the 'human element' - even w/o terrorists there're plenty of small causes that add up to big effects. & was the theory of terrorism recounted above necessarily paranoid? Think of the leak of "The Freedom Fighter's Manual", the CIA training manual for conducting sabotage in Nicaragua. It included instructions like "Put nails on roads and highways" & gave instructions for being an arsonist. This 'Freedom Fighter's Manual' was specifically aimed against the Marxist government, an ally of the USSR.

"In 1983, the Central Intelligence Agency released two manuals to the Nicaraguan Contra rebels. The first, the Freedom Fighter's Manual, was a manual that was airdropped over known Contra camps. This was a 15-page manual that was illustrated with captions to educate the (mostly illiterate) Contras on how to cause civil disruptions in the face of the Sandinista government. The manual starts off with simple instructions including calling into work as to decrease production. Soon after, the instructions become more destructive by educating Contras on how to perforate fuel tanks with ice picks and how to create Molotov Cocktails to burn the fuel supplies." - https://en.wikipedia.org/wiki/CIA_activities_in_Nicaragua#Freedom_Fighter.27s_Manual

That's 1983, during the Reagan administration, 3 yrs before Chernobyl, also during the Reagan administration. Of course, taking on the USSR wd've been a much bigger task & they had nuclear weapons. Reagan preferred invading Grenada, a country that had done nothing to the US & didn't stand a chance against US invasion. Bullies prefer picking on targets that can't fight back.

"We had a great empire! And now how do you like that, coming around here. A great empire! Fuck!

"Until Gorbachev came along. A devil with a birthmark. Gorbie, Gorbie was working for them, for the CIA. What are you trying to tell me, huh? Coming around here. They're the ones who blew up Chernobyl. The CIA and the democrats. I read it in the papers. If Chernobyl hadn't blown up, the empire wouldn't have collapsed. A great empire! Fuck! Now they're coming around here. You could buy a loaf of bread for twenty kopeks under the Communists, now it costs two thousand rubles. What did the democrats do? They sold everything off! Our grandchildren won't have shit." - p 205

Don't throw out the baby w/ the bathwater. Trying to look for pravda in anyone's statement can be like being a starving man trying to pick out edible pieces of corn from another man's shit. Let's say the CIA aren't 'clever' enuf to've caused Chernobyl, that it really was caused by things like poor concrete manufacture caused by unrealistic quotas & other corruptions. Maybe Chernobyl was still largely responsible for the fall of the USSR.

A kopek is 1/100th of a ruble. If the defender of the soviets above isn't exaggerating too much that means that the cost of bread went up 10,000 times.

"After the fall of the Soviet Union, the Russian Federation introduced new coins in 1992 in denominations of 1, 5, 10, 20, 50 and 100 rubles."

[..]

"As high inflation persisted, the lowest denominations disappeared from circulation and the other denominations became rarely used." - https://en.wikipedia.org/wiki/Russian_ruble#Post-Soviet_ruble_.281992.E2.80.931997.29

"In Kiev they gave us some money, but we couldn't buy anything: hundreds of thousands of people had been uprooted and they'd bought everything up and eaten everything. Many had heart attacks and strokes, right there at the train stations, on the buses." - p 157

Absorb that for a moment. "hundreds of thousands of people had been uprooted" - that's not an insignificant number. Imagine having to leave yr home, to leave behind all the things of yr life, never to be able to return to them. What do you do next? You can't go back to yr home, yr job, most or all of yr support system is gone. How many times has such a catastrophe happened b/c of a volcano? An earthquake? A Tsunami? Now how many times has it happened b/c of human stupidity & malevolence? Wars? These things like nuclear power that make some people very wealthy in order to ostensibly provide greater electricity for everyone have a hazard potential that's pretty damned serious. Then imagine being one of the people who tries to correct the mess only to be lied to & destroyed:

"So here's how it was: I received a notice, and, as I'm a disciplined person, I went to the military recruiter's office the next day. They went through my file. "You," they tell me, "have never gone on an exercise with us. And they need chemists out there. You want to go for twenty-five days to a camp near Minsk?" And I thought: Why not take a break from my family and my job for a while? I'll march around a bit in the fresh air." - p 158

""Are you kidding?" says the captain who came with us, laughing. "Twenty-five days? You'll be in Chernobyl six months." Disbelief. Then anger. So they start convincing us: anyone working twenty kilometers away gets double pay, ten kilometers means triple pay, and if you're right at the reactor you get six times the pay. One guy starts figuring that in six months he'll be able to roll home in a new car, another wants to run off but he's in the army now. What's radiation? No one's heard of it." - p 159

Sure if the guy can still drive or see he can drive his new car to the hospital where his skin will fall off. The carrot on the stick is radioactive. But what's radiation? What you can't see can't hurt you. NOT. People want life to just go on normally so they act like nothing's changed. This too shall pass.. & it shall.. but not before un-tolled misery occurs. What wd you do if you were caught in such a mess? What wd I do? The sensible thing to do wd be to get away as far as possible as soon as possible - but how easy is that to do if you're poor?

"They brought us the insides of domestic and undomesticated animals. We checked the milk. After the first test it became clear that what we were receiving couldn't properly be called meat—it was radioactive byproducts. Within the zone the herds were taken care of in shifts—the shepherds would come and go, the milkmaids were brought in for milking only. The milk factories carried out the government plan. We checked the milk. It wasn't milk, it was a radioactive byproduct.

"For a long time after that we used dry milk powder and cans of condensed and concentrated milk from the Rogachev milk factory in our lectures as examples of a standard radiation source. And in the meantime, they were being sold in the stores. When people saw that the milk was from Rogachev and stopped buying it, there suddenly appeared cans of milk without labels. I don't think it was because they ran out of paper." - pp 165-166

"I worked at the inspection center for environmental defense. We were waiting for some kind of instructions, but we didn't receive any. There were very few professionals on our staff, especially among the directors: they were retired colonels, former Party workers, retirees or other undesirables. If you messed up somewhere else, they'd send you to us. Then you sit there shuffling papers. They only started making noise after our Belarussian writer Aleksei Adamovich spoke out in Moscow, raising the alarm. How they hated him! It was unreal. Their children live here, and their grandchildren, but instead of them it's a writer calling to the world: save us! You'd think some sort of self-preservation mechanism would kick in." - p 168

But, of course, the "self-preservation mechanism" of the robopath is to *not rock the boat* b/c that's what's enabled them to get this far in life w/o having to have much energy or vision. Kill the messenger, right?! Esp when the messenger points out to them that the boat has sunk so rocking it one way or the other is irrelevant.

"During World War II Ales Adamovich, a teenager, still a school student, became a partisan unit member in 1942-1943. During this time, the Nazis systematically torched hundreds of Belarusian villages and exterminated their inhabitants. Later, he wrote one of his most recognized works, *The Khatyn Story*, and the screenplay for the film *Come and See*, which was based on his real-life experiences as a messenger and a guerilla fighter during the war.

"Starting in 1944, he resumed his education. After the war, he entered the Belarusian State University where he studied in the philology department and where he completed graduate course; he later studied in Moscow at the Higher Courses for Screenwriters and in the Moscow State University. Starting in the 1950s in Minsk, he worked in the field of philology and literary criticism; later also in cinematography. Was a member of the Union of Soviet Writers since 1957. In 1976 was awarded the Yakub Kolas Belarus State prize in literature for *The Khatyn Story*. He lived and worked in Moscow since 1986 and was an active member of the Belarusian community of that city.

"After the Chernobyl accident in 1986, of which Belarus has suffered more than any other country, Adamovich started actively raising awareness of the catastrophe among the Soviet ruling elite.

"In late 1980s Ales Adamovich supported the creation of the Belarusian Popular Front but did not become a member of the movement. In 1989 Adamovich became one of the first members of the Belarusian PEN center (Vasil Bykaŭ was founder and president of the Belarusian PEN). In 1994 the Belarusian PEN Center instituted the *Ales Adamovich Literary Prize*, a literary award to the gifted writers and journalists. The prize is awarded annually on September 3 (Ales Adamovich's birthday) at the award ceremony that is usually part of the annual international conference." - https://en.wikipedia.org/wiki/Ales_Adamovich

That doesn't tell us much but it's better than nothing. I highly recommend *Come and See*.

While we're on the subject of Pravda, of 'truth', where do YOU get yr 'truth' from? I trust direct experience the most. I was raised by a robopath who trusted propaganda more than direct experience. When she read in the newspapers that all men w/ long hair were homosexuals she sd that that was right - despite the fact that her long-haired heterosexual son was sitting right there.

"I'm a historian. I used to work on linguistics, the philosophy of language, but language thinks us. When I was eighteen, or maybe a little earlier, when I began to read samizdat and discovered Shalamov, Solzhenitsyn, I suddenly understood that my entire childhood, the childhood of my street, even though I grew up in a family that was part of the intelligentsia (my grandfather was a minister, my father a professor at the university of St. Petersburg), all of it was shot through with the language of the camps." - p 175

Samizdat: underground publications forbidden by the Soviets. I contributed to at least one of them in Yugoslavia, SECOND MANIFESTo no: 3 (July 1985), w/ the "Monty Cantsin performing w/ White Colours Transparent Smile proposal". It's nice for me to imagine that I might've contributed to anyone's revelation.

One of the jokes that didn't make it into my April 29th cabaret presentation was this:

"An American robot is on the roof for five minutes, and then it breaks down. The Japanese robot is on the roof for five minutes, and then—breaks down. The Russian robot is up there two hours! Then a command comes in over the loudspeaker: "Private Ivanov! In two hours you're welcome to come down and have a cigarette break."" - p 191

& who were the 'volunteers' who did this clean-up work? Of course, there are many stories but this next one's no joke: "The commander says: "Only volunteers go up on the roof. The rest of you step aside, you'll have a talk with the military prosecutor." Well, those guys stood around, talked about it a little, and then agreed. If you took the oath, then you should do what you have to do. I don't think any of us doubted that they'd put us in jail for insubordination. They'd put a rumor that it would be two or three years." (p 192) Hhmm.. let's see.. dose of radiation that might make you impotent & will probably lead to a very early death in horrible conditions.. or 2 or 3 yrs in jail only to maybe be let out & forced into working in radioactive conditions anyway. Talk about being between a rock & a hard place!

"People ask me: "Why don't you take photos in color? In color!" But Chernobyl literally means *black event*. There are no other colors there." - p 193

So, what is it it? Chernobyl = "Wormwood" or "Black Event"? Or both? Wchever way you turn it it's not a very auspicious naming. However, never trust just one source. According to Yahoo Answers:

"Best Answer: Actually, Chernobyl is an altered form of the word Chornobyl, which means mugwort, which is a grass with black color. Chernobyl was mistakenly translated to mean "wormwood" some time ago.

"Taken from WordIQ Dictionary/Advanced Encyclopedia:

"Name Origin: "The city is named after the chornobyl' grass, or mugwort. The word itself is a combination of chornyi (чорний, black) and byllia (билля, grass blades or stalks), hence it literally means black grass or black stalks. "" - https://answers.yahoo.com/question/index?qid=20070126134416AA12Bsd

So maybe it's neither "wormwood" or "black event" but black grass. Maybe Pohl was smoking some black grass when he believed that the bible cd be relevant to anything other than stupidity. Just like the whole business of what shd've been done & why it wasn't:

"I'm calling over a government line, but they're already blocking things. As soon as you start talking about the accident, the line goes dead. So they're listening, obviously! I hope it's clear who's listening—the appropriate agency. The government within the government. And this is despite the fact that I'm calling the First Secretary of the Central Committee. And me? I'm the director of the Institute for Nuclear Energy at the Belarussian Academy of Science. Professor, member-correspondent of the Academy. But even I was blocked.

"It took me about two hours to finally reach Slyunkov. I tell him: "It's a serious accident. According to my calculations"—and I'd had a chance by then to talk with some people in Moscow and figure some things out—"the radioactive cloud is moving toward us, toward Belarus. We need to immediately perform an iodine prophylaxis of the population and evacuate everyone near the station. No man or animal should be within 100 kilometers of the place."

""I've already received reports," says Slyunkov, "There was a fire and they've put it out."

"I can't hold it in. "That's a lie! It's a blatant lie! Any physicist will tell you that graphite burns at something like five tons per hour. Think of how long it's going to burn!"" - pp 210-211

So who's the demon here? Who's the scapegoat? Who's truly responsible? No doubt there were some people who saved their own careers by towing a line they knew to be disastrous for the multitude. No doubt there were thieves who only gave a shit about making money off of the temporary gains of looting w/o caring about the long-term effects of spreading the radioactivity, wch they might not've believed in to begin w/. But in the long run, this was business-as-usual in an unusual situation that people cdn't cope w/. The primary foolishness was in the planners thinking they cd cope w/ such a disaster if & when it were to occur. Oops! The 'masters of nature' just had a slave revolt.

"Well, I have maps and figures. What do they have? They can put me in a mental hospital. They threatened to. And they could make sure I had a car accident—they warned me about that, also. They could drag me into court for anti-Soviet propaganda. Or for a box of nails missing from the Institute's inventory.

"So they dragged me into court.

"And they got what they wanted. I had a heart attack." - p 216

The status quo can be seen as being for the good of the majority when that majority is housed & fed & dependent on social order for continuance of conditions. Is it possible for the social order to continue under conditions where pravda *isn't suppressed*? I'd like to think so.

"We always say "we," and never "I." "We'll show them Soviet heroism," "we'll show them what the Soviet character is made of." "We'll show the whole world! But this is me, this is I. I don't want to die! I'm afraid.

"It's interesting to watch oneself from here, watch one's feelings. How did they develop and change? I've noticed that I pay more attention to the world around me. After Chernobyl, that's a natural reaction. We're beginning to learn to say "I." I don't want to die! I'm afraid.

"That great empire crumbled and fell apart. First Afghanistan, then Chernobyl." - p 219

When I was a kid, my mom always talked about how the communists only had propaganda & how the Americans had real news. When I asked her what communism was she didn't know. To me, that showed that she was the victim of US propaganda, that there *was* propaganda in America.

In some ways, for me, the USSR *was* a noble experiment that came entirely too close for comfort to succeeding for the greedy bosses of the world. Look at how much energy the US alone has expended on trying to suppress the spread of communism to Asia & South America.

But I'm an anarchist, not a communist. For me, the USSR failed partially b/c its way of spreading the 'revolution' just became tantamount to same-old, same-old imperialism. It *was* an "empire" in ultimately a way entirely too similar to what it was ostensibly revolting against.

Making matters worse, I think the USSR, esp under Stalin, suppressed the development of its own culture w/ the idiotic notion of Socialist Realism. The same mistake is made that's made everywhere else: that the working class is intrinsically stupid & must have its culture kept at a consistently moronic level to constantly bombard them w/ oversimplistic notions of who they supposedly 'are'.

The US has the PR reputation of being the great nation for individualists, the great nation where individual effort & vision is rewarded. In my own personal experience that's utter bullshit. It's also utter bullshit to me that individualism is by definition against the good of the community - wch is, perhaps, a more communist take on things. Why can't the individual stay true to personal integrity at the same time as recognizing that mutual aid is mutually beneficial? It seems obvious to me that it's healthier for the individual to have many friends than it is to have many enemies.

As such, the person quoted above who's "beginning to learn to say "I"" is someone I can relate to at an individual level enuf to feel the tendency for friendship w/ them. I wdn't feel that if they presented themselves as just another statistic in the herd. I don't befriend herds, I befriend individuals.

As for the "great empire" crumbling & then falling apart? "First Afghanistan, then Chernobyl"? The US cd probably learn a thing or 2 from that lesson. "["]The papers are saying that it's not just Chernobyl, all of Communism is blowing up. ["]" (p 235) Some people act like now that 'communism is dead' (wch is delusional enuf in itself) 'capitalism is the winner' as if it, too, can't die.

"Now people come here from other wars. Thousands of Russian refugees from Armenia, Georgia, Abkhazia, Tajikistan, Chechnya—from anywhere where there's shooting, they come to this abandoned land and the abandoned houses that weren't destroyed and buried by special squadrons. There are over 25 million ethnic Russians outside of Russia—a whole country—and there's nowhere for some of them to go but Chernobyl. All the talk about how the land, the water, the air can kill them sounds like a fairy tale to them. They have their own tale, which is a very old one, and they believe in it—it's about how people kill one another with guns." - p 239

Will humanity ever move forward? Will arms manufacturers & dealers ever be recognized as criminals far more harmful than someone who steals a piece of meat from a supermarket? Stay tuned! Detune.

"The experience of Chernobyl is not unique, but follows the secrecy pattern used at many lesser accidents which were mishandled in the same way. This has occurred both in the developed and developing world. In particular, I would note the radioactive pollution of the Mitsubishi Asian Rare Earth facility in Bukit Merah, Malaysia, the radioactive waste dumped in Nigeria and the contaminated food distributed to Egypt, Papua New Guinea, India and other countries during the Chernobyl disaster clean up.

"However, the health problems due to Chernobyl continue to be very acute right now, and demand international attention and action. Scientists and physicians are deprived of their freedom, and the people, especially the children, are suffering. This crisis can serve to point out the serious secrecy, vested interest and collusion of international agencies protecting nuclear technologies. The public face of the nuclear industry has been "clean and safe". It is important to unmask this public face, serving as a warning to economically developing countries deciding on energy technologies and bringing needed humanitarian aid to the victims. Preserving the false image of nuclear technology keeps the industry and nuclear agencies in business." - "Avoidable Tragedy post-Chernobyl - A Critical Analysis" by Rosalie Bertell, Ph.D., G.N.S.H., President Emerita of the International Institute of Concern for Public Health, Member of the Board of

Regents, International Association of Humanitarian Medicine - Journal of
Humanitarian Medicine, Vol. II, No. 3, pp 21 - 28.

"Belarusian writer and journalist Svetlana Alexievich who writes about Chernobyl
has won the 2015 Nobel Prize for literature." - http://
fromchernobyltofukushima.com/nobel/

The Truth About the Drug Companies

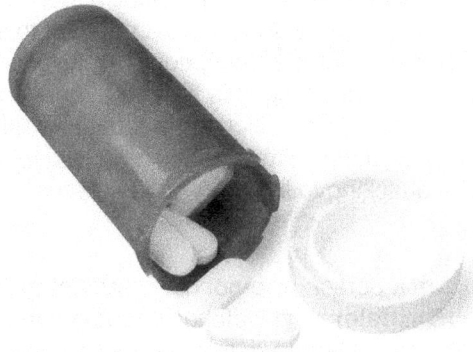

HOW THEY DECEIVE US
AND WHAT TO DO ABOUT IT

MARCIA ANGELL, M.D.

Former editor in chief of The New
England Journal of Medicine
Winner of the Polk Award

review of
Marcia Angell, M.D.'s **The Truth About the Drug Companies -
How They Deceive Us and What to do About It**
by tENTATIVELY, a cONVENIENCE - December 25-27, 2020

One of the many things that I've found annoying about the pseudo-dialog around what I call the PANDEMIC PANIC, the discussion about what's 'real' & what's a media-fabrication regarding COVID-19, has been some people's asking for "the science" that supports any position taken contrary to the mainstream narrative. This isn't because I'm opposed to science, although I do find it as potentially fallible as anything else, but because the people asking for it haven't generally, in my experience, much notion of what science is - nor wd they truly understand any science that they might encounter.

In other words, again in my personal experience, the people asking for "the science": 1. aren't scientists, 2. aren't intellectuals, 3. don't even read bks - except for, perhaps, the occasional thriller or bk relevant to some subcultural concern such as bike-riding. Nor are they people likely to've ever asked for "the science" to support much of anything else they've ever encountered in their life. Nor wd they be able to explain "the science" that backs what're hypothetically 'their own' positions on anything. The responsibility is solely on the person whose opinion they're attacking to 'prove' w/ "the science" that what they're saying is 'true'.

I, on the other hand, am a person who not only reads bks (thousands of them), but also *writes bks* (15 to date); who watches documentaries, & also *makes documentaries* (hundreds of them to date); & someone who writes & publishes criticism (something like 1,500 pieces to date). As such, I can easily demonstrate actual experience w/ critical thinking that the people asking for "the science" can't. At best, they can quote talking points from a radio program that they heard. Because they have other friends who heard the same program or something similar & because these friends can also paraphrase from these programs this parroting takes on a 'reality' to them.

W/ all this, & more, in mind, I've been accumulating bks that address medical science issues w/ the intention of actually reading them & quoting them & writing about them. Some of these bks, such as this one, are too based in commonly acccepted scientific legitimacy for most people to be able to easily dismiss them as somehow 'lunatic fringe' or 'conspiracy theorist'. Others are bks written by people so widely lambasted by what I call *Fact Chokers* (censors) that I'm curious about what they *actually say* instead of what people are being told they say in an attempt to discourage readers from finding out for themselves. I may or may not agree w/ them, I won't know until I actually read one of their bks. Finally, at least a few may say things that I find completely egregious & full of hidden agendas.

I decided to start reading these bks w/ this one b/c the title promised to support opinions & observations I already have AND b/c the author is fully credited in the area she's criticizing & is, therefore, difficult for people wanting "the science" to easily write off (w/o being told to do so by the people who tell them what 'to think' in the 1st place).

The author's bio in the back of the bk informs us of the following:

"The former editor in chief of *The New England Journal of Medicine* and a physician trained in both internal medicine and pathology, Marcia Angell is a nationally recognized authority in the field of health care and an outspoken proponent of medical and pharmaceutical reform. *Time* magazine named her one of the twenty-five most influential people in America. Dr. Angell is the author of *Science on Trial*." - p 307

TO BEGIN: ***READ THIS BOOK, IT'S ABSOLUTELY IMPORTANT***.

"Prescription drug costs are indeed high—and rising fast. Americans now spend a staggering $200 billion a year on prescription drugs, and that figure is growing at about 12 percent per year (down from a high of 18 percent in 1999).1" - p xii

"1. There are several sources of statistics on the size and growth of the industry. One is IMS Health (www.imshealth.com), a private company that collects and sells information on the global pharmaceutical industry. See www.imshealth.com/ims/portal/front/articleC/0,2777,6599_3665_41336931,00.html for the $200 billion figure." - p 267

It's important to inform you that everything Angell refers to is reinforced by endnotes that one can use to follow up. Alas, I DID just follow up on that one & got this message: "The page you requested was removed.". Given that this bk was published in 2004, it's no wonder that links might be broken. It's also possible that the recent spate of censorship (worse than any I've previously noted in my life) has something to do w/ it as might litigious behaviors of Big Pharma.

"I witnessed firsthand the influence of the industry on medical research during my two decades at *The New England Journal of Medicine*. The staple of the journal is research about causes of and treatments for disease. Increasingly, this work is sponsored by drug companies. I saw companies begin to exercise a level of control over the way research is done that was unheard of when I first came to the journal, and the aim was clearly to load the dice to make sure their drugs looked good. As an example, companies would require researchers to compare a new drug with a placebo (sugar pill) instead of with an older drug. That way the new drug would look good even though it might actually be worse than the older one." - p xviii

It's also important to emphasize that this bk is very solid in its presentation of the objectionable practices of Big Pharma. There are, in fact, so many issues brought to light & explained so clearly that this review can only hint at a few that I found most compelling. Again, I encourage the reader of this review to *read the entire bk from front-to-back* in order to thoroughly understand its well-developed points.

It might help the reader to understand my position here to explain that I don't take medicine except under truly extreme circumstances. I've taken many illegal drugs, esp important being consciousness-expansion drugs (a term I prefer to "psychedelics"). I'm particularly in favor of LSD & mushrooms — but I don't recommend them for everyone & I don't recommend using them frivolously. I also essentially stopped using those decades ago. Otherwise, I don't even take aspirin. I also rarely get headaches, & the worst headaches I've ever gotten have been from stupid excessive use of alcohol (I strongly warn people against hangovers where it hurts to think or move!). It used to be a joke of mine that all drugs that keep politicians alive shd be illegal. That upset some people b/c the implication was that I think medicines shd be illegal & many people I know are very dependent on them.. or at least think they are. Given my objection to a medicated society it was very welcome to me to read Angell's critique of the drug industry. Heroin is definitely a problem (& we can 'thank' Bayer for the early days of that) but pharmaceutical pushers are at least as bad — *& they're legal!*

"From 1960 to 1980, prescription drug sales were fairly static as a percent of U.S. gross domestic product, but from 1980 to 2000, they tripled. They now stand at more than $200 billion a year.1" - p 3

"1. These figures come from the U.S. Centers for Medicare & Medicaid Services, Office of the Actuary, National Health Statistics Group, Baltimore, Maryland. They were summarized in Cynthia Smith, "Retail Prescription Drug Spending in the National Health Accounts," *Health Affairs*, January-February 2004, 160." - p 268

That probably wdn't've been online as of the writing of the bk but there's some sort of gateway to it online now: https://www.healthaffairs.org/doi/abs/10.1377/hlthaff.23.1.160 .

Angell starts off w/ some historical philosophizing about how the Reagan presidency inaugurated much of the unrestrained greed of Big Pharma as we know it today. She doesn't however, blame the problem entirely on Republicans, she's quite frank in her look at similarly acting Democrats.

"You could choose to do well or you could choose to do good, but most people who had any choice in the matter thought it difficult to do both. That belief was particularly strong among scientists and other intellectuals. They could choose to live a comfortable but not luxurious life in academia, hoping to do exciting cutting-edge research, or they could "sell out" to industry and do less important but more remunerative work. Starting in the Reagan years and continuing through the 1990s, Americans changed their tune. It became not only reputable to be wealthy, but something close to virtuous. There were "winners" and there were "losers," and the winners were rich and deserved to be." - p 6

Of course, the author is referring to her own professional class here; simultaneously there were punks & anarchists & other 'lunatic fringe' types whose priorities were definitely not w/ getting rich but were instead w/ Truth, Justice, & the Unamerican Way. I was solidly in that camp. How many of us were following legal developments such as what Angell details next I don't know,

I certainly wasn't. But the Reagan administration in general was definitely high on the shit list.

"The most important of these laws is known as the Bayh-Dole Act, after its chief sponsors, Senator Birch Bayh (D-Ind.) and Senator Robert Dole (R-Kans). Bayh-Dole enabled universities and small businesses to patent discoveries emanating from research sponsored by the National Institutes of Health (NIH), the major distributor of tax dollars for medical research, and then to grant exclusive licenses to drug companies. Until then, taxpayer-financed discoveries were in the public domain, available to any company that wanted to use them." - p 7

Hhmm.. Taxpayer money pays for research, results enter Public Domain. That seems reasonable to me. But it also seems reasonable for researchers to benefit from their hard work above & beyond just salaries. Surely, a compromise solution cd be reached in wch the research stays in the public domain but the researchers are still rewarded for their exceptional accomplishment. At any rate, the Reagan admin was about benefitting big business, not the public. & the following is still from his January 20, 1981 – January 20, 1989 reign.

"Starting in 1984, with legislation known as the Hatch-Waxman Act, Congress passed another series of laws that were just as big a bonanza for the pharmaceutical industry. These laws extended monopoly rights for brand-name drugs. Exclusivity is the lifeblood of the industry because it means that no other company may sell the same drug for a set period. After exclusive marketing rights expire, copies (called generic drugs) enter the market, and the price usually falls to as little as 20 percent of what it was." - p 9

A justification for the original drug's high price is basically that the drug company had to spend a fortune on R&D (Research & Development). A significant part of this bk is spent debunking that as a PR myth.

"By 1990, the industry had assumed its present contours as a business with unprecedented control over its own fortunes. For example, if it didn't like something about the FDA, the federal agency that's supposed to regulate the industry, it could change it through direct pressure or through its friends in Congress." - p 10

Bypass democratic process anyone? The good ole 'merican way being pay-offs-every-wch-way. Profits before People, eh?

"The fact that Americans pay much more for prescription drugs than Europeans and Canadians is now widely known. As estimated 1 to 2 million Americans buy their medicines from Canadian drugstores over the Internet, despite the fact that in 1987, in response to heavy industry lobbying, a compliant Congress had made it illegal for anyone other than manufacturers to import prescription drugs from other countries." - p 15

I'm reminded of my friend Vermin Supreme (https://archive.org/details/

VerminSupremeHisHumbleBeginnings), a perpetual candidate for just about any
political office that he might be had by, & his proposed Health Plan shd he get
into power: a bus ticket to Canada. Yes, for some reason, the Canadian medical
system doesn't seem hell-bent on sucking every last asset out of its patients
before drugging & starving them to death in a hospice.

Every once in a while, one of these greedy big companies gets caught
committing a crime in pursuit of the Great American Dream (getting rich as fuck &
not giving a damn about who gets hurt by it) & has to pay the piper - but like all
big corporations busted in similar manner they've made so much profit off their
crime that the fines, enormous tho they may be, just come out of the profits as an
unfortunate expense.

"TAP Pharmaceuticals, for instance, paid $875 million to settle civil and criminal
charges of Medicaid and Medicare fraud in the marketing of its prostate cancer
drug, Lupron." - p 19

But there're all sorts of shenanigans going on that you're probably not aware of.
Have you ever been unwittingly used by a dr in a study w/o realizing that you're a
cash cow?

"To get human subjects, drug companies or contract research organizations
routinely offer doctors large bonuses (averaging about $7000 per patient in 2001)
and sometimes bonuses for rapid enrollment. For example, according to a 2000
Department of Health and Human Services inspector general's report, physicians
in one trial were paid $12,000 for each patient enrolled, plus another $30,000 on
the enrollment of the sixth patient. One risk of this bounty and bonus system is
that it can induce doctors to enroll patients who are not really eligible. For
instance, if it means an extra $30,000 to you to enroll a patient in an asthma
study, you might very well be tempted to decide your next patient has asthma,
whether he does or not ("Sounds like a little wheeze you have there. . . .").
Obviously, if the wrong patients are enrolled, the results of a trial are unreliable,
and that is probably often the case." - pp 30-31

Now you don't think that 6th patient enrolled is getting $30,000 too do you? Of
course not.. & they're getting hoodwinked into thinking that they're advancing
science & not being used for profiteering at the possible expense of their health.

"*The Wall Street Journal* urged the FDA to "reform its slow and blinkered
approach to potentially life-saving therapies" and "view itself not as a gatekeeper
but as a facilitator." The Washington Legal Foundation warned in one of its
advertisements in *The New York Times*, "Make no mistake, unneccessary
approval delays have human costs. Rigid procedures, endless data requests,
and the pursuit of absolutely risk-free products keep new treatments bottled up at
FDA while radically ill patients wait, suffer, and often die."" - pp 34-35

Oh, really? Where's 'the science'? In other words, if we're to "Make no mistake"
it might be a good idea to have some proof of the assertions made. It seems just

as likely, if not more so, that people die from rushing thru approval for big business profiteering. "[P]otentially life-saving therapies" might also be killers.

"Furthermore, all the data on the costs of those drugs were supplied by the companies to the Tufts group confidentially, and as far as I can tell, the authors were not able to verify the information. They were supposed to take the companies' word, and we were supposed to take theirs. That situation is extremely unusual in scientific publishing, where it is understood that the salient data will be made available to readers so they could evaluate the analysis for themselves." - p 42

In other words, it's like having the police investigate themselves or asking the Mafia if they've committed any crimes: *What me?!* A similar situation existed w/ the 'discrediting' of Hydroxychloriquine as a potentially helpful drug in connection w/ COVID-19: HCQ isn't expensive b/c it's been in use for a long time & long-since fallen out of patent, the drugs being developed are expensive & promise to make Big Pharma a bundle. I quote from my own bk, quoting, in turn, another source:

"& pull out of this, the Lancet and New England Journal of Medicine's retraction of their falsified HCQ discrediting studies

OK! Here's a relevant paragraph of their retraction:

"Our independent peer reviewers informed us that Surgisphere would not transfer the full dataset, client contracts, and the full ISO audit report to their servers for analysis as such transfer would violate client agreements and confidentiality requirements. As such, our reviewers were not able to conduct an independent and private peer review and therefore notified us of their withdrawal from the peer-review process."" - p 932, Unconscious Suffocation - A Personal Journey through the PANDEMIC PANIC

"One would expect a multinational database such as this to be a treasure trove coveted by researchers. Strangely, this is not so. Surgisphere has a razor thin folder of contributions to past publications. Besides the Lancet publication, Surgisphere's only other peer-reviewed publication is one entitled Cardiovascular, Drug Therapy, and Mortality in Covid-19 that was published on May 1, 2020 in The New England Journal of Medicine.

The Research section of Surgisphere's website features twenty-three "Case Studies from Around the World" as evidence of their prior work and product features. The vast majority of these "case studies" lack scientific substance and actually consist of short letters, press releases or potential use-cases for its database.

In place of actual research, the website appears primarily promotional and gives the impression of an immature tech company with lofty goals as opposed to a global database with real-time data on millions of patients.

A company with only five employees, most of which joined only two months ago.

According to LinkedIn, Surgisphere has five employees, only one of which has a medical degree—the founder Dr. Sapan Desai. The remaining four employees appear to have little to no science or medical background, but with a plethora of experience in business development and sales & marketing. The team's personnel consist of a VP of Business Development and Strategy, VP of Sales and Marketing and two freelance writers creating content for Surgisphere." - https://www.medicineuncensored.com/a-study-out-of-thin-air - quoted on p 1030, Unconscious Suffocation - A Personal Journey through the PANDEMIC PANIC

I don't mean to diss either the *Lancet* or the *New England Journal of Medicine* here, I'm just pointing out that this type of 'take-my-word-for-it' pseudo-science is exactly the type of 'science' that people asking for "the science" seem to accept w/o question. I *do mean to diss* **Surgisphere** who appear to've been formed just for the purpose of publishing fake studies that benefit Big Pharma's financial interests. But I digress.

"The NIH has been friendly to big pharma as well. (As we will see, some senior scientists at the NIH have had extensive financial dealings with drug companies.) Under considerable pressure from industry, in 1995, the agency completely abandoned its 1989 policy requiring "a reasonable relationship between the pricing of a licensed product, the public investment in that product, and the health and safety needs of the public." According to an NIH report, "Shortly after the policy of 'reasonable pricing' was introduced, industry objected to it, considering it a form of price control." And so it was! A well-intended but doomed effort to hold the industry accountable. As a result, companies like Bristol-Myers Squibb could charge whatever they liked for drugs like Taxol." - p 70

Alas, money talks & just about everybody walks. How many of us out here *can't be bought?!* Not many, that's for sure. Personally, I think my integrity, such as it is, is *priceless*. Big Pharma, on the other hand, seems to have little or no integrity at all & more greed than you can count on a billion sticky fingers. Their PR is that they need huge amts of moolah to support their R&D when in actuality they're just cranking out the same old drugs in new packages so they can extend their exclusive rights to them.

"So it is with big pharma. Every now and then, drug companies bring an innovative drug to market, but mainly they turn out a seemingly inexhaustible supply of leftovers—"me-too" drugs that are versions of drugs in the distant past." - pp 74-75

"Sometimes it's simply a matter of extending the life of a blockbuster drug that is going off patent by making a virtually identical drug and shifting users to the new one. The drug just has to be different enough to qualify for a new patent. Take the case of Nexium. Nexium is a heartburn drug of the proton pump inhibitor type made by the British company AstraZeneca. It came out on the market in

2001, just as the company's blockbuster drug for heartburn, Prilosec, was scheduled to go off patent. That was no coincidence. Unless there was a replacement, the loss of the Prilosec patent would have been a devastating blow to AstraZeneca. At $6 billion in annual sales, Prilosec was once the top-selling drug in the world. When the patent expired, it would face competition from generic manufacturers, and its sales would plummet." - pp 76-77

Poor babies, right? I'm sure that $6 billion a year was barely enough to make ends meet so the company wasn't even able to set any money aside for an acid-rainy day.

"Second, the market has to consist of *paying* customers. It doesn't help the bottom line to turn out drugs for nonpaying customers. That is why the pharmaceutical industry is supremely uninterested in finding drugs to treat tropical diseases, like malaria or sleeping sickness or schistosomiasis (an extremely common Third World disease caused by parasitic worms). Although these diseases are widespread, they are not important to the industry, since those who suffer from them are in countries too poor to buy drugs." - p 84

But.. but.. these Big Pharma companies are in CHRISTIAN countries!! Don't tell me that CHRISTIANS ARE HYPOCRITES?! Why, my whole world-view wd crumble if I had to accept that money is more important than morals to the people in Big Pharma. For those of you who're sarcasm-challenged, yes, I'm mocking capitalism & religion again.

The Truth About the Drug Companies - How They Deceive Us and What to do About It touches on & reinforces many of the things that I've been expressing dissatisfaction w/ for a looonnnnnggggg time: one of these being what I consider to be the excessive amt of 'disorder' diagnosing & the subsequent overkill of medications.

"Prozac, you will remember, was approved not only for depression but also for a series of related disorders. The me-too manufacturers simply expanded the list of psychiatric disorders. GlaxoSmithKline's Paxil, for instance, was approved to treat something called "social anxiety disorder"—said to be a debilitating form of shyness."

[..]

"The fact that few psychiatric disorders have objective criteria for diagnosis makes these disorders easier to expand than most physical illnesses." - p 88

I was shy once. Y'know what I did? I started working as a nude model. That helped. I started going out & getting drunk. I started pretending that I was drunk when I danced even tho I wasn't. Remember when people 'cured' their own shyness by having friends who were nice? That sort of thing? I mean, c'mon, do we really *need* a fucking prescription to help us deal w/ every little thing?! I think not. But it's in the interest of THE GOD ABOVE ALL OTHER GODS, **PROFIT**, to literally define every human characteristic as a 'disorder' & then prescribe drugs for it. Have you got a personality? *You must have Personality Disorder*, we can do away w/ that for you.

"Paxil was also approved for "*generalized* [my italics] anxiety disorder," and shortly after September 11, 2001, the company launched an ambitious campaign promoting the drug for this use. Commercials showed images of the World Trade Center towers collapsing. And who didn't feel anxious about that? But the implication was that even this perfectly appropriate (and, for most people, temporary) anxiety should be treated with drugs. The *New York Times* columnist Maureen Dowd summed it up best: "The more anxious the companies feel about profits, the more generalized the generalized anxiety disorders get."" - pp 88-89

&, gee, guess when diagnosis of GAD has been on the rise? During the QUARANTYRANNY. What a surprise. NOT. Not happy b/c you're being constantly force-fed a fear of death AND being severely restricted in all healthy social activities? *We've got a drug to make you pretend reality doesn't exist, it's a 'recession' & some politician is going to make it all better if you just go along with the program.* Even if you don't believe that one they'll tell you another one.. & another one.. ad infinitum.

But, but.. Won't those good scientists at the NIH get it all straightened out for us? No doubt there are scientists all over the place w/ integrity & deep knowledge, some of them may even rise to the top (& I'm rooting for them) — but, generally, the scum rises to the top b/c that's what they're aiming for, b/c that's what enables them to make. the. most. money. To quote myself: "When Money's God, Poor People are the Human Sacrifices."

"Some NIH scientists made hundreds of thousands of dollars in consulting fees. The deputy director of the Laboratory of Immunology, for instance, whose salary was $179,000 in 2003, was reported to have collected more than $1.4 million in consulting fees over eleven years and received stock options valued at $865,000.

"It is impossible to know to what extent these financial deals influenced NIH judgments about grants, research priorities, or the interpretation of results, but they certainly are cause for concern." - p 105

Beware of innocuous seeming euphemisms. "Consulting"? What's wrong w/ that? Someone's pd a fee to give their expert opinion - but what if that's not what's really happening? What if people in influential positions are being given all expenses pd vacations as "consultants" & expected to absorb a few hrs of Big Pharma propaganda in the process? Non-euphemistically, this is known as bribery. Gosh, it's so depressing, I think I'll drug myself into oblivion & make it all go away.

"Take the case of antidepressants of the SSRI type—Zoloft and Paxil. It is generally accepted that SSRIs are highly effective drugs. Millions of Americans take them, and many psychiatrists and primary care doctors swear by them. But a recent study throws doubt on the general enthusiasm. Using the Freedom of Information Act (a law that allows citizens to obtain government documents), the authors obtained FDA reviews of every placebo-controlled clinical trial submitted for initial approval of the six most widely used antidepressant drugs approved between 1987 and 1999—Prozac, Paxil, Zoloft, Celexa, Serzone, and Effexor (all but the last two are SSRIs). As is typical, most of the forty-two clinical trials lasted for just six weeks.

"Their findings were sobering. On average, placebos were 80 percent as effective as the drugs." - pp 112-113

Ha ha! Get a life, people!! Stop taking the antidepressants & develop a sense of humor.. or at least find a friend who has one. Or stop believing the hype that convinces you to be a slave to medications.

"Until 1997, drug companies didn't advertise much on television, because the Food and Drug Administration (FDA), which has jurisdiction over all prescription drug advertising, requiring them to include full information about side effects in their ads. This made thirty-second spots difficult—and even counterproductive. A drug could sound pretty scary with a rapid-fire listing of side effects. But in 1997, the FDA announced it would change the rules for broadcast ads. Instead of including a complete rundown of risks, companies would merely have to mention the major ones and refer viewers to a source of additional information" - pp 123-124

Give them an inch & they'll take your whole foot. Only a researcher wd be likely to follow-up on checking out the side-effects - & they're already questioning.

"Doctors don't want to alienate their patients, and too many of them find it faster and easier to write a prescription than to explain why it isn't necessary. That is why DTC ads are prohibited in every other developed country (except New Zealand)." - p 125

DTC = Direct To Consumer.

"Vanessa Fuhrmans and Gauram Naik, "In Europe, Prescription-Drug Ads Are Banned—and Health Costs Lower," *Wall Street Journal*, March 15, 2002, B1." - p 280

Given that the purpose of advertising is to convince the consumer that they want & need the object or service advertised then advertising for medications is geared to convince people that they're sick. Any sort of discomfort or awkwardness or minor pain that a person might feel becomes elevated to a DISORDER that 'must' be treated by advertising. &, of course, serious psychology is behind the ads so they're expert at knowing how to play on people's fears.

"Finally, whether the public benefits from taking more and more medicines for increasingly broadly defined diseases is open to serious question. One could make a strong argument that Americans with minor ailments suffer more from overmedication, and all the side effects and drug interactions that go with it, than from undermedication." - p 126

My position exactly. But there's so much bribery going on.

"And TAP didn't stop there. In 1996, the company also tried to persuade a large Massachussetts HMO, Tufts Health Plan, to stay with Lupron by offering its medical director of pharmacy programs a $25,000 "educational" grant that he could use for anything he wanted. The company couldn't have chosen a worse target: The medical director of pharmacy programs, Joseph Gerstein, is someone I know to be among the least likely people to take a bribe. When Gerstein refused, the company upped the offer to $65,000. But this time Gerstein, who with the support of Tufts had alerted federal authorities, taped the conversation, and that led to the unraveling of the company's illegal activities." - p 131

Ha ha ha. Oh, how I love seeing white collar criminals taken down. Of course, no-one goes to jail - while a friend of mine can get sentenced to 2 yrs in prison when he tries to legally purchase a case of beer not having an ID on him & gets into an altercation. HE goes to prison - but these corrupt shits sail clear as usual, a little bruised but still straight on course for more of the same. Nonetheless, thank you Joseph Gerstein for not beng bought.

"Many doctors become indignant when it is suggested that they might be swayed by all this industry largesse. But why else would drug companies put so much money into them? As Stephen Goldfinger, chairman of the APA's Committee on Commerical Support, said, "The pharmaceutical companies are an amoral bunch. They're not a benevolent association. So they are highly unlikely to donate large amounts of money without strings attached. Once one is dancing with the devil, you don't always get to call the steps of the dance."" - p 147

Again, beware of innocuous seeming euphemisms: in this case "education". Everyone's in favor of education, right? But what if the education has a subtext, a sales subtext? As always, I encourage people to be critical readers: even beyond the context of the issues at the heart of this bk there's 'education' going on all the time that has ulterior motives. Who defines what? How many history classes in universities are seeking out future CIA agents? The expression is that history belongs to the victor so if someone in a class consumes that history uncritically they might just be perfect for the Company, eh?

"In 2002, General Electric, with funding from big pharma, launched The Patient Channel, which shows medical programming interspersed with drug ads to patients in hospitals and waiting rooms across the country. Within a year, some eight hundred hospitals were carrying the network twenty four hours a day, seven days a week. Supported entirely by its advertisers, The Patient Channel cost hospitals nothing. Patients could choose among half-hour segments, such as

"Cancer Related Fatigue" or "Breathe Easy: Allergies and Asthma." Hospitals liked the idea, because they were told it would satisfy accreditation requirements that they educate patients about their illnesses. But the Joint Commission on Accreditation of Healthcare Organizations, the accrediting body, disagreed." - p 150

"It would deliver vulnerable, captive customers right to the companies' doorstep —or more precisely, bring the companies' doorstep to them." - p 151

"vulnerable, captive customers" is right. Patients desperately in search of a solution to their problem(s) will be presented w/ things posing as just that & will eagerly reach out for the product - but what if it's not *really* the solution?

"One of the least savory marketing efforts is Wyeth's campaign to "educate" college students about depression. What is really being marketed is the condition. If students can be convinced they have a treatable depression, selling the company drug Effexor is easy. To that end, Wyeth sponsors a ninety-minute forum on college campuses called "Depression in College: Real World, Real Life, Real Issues." It features doctors, psychologists, and Cara Kahn of the MTV reality show *Real World Chicago* who takes Effexor." - pp152-153

Now, here's a little story from my own life from my own bk:

"Living in continual poverty, despite working for a living, led to my finding out about clinics. Many of these had sliding scales, were otherwise cheap, or could be free if one pleaded penury. I engaged with 3 of them. At one of them the hours were very limited, maybe from 1:30PM to 4:30PM. I went there, it was cold out & raining, but the doctor hadn't arrived by 1:30 so the prospective patients were waiting in line out in the rain. That couldn't have helped their health much. Finally, the doctor arrived, an hour later, at 2:30 & I eventually got to see him in the company of 2 oncology med students. I told him that I was worried about my kidneys because they'd been hurting me lately & because my urine was particularly foul-smelling. He punched the skin over where my kidneys were, asked if that hurt & when I told him that it didn't he told me that my kidneys were fine. The doctor, however, was obviously far from fine. He told me that he'd recently had back surgery, he seemed very depressed. He also told me that he was a doctor at a university where he prescribed drugs for the students. I got the impression that that was pretty much it, a student would come in & say that they were depressed &/or stressed & he'd prescribe some pills. I wasn't impressed.

The doctor then informed me that I should go to the psychiatrist in the cliniic & get drugs prescribed for me because I'd been moving my hands too much while we were talking. Apparently, he took this to be a symptom of psychomotor agitation or some such. I didn't bother to explain that I'm a drummer & that when I'm bored & have nothing better to do I often practice on my legs. Instead, I explained to him that there's nothing wrong with my mind & that I didn't require any psychiatric prescriptions. It was obvious to me that he was there at the clinic to get a tax write-off, that he had close to no medical knowledge, & that what he had to offer was the chance to become addicted to drugs that would enable him to get kickbacks from the pharmaceutical companies. This man was more of a menace to patients than anything else — but poor people are expected to grovelingly accept whatever crap is handed us & be 'grateful' for it — but, as I like to say, **Before you decide against biting the hand that feeds you ask why it has so much food in the 1st place.**" - pp 1159-1160, <u>Unconscious Suffocation - A Personal Journey through the PANDEMIC PANIC</u>

"Phase I through III clinical trials are directed toward getting initial FDA approval, and they must meet the agency's scientific standards. Phase IV trials, in contrast, are studies of drugs already on the market, and many of them don't have to meet any standards at all." - p 161

Hence, there's a loophole. CRO = Contract Research Organization.

""Phase IV studies are the fastest growing segment of clinical spending," CenterWatch wrote. "This sweet spot in the market is being actively pushed by CRO's and offers unique opportunity for experienced, community-based clinical investigators." It's a "sweet spot" for the doctors, too. They usually make more working for CROs than spending the same time caring for patients. There are now tens of thousands of private doctors doing this work—many of them essentially being paid to prescribe a company drug." - p 165

"I find it hard to imagine that a system this corrupt can be a good thing, or that it is worth the vast amounts of money spent on it. But in addition, we have to ask whether it really is a net benefit to the public to be taking so many drugs. In my view, we have become an overmedicated society." - p 169

"The result is that too many people end up taking drugs when there may be better ways to deal with their problems.

"This conclusion was underscored by a large trial sponsored by the National Institutes of Health of ways to prevent adult-onset diabetes in people at high risk for the disease. One group in a trial received a placebo, and 29 percent of patients in that group developed diabetes over a three year period. The second group received a drug called metformin (the generic form of Bristol-Myers Squibb's blockbuster Glucophage), and they did somewhat better—22 percent developed diabetes. But the third group did much better than the other two. They were placed on a moderate diet and exercise program, and only 14 percent got diabetes. In other words, diet and exercise were better than the drug." - p 170

Once again, I share a personal story from my own bk:

"Since I'd suspected for some time that I was probably Diabetic Type 2 I decided to make an appointment to see a doctor. I had just turned 66 & had gone out with 2 friends in succession the night before the appointment. With the 1st friend I'd drunk 2 beers & I have a taste for strong beer. With the 2nd friend I was drinking wine & we went for a walk. The 2nd friend is an alcoholic so our walk was basically a beeline to her favorite bar. I told her that I didn't want another drink but she insisted & since I rarely have a chance to hang out with her I acquiesced & drank one more beer before I went home. The next morning when I woke up my legs were numb, it was horrible, it was worse than ever.

I walked to the doctor's office as a part of my intention to get more exercise. When I got there, I explained that I had Medicare, parts A & B, but that I didn't know what it covered. I told the receptionist that if it didn't cover this doctor's visit that I wouldn't go, that I'd turn around & leave because I was completely broke & couldn't afford it if my Medicare didn't cover it. She said that she was pretty sure that it covered it. I should've turned around & left. There was one other receptionist I had to see & I went through the same routine with her. She, also, was pretty sure my Medicare would cover it.

A doctor's assistant of some sort saw me next in the examination room & put me through the intial process of getting weighed & getting my blood-pressure taken, etc. The doctor appeared & the assistant left. The doctor asked questions & I explained what I was there for & described what my diabetes symptoms were. She agreed that I probably was diabetic. I told her that I didn't want medication & that I wanted to solve the problem with diet & exercise. She told me that that was "IMPOSSIBLE". I had told her that I don't use condoms so she insisted that I get all STD blood-tests. I explained that I'm a serial monogamist, that I'd been tested before my last girlfriend, & that I knew that I only have herpes. She continued to pressure me to get STD bloodtests, I replied that as long as it was paid for by my Medicare that I would do it. She said she thought it would be. When I explained my aversion to drugs & made further philosophical explanations such that I'm a naturist she rolled her eyes in mockery of my opinions. This didn't endear her to me OR work as peer pressure. She was most definitely NOT my peer or anywhere close.

She spent the entire time trying to browbeat me into shooting insulin, which I was told I had to do IMMEDIATELY, & to get other tests. Her whole approach was to try to make me be terrified of my apparent imminent death. I explained to her that I'm not particularly afraid of dying but that I'm worried about having the quality of my life lowered. Having to inject insulin every day would definitely do that. She impressed me as basically just a drug pusher who was trying to extract money from me, I didn't think she really had my health in mind."

[..]

"Within FIVE DAYS I got rid of my symptoms. Within TWO WEEKS I got my blood sugar down to an acceptable level. This was what the doctor had told me was "impossible". When I communicated this to the medical center I was told that I must've been "pre-diabetic", a fine bit of back-pedaling to someone who was told they were close to death before. In the meantime, when I tried to get more sticks for pricking myself when I was checking my blood-sugar I was told that there was a problem with my Medicare paying for it & that the doctor was trying to work it out. The pharmacist told me that I DID have needles there (for injecting insulin) & that there were also pills that were a substitute for injectable insulin. I explained that I hadn't asked for OR wanted either of them. Nonetheless, every time I went back I was reminded that they were there waiting for me to pick them up.

Roughly 6 months later I decided to cancel my Medicare part B assistance because I was having to pay for it and it was reducing my Social Security to something like $450 to $500 a month. As soon as I did this I was informed that, in fact, Medicare had not covered my visit & that I would possibly be billed for something like $1,300. Much or most of that was for the STD blood tests that I never wanted in the 1st place. I haven't been billed for it yet, perhaps the doctor had a crisis of conscience. Whatever the case, I will NEVER go back there, certainly NOT to that doctor & I've repeatedly informed representatives of the medical center who've called to harass me that this is the case. I did make a movie about my diabetes & my treatment of it that you can find here:

https://youtu.be/2GLu66dgpKl " - pp 1171-1173, Unconscious Suffocation - A Personal Journey through the PANDEMIC PANIC

If such conscienceless drug-pushing isn't foul enough for you, consider this:

"And finally, the Food and Drug Administration Modernization Act of 1997 added six months of" [patent] "protection if drug companies test their drugs in children. One might think that drugs to be used in children should be tested in them anyway, as a condition of FDA approval. But while the agency can require such testing, it seldom does. Instead, Congress offered the industry a gigantic bribe. The result is that drug companies now test their blockbusters, including drugs to treat primarily adult diseases like high blood pressure, in children, just because the extra protection is so lucrative." - p 182

"*The law granting drug companies an extra six months of exclusive marketing rights for testing drugs in children should be repealed.* That law is virtual bribery, and it doesn't even accomplish its stated purpose. Drug companies take advantage of the law to test blockbuster drugs in children whether the drugs are meant for this age-group or not. For an investment of a few million dollars or less, they can increase their revenues by hundreds of millions." - p 248

Surely, testing drugs on children for non-childhood diseases can't be good for the kids?!

"The most ingenious move to extend the life of Prozac was the creation of Sarafem—the identical drug in the identical dose, but colored pink and lavender instead of green, and taken for a new indication." - p 188

"Lilly renamed Prozac "Sarafem" and got FDA approval to market it for "premenstrual dysphoric disorder (PMDD)""

[..]

"Lilly received a six-month extension because it tested the drug on children—
something I can't imagine would be scientifically illuminating, since these
"children" must have been very nearly adults if they were experiencing PMDD." -
p 189

I mean, c'mon people, how low can you go? & w/ all this evidence of the
extremes of dishonesty that Big Pharma will go to to get even more fabulously
wealthy is it any wonder that some of us distrust the heavy push for mandatory
vaccines? That wd be a Cash ELEPHANT putting all previous Cash Cows into
miniature perspective.

"The pharmaceutical industry has by far the largest lobby in Washington—and
that's saying something. In 2002 it employed 675 lobbyists (more than one for
each member of Congress)—many drawn from 138 Washington lobbying firms—
at a cost of over $91 million." - p 198

I don't think $91 million spent on 675 lobbyists is just so they can twiddle their
thumbs, do you?

"Torricelli introduced a bill to make it easier to extend the patents of Schering-
Plough's blockbuster Claritin and a few other drugs. According to Common
Cause, this bill was introduced a day after Schering-Plough made a $50,000
contribution to the Democratic Senatorial Campaign Committee, which Torricelli
chaired. Senator Hatch held hearings on the bill." - pp 200-201

Yes, no matter how you spin it, it's bribery: the politicians go to the highest bidder
- *they're there to be bought.* Even though I think people generally understand
this they still act as if it's somehow not true. Actually facing the reality of this is
too much for people's peace of mind. But there are those of us who pay attn to
these things, some of us even have that increasingly rare quality: *a memory* - &
while the rest of you blunder along demanding "the science" that you're incapable
of paying attn to or understanding in your drug-addled way, some of us even
have enough sense to put 2 & 2 together in the puzzle of who's got the money &
how are they getting it.

"When the World Trade Organization was formed in 1995, members were
required to honor twenty-year patents on drugs. At the time, many countries did
not even consider drugs patentable. Exceptions were to be allowed for public
health emergencies. (In that case, governments could issue "compulsory
licenses" to have needed drugs produced by manufacturers.) Poor countries
were given until 2005 to comply. It was in this context that in the late 1990s
South Africa—desperate to control its HIV/AIDS epidemic—threatened to
produce or import drugs to fight it. The pharmaceutical industry adamantly
opposed any such move and the Clinton administration, no doubt reflecting the
industry's influence on Washington, warned of trade sanctions." - p 206

"The point is that the United States is generally seen as siding with drug company interests against the needs of millions of HIV/AIDS victims in the Third World." - p 207

For those of you who still wonder why some of us oppose the actions of the WTO, the WEF, the IMF, etc, consider the above 2 quotes. I, personally, think & feel that humanitarian values are more important than getting rich. But that's just me.

"The United States is the only developed nation that does not regulate drug prices. All the others—Australia, Canada, France, Germany, Italy, Japan, the Netherlands, Spain, Switzerland, Sweden, Britain, and so on—do." - p 219

Spokespeople for US government policy have already come out in favor of torture, how long will it be before this notion of 'Free Trade' blatantly includes slavery?

"Between 2000 and 2003, according to Michael Loucks, chief of the Health Care Fraud Unit in the U.S. Attorney's Office for the District of Massachusetts, eight companies paid out a total of $2.2 billion in fines and settlements. Four of those companies—TAP Pharmaceuticals (discussed in Chapter 7), Abbott, AstraZeneca, and Bayer—pleaded guilty to criminal charges. TAP, the champion so far, paid a total of $885 million, of which $290 million were criminal fines. Loucks pointed out in a speech that the company received $2.7 billion in revenues from Medicare during the 1990s, so it still came out well ahead." - pp 232-233

Aye!, there's the rub! That's the way the accumulation of wealth has always worked the whole world round: if you steal more than can ever be taken back from you your position for doing it more & more becomes more & more secure. People who think that billionaires don't consistently have a history like this are naive. When I was a teenager it was common for people to say "You're an idealist" to anyone who might point out things like what I've just written. This meant that an "idealist" doesn't live in 'reality' where a complete disregard for anything but getting rich is considered 'The Way of the World.' Bullshit. An idealist, if one accepts the term, is a person who has a clear vision of what they ethically prefer & chooses to not be bribed into muddying that vision.

"But there is value in trying to define the ideal system, so that we can move toward it in the best way possible—unevenly and incompletely, if necessary—but at least with an understanding of where we want to go." - p 238

If the goal is good health for as many as possible why let greed corrupt it?

The author is a reformer, there's always the question of whether reform is more to the heart of the matter than larger scale possibilities. That sd, I'm happy to consider her reform proposals & hope that people more in a position to make them happen consider them. In the meantime, I think I'll avoid the Medical

Industry as much as possible.

"*Food and Drug Administration regulations should require that new drugs be compared not just with placebos but with old drugs for the same conditions.* Approval should depend on whether the new drug adds something useful in terms of greater effectiveness, greater safety, fewer side effects, or substantially greater convenience." - p 240

That seems like an obviously good idea to me.

"When your doctor prescribes a new drug, ask him or her these questions: *What is the evidence that this drug is better than an alternative drug or some other approach to treatment? Has the evidence been published in a peer-reviewed medical journal? Or are you relying on information from drug company representatives?* Insist on getting a straight answer and, if necessary, a reference to a journal article or a medical textbook." - pp 261-262

Another good idea - unfortunately it's one that's easier sd than done. For one thing, a patient visiting a dr is usually at a low point, for another, insisting on getting a straight answer won't necessarily yield the truth.

All in all, I think this is an extremely important bk - esp in this time when a belief in the Medical Industry is justifying running roughshod over civil liberties. Read it. Even those of you who demand "the science" w/o actually having an intellect capable of understanding what you're asking for might be able to understand the clear presentation of Big Pharma corruption as laid out here. Thank you, Marcia Angell, M.D..

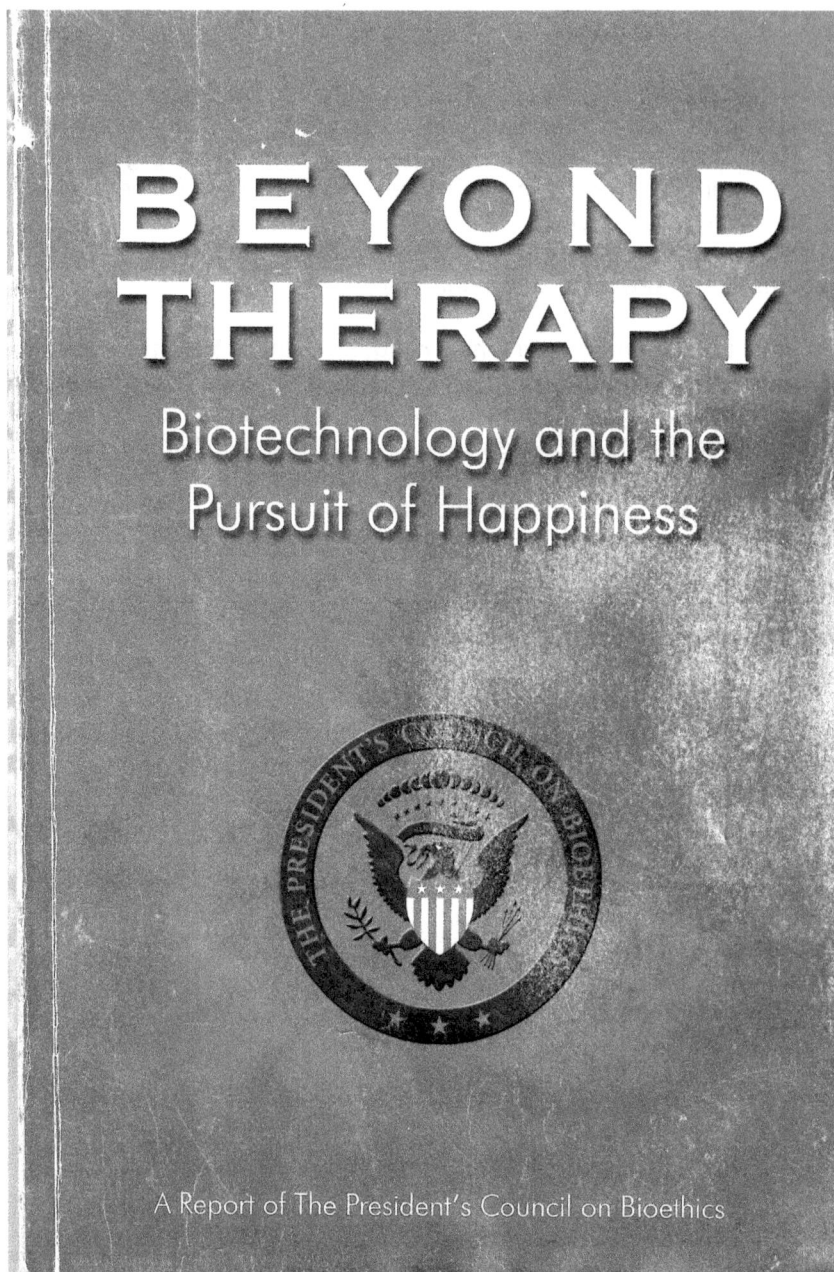

BEYOND THERAPY

Biotechnology and the Pursuit of Happiness

A Report of The President's Council on Bioethics

review of
The President's Council on Bioethics's **Beyond Therapy -
Biotechnology and the Pursuit of Happiness**
by tENTATIVELY, a cONVENIENCE - May 20-28, 2021

"The Future is sooooo Yesterday"

As I've commented previously, (review of Marcia Angell, M.D.'s <u>The Truth About the Drug Companies - How They Deceive Us and What to do About It</u>),

"One of the many things that I've found annoying about the pseudo-dialog around what I call the PANDEMIC PANIC, the discussion about what's 'real' & what's a media-fabrication regarding COVID-19, has been some people's asking for "the science" that supports any position taken contrary to the mainstream narrative. This isn't because I'm opposed to science, although I do find it as potentially fallible as anything else, but because the people asking for it haven't generally, in my experience, much notion of what science is - nor wd they truly understand any science that they might encounter.

"In other words, again in my personal experience, the people asking for "the science": 1. aren't scientists, 2. aren't intellectuals, 3. don't even read bks - except for, perhaps, the occasional thriller or bk relevant to some subcultural concern such as bike-riding. Nor are they people likely to've ever asked for "the science" to support much of anything else they've ever encountered in their life. Nor wd they be able to explain "the science" that backs what're hypothetically 'their own' positions on anything. The responsibility is solely on the person whose opinion they're attacking to 'prove' w/ "the science" that what they're saying is 'true'."

In OTHER other words, I decided to read the occasional science bk. These people who want "the science" are basically just looking for sound bites to quote that come from their favorite propaganda sources - usually NPR. Such sound bite quoting presents a completely false front of being intellectually informed. I decided to go a step further & to read some medical science bks, admittedly targetted somewhat at the layperson, & to write reviews that wd express what I've learned from the bks & my opinions regarding these works. The 1st bk I chose was the above-mentioned <u>The Truth About the Drug Companies</u> wch completely validated positions that I'd already had for yrs.

This 2nd bk I chose as medical science to read was this one - wch I expected to be in an adversarial position in relation to. When I saw it for sale I practically groaned out loud at what I thought it might present. After all, "The President's Council on Bioethics" was formed during the George W. Bush administration. "W" was one of my most hated presidents, someone who felt free to advocate torture as AOK as USA practice. That was despicable almost beyond belief.

Imagine my surprise when I found <u>Beyond Therapy</u> to be thoughtful &, once again, essentially reinforcing many of the my own criticisms. As w/ so many non-fiction bks I choose to read, there's so much here of value that writing this review will be quite a task - but one that's very worthwhile.

The Council was created on November 28, 2001, by means of Executive Order 13237. There were 17 members of The President's Council on Bioethics. Their names & credentials are listed on pp xi-xiii. I doubt that people wanting "the science" wd find them lacking or 'debunked' or quacks or whatever:

Leon R. Kass, M.D., PH.D., Chairman
Elizabeth H. Blackburn, Ph.D.
Rebecca S. Dresser, J.D., M.S.
Daniel W. Foster, M.D.
Francis Fukuyama, Ph.D.
Michael S. Gazzaniga, Ph.D.
Robert P. George, J.D., D.Phil.
Mary Ann Glendon, J.D., M.Comp.L.
Alfonzo Gómez-Labo, Dr. phil.
William B. Hurlbut, M.D.
Charles Krauthammer, M.D.
William F. May, Ph.D.
Paul McHugh, M.D.
Gilbert C. Meilaender, Ph.D.
Janet D. Rowley, M.D.
Michael J. Sandel, D.Phil.
James Q. Wilson, Ph.D.

I'd never heard of any of these people - but there's no reason why I wd've.
They're mostly professors. Meilaender is a "Professor of Christian Ethics".
Wilson is a "Reagan Professor of Public Policy". While their credentials are of
the type widely respected in our society they're not necessarily ones I entirely
trust: Law, Government, Journalism, Religion, & Medicine are all hypothetically
based in deep ethical principles but how often is that actually the case? W/
reservations in mind, I approached this bk wondering what duplicitous logic might
be presented to justify crimes against humanity. NONETHELESS, I found it to
actually present its case(s) w/ calm, considered integrity.

Readers of this review might be wondering what any of this has to do w/ the so-
called Covid-19 pandemic & the various actions that've been hyped as protecting
the general public. After all, this report was presented on October 15, 2003 - 16
yrs *before* Covid. The importance of this report lies in its analysis of the ethical
problems to be encountered w/ future medicine. That future was in-progress
then & that future is in-progress now. Mark Lynas published a blog entry on
December 17, 2020, entitled "Yes, some COVID vaccines use genetic
engineering. Get over it." (https://allianceforscience.cornell.edu/blog/2020/12/yes-
some-covid-vaccines-use-genetic-engineering-get-over-it/) As the title
indicates, the author thinks that this genetic engineering is fine. Nonetheless,
there are other people who are concerned about the ethical implications of some
medical directions - wch include genetic engineering.

The beginning of the Council's October 15, 2003, "Letter of Transmittal to the
President", written by the Council's Chairman, Leon R. Kass, M.D., is as follows:

"I am pleased to present to you Beyond Therapy: Biotechnology and the Pursuit
of Happiness, a report of the President's Council on Bioethics.

"The product of more than sixteen months of research, reflection, and deliberation, we hape this report will prove a worthy contribution to public understanding of the important questions it considers. In it, we have sought to live up to the charge you gave us when you created this Council, namely, "to undertake fundamental inquiry into the human and moral significance of developments in biomedical and behavioral science and technology" and "to facilitate a greater understanding of bioethical issues."

"Biotechnology offers exciting and promising prospects for healing the sick and relieving the suffering. But exactly because of their impressive powers to alter the workings of body and mind, the "dual uses" of the same technologies make them attractive also to people who are not sick but who would use them to look younger, perform better, feel happier, or become more "perfect."" - p xv

Later in the same letter, some reservations appear:

"We want better children—but not by turning procreation into manufacture or by altering their brains to gain them an edge over their peers. We want to perform better in the activities of life—but not by becoming mere creatures of our chemists or by turning ourselves into tools designed to win or achieve in inhuman ways. We want longer lives—but not at the cost of living carelessly or shallowly with diminished aspiration for living well, and not by becoming people so obsessed with our own longevity that we care little about the next generations. We want to be happy—but not because of a drug that gives us happy feelings without the real loves, attachment, and achievements that are essential for true human flourishing." - p xvii

How many people have given much thought to the above considerations? Entirely too few IMO (In My Opinion) - including those who clamor for "the science". Yes, let's have "the science" - but let's not act as if all science inevitably generates change for the better.

Definitions of "biotechnology" are given in a footnote on p 1:

"These range from "engineering and biological study of relationships between human beings and machines" (*Webster's II New Riverside University Dictionary*, 1988), to "biological science when applied especially in genetic engineering and recombinant DNA technology" (*Mirriam-Webster OnLine Dictionary*, 2003), to "the use of biological processes to solve problems or make useful products" (Glossary provided by BIO, the Biotechnology Industry Organization, www.bio.org, 2003). In the broader sense of the term that we follow here, older technologies would include fermentation (used to bake bread and brew beer) and plant and animal hybridization. Newer biotechnologies would include, among others, processes to produce genetically engineered crops, to repair genetic defects using genomic knowledge, to develop new drugs based on knowledge of biochemistry or molecular biology, and to improve biological capacities using nanotechnology. They include also the products obtained by these processes: nucleic acids and proteins, drugs, genetically modified cells, tissues derived from stem cells, biomechanical devices, etc." - p 1

"In this sense, it appears as a most recent and vibrant expression of the technological spirit, a desire and disposition rationally to understand, order, predict, and (ultimately) control the events and workings of nature, all pursued for the sake of human benefit." - p 2

This was the type of rhetoric that I was expecting, one untempered by acknowledgment of *things that can go wrong*. Note no mention of bioweaponry, e.g.. Note no mention of "control" as imprisonment. But it isn't long before the authors redeem themselves:

"Biotechnologies are already available as instruments of bioterrorism (for example, genetically engineered super-pathogens or drugs that can destroy the immune system or erase memory), as agents of social control (for example, tranquilizers for the unruly or fertility-blockers for the impoverished)" - p 6

or look at endnote 37 on p 100:

"See, for example, Eberstadt, M., "Why Ritalin Rules," *Policy Review* 94, April/ May 1999; DeGrandpre, R. *Ritalin Nation: Rapid-Fire Culture and the Transformation of Human Consciousness*, New York: Norton, 1999; and Hancock, L., "Mother's Little Helper," *Newsweek*, 18 March 1996."

I'm sure biotechnology has come a long way in the 18 yrs since this bk was written. Even then, many uses were somewhat 'normalized':

"The already widely accepted "beyond therapy" uses of biomedical technologies include: pills for sleep and wakefulness, weight loss, hair growth, and birth control; surgery to remove fat and wrinkles, to shrink thighs, and to enlarge breasts; and procedures to straighten teeth and select the sex of offspring. These practices are already big business. In 2002 Americans spent roughly one billion dollars on drugs used to treat baldness, about ten times the amount spent on scientific research to find a cure for malaria, a disease that afflicts hundreds of millions of people worldwide." - footnote, pp 8-9

I'm reminded that I met a Danish guy (he was filled w/ cheese) in 1994 in Europe who told me a story about camping out in Africa & getting malaria. He became so debilitated that he was too weak to try to go somewhere for help. He was alone. Finally, some local tribespeople came by, saw what was wrong & showed him a nearby plant to eat. That healed him enuf to recover. He's suffered from return bouts since but if it hadn't been for that plant he probably wd've died. Is that plant known about by 'Western Medicine'?

"We shall not here consider biotechnologies as instruments of bioterrorism or of mass population control. The former topic is highly specialized and tied up with matters of national security, an area beyond our charge and competence. Also, although the practical and political difficulties they raise are enormous, the ethical and social issues are relatively uncomplicated. The main question about bioterrorism is not what to think about it but how to prevent it. And the use of tranquilizing aerosols for crowd control or contraceptive additions to the drinking water, unlikely prospects in liberal democratic societies like our own, raise few issues beyond the familiar one of freedom and coercion." - p 10

Then again, this Council was created by President George W. Bush. "Liberal democratic societies" don't advocate & use torture either but Bush did &, as far as I know, it didn't start or stop w/ him. While "the ethical and social issues" may be "relatively uncomplicated" for people who HAVE ethics - that doesn't mean that the people calling the shots in a so-called 'liberal democratic society' *have those ethics*. While I don't know of "tranquilizing aerosols for crowd control" that doesn't mean that militarized police & weapons that can damage crowd hearing aren't just as insidious, if not more so. There are those who argue that Margaret Sanger was a eugenicist whose Planned Parenthood program was designed to reduce breeding amongst the poor. I'm pro- Planned Parenthood but I still think the criticism is valid.

"It reflects humankind's deep disatisfaction with natural limits and its ardent desire to overcome them." - pp 10-11

"When René Descartes, in his famous *Discourse on Method*, set forth the practical purpose for the new science he was founding, he spoke explicitly of our becoming "like masters and owners of nature"" - p 11

I don't like that *enslaving* attitude. We are a part of nature, nature is much bigger than any individual intellect or even the collective human intellect. What if instead of enslaving something that we deny being a part of we tried to learn how to appreciate & work w/ its vast capabilities?

The authors address the important distinction between therapy & enhancement:

"Gene therapy for cystic fibrosis or Prozac for major depression is fine; insertion of genes to enhance intelligence or steroids for Olympic athletes is, to say the least, questionable." - p 14

But even Prozac is subjected to further scrutiny - & I think rightfully so.

As a literate person, I enjoy & appreciate that the Council quotes literary sources. For one thing, by doing so, they're acknowledging writing as a thing that speaks of & to the culture it's produced in. Sometimes I wonder whether that respect is vanishing from our society.

"Dreams of human perfection—and the terrible consequences of pursuing it at all costs—are the themes of Greek tragedy, as well as of "The Birth-mark," the Hawthorne short story with which the President's Council on Bioethics began its work." - p 18

& what about death? This is another subject that I've been somewhat philosophically preoccupied w/ for the last 2 yrs. When I went to a dr about Diabetes Type II she immediately tried to terrorize me w/ extreme fears of impending death. I explained to her, truthfully, that I'm not afraid of death as much as I am of having my quality of life degrade further as I approach it. If I had followed the dr's attempted orders the quality of my life wd've deteriorated substantially & her income wd've increased proportionately.

"Curiously, we may even be more afraid of death than our forebears, who lived before modern medicine began successfully to do battle with it." - p 18

"In Chapter Two, we consider the pursuit of "better children," using techniques of genetic screening and selection to improve their native endowments that might make them more accomplished, attentive, or docile." - p 21

Docility might be more convenient for control freaks but I wdn't put it in the same category as "accomplished" & "attentive".

Adding to the overall conventional credibility of this report:

"we commissioned presentations from a wide array of scientists working or writing in the pertinent fields of biology and biotechnology: preimplantation genetic diagnosis and genetic enhancement (Gerald Schatten and Francis Collins); choosing sex of children (Arthur Haney and Nicholas Eberstadt); drugs to modify behavior in children (Lawrence Diller and Steven Hyman); genetic enhancement of muscle strength and vigor (H. Lee Sweeney); genetic enhancement of athletic performance (Theodore Friedmann); aging and longevity research (Steven Austad and S. Jay Olshansky); memory, and drugs that might improve or blunt it (James McGaugh and Daniel Schacter); and mood-brightening drugs (Peter Kramer and Carl Elliott)." - p 22

I wonder, however, if it might not be useful to study what cultures that don't use 'Western Medicine' have to show us. E.G.: are there memory problems in Amazonian tribes? It seems to me that memory problems in the society I live in are fantastically widespread & getting worse every day. Is that the case everywhere or are the drugs commonly used & the devices relied upon not a major cause of memory dysfunction? & what about HAPPINESS?, another subject that I'm preoccupied w/. Is it in such short supply in all cultures? If not, what distinguishes the cultures where there's greater happiness from the culture I see around me?

The bk is structured to address central purposes for biotechnology. Chapter 2 is entitled "Better Children".

"To help our children on their way and make them strong in body and mind, we clothe and feed them, so that they get rest, fresh air, and exercise, and take great pains regarding their education. Beyond ordinary schooling, we give them swimming and piano lessons, enroll them in Scouts or Little League, and help them acquire a variety of skills—artistic, intellectual, and social."

[..]

"Needless to say, the thing is easier said than done" - p 27

"For what, exactly, is a good or a better child?

"Is it a child who is more able and talented? If so, able in what and talented how? Is it a child with better character? If so, having which traits or virtues? More obedient or more independent? More sensitive or more enduring? More daring or more measured? Better behaved or more assertive? Is it a child with the right attitude and disposition toward the world? If so, should he or she tend toward more reverence or skepticism, high-mindedness or toleration, the love of justice or the love of mercy?" - pp 10-11

I, of course, think of my own childhood. My mother was an OCD robopath & untertan; my father was a completely selfish con artist who had almost no feeling for or interest in his family. They were separated by the time I was 9 & divorced by the time I was 10. My mom was completely unsuited to making her way in the world b/c she'd been a firm believer in having a solid family unit w/ a father who took care of the financial end. My dad never pd alimony or child support &, according to my mom, stole her mother's life savings. Were either of these people capable of making good life decisions for their children? I think not. Nonetheless, they were what I was stuck w/.

I *did* go swimming, I loved that. I was forced to take piano lessons, I hated them but I still went to them for 3 or 4 yrs. I'm a pianist as an adult but I have my doubts that the lessons contributed much other than aversion. I was forced into the Cub Scouts, wch I also hated. I was told that it wd get good once I was in the Boy Scouts. They were even worse so I got to drop out. I was forced into Little League softball, something I had no aptitude for, or interest in, whatsoever. After my ineptitude there became obvious I was allowed to drop out. I had art lessons in the summer. I enjoyed those & had the 'talent'. My teacher got arrested as a bank robber. Teaching art to kids apparently wasn't paying the bills. Surprise, surprise. The point is that the things that my mom did 'for me' were all attempts to make me a 'normal' kid & to get me out of the house under adult supervision. In at least one instance, my 'adult supervision' was quietly gotten out of the way as a child molestor.

THEREFORE, the above questions about what constitutes "a good or a better child" strike me as very pertinent. From the perspective of my dysfunctional & mostly selfish parents, the only role for me was to follow orders & keep my mouth shut. My mom wanted me to go to the Vietnam War. She had no idea what the war was about but she was brainwashed enuf to accept that it was obviously necessary. I was opposed to the war & was willing to risk 10 yrs imprisonment for refusing to register fo the draft. Since it was the 'normal' thing to do, my mom absolutely cd not accept my choice but was fine w/ my going thousands of miles away to kill people who'd never done anything to me & to risk being mutilated or killed in the process. I often joke that if it had been the 'normal' thing to do my mom wd've had me castrated & lobotomized. As for my dad? If I'd died in front of him maybe he wd've checked my pockets to see if I had any money.

My point is obvious: if my parents had had even more power over my life than they already did I wd've been a vegetable - so I'm not in favor of biotechnology that indiscriminately enables that. How about fostering a society where people who become parents are less naive & more competent instead? Of course, that's a somewhat 'impossible' dream.

"We give our children supplementary vitamins, fluoridated toothpaste, and, where necessary, corrective lenses or hearing aids. We even use biological means of improving their limited human capacity to resist disease: we immunize our children against polio, diptheria, and measles, among other infectious diseases, by injecting them with attenuated viruses and bacteria in the form of vaccines." - p 30

But there are those of us who think that not all of those things are healthy. Let's take the example of "corrective lenses" "where necessary": what are the criteria for necessity & is that criteria infallible? Where there's a profession that promotes glasses & a profession that manufactures them, etc, it seems almost inevitable to me that the 'necessity' for them will (& has) become exaggerated. I'm 67, I've never worn glasses, my eyesight is still excellent - not as good as it was when I was younger, but I can still read & drive at night, etc.. Once a person's eyes become dependent on glasses it's all downhill. Their eyesight is only going to be functional while they're wearing them & stronger glasses will gradually be needed as their natural unassisted eyesight gets weaker. So what if some sort of eye exercise were promoted for young children w/ weak eyesight? Maybe their eyesight wd improve & they wdn't develop a lifelong dependency on an object.

"It is merely to insist that the use of equipment in sports, as in the rest of life, changes and even binds the human users, often without their knowing it.

"The point was beautifully made by Rousseau, commenting on how even the earliest human inventors of artful aids to better living "imposed a yoke on themselves without thinking about it":

"For besides their continuing thus to soften body and mind, as these commodities had lost almost all pleasantness through habit, and as they had at the same time *degenerated into true needs*, being deprived of them became much more cruel than possessing them was sweet; and people were unhappy to lose them without being happy to possess them. (Emphasis added.)" - p 125, Beyond Therapy

As for vaccines? These days, in the New Dark Ages, anyone who questions vaccines in any way is pejoratively pigeonholed as a 'bonehead' or whatever. I was vaccinated against polio as a child, I didn't get polio, maybe the vaccine helped, maybe it didn't. I was vaccinated against avian flu at 14, I really doubt that that was necessary. As far as I know I haven't been vaccinated in the ensuing 53 yrs *except by nature*. Many people will take exception to the latter notion but I find it completely reasonable. By natural exposure to my environment, I become innoculated to it & by it. If I contracted COVID-19 in the last 16 mnths it had the effect of making me feel very tired for ONE DAY. I didn't cough, my nose didn't run. I felt a bit strange the next day but not sick. To me, that was my vaccination from nature. If my body had been more worn out than it is, the C-19 might've pushed me over the edge to death - but I wdn't blame the Covid as much as I wd my deteriorated body. I'm much more likely to have my body defeated by human-made toxicities: polluted water, polluted air, radioactivity, sugar, *drugs*. I'm considerably less concerned w/ being hit by lightning than I am w/ being hit by a car.

A footnote on p 33 can give us a little comic relief:

"The Repository for Germinal Choice, a California sperm bank accepting deposits only from Nobel Laureates or other comparably accomplished donors, recently closed its doors, having done only minimal business in the roughly twenty-five years of its existence."

Maybe they shd've tried pop star donors instead - or how about doner kebabs?

"Yet, second, to achieve these benefits prenatal dignosis adopts a novel approach to preventative medicine: it works by eliminating the prospective patient before he can be born. This kind of preventative medicine is thus in fact a species of negative eugenics—elimination of the genetically unfit and a reduction in the incidence of their genes—albeit carried out voluntarily and on a case-by-case basis." - p 36

Voluntarily? The mother volunteers, the baby that doesn't get to be born doesn't. Imagine if the joke were on the drs if it were eventually discovered that something about those genes was actually very special & useful, an unusual talent. Oh well, extinct. As the footnote on p 38 says:

"Growing recognition of the complexity of gene interactions, the importance of epigenetic and other environmental influences on gene expression, and the impact of stochastic events is producing a strong challenge to strict genetic determinism. Straightforward genetic engineering of better children may prove impossible, not only in practice but even in principle."

In fact, all human endeavor can backfire - but the more directly one fiddles w/ the complexity of genetics the higher the likelihood of sd complexity being far beyond the perceptual limitations of people who may be more pompous than they are actually brilliant.

"Insertion of genes into the host genome can cause abnormalities, either by activating harmful genes or by inactivating useful ones. Recently, for example, children undergoing experimental gene therapy for immune system deficiencies have developed leukemia after retroviral gene transfer into bone marrow stem cells, very likely the result of activation of a cancer-producing gene by the virus used to transfer the therapeutic genes into the cell." - pp 38-39

IMO, the human immune system is amazing, far superior to medicine. So what happens to it when it goes awry? My immediate suspect is *human intervention*. Why did these children have immune system deficiencies in the 1st place? Cd there be something in their parents' history that caused it? & was that something *medically or environmentally induced?!*

These days I wonder when certain diseases came into being & whether they're a side-effect of human-created environment changes. Take, e.g., Multiple Sclerosis: I have at least 3 friends w/ MS. SO, I ask online about MS's history & read this:

"Multiple sclerosis was first recognised as a condition in the middle of the 19th century. Prior to this time, there are reports of a few instances of what may have been MS, although the variety of symptoms, the range of other possible causes and the incompleteness of records make these impossible to confirm." - https://mstrust.org.uk/a-z/history-ms

When I asked: "What are the causes of MS?" I found this:

"The cause of multiple sclerosis is unknown. It's considered an autoimmune disease in which the body's immune system attacks its own tissues. In the case of MS, this immune system malfunction destroys the fatty substance that coats and protects nerve fibers in the brain and spinal cord (myelin).

"Myelin can be compared to the insulation coating on electrical wires. When the protective myelin is damaged and the nerve fiber is exposed, the messages that travel along that nerve fiber may be slowed or blocked.

"It isn't clear why MS develops in some people and not others. A combination of genetics and environmental factors appears to be responsible."

- https://www.mayoclinic.org/diseases-conditions/multiple-sclerosis/symptoms-causes/syc-20350269

The same article goes on to list these "risk factors":

"These factors may increase your risk of developing multiple sclerosis:

"**Age.** MS can occur at any age, but onset usually occurs around 20 and 40 years of age. However, younger and older people can be affected.

"**Sex.** Women are more than two to three times as likely as men are to have relapsing-remitting MS.

"**Family history.** If one of your parents or siblings has had MS, you are at higher risk of developing the disease.

"**Certain infections.** A variety of viruses have been linked to MS, including Epstein-Barr, the virus that causes infectious mononucleosis.

"**Race.** White people, particularly those of Northern European descent, are at highest risk of developing MS. People of Asian, African or Native American descent have the lowest risk.

"**Climate.** MS is far more common in countries with temperate climates, including Canada, the northern United States, New Zealand, southeastern Australia and Europe.

"**Vitamin D.** Having low levels of vitamin D and low exposure to sunlight is associated with a greater risk of MS.

"**Certain autoimmune diseases.** You have a slightly higher risk of developing MS if you have other autoimmune disorders such as thyroid disease, pernicious anemia, psoriasis, type 1 diabetes or inflammatory bowel disease.

"**Smoking.** Smokers who experience an initial event of symptoms that may signal MS are more likely than nonsmokers to develop a second event that confirms relapsing-remitting MS."

I find that interesting. Is it that the person is white & living in a temperate climate or is there something else about that combination that leads to the body attacking itself?

"The precise cause of MS still remains unknown. Genetic, environmental, and immunological factors have been implicated in the etiology of this complex and heterogeneous disease. The rise in the incidence and prevalence of MS in the world in the past decades paralleled the rapid socioeconomic development, urbanization, and westernization, which was marked by radical change in dietary and lifestyle habits. The industrial revolution and the contemporary age in Western countries gave rise to the fast-food industry and widespread consumption of excessive salt, refined vegetable oils, and sugars and also led to reduced physical activity, exposure to artificial light at atypical biological times, and insufficient and poor-quality sleep. The influence of other environmental factors, such as Epstein–Barr infection, vitamin D levels, smoking, obesity, and geographical location, has already been extensively reviewed. Here, we mainly focus on how Western diet and sleep-circadian disruption may contribute to our understanding of MS etiology." - https://www.ncbi.nlm.nih.gov/pmc/articles/PMC5947729/

"Even more dauntingly, any gene introduced on such a chromosome would now be present in three copies (one from mother, one from father, and one on the extra chromosome) instead of the usual two, throwing off the normal balance of gene copies among all the genes. The consequences of such "triploidy" can be deleterious (for example, Down syndrome)." - p 39

"Down syndrome, which arises from a chromosome defect, is likely to have a direct link with the increase in stress levels seen in couples during the time of conception, say Surekha Ramachandran, founder of Down Syndrome Federation of India, who has been studying about the same ever since her daughter was diagnosed with the syndrome, 37 years ago." - https://www.hindustantimes.com/pune-news/stress-in-couples-during-conception-could-be-a-likely-cause-of-down-syndrome-says-researcher-in-pune/story-g63YeZRW59I23tU6qooJpK.html

"Between 1979 and 2003, the number of babies born with Down syndrome increased by about 30%." - https://www.cdc.gov/ncbddd/birthdefects/downsyndrome/data.html

What factors led to this increase? Are any of them of human origin? Is something as simple as more women giving birth at later ages the primary factor?

"preimplantation genetic diagnosis (PGD)" - p 40

"Over the years, more and more traits will presumably become identifiable with the aid of PGD, including desirable genetic markers for intelligence, musicality, and so on, as well as undesirable markers for obesity, nearsightedness, color-blindness, etc." - p 41

I've always found color-blindness interesting. As I understand it, it's only a trait that men can have. If it were a trait that only women can have wd referring to it negatively be considered sexist?

"In addition to welcome consequences for the health of children, such practices may have more ambiguous or worrisome consequences for our ideas about the relation of sex and procreation, parents and children, the requirements of responsible parenthood, and beliefs in the equal worth of all human beings regardless of genetic (or other) disability." - p 45

Call me old-fashioned but there's something about Old School Fucking that I'm rather fond of - some might say excessively fond of. I'm rooting for the old in-&-out - & I don't mean a hypodermic.

"No one would *wish* to be afflicted, or to have one's child be afflicted, by a debilitating genetic disorder, and the new technologies hold out the prospect of eliminating or reducing the prevalence of some of the worst conditions.**"

[..]

"**We know of at least one exception: the case of a deaf couple using genetic screening to produce a deaf child." - p 46

"In the kingdom of the blind, the one-eyed man is king." - Desiderius Erasmus
(28 October 1466 – 12 July 1536)

& what if the one-eyed king is the one calling the genetic shots? Maybe he
doesn't want any sighted contenders for the throne.

"When biomedical technology permits the substitution, for natural procreation and
the rule of chance, of a procedure in which parents begin to control their child's
genotype, reproduction becomes to some extent like obtaining or making a
product to selected specifications. Even if the parents are guided by their own
sense of what would be a good or perfect baby, their selection may serve to
satisfy their own interests more than that of the child. The new technologies,
even when used only to screen out and get rid of the sick or "imperfect," imply a
changed attitude of parents toward their children, a mixture of control and tacit
expectations of perfection, an attitude that might grow more pronounced as the
relevant techniques grow more sophisticated." - p 51

I don't know about YOU, not knowing who YOU are, but I find the above concern
to be extremely important. I was raised by someone w/ OCD (Obsessive-
Compulsive Disorder), she found it almost unbearable for other people to do
anything that wasn't done under her very specific rules. The Living Room, the rm
that one entered thru the front door of the house I grew up in, was off-limits for
children - we weren't allowed to sit on the furniture. What wd her 'perfect' child
have been? One that absolutely, unquestioningly, followed orders. Is it really
healthy for medical science to *enable* people like that?!

"First, the goal of eliminating embryos and fetuses with genetic defects carries
the unspoken implication that certain "inferior" kinds of human beings—for
example, those with Down syndrome—do not deserve to live. The assumption
that the genetically unfit ought to be prevented from being born embodies and
invites a profoundly denigrating and worrisome attitude toward those who *do* get
to be born." - p 52

"Dr. Gerald Schatten, a leading researcher in the field of reproductive biology,
stated that the overall goal of assisted reproductive technology is "to help
prospective parents *realize their own dreams* of having a *disease-free
legacy*" (emphasis added). - p 54

What about *all sunny days*? Maybe the rain serves a purpose beyond whether
one gets a good tan at the beach. I don't want any diseases either but I have to
wonder whether people working toward a "*disease-free legacy*" have a myopic
vision - I mean, nature is a huge living entity, humans are but a small part of that;
what might be good for humans might be bad for the larger ecology. **What if
diseases are nature's antibodies? What if all these human autoimmune
deficiencies are mirrors of the autoimmune deficiencies humans wreak on
nature?** Of course, I favor the human side, being one myself, but that doesn't

mean I can't at least consider the possibility of a bigger picture - such as one in wch people who die of COVID-19 have reached a point in a natural cycle where their time has come.

"Over the past several decades, disturbing evidence has accumulated of the widespread use of various medical technologies to choose the sex of one's child, with a strong preference for the male sex. The natural sex ratio at birth is 105 baby boys born for every 100 baby girls. But in several countries today the ratio approaches or even exceeds 120 baby boys born for every 100 girls. There is also evidence that the ratio at birth of boys to girls is rising among certain ethnic groups in the United States." - p 58

All you cultures that prefer male offspring to female can *kiss my ass* - but, NO!, Really!, I don't want you anywhere *near* my ass. As a hetereosexual I'll put in an order for 50/50, ok?

"In 1999, the American Society for Reproductive Medicine (ASRM) criticized the use of PGD and sperm sorting for sex selection, fearing that such practices might contribute to gender stereotyping and discrimination." - p 64

There're multiple SciFi novels where one thing leads to another & there's only one sex left in the world. (https://en.wikipedia.org/wiki/Single-gender_world) I think of Monsanto's seeds that don't reproduce so that they have to be bought again & again. The *what-if-something-goes-wrong?* possibilities are enormous. As I like to say: I hate women but I hate men even more. An all-male world wdn't be one I'd want to live in.

"In a previous Council report, on human cloning, we emphasized how cloning-to-produce-children alters the very nature and meaning of human procreation, implicitly turning it (at least in concept) into a form of manufacture and opening the door to a new eugenics. Sex selection raises related concerns." - p 70

Indeed. We already have, to use Noam Chomsky's famous expression, *Manufacturing Consent* &, during the time of the QUARANTYRANNY, an almost completely successful *Manufacturing Obedience & Conformity* has been achieved - What better, from the POV of the manufacturers, than simply churning out slaves? 'Perfect workers' & sex slaves - people who can't & don't reproduce themselves but can be *reproduced* when needed. No doubt, the rich will be able to buy custom-made playmates & servants. Who needs those messy old school humans anymore? People like myself ask too many questions & *just don't cooperate w/ our destruction*. That makes us 'obsolete' as the Juggernaut of Pseudo-Progress marches us into the New Dark Ages. Let's get everybody on medications, that'll straighten 'em out.

"In recent years the rate at which children are diagnosed with ADHD and treated with stimulants has risen dramatically. Although it is difficult to get precise figures, it is estimated that up to four million American children are taking Ritalin or related drugs on a daily basis." - p 75

If you don't think that's a Medical Industry created problem, iatrogenesis, then you're probably going to be fine w/ a society in wch very few people can concentrate w/o being on drugs. Personally, I can concentrate just fine w/o them. How many people do you know who have dysfunctional memories? That's a related problem. It's fairly widely recognized that speed, wch wd include Ritalin, provides a temporary boost that depletes a person & results in premature aging.

"Among their effects are diminished fatigue, improved concentration, decreased distraction and restlessness, and enhanced effort on demand, as well as increased blood pressure and greater physical strength, speed, and endurance."
- p 78

Right. & how many users *burn-out* like a light-bulb w/ too much electricity coursing thru it?

"Because of their addictive effects in adults, stimulants like Ritalin and Adderall are not only prescription drugs; since 1971 they have been classified as Schedule II controlled substances. This means their production is strictly monitored and regulated by the federal Drug Enforcement Administration (DEA). Yet, closer to the ground of action, their prescription and actual use by pediatricians and other physicians are unregulated, and there is no scrutiny of off-label uses." - pp 78-79

SOoooo.., drs are free to prescribe ANY drug for ANY purpose, regardless of whether it's approved for that purpose by the FDA. That's what off-label means. Hence, Big Pharma can bribe drs to prescribe a drug they manufacture for purposes other than the FDA approved one so that they can boost sales. & the bribes won't even be recognized as such.

Back to ADHD: "The criteria further require that at least some of the symptoms must have begun before the age of seven; as defined, ADHD is a childhood disorder.*"

[..]

"*Notwithstanding this conclusion, there has been much recent discussion about "adult ADHD," and pharmaceutical companies are aggressively advertising remedies for this "disorder" on television." - p 79

In fact, I once briefly dated a woman who was in her mid-30s who was prescribed dextroamphetamine for her "ADHD". At the time, the effects were fairly intense. It's now 16 yrs later, I wonder what the state of her health is in her early 50s. If she's showing advanced speed-induced deterioration her prescribing dr shd be held legally accountable.

"Yet despite the generic genetic and environmental correlations, there is at present no clear biological marker or physiological test for ADHD. The disorder is dignosed solely on the basis of observed and reported symptoms.

"In florid cases, a symptom-based diagnosis is easy to make. But the symptoms themselves shade over along a continuum into normal levels of childish distractibility or impulsiveness, and, in all cases, evaluation is unavoidably subjective."

[..]

"**Dr. Lawrence Diller, a pediatrician specializing in behavior problems whose referral practice gets mostly hard-to-diagnose cases, estimates that in his experience less than half of the children for whom he prescribes Ritalin are genuine cases of ADHD. See Diller, L. "Prescription Stimulant Use in Children: Ethical Issues," presentation to the President's Council on Bioethics (www.bioethics.gov)" - p 80

"What is clear, however, is that stimulant prescriptions have skyrocketed in recent years. The DEA attempts to callibrate its production quotas to meet demand, so that production levels roughly correlate with prescription levels. In the decade between 1990 and 2000, annual production of methylphenidate increased by 730 percent, and annual production of amphetamine increased by an even more astounding 2,500 percent. The overwhelming majority of those taking these medications are children, although adult use has been growing rapidly. Estimates of the number of American children taking Ritalin-like stimulants hover around three to four million.**"

[..]

"**We lack comparable data for other countries. In this presentation to the Council, Dr. Lawrence Diller reported that the United States uses 80 percent of the world's Ritalin." - p 82

Do you think that's a problem?! I do. What are those 3 to 4 million adults doing today? Are they dependent on speed to be able to focus? What other side-effects are they suffering from if there's been prolonged use?

"Stimulants such as Adderall and meth cause the salivary glands to dry out which allows the mouths acids to eat away at the tooth enamel causing cavities. Teeth can become translucent. Hair loss, malnourishment, and hypertension were just a few of the many side effects of Adderall abuse. Also, psychosis."

"**Adverse Effects of Methamphetamine**
  •   Psychosis, including hallucinations and extreme paranoia
  •   Change in brain structure and function
  •   Deficits in thinking and motor skills
  •   Increased distractibility
  •   Insomnia
  •   Memory loss

- Aggressive or violent behavior
- Depression
- Severe dental problems
- Weight loss
- Meth Addiction

Adderall abuse can create feelings of euphoria and increased alertness and energy, similar to the effects of other stimulants such as cocaine and methamphetamine.

Short Term Effects of Adderall
- Irritability
- Loss of Appetite
- Weight Loss
- Insomnia
- Restlessness
- Cardiac Issues

Long Term Effects of Adderall
- Paranoia
- Depression
- Hostility
- Adderall Addiction"

- https://clearrecoverycenter.com/similarities-between-adderall-and-methamphetamine/

"We have no doubt that, in most cases, parents, teachers, and physicians are acting in what they sincerely deem the best interest of the child. But anecdotes abound of schools and teachers pressuring parents to medicate their children, often as a condition of continued enrollment; of doctors, pushed by hectic schedules and distorted insurance rules, prescribing stimulants to children they have not fully examined; and of parents seeking a quick way to calm their unruly child or pressuring their doctors to give their son the same medication that is helping his schoolmates." - p 84

Iatrogenesis.

"Special safety concerns have been raised about the growing practice of prescribing stimulants "off-label" to toddlers as young as two years old. One concern is that, between the ages of two and four, the brains of children are still undergoing important biological development that might be adversely affected by the use of psychotropic drugs. At present, stimulants are approved by the FDA only for treatment of children age six and above." - footnote, p 87

One online commentator, James Evan Pilato, has recently expressed shock that prominent punk rockers who embraced heavy alcohol & other drug use are also advocating Covid-19 vaccination. To me, that fits. What was once our youth culture broke away from 'straight' culture by drug use, by getting 'high'. This was legally oppressed in a heavy-handed way that added to the problem, making drug-users sure that they were right in contrast to the obvious overkill of the laws.

There was a time when a person cd get sentenced to 10 yrs in prison for 2 joints of marijuana. That's insane. But, now, marijuana is almost completely legal, recognized as having medical value. What if the same system that pushed the Draconian laws is now co-opting what were once illegal consciousness expansion &/or disinhibiting tools by exploiting counter-culture's belief in drugs to get the people to accept *any drug* as for their betterment?

"This enhanced ability to make children conform to conventional standards could also diminish our openness to the diversity of human temperaments. As we will find with other biotechnologies with a potential for use beyond therapy, behavior-modifying drugs offer us an unprecedented power to enforce our standards of normality." - p 90

Exactly. & as a person who highly values original thought, I see that as a threat. *& it's very relevant to trends in QUARANTYRANNY.*

"Here let me try to make everyone feel even worse. Check out this bioethics professor's article on "Morality Pills". By the end he suggests that we need to sneak this into the water supply to 'fix' those noncompliant little imps (like us).

https://theconversation.com/amp/morality-pills-may-be-the-uss-best-shot-atending-
the-coronavirus-pandemic-according-to-one-ethicist-142601?
fbclid=IwAR0RCxMruaA2J-4WAzTu2iAt93Z-2vFTpB6roifGXXWUU2U9DtROKml
RB1k

****

'Morality pills' may be the US's best shot
at ending the coronavirus pandemic,
according to one ethicist
Parker Crutchfield, Western Michigan University
August 10, 2020 8.07am EDT"

- Unconscious Suffocation - A Personal Journey through the PANDEMIC PANIC,
by Amir-ul Kafirs & fellow HERETICS, p 1083

That article about 'Morality Pills' appeared in USA Today, no less. That's what passes for academic credibility for some people these days. I remember when USA Today 1st appeared, on September 15, 1982, that literate people were horrified at how dumbed-down it was! It's listed online as being aimed at a 10th-grade reading level (https://www.attorneysync.com/blog/grade-level-writing/). Other 'news'papers aim even lower.

Once upon a time, Yippies scared the shit out of straight society by saying they had plans to put LSD in the water supply. I think they were just fucking w/ people. Now, we actually have a professor of bioethics proposing putting a psychoactive substance in the water supply to make everyone go along w/ the

QUARANTYRANNY program & his article to this effect gets printed in a widely circulated mainstream 'news'paper! 50 yrs ago such a proposal wd've been rc'vd w/ horror, now it's 'AOK'. What idiocy. Any responsible user of LSD, e.g., wdn't spike a punch w/ it at a party.

If you don't distrust Big Pharma, I reckon you're naive or too far gone down the addict road. As an aside, consider this article:

"Only The Illiterate Trust Big Pharma"
MAY 22, 2021 BY GABRIELLE LAFAYETTE
- https://outerlimitsradioshow.com/2021/05/22/only-the-illiterate-trust-big-pharma/

But back to Beyond Therapy:

"Children learn by their elders' example, and in this instance they may learn from those whose opinions matter most to them that behavior is simply a matter of chemistry, and that responsibility for their actions falls not to themselves but to their pills. They may behave better, but they will not have learned why, or even quite how." - p 92

That seems like 'common sense' to me so why does it seem like 'uncommon sense' these days? My answer to that question is somewhat put forth in the preceeding passages.

Chapter 3's title & theme is "Superior Performance".

"Our motives for seeking superior performance are varied and complex, as human desire and human aspiration always are. We seek to win in competition, to advance in rank and status, to increase our earnings, to please others and ourselves, to gain honor and fame, or simply to flourish and fulfill ourselves by being excellent in doing what we love." - p 101

Is that true any more? It seems that most people just don't give a shit - but maybe that's just among my friends. The lack of ambition for accomplishment around me is pretty grim from my POV. Then again, people who're mainly motivated by advancing in rank or status & increasing earnings are usually pretty dull.

"recombinant viruses* containing a rat IGF-1 gene were injected into the anterior compartment of the rear legs of young mice containing the extensor digitotum longus (EDL) muscle. The resulting increased production of IGF-1 promoted an average increase of about 15 percent in EDL muscle mass and strength in young adult mice. Strikingly, such injections led to a 27 percent increase in the strength of the EDL muscles when the mice approached the average lifespan of 27 months. In fact, the continued presence of additional (rat) IGF-1 genes essentially prevented the decline in muscle size and strength observed in untreated old mice."

"*Recombinant viruses, engineered to express a specific foreign gene, are frequently used to stimulate the production of functionally effective amounts of the foreign protein to treat disease. Recombinant viruses created from genetically engineered human Adenovirus-associated Virus (AAV) have proved to be efficient delivery systems of foreign genes into muscle cells. As AAV is a small virus, only small foreign genes can be used effectively with this virus. Fortunately, the DNA sequence encoding IGF-1 is small enough to function well in AAV-based recombinant viruses." - p 116

So that's where such research was at by 2003. Imagine where it's at now, 18 yrs later. Wd it be possible, e.g., to inject a bear's ability to recycle its own urine & to, therefore, hibernate for long periods, into a human?

"It has been suggested that along with the regular Olympics and the Special Olympics, we have the "Bio-Olympics," where the competition is unconstrained and the athletes are free to use any legal form of pharmaceutical or physiological enhancement." - footnote, p 123

Naturally, the above evokes imagining the 'Extreme Sports' version of the Bio-Olympics in wch illegal forms of enhancement are also fair game. Regardless of how laws develop to govern such developments it seems almost 'inevitable' that the illegal enhancements will flourish somewhere. Cyberpunk novels are full of such things.

"When and if we use our mastery of biology and biotechnology to alter our native endowments—whether to make the best even better or the below-average more equal—we paradoxically make improvements to our performance less intelligible, in the sense of being less connected to our own self-conscious activity and exertion. The improvements we might have once made through training alone, we now make only with the assistance of artfully inserted IGF-1 genes or anabolic steroids. Though we might be using rational and scientific means to remedy the mysterious inequality or unchosen limits of our native gifts, we would in fact make the individual's agency *less* humanly or experientially intelligible to himself." - p 128

"With interventions that bypass human experience to work their biological "magic" directly—from better nutrition to steroids to genetic muscle enhancements—our silent bodily workings and our conscious agency are more alienated from one another.

"The central question becomes: Which biomedical interventions for the sake of superior performance are consistent with (even favorable to) our full flourishing as human beings, including our flourishing as active, self-aware, self-directed agents? And, conversely, when is the alienation of biological process from active experience dehumanizing, compromising the lived humanity of our efforts and thus making our superior performance in some way false—not simply our own, not fully human?" - pp 130-131

I take the above issues very seriously. Not surprisingly, to me, it's also a class struggle matter: if a rich person were to increase muscular ability thru an IGF-1 injection wd that justify as full a feeling of accomplishment as the increased muscular ability of someone who physically works? OF COURSE NOT. Things rc'vd thru privilege are empty of important experiential content.

"A second source of disquiet centers on issues of freedom and coercion, both overt and subtle. The pride most nations (and schools) take in their athletes is often far from benign, and there are well-known cases in which countries and coaches have forced athletes to use performance-enhancing drugs. In East Germany before the fall of communism, to take just a single example, the young members of the women's Olympic swim team took regular doses of the anabolic steroid known as Oral-Turinabol. This improved their strength and endurance, but it also caused terrible masculinizing side effects (severe acne, uncontrollable libido, gruff voices, abnormal hair growth). Those women who were brave enough to inquire about what they were taking were told that the drugs were simply "vitamin tablets." As one of the swimmers testified years later: "I was fifteen years old when the pills started. . . . The training motto at the pool was, 'You eat the pills, or you die.' It was forbidden to refuse.""

[..]

"In professional sports, where not only victory but big money is at stake, the pressures not to disarm oneself pharmacologically will be—are already— enormous." - p 135

The parallel to today's QUARANTYRANNY world seems glaringly obvious to me. *The pressure is on to conform.* Oh, a person can be this, that, or the other thing (b/c those 'freedoms of choice' have become trivialized) *as long as they all take their "vitamin tablets"* - if they know 'what's good for them'. But what about those of us who *do know what's good for us* & know that it's not what we're being pressured into? We face demonization, we face becoming pariahs. Resisting takes strength of character - something most people seem to lack.

"With drugs like steroids, the grave long-term health risks are well known: they include, among others, liver tumors, fluid retention, high blood pressure, infertility, premature cessation of growth in adolescents, and psychological effects from excessive mood swings to drug dependence." - p 137

&, yet? How often are they still prescribed? Take this story from my own life:

"In 2014 I was working on my house & my backyard extensively. In the process of weed-whacking the jungle, I got poison ivy for the 1st time in my life. For me, it was as if the plants were fighting back with chemical warfare. I applied cream to my hands but I seemed to have poison ivy on my eyelid too & I was afraid to rub cream on that. I went to a medical place & explained the problem, my poison ivy wasn't that bad but I was afraid for my eyesight & afraid of getting the oil on a valuable object at one of my museum jobs. I was told that I could get a steroid injection & then be prescribed steroids in pill form. I'd never had steroids before & didn't have any opinion about them.

"When it was time to give me the injection I was told that it was best to get it in the hip, in this case, my right hip. When I asked why I was told that "it's not as painful this way." I got the injection & went home & started taking the pills in the recommended dosage, it might've been twice a day. I'd been warned that the steroids might make me sleepy. Instead, I was very wide awake all night, I felt great. Within a few days, however, I started getting severe pains radiating out from where the injection had been given. It wasn't long before any pressure on my hip & leg joints, such as by standing, was excrutiatingly painful. I started crawling around my house, standing was too painful — even then I was often in so much pain that I screamed uncontrollably.

"At the same time, a friend came to visit that I hadn't seen in decades. I warned him over the phone of my situation & once he got here he drove me to the ER. Once there, I was put in a wheelchair & taken to an examination room where a doctor told me he didn't know what was wrong with me but thought I should take some drugs for it. I explained to him that I thought that the steroids were the problem & that I didn't want any more drugs. He told me that was "impossible". I was given many X-Rays & when the doctor looked at them he said he still didn't know what was going on but speculated that I have Reactive Arthritis but that he didn't know what I was reacting to. That made sense to me & I told him that I was reacting to the steroids. Once again, he told me that that was impossible. Then he asked me if I still needed the wheelchair. I found that astounding since nothing whatsoever had been done to alleviate my need to not walk. I made it back to my friend's car in the wheelchair & we left. I immediately STOPPED taking the steroids. By the next day my pain was down to about 50% of what it had been. It was still extremely painful to walk but I wasn't screaming anymore. By the day after that the pain had halved again. It was now bearable. By the next day I could walk enough to go on another political trip to Harrisburg. My opinion that the steroids were triggering the reactive arthritis was vindicated for me. I'll never get a steroid injection again & I doubt that I'll ever take the pills again either. The bill for the trip to the ER was astronomical, of course, but the insurance that I had that 1st year was good enough to make my part affordable (barely)." - Unconscious Suffocation - A Personal Journey through the PANDEMIC PANIC, by Amir-ul Kafirs & fellow HERETICS, pp 1168-1169

"No biological agent powerful enough to achieve major changes in body or mind is likely to be entirely safe or without side effects. Moreover, targeted interventions aimed at enhancing normally functioning capacities, not repairing broken parts, could produce lop-sided "improvements" that throw whole systems out of kilter: monster muscles could threaten unenhanced bones and ligaments." - p 137

Yet-another thing that seems obvious to me at the same time that it's largely overlooked. In a way, it's the story of humanity: we're great at enhancing our abilities but deficient in developing what I sometimes call the Martial Artist

Temperament that shd provide guidance. I'm reminded of an incident in Pittsburgh sometime in the last 20 or so yrs: someone w/ a powerful SUV was running smaller cars off the road. In one instance they forced a smaller car into a dead-end alley & repeatedly rammed it. When the police chief was interviewed he sd he cdn't understand why someone wd do something like that. It seemed obvious to me: they'd suddenly acquired enormous power w/o any maturity tempering it so they went wild. Imagine giving a child an automatic weapon. Then they get upset w/ someone over something minor & shoot them. B/c they can. Humanity is that child.

"how people exploit the relatively unlimited uses of biotechnical power will be decisively determined by the perhaps still more unlimited desires of human beings, especially—and this is a vital point—as these desires themselves become transformed and inflated by the new technological powers they are all the while acquiring." - p 277

"We discipline our gifts through choice and effort in the service of enabling them to shine forth in our own beautiful and splendid activity. We take pleasure in our own performance and achievement. The added bonus of victory and the recognition that follows from it we esteem largely because they confirm that our own embodied excellence has been attained and that our desire for superior performance has been satisfied.

"In trying to achieve better bodies through muscle-enhancing agents, pharmacological or genetic, we are not in fact honoring our bodies or cultivating our individual gifts. We are instead, whether we realize it or not, voting with our syringes to have a *different* body" - p 148

& maybe that doesn't matter to many people. To me, it's important. A quantatative increase *does not equal* a qualitative one.

"The ironies of biotechnological enhancement of athletic performance should now be painfully clear. First, by turning to biological agents to transform ourselves in the image we choose and will, we in fact compromise our choosing and willing identity itself, since we are choosing to become less than normally the source or the shapers of our own identity. We take a pill or insert a gene that makes us into something we desire, yet only by seeming to compromise the self-directed path toward its attainment. Second, by using these agents to transform our bodies for the sake of better bodily performance, we mock the very excellence of our own individual embodiment that superior performance is meant to display. Finally, by using these technological means to transcend the limits of our natures, we are deforming also the character of human desire and aspiration, settling for externally gauged achievements that are less and less the fruits of our own individual striving and cultivated finite gifts." - p 150

In other words, winning is no longer an accomplishment, it's just another purchase.

Chapter 4 is "Ageless Bodies".

"*Maximum Lifespan*: The longest lifespan ever recorded for a species—in humans today it is 122.5 years." - p 164

Now, how cd I not be curious about who that was?

"**Jeanne Louise Calment** (21 February 1875 – 4 August 1997) was a French supercentenarian and the oldest human whose age is well-documented, with a lifespan of 122 years and 164 days."

[..]

"On 8 April 1896, at the age of 21, she married her double second cousin, Fernand Nicolas Calment (1868–1942). Their paternal grandfathers were brothers, and their paternal grandmothers were sisters. He had reportedly started courting her when she was 15, but she was "too young to be interested in boys". Fernand was heir to a drapery business located in a classic Provençal-style building in the center of Arles, and the couple moved into a spacious apartment above the family store. Jeanne employed servants and never had to work; she led a leisurely lifestyle within the upper society of Arles, pursuing hobbies such as fencing, cycling, tennis, swimming, rollerskating ("I fell flat on my face"), playing the piano and making music with friends."

[..]

"Calment's remarkable health presaged her later record. On television she stated "J'ai jamais été malade, jamais, jamais" (I have never been ill, never ever). At age 20, incipient cataracts were discovered when she suffered a major episode of conjunctivitis. She married at 21, and her husband's wealth allowed her to live without ever working. All her life she took care of her skin with olive oil and a puff of powder. At an unspecified time in her youth, she had suffered from migraines. Her husband introduced her to smoking, offering cigarettes (or cigars) after meals, but she did not smoke more. ("After the meal, after just one, I'd had enough of it".) Calment continued smoking in her elderly years, until she was 117. At "retirement age" she broke her ankle, but before that had never suffered any major injuries. She continued cycling until her hundredth birthday. Around age 100, she fractured her leg, but recovered quickly and was able to walk again.

"After her brother, her son-in-law and her grandson died in 1962–63, Calment had no remaining family members. She lived on her own from age 88 until shortly before her 110th birthday, when she decided to move to a nursing home. Her move was precipitated by the winter of 1985 which froze the water pipes in her house (she never used heating in the winter) and caused frostbite to her hands. According to one of her doctors, she had been quite healthy until she moved to the nursing home, and only began showing signs of ageing during her stay."

[..]

"Medical student Georges Garoyan published a thesis on Calment when she was 114 years old in January 1990. The first part records her daily routine, and the second presents her medical history. She stated that she had been vaccinated as a child but could not remember which vaccine(s). Apart from aspirin against migraines she had never taken any medicine, not even herbal teas. She did not contract German measles, chickenpox, or urinary infections, and was not prone to hypertension or diabetes. In April 1986, aged 111, she was sent to a hospital for heart failure and treated with digoxin. Later she suffered from arthropathy in the ankles, elbows, and wrists, which was successfully treated with anti-inflammatory medication. Her arterial blood pressure was 140mm/70mm, her pulse 84/min. Her height was 150 cm (4 ft 11 in), and her weight 45 kg (99 lb), showing little variation from previous years. She scored well on mental tests, except on numeric tasks and recall of recent events."

- https://en.wikipedia.org/wiki/Jeanne_Calment

Amazing. She wd've lived thru the famous "Spanish Flu" era. Did she wear a mask? I wonder if she wd've realized that she'd start to age in the nursing home if she wd've decided against it. I mean, why stop at 122? Personally, I figure that once I pass 100 I'll just keep going indefinitely.

"Life-extension may take three broad approaches: (1) efforts to allow more individuals to live to old age by combating the causes of death among the young and middle-aged;"

[..]

"The first, particularly in the form of combating infant mortality (mostly through improvements in basic public health, sanitation, and immunization), is largely responsible for the great increase in lifespans in the twentieth century, from an average life expectancy at birth of about 48 years in 1900 to an average of about 78 years in 1999 in the United States" - p 165

"For instance, if diabetes, all cardiovascular diseases, and all forms of cancer were eliminated today, life expectancy at birth in the United States would rise to about 90 years, from the present 78." - p 166

It seems like a change of diet wd take c/o most of those. It's interesting to know that if I'm average I only have 11 more yrs to go. I've never been particularly average but I'm not sure wch way that makes my likelihood of death go, earlier or later? Lately, I've been telling people I'm already dead. That way, if they complain that I smell funny I can act offended by their deathism.

""memory drugs" have a significant effect on other bodily functions. So, for example, amphetamines, Ritalin, and dunking one's hand in freezing water have a "positive effect" on the capacity to remember new information, at least over the short term. But these drugs or experiences work on memory only indirectly, affecting not the specific memory systems but the other systems of the body that influence how the different memory systems function.*

"*The above description draws heavily on Steven Rose (Rose, S., "'Smart drugs': do they work, are they ethical, will they be legal?," *Nature Reviews Neuroscience* 3: 975-979, 2002). As Rose has said: "[M]emory formation requires, amongst other cerebral processes: perception, attention, arousal. All engage both peripheral (hormonal) and central mechanisms. Although the processes involved in recall are less well studied it may be assumed that it makes similar demands. Thus agents that affect any of these concomitant processes may also function to enhance (or inhibit) cognitive performance." - p 170

"In 1999, another group of researchers succeeded in genetically engineering mice to learn tasks much more readily. They inserted into a mouse embryo a gene that caused over-expression of a specific receptor in the outer surface of certain brain cells, "long suspected to be one of the basic mechanisms of memory formation" because it allows to make an association between two events."*"

[..]

"* The difficulty of simple and direct improvement in complex neurological processes is underscored by the results of this experiment. Together with some improvements in memory the mice experienced other neurological changes, including hypersensitivity to inflammatory pain." - p 171

Wdn't it be interesting if the "hypersensitivity to inflammatory pain" were akin to immersion of the hand into freezing water as a memory aid? What if instead of churning out Adderal addicts at the universities there was a fad of students carrying around ice cubes in small coolers?

I like to fast, I've done it many times from one wk to 4. I think it's good for me. But when I was diagnosed w/ Diabetes Type II & I told the medical dietician that I planned to fast every other day she seemed alarmed & didn't think it was a good idea. I did it anyway, it seemed to work just fine. Consider this:

"It has been known since the mid-1930s that substantial reductions in the food intake of many animals (combined with nutritional supplements to avoid malnutrition) can have a dramatic effect on lifespan. With nearly seven decades of laboratory research, this is by far the most studied and best-described avenue of age-retardation, though scientists still lack a clear understanding of how it works. What is clear, however, from numerous studies in both invertebrates and vertebrates (including mammals), is that reduction of food intake to about 60 percent of normal has a significant impact not only on lifespan but also on the rate of decline of the animal's neurological activity, muscle functions, immune response, and nearly every other measurable marker of aging." - p 173

In other words, it's 'normal' to overeat. Too bad I'm about to go out to a restaurant. I probably won't even remember who I am after I finish gorging. I won't even remember who YOU are.. *wait!*, I don't know who you are now! Sheesh, I really need to lose some pounds.. or is it Euros?

"A useful review of caloric restriction work in animals is Weindruch, R., et al., *The Retardation of Aging and Disease by Dietary Restriction.* Springfield, IL: Charles Thomas Publishers, 1998." - endnote, p 203

"To reduce food consumption to 60 percent of normal, the average active adult human being would have to lower his daily caloric intake from 2,500 calories a day to 1,500. By any standard, that is a severely restricted diet that few people would want to sustain for long periods. Accordingly, much research is being devoted to the search for pharmaceuticals (known as "caloric restriction mimetics") that might mimic the benefits of caloric restriction without actually forcing people to go hungry. See Lane, M., et al., "The Serious Search for an Anti-Aging Pill," *Scientific American* 287(2): 36-41, 2002." - footnote, p 174

Pills, pills, & more pills. If people developed more self-discipline & realized that their desires & 'needs' can be reshaped in a way ultimately more beneficial for them w/o being painful or even unpleasant they cd do w/o the pills. Good fucking luck. Good lucking fuck. That sd, I still ate too much when I went out to the restaurant tonight.

"Some single-gene mutations do, however, have serious side effects, including, most commonly, sterility or reduced fertility—problems also observed with other techniques of age-retardation"

[..]

"The effects of induced age retardation on fertility and reproductive fitness invite interesting speculation on the possible connection between longevity and reproduction: prolongation of life for the individual may be in tension with renewal of life through generation; conversely, fitness for reproduction is correlated with the process of decline leading to death." - p 176

Makes sense to me.

Think of mayflies: after maturing to winged creatures they only live for a few hrs to 2 days. During that time, the female can produce between 50 or more than 100,000 eggs. "Mating takes place soon after the final molt. In most species death ensues shortly after mating and oviposition (egg deposition). Winged existence may last only a few hours, although *Hexagenia* males may live long enough to engage in mating flights on two successive days, and female imagos that retain their eggs may live long enough to mate on either of two successive days. Groups of male imagoes perform a mating flight, or dance, over water as dusk approaches, flying into any breeze or air current. Individuals may fly up and forward, then float downward and repeat the performance. Females soon join the swarm, rising and falling as the dance continues. The male approaches the female from below and behind and grasps her thorax with his elongated front legs. Mating is completed on the wing. After her release by the male, the female deposits her eggs and dies. A few species are ovoviviparous—i.e., eggs hatch within the body of the female generally as she floats, dying, on the surface of a stream or pond." (https://www.britannica.com/animal/mayfly)

Let's anthropomorphize that, shall we? A guy goes to a bar looking to get laid. He's wearing clothes that he thinks the girls will like, he's sharpened his pick-up line wit. He finds a girl that wants his action, after a few drinks they go out to his car & fuck before they can even manage to get to one or the other of their homes. He dies. She dies - but not before popping out 100,000 babies who mature very quickly & go into the bar looking for some action (It's a big bar, some sort of outdoor beer hall).

Now, let's consider sea turtles: "The oldest sea turtle is said to be 400 years old. It is in captivity in the Guangzhou Aquarium in China and weighs 300 kilograms. It is not possible, however, to determine the age of the oldest sea turtle in the wild." (https://www.reference.com/pets-animals/old-oldest-sea-turtle-111ccf2e495421d3) But what of its reproductive habits? "Most marine turtles take decades to mature—between 20 and 30 years—and remain actively reproductive for another 10 years." (https://www.worldwildlife.org/stories/how-long-do-sea-turtles-live-and-other-sea-turtle-facts) Let's humanize again: The girl has heard of the bar but she's not really in any hurry to go there. She moves to the city that the bar's in.. but she doesn't go to the bar. 5 yrs later, she moves to the neighborhood the bar's in.. but she doesn't go in, she just walks past it a few times. Finally, at age 30, she goes in. She's not really in a hurry to pop out the kids but it happens eventually. She lives a really long life. You see? That's why I've decided to not have any kids until I'm 100.

"It is possible that age-retarding techniques, like many medical interventions, will have unseen effects: they might work well for some, not well for others, and cause serious side effects in yet others. For example, for some recipients of greater longevity, the result might include a much longer period of decline and debility." - p 183

"Would people in a world affected by age-retardation be more or less inclined to swear lifelong fidelity "until death do us part," if their life expectancy at the time of marriage were eighty or a hundred more years, rather than, as today, fifty? And would intergenerational family ties be stronger or weaker if there were five or more generations alive at any one time?" - p 190

I find those questions interesting - not b/c I particularly identify w/ them. I've never pledged "until death do us part", I'm not even sure I think that's healthy, if it happens, it happens. As for "intergenerational family ties"? I've always had next-to-nothing in common w/ my family, so I've had essentially *no* "intergenerational family ties" worth speaking of under what are currently 'normal' circumstances. People have strong family ties b/c *that's what they're born w/* - but when they're born into a family that basically doesn't like anything about them, an intellectual amongst anti-intellectuals, a free-thinker amongst puppets, an atheist amongst fervent believers, those ties are more like nooses.

"Yet it is possible that an individual commited to the technological struggle against aging and decline would be less prepared for and less accepting of death, and the least willing to acknowledge its inevitability." - p 190

Is that where we are now? It seems to me that the irrational terror of Covid-19 that I've seen among healthy people in their 20s in the past 14 mnths might be the mindset of a generation that feels like an immortality breakthru is looming, maybe they'll be the 1st generation to escape death? Sometimes it seems like people are acting as if death's a new thing that boogieman Rump invented.

"Given that these technologies would not in fact achieve immortality, but only lengthen life, they could in effect make death even less bearable, and make their beneficiaries even more terrified and obsessed with it." - p 190

"And in the absence of fatal illnesses to end the misery, pressures for euthanasia and assisted suicide might mount." - p 191

Maybe. It might seem like the passages I'm quoting are excessively pessimistic, maybe it's just that those are the parts I pick. I pick them b/c I think they're thoughtful responses to possibilities that I think many people might embrace unquestioningly. 'Prolonged lifespan?, FABULOUS!' How many of such people are going to consider the possibility that the last 20 or 30 yrs of that prolonged lifespan might be spent blind, deaf, & in a wheelchair? The authors of this report at least have enuf sense to consider such possibilities.

"A lifespan of approximately 150 years could reasonably be expected to allow one to see his or her great-great-great-great-grandchild. But this child would have as many as 63 other such great-great-great-great-grandparents, along with 32 great-great-great-grandparents, 16 great-great-grandparents, 8 great-grandparents, 4 grandparents and two parents—and, if certain demographic trends continue, few if any siblings, uncles and aunts, or cousins." - footnote, p 194

That's an interesting thing to think about, isn't it? Is it something you considered when thinking of prolonged lifespan? Imagine what Thanksgiving dinner or Christmas dinner might look like!! 127 people - wd the family have to rent a hall someplace? Wd there be intense competition for available rentals? How many turkeys & pigs wd be killed?

Chapter 5 is called "Happy Souls":

"For it is ultimately our desire for happiness—for the fulfillment of our aspirations and the flourishing of our lives—that leads us to seek, among other things, better children, superior performance, and ageless bodies (and minds)." - p 205

Is that true anymore? I don't know. Sometimes it seems to me like most people are in a frenzy to just be holier-than-thou - w/o happiness even being a consideration. But, then, I feel like I'm living in the New Dark Ages (to paraphrase my friend Dick Turner).

"The currently available drugs to alter memory and mood, and the new drugs and their uses that may be just around the corner, invite other large questions about the character of human life. By using drugs to satisfy more easily the enduring aspirations to forget what torments us and approach the world with greater peace of mind, what deeper human aspirations might we occlude or frustrate? What qualities of character may become less necessary and, with diminished use, atrophy or become extinct, as we increasingly depend on drugs to cope with misfortune?" - p 208

Exactly. W"hat deeper human aspirations might we" be "occlud"ing now? Why does it seem to me that so few people have many aspirations at all?, why so few people seem to have almost ZERO curiosity? There're a set of 'politically correct' aspirations that I support: anti-racism, anti-homophobia - but it seems that those're now rolled up w/ anti-free-thinking (to my dismay)!!

"But if enfeebled memory can cripple identity, selectively altered memory can distort it. Changing the content of our memories or altering their emotional tonalities, however desirable to alleviate guilty or painful consciousness, could subtly reshape who we are, at least to ourselves. With altered memories we might feel better about ourselves, but it is not clear that the better-feeling "we" remains the same as before. Lady Macbeth, cured of her guilty torment, would remain the murderess she was, but not the conscience-stricken being even she could not help but be."

[..]

"Many stoop to fraud to obtain happiness, but none want their feeling of flourishing itself to be fraudulent. Yet a fraudulent happiness is just what the pharmacological management of our mental lives threaten to confer upon us." - p 212

I don't want people to be traumatized, but I do want to reduce the circumstances under wch trauma occurs as much as possible rather than erase memories. Is that possible *at all*?! Maybe not. Alas, memories are too crucial to be cavalier about them.

"If happiness requires better memories, how would we improve them if we could? What would be an excellent or perfect memory?

"The most obvious answer is "perfect recall." An individual with a perfect memory, forgetting nothing, would remember every fact, face, and encounter, every mistake he ever made, every injury suffered at the hands of others. But even a little reflection shows that indiscriminate and total recall is not a blessing but a curse. Those who have it suffer like the Jorge Luis Borges character, "Funes, the Memorious," who describes his "all-too-perfect" memory as "a garbage disposal" or like the famous memory patient Shereshevskii, whose photographic memory prevented him from forming normal human relationships. "Perfect memory" makes those who possess it miserable and dysfunctional." - p 218

"Luria, A. R., *The Mind of a Mnemonist: A little book about a vast memory* (Solotaroff, L. trans.). New York, Basic Books, 1968." - endnote, p 271

The Mind of a Mnemonist, about Shereshevskii, is in my Top 100 Books (http://idioideo.pleintekst.nl/Top100Books.html). I'm going to have to expand that list to Top 200 Books eventually b/c there's just too much left off it. Reading about Shereshevskii is utterly fascinating - esp for those of us who're concerned w/ memory.

"People who have suffered damage to the amygdala typically have no difficulty remembering recent mundane events, but they do not exhibit the enhanced long-term memory normally produced by emotionally arousing experiences." - p 222

"Worse, in still other cases, the use of such drugs would inoculate individuals in advance against the psychic pain that *should* accompany their commission of cruel, brutal, or shameful deeds." - p 228

"The **Speed Freak Killers** is the name given to serial killer duo **Loren Herzog** and **Wesley Shermantine**, together initially convicted of four murders — three jointly — and suspected in the deaths of as many as 72 people in and around San Joaquin County, California. They received the "speed freak" moniker due to their methamphetamine abuse." - https://en.wikipedia.org/wiki/Speed_Freak_Killers

It's my opinion that prolonged speed use creates a sociopathic personality. The above Wikipedia entry isn't what I was looking for. I had a vague memory that there were more California-based serial killers who were speed users but I didn't find anything about that either. I had a vague memory that one serial killer duo were speed freaks & made the claim that they were essentially sleep-walking when they commited their murders. I thought that that might be the so-called "Hillside Strangler" but I didn't find anything confirmational online. I'm sure I read that somewhere in a bk. What horrified me / interested me about that was the possibility that long-term speed abusers might essentially go into a walking trance from sleep-deprivation during wch they might commit acts uninhibited by 'normal' conscious behaviors.

"These symptoms are observed especially among combat veterans: indeed, PTSD is the modern name for what used to be called "shell shock" or "combat neurosis."" - footnote, p 225

It's not my intention to draw a parallel between PTSD & speed abuse consequences. I'm simply moving on in the bk. If anything, the 2 might be polar opposites: PTSD produced by a sensitivity & speed sociopathy (in my formulation) produced by a lack of affect & empathy.

"As Daniel Schacter has observed, "attempts to avoid traumatic memories often backfire":

"Intrusive memories need to be acknowledged, confronted, and worked through, in order to set them to rest for the long term. Unwelcome memories of trauma are symptoms of a disrupted psyche that requires attention before it can resume healthy functioning. Beta-blockers might make it easier for trauma survivors to face and incorporate traumatic recollections, and in that sense could facilitate long-term adaption. Yet it is also possible that beta-blockers would work against the normal process of recovery: traumatic memories would not spring to mind with the kind of psychological force that demands attention and perhaps intervention. Prescription of beta-blockers could bring about an effective trade-off between short-term reductions in the sting of traumatic memories and long-term increases in persistence of related symptoms of a trauma that has not been adequately confronted." - p 227

In other words, our pill-popping society is always trying to find the 'easy' way out, the shortcut - but in order to truly solve the problem *there is no shortcut, no easy way out.* Then again, there's this counterargument:

"If there are some things it is better never to have known or seen, why not use our power over memory to restore a witness's shattered peace of mind? There is great force in this argument, perhaps especially in cases where children lose prematurely that innocence that is rightfully theirs." - p 230

Nonetheless, I can imagine a future in wch during waking hrs the treated 'witness'/victim isn't troubled by bad memories.. but then they go to sleep & are haunted by nightmares. What if trauma simply can't be erased? What if it can only ultimately be satisfactorily dealt w/, for better or worse, by coping?

"There is already ongoing controversy about excessive diagnosis of PTSD." - footnote, p 228

What about excessive diagnosis of all-things-psychological? At the same time that I find it fascinating that the DSM (Diagnostic and Statistical Manual of Mental Disorders, of wch I have the most recent, 5th, edition) lists so many disorders & is so meticulous in doing so, I wonder whether it might not be healthier for people to just learn to be a bit more comfortable w/ just *being who they are* instead of having so much diagnosed as 'disorders'. I don't mean to downplay the suffering that people go thru & the dedicated attempts to ameliorate it, &, of course, there're so many extreme cases, but sometimes I think that there's *overdiagnosis in general* & that that can be more harmful than the condition diagnosed if it gives the patient something to brood & obsess over.

There are so many passages in this bk that I have marked as "important", this is another one of them:

"Without truthful memory, we could not hold others or ourselves to account for what we do and who we are. Without truthful memory, there could be no justice or even the possibility of justice; without memory, there could be no forgiveness or the possibility of forgiveness—all would simply be *forgotten*.' - p 232

That might be easy for the Council to say - but maybe none of them were traumatized as children; presumably none of them have had the experience of getting a life-w/o-parole sentence for a crime they didn't commit. In other words, not everyone thinks that highly of the 'justice system' as it is, there are entirely too many who see it as an 'injustice system'. Still, I'm an advocate of "truthful memory".

"To have only happy memories would be a blessing—and a curse. Nothing would trouble us, but we would probably be shallow people" - p 234

& that's the general philosophical rub, isn't it? As my friend Dick Turner told me: "John Rechy a gay beat writer" [..] "said "if you don't think about killing yourself at least once a day there's probably something wrong with you"" & I think that that's somewhat the attitude of depressed people who've learned to cope w/ it. In other words, their depression is a sign that they're sensitive & still alert & alive - even if it is painful from time-to-time. I'm an extreme depressive but then I write reviews like this & they make me laugh so hard I shit myself.. & then I have to clean up the mess & I get depressed all over again. See, it's a natural cycle. Man, that's deep.

"Given the wide variety of mood-altering agents, present and projected, and given our ignorance of the precise effect any particular drug will bring about in any given person, we are somewhat at a loss about what to call these chemicals: "antidepressants" seems too narrow, "mood-altering agents" too non-specific, "mood-elevators" or "mood brighteners" too specialized, "euphoriants" inaccurate." - p 239

I call them "happy pills" but, then, I'm being sarcastic. I figure that instead of taking pills people shd determine what wd make them happy in their lives & then try to bring it about. I used to say that I stayed fit just by living my life: w/ lots of sex & walking & general busyness. Then my sex life deteriorated & I got a little flabby. I still don't work out in gyms, that just seems like the wrong way to go about it.

"We already have at our disposal a wide range of newer psychotropic agents useful in altering mood, some named above. But selective serotin reuptake inhibitors (SSRIs), such as Prozac, Paxil, Celexa, Lexapro, and Effexor stand out. SSRIs are the newest and most advanced mood-brigtheners available. There is nothing futuristic about them—a recent poll suggests one in eight adult

Americans use them today, mostly as treatment for diagnosed illness" - p 240

Ok, let's simplify & say that that means that at least 1 in 8 Americans are so depressed that they think they need Happy Pills to be able to endure life. What's wrong w/ this picture?! What's wrong w/ their lifestyle? Why can't they be happier just w/ their lives as they are? Like so much of pill culture, the Happy Pills cover over the symptoms w/o addressing or solving the problem. It's just like w/ pain killers: you hurt yourself, yr body tells you where it hurts so that you'll be careful about not hurting that part of yr body anymore until it has time to heal. Instead, you take a pain-killer & continue to make yr injury worse since the pain isn't there to slow you down.

"The millions of Americans now taking SSRIs are probably only the beginning. Epidemiologists widely consider depression to be *under*treated in America: according to recent studies, between 9.5 percent and 20 percent of Americans suffer from some form of depression. If all were treated with mood brighteners, one out of every five to ten people would use them. Moreover, the rate of diagnosed depression appears to be climbing in the United States, as in all developed countries—probably due not just to greater reporting, but to real increase." - p 242

What if that's b/c people are *avoiding their problems* instead of facing them?

"Neurologically, what SSRIs do is alter the brain's handling of serotonin. Like other neurotransmitters, serotonin is released from one neuron to bind with and thereby activate another. The brain recycles serotonin after each release, gathering it up again by means of a "reuptake system." SSRIs inhibit the serotonin reuptake system, thus increasing the concentration of serotonin available to the receiving neurons—hence the name, "serotonin reuptake inhibitor." (Since SSRIs inhibit serotonin reuptake without interfering with reuptake of other neurotransmitters, we get the full name , "selective serotonin reuptake inhibitor.")" - p 243

Does anyone else worry that interfering w/ the brain's natural processes, formed over x-number of millenia thru evolution, might have short or long-term effects that the experimenters can't envision?

"One effect of SSRIs is clear: they relieve a number of disorders of mood, particularly depression. Yet the nature of these disorders is complicated and their causes remain largely unknown. In DSM-IV the lengthy discussion of depression (like the discussions of other psychiatric disorders) is essentially a compendium of symptoms, with no attempt at a coherent account of the nature or causes of an illness." - p 244

Well, to not completely randomly pick something relevant from DSM-V I picked the following:

"A large number of substances of abuse, some prescribed medications, and several medical conditions can be associated with depression-like phenomena. This fact is recognized in the diagnoses of substance/medication-induced depressive disorder and depressive disorder due to another medical condition." - DSM-V, p 155

There's that iatrogenesis again. Any user of cocaine, e.g., knows about "coming down", the depressing crash from the exhilirating high: is there a "coming down" from artificially increased serotonin?

"Yet "calmness" is not the only way to understand the effects of SSRIs on mood and psychic experience. For one thing, the calmness explanation stumbles on the example of MDMA (ecstasy), which also makes more serotonin available and induces not calm but bliss, social and sensory openness, and feelings of intense affection." - p 246

[..]

"MDMA functions differently from SSRIs: rather than inhibiting serotonin reuptake, it increases serotonin production, causing massive dumps of serotonin into the synapses. Yet to the receiving neuron, more serotonin is available either way. Whether the difference between SSRIs and MDMA is one of degree or of kind, and what the example of one means for the other, is not clear." - footnote, p 246

I took ecstasy a few times in 1986 when I cd get the pure stuff complete w/ health advice for how to properly take c/o yourself so as to not suffer any health problems as a result. That was one responsible dealer, you don't meet them too often. I don't even know if one can get it anymore or if what's now called "Molly" is what passes for ecstasy or if it's all cut w/ drugs that have no business being there, like heroin - deliberately added to foster addiction by dealers considerably less scrupulous than the one I bought from. At any rate, The description of "bliss, social and sensory openness, and feelings of intense affection" is accurate enuf in my experience for this bk to be trustworthy.

"These three accounts of what SSRIs fundamentally do—induce calmness, provide a background sense of well-being, change personality"

[..]

"As the example of SSRIs shows, even though we are ignorant, even though we suspect that the unknown effect of the drugs are subtle and deep, we make substantial use of them nonetheless." - p 250

But, but, what about the reduced sexuality?!

"Yet further reflection gives rise to questions—about both ends and means—that ought, at the very least, to give pause to anyone tempted by the pharmacological road to happiness. For we care that our children—and that we ourselves—have not only the sense or feeling of well-being, but well-being itself. We desire not simply to be satisfied with ourselves and the world, but to have this satisfaction as a result of deeds and loves and lives worthy of such self-satisfaction. We do not want to kill our aspiration for a better life by drowning in a self-absorbed contentment those experiences of lack and self-discontent that serve as an aspiration's source, or those engagements with the world and other people that serve as aspiration's vehicle.*

"*Consider the analogy of "treating" the anxiety and disproportionate urgency (and associated danger) of adolescent sexuality by extinguishing it at its biological source (note that in some patients Prozac will diminish libido). This fundamental biological drive, and its attendant discontent, is inextricably related to the larger longings of romantic love and in turn to some of life's highest aspirations and achievements." - p 251

Is that what "I'm gonna fuck you up!' refers to?

"To what extent is the happiness of the happy person attributable to the drug and to what extent is it "her own"? To what extent are drug-induced psychic states connected with or disconnected from life as really lived?"

[..]

"With a drug like Ecstasy, the answers to such strange and difficult questions—about the identity of the person taking such drugs and the status of the positive feelings they induce—are more obvious, if no less disquieting. People high on Ecstasy routinely profess their love for perfect strangers. Imagine that a young party-goer, under the influence of the drug, tells a young woman that he loves her and wants to marry her." - p 252

At least these hypothetical strangers are perfect. I seem to only meet imperfect ones.

"Should the fact that his feelings are produced by the drug, rather than inspired by the woman, matter? It should of course matter to *her*. His drug-based professions of love cannot be taken seriously." - p 253

They might still get him laid.

"A central concern with mood-brightening drugs is that they will estrange us emotionally from life as it really is, preventing us from responding to events and experiences, whether good or bad, in a fitting way." - p 255

Of course, it's not just drugs that do this, the mass media is at least as culpable, if not more so - *AND* the mass media is pushing the drugs all the time too. What a double whammy. IMO this is a calculated strategy to keep people away from reality - &, judging by the last 14 mnths, it's working wonderfully. The drugs make people more suceptible to suggestion & the mass media tells them what to 'think'. Presto, Chango, a nice little robopathic population is ready & willing to fall into lockstep. Hebephrenia: inappropriate reactions.

""Feeling good" may not always be good or good for us. Never to suffer loss may mean never to love deeply; never to feel ashamed may mean that our standards with ourselves are too low; never to be dissatisfied with ourselves may mean that we aspire to too little." - p 258

What about never being inspired? Never being curious? Having a generally flattened affect? Never being friendly? These are contemporary problems to me that're deeply rooted in our over-medicated & over-media-controlled excuse for an existence.

"In sum, a mood-brightening drug that always made us pleased with ourselves no matter what we did—a drug that guaranteed our self-esteem when such esteem is not warranted—might shrink our capacity for true human flourishing.*"

[..]

"* The cultivation and corruptions of a spurious self-esteem are, of course, possible without using drugs. Examples abound in our current cultural climate." - p 260

Indeed, in fact, for me, these last 14 mnths of the QUARANTYRANNY have been times of almost constant subjection to pompous condescension from people who have never & *will never* accomplish even 1% of what I've accomplished in my life & yet these people feel entirely justified in telling me how to do just about *everything* w/o, apparently, the slightest self-consciousness about it at all. Is it mostly drug-induced? I think the general cultural climate is more responsible but maybe the drugs are a more important factor than I've considered.

"Thanks to the efficacy of mood brightening agents, and of psychotropic drugs more generally, there may well be a temptation to redefine and to *treat* what are currently considered normal emotions, moods, and temperaments on the model of mental illness, and mental illness as a matter purely of bodily—ultimately, of molecular—character and causation." - p 262

Chemistry matters - but to downplay the social is very dangerous b/c it essentially acts as if any social conditions are AOK as long as the person in them is 'properly medicated'.

"Living in continual poverty, despite working for a living, led to my finding out about clinics. Many of these had sliding scales, were otherwise cheap, or could be free if one pleaded penury. I engaged with 3 of them. At one of them the hours were very limited, maybe from 1:30PM to 4:30PM. I went there, it was cold out & raining, but the doctor hadn't arrived by 1:30 so the prospective patients were waiting in line out in the rain. That couldn't have helped their health much. Finally, the doctor arrived, an hour later, at 2:30 & I eventually got to see him in the company of 2 oncology med students. I told him that I was worried about my kidneys because they'd been hurting me lately & because my urine was particularly foul-smelling. He punched the skin over where my kidneys were, asked if that hurt & when I told him that it didn't he told me that my kidneys were fine. The doctor, however, was obviously far from fine. He told me that he'd recently had back surgery, he seemed very depressed. He also told me that he was a doctor at a university where he prescribed drugs for the students. I got the impression that that was pretty much it, a student would come in & say that they were depressed &/or stressed & he'd prescribe some pills. I wasn't impressed.

"The doctor then informed me that I should go to the psychiatrist in the cliniic & get drugs prescribed for me because I'd been moving my hands too much while we were talking. Apparently, he took this to be a symptom of psychomotor agitation or some such. I didn't bother to explain that I'm a drummer & that when I'm bored & have nothing better to do I often practice on my legs. Instead, I explained to him that there's nothing wrong with my mind & that I didn't require any psychiatric prescriptions. It was obvious to me that he was there at the clinic to get a tax write-off, that he had close to no medical knowledge, & that what he had to offer was the chance to become addicted to drugs that would enable him to get kickbacks from the pharmaceutical companies. This man was more of a menace to patients than anything else — but poor people are expected to grovelingly accept whatever crap is handed us & be 'grateful' for it — but, as I like to say, **Before you decide against biting the hand that feeds you ask why it has so much food in the 1st place.**

"A partial take-away from the above is that one should not automatically accept the diagnosis of a doctor. Doctors have a mindset that defines everything in terms of disorders & diseases. This is a form of morbidity. Think of something like "restless legs syndrome": how easy it would be to misdiagnose this!

""**Restless legs syndrome (RLS)** is generally a long-term disorder that causes a strong urge to move one's legs. There is often an unpleasant feeling in the legs that improves somewhat by moving them. This is often described as aching, tingling, or crawling in nature. Occasionally, arms may also be affected. The feelings generally happen when at rest and therefore can make it hard to sleep. Due to the disturbance in sleep, people with RLS may have daytime sleepiness, low energy, irritability, and a depressed mood. Additionally, many have limb twitching during sleep.

""Risk factors for RLS include low iron levels, kidney failure, Parkinson's disease, diabetes mellitus, rheumatoid arthritis, and pregnancy. A number of medications may also trigger the disorder including antidepressants, antipsychotics, antihistamines, and calcium channel blockers." - https://en.wikipedia.org/wiki/Restless_legs_syndrome

"I sometimes move my legs quickly while I'm sitting or laying down. Is this a "syndrome", in other words does it have to be diagnosed in terms of bad health? In my opinion, NO! Perhaps my body just isn't ready for repose, maybe it feels a need to go walking or is otherwise just manifesting a state of energeticness.

"And what about SAD?, Seasonal Affective Disorder? Is it symptomatic or a "disorder" if we feel depressed when there's very little sun & it's cold? Or is that just a natural part of an instinct for hibernation that we defy with electricity, heat, & the rat race? HPD?, Histrionic Personality Disorder? That's a perfect one for natural-born performers. Instead of being praised for their performative abilities, they're chastised for their attention-grabbing."

- Unconscious Suffocation - A Personal Journey through the PANDEMIC PANIC, by Amir-ul Kafirs & fellow HERETICS, pp 1159-1161

"Especially as health comes to be regarded less as the absence of disease but as some positive state of well-being, ever open-ended and unlimited in its boundaries, the incentive increases to medicalize not only health but all human activities, psychic and social. One need not philosophically embrace the World Health Organization's notorious definition of health—as "complete physical, mental, and social well-being" - p 263

"Around 2013, when I 1st started having to deal with consequences of the AHCA (Affordable Health Care Act) I started to realize how invasive it was into one's private life. It seemed to me that what hadn't been accomplished with the post-9/11 Department of Homeland Security violations of Civil Liberties was now being pushed further with the AHCA. The criminalization of NOT having health insurance gave the 'perfect excuse' for intimate data gathering. Because of this, I postulated that future police state actions would be justified as for 'our own good' & executed in the name of 'Public Health'." - Unconscious Suffocation - A Personal Journey through the PANDEMIC PANIC, by Amir-ul Kafirs & fellow HERETICS, p 7

"The second danger is that social goals or expectations—the external pressure to be productive, to gain status and recognition, to get ahead—will produce a "mood-brightened society," where pharmacological interventions in our psyches become normal or expected for students, employees, and ultimately everyone."

[..]

"The second danger involves the slavish self, whose worth is measured only in the eyes of others or according to his success in the rat race, and who takes mood-brightening (or other) drugs to assert himself or to increase his chances of meeting society's demands." - p 267

There're already workplaces where people carry apps to check whether the people around them have had COVID-19 or whatnot, there're devices to measure social distancing - after that invasive idiocy is accepted, why not apps that tell you whether the people around you are on the 'right' drugs to give them the 'right' attitude?

"the willingness to use or forgo medication for various sorts of psychic distress will be affected by the poverty or richness of private life, and the degree to which strong family or community support is (or is not) available for coping with that distress directly." - p 277

It helps to have friends you can trust. So, hey!, you want to get more people on drugs? Socially isolate them 'for their own good', of course, so that their psychic distress goes thru the roof & they can't have a supportive social environment. QUARANTYRANNY.

Heed the warnings in <u>Beyond Therapy</u>.

"Athletes who take steroids to boost their strength may later suffer premature heart disease. College students who snort Ritalin to increase their concentration may become addicted. Melancholics taking mood-brighteners to change their outlook may experience impotence or apathy. To generalize: no biological agent used for purposes of self-perfection or self-satisfaction is likely to be entirely safe. This is good medical common sense: anything powerful enough to enhance system A is likely enough to harm system B (or even system A itself)." - p 280

The authors of this bk are even class-conscious.

"If, as is now often the case with expensive medical care, only the wealthy and privileged will be able to gain easy access to costly enhancing technologies, we might expect to see an ever-widening gap between "the best and the brightest" and the rest. The emergence of a biotechnologically improved "aristocracy"— augmenting the already cognitively stratified structure of American society—is indeed a worrisome possibility, and there is nothing in our current way of doing business that works against it." - pp 281-282

"What is freely permitted and widely used may, under certain circumstances, become practically mandatory. If most children are receiving memory enhancement or stimulant drugs, failure to provide them for your child might be seen as a form of child neglect."

[..]

"And, a point subtler still, some critics complain that, as with cosmetic surgery, Botox, and breast implants, many of the enhancement technologies of the future will very likely be used in slavish adherence to certain socially defined and merely fashionable notions of "excellence" or improvement, very likely shallow and conformist." - p 284

I dated someone in the early 1990s who was from Scarsdale, NY, a very rich area. She told me that she was the *only* girl at her high school who didn't get a nose job as a teenager. Maybe she was exaggerating but I'm sure there was mostly truth to it.

"Over the past few decades, environmentalists, forcefully making the case for respecting Mother Nature, have urged us a "precautionary principle" regarding all our interventions into the natural world. Go slowly, they say, you can ruin everything. The point is certainly well taken in the present context. The human body and mind, highly complex and delicately balanced as a result of eons of gradual and exacting evolution, are almost certainly at risk from any ill-considered attempt at "improvement."" - p 287

"One revealing way to formulate the problem of hubris is what one of our Council Members has called the temptation of "hyper-agency," a Promethean aspiration to remake nature, to serve our purposes and to satisfy our desires. This attitude is to be faulted not only because it can lead to bad, unintended consequences; more fundamentally, it also represents a false understanding of, and an improper disposition toward, the naturally given world." - p 288

"It is worrisome when people speak as if they were wise enough to redesign human beings, improve the human brain, or perhaps the human life cycle. In the face of such hubristic temptations, appreciating that the given world—including our natural powers to alter it—is not of our own making could induce a welcome attitude of modesty, restraint, and humility." - p 289

That's what I'm sayin'. I cd really live w/o this contemporary arrogance.

"Earlier this week, *Atlantic* magazine – fast becoming the favored media outlet for self-styled intellectual elites of the Aspen Institute type – ran an in-depth article of the problems free speech pose to American society in the coronavirus era. The headline:
Internet Speech Will Never Go Back to Normal'

[..]

"The new piece by Goldsmith and Woods says we're there, made literally sick by our refusal to accept the wisdom of experts. The time for asking the (again, literally) unwashed to listen harder to their betters is over. The Chinese system offers a way out. When it comes to speech, don't ask: tell."

[..]

"My obvious question is: Are Goldmith and Woods among those hypothetically 'qualified' to be future censors because of their educational affiliations and their status as lawyers? If so, I quote from a billboard I saw 25 years ago in rural Canada to put things in perspective: "If lawyers are so smart, why are there so many of them?""

- Unconscious Suffocation - A Personal Journey through the PANDEMIC PANIC, by Amir-ul Kafirs & fellow HERETICS, pp 194-196

"The "giftedness of nature" also includes smallpox and malaria, cancer and

Alzheimer disease, decline and decay." - p 289

Exit thru God's little giftshop.

Let's summarize:

"As we have already noted, healthy people whose disruptive behavior is "remedied" by pacifying drugs rather than by their own efforts are not learning self-control: if anything, they may be learning to think it unnecessary. People who take pills to block out from memory the painful or hateful aspects of a new experience will not learn how to deal with suffering or sorrow. A drug that induces fearlessness does not produce courage." - p 291

A billionaire who's accustomed to always getting what they want isn't going to take NO for an answer. Therefore, they don't learn humility or sensitivity to the needs of others.

The crux:

"As the power to transform our native powers increases, both in magnitude and refinement, so does the possibility for "self-alienation"—for losing, confounding, or abandoning our identity. I may get better, stronger, and happier—but I know not how. I am no longer the agent of self-transformation, but a passive patient of transforming powers. Indeed, to the extent that an achievement is the result of some extraneous intervention, it is detachable from the agent whose achievement it purports to be. "Personal achievements" impersonally achieved are not truly the achievements of persons. That I can use a calculator to do my arithmetic does not make *me* a knower of arithmetic; if computer chips in my brain were to "download" a textbook of physics, would that make *me* a knower of physics?" - p 294

"The significance of these two prominent features of American life—the power of free markets and the prestige of medicine—points us also toward a greater understanding of the implications of our new biotechnical powers for our American ideals." - p 308

"Had we looked only at the perils of the technologies that seem to lie in our future, and had we sought to imagine the worst, it would not have been difficult to raise up specters of terrifying and inhuman violations, or of an unprecedented despotism of man over man, with powerful new technologies serving as the whips of the new slave masters. The recent history of the human race offers no dearth of sources for such nightmarish visions." - p 309

Alas, as far I'm concerned, we're there, there in the New Dystopia disguised as a medical concern for our own good. Take, e.g., the actions of Pennsylvania Governor Tom Wolff: he's been using emergency powers in the name of COVID-19 to be despot for over a yr now. The voters, finally given an opportunity to fight back, approved 2 constiutional amendments that wd limit the governor's emergency powers. However, before the state had a chance to certify this vote, Wolff rushed to extend his 'emergency' even further:

"HARRISBURG, PA — Pennsylvania Gov. Tom Wolf on Thursday signed a renewal of the Proclamation of Disaster Emergency for the COVID-19 pandemic. It is the fifth renewal of the disaster emergency. The first proclamation was signed on March 6, 2020." - https://patch.com/pennsylvania/ardmore/s/hkxvp/gov-wolf-extends-covid19-disaster-declaration-for-5th-time?utm_source=alert-breakingnews&utm_medium=email&utm_campaign=alert&fbclid=IwAR05ygDlnXL0skovm4njh_9eglQrAVSKoWSAwy6c1sCKaYs7kvSghlJJhqo

& who am I to question the great wisdom of government allied w/ medical 'science' you might ask?:

"While academic science is traditionally a system for producing knowledge within a laboratory, validating it through peer review, and sharing results within subsidiary communities, anti-maskers reject this hierarchical social model. They espouse a vision of science that is radically egalitarian and individualist. This study forces us to see that coronavirus skeptics champion science as a personal practice that prizes rationality and autonomy; for them, it is not a body of knowledge certified by an institution of experts." - https://arxiv.org/pdf/2101.07993.pdf

Yes, I'm "egalitarian and individualist" & I'm looking at the science thru the POV of my own mind, not the fake mind that's media fed me.

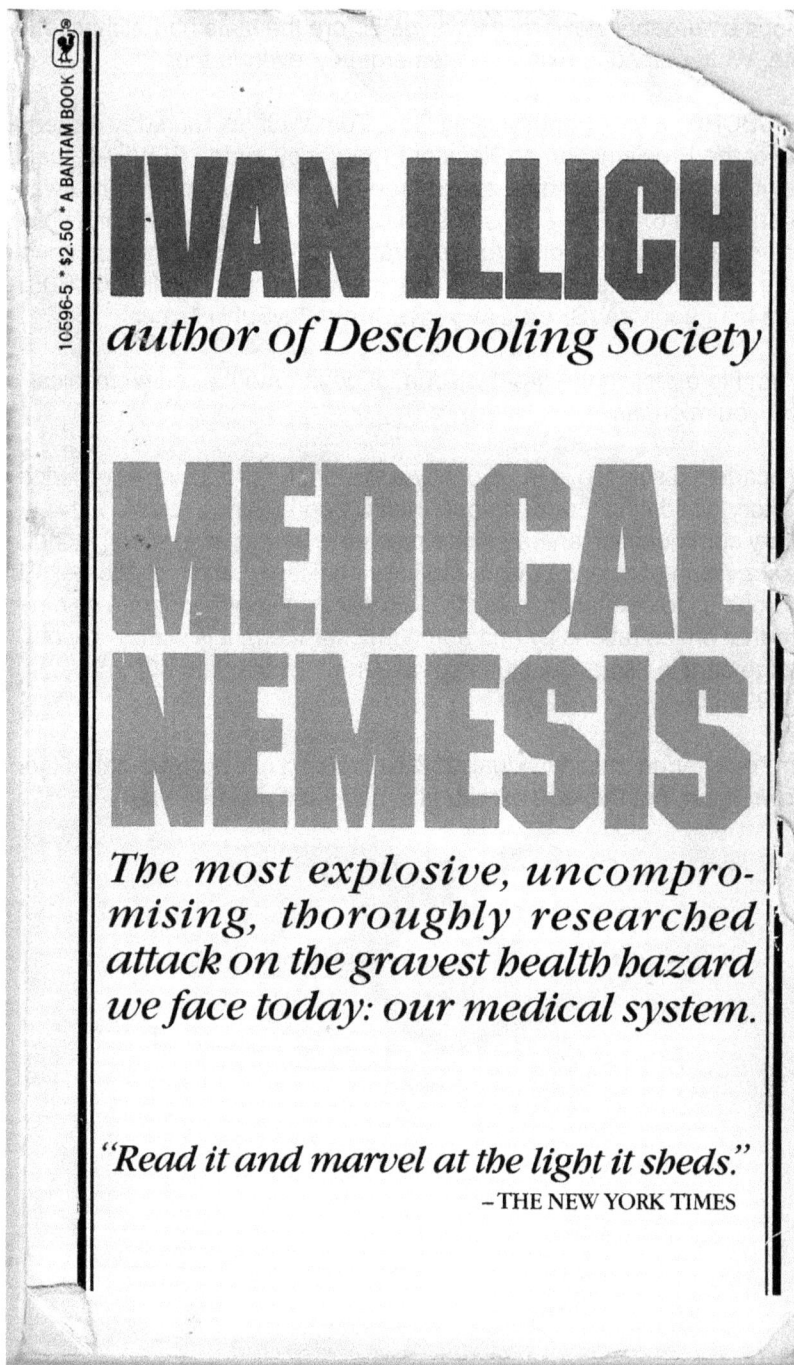

IVAN ILLICH
author of Descbooling Society

MEDICAL NEMESIS

The most explosive, uncompromising, thoroughly researched attack on the gravest health hazard we face today: our medical system.

"Read it and marvel at the light it sheds."
— THE NEW YORK TIMES

review of
Ivan Illich's **Medical Nemesis**
by tENTATIVELY, a cONVENIENCE - July 15-24, 2021

Backlash to Arrogance: Let Illich, Be Heard

1st, this bk is truly brilliant. I've been hearing about Illich for a long time, mainly in connection w/ his <u>Deschooling Society</u>, but it wasn't until I finally read this that I realized that he's a thinker who seriously rethinks major institutions in society in a way that's incredibly well-researched & very, VERY thorough. <u>Medical Nemesis</u> was published in 1976. I remember having revelations about things like preplanned obsolescence in the early '70s - thinking that it was such an obviously bad thing that corporations wd be forced to stop structuring their profits around it. No such luck, preplanned obsolesence is worse than ever now BY FAR. The same applies to all of the problems that Illich points out about medicalization & iatrogenesis. Not only have they not gone away, they've reached a sort of melting point in today's day & age.

I read recently someone's opinion that the saying 'Those who don't know history are doomed to repeat it' (credited in various forms to various people) isn't true b/c even people who *do know history* are doomed to repeat it anyway - & that's the way I feel about the times I'm currently living in. I, e.g., understand the way that the mass media manipulates public opinion as a form of mind control but I also see it working time & again regardless of how much I point out things that I, at least, think are obvious. The result is that I'm stuck in a 'reality' where history repeats itself not b/c I don't remember but b/c the majority of the people around me are so easily kept in a destructive loop that goes unquestioned. Illich has done the questioning about as intelligently as any human being can be expected to - but very few people I know wd ever read one of his bks & even if they did their minds wd gloss over at the lack of sound-bytes & simple talking points. The absence of oversimplification in <u>Medical Nemesis</u> is just too challenging for people whose mindset is Good Guys vs Bad Guys - if everything isn't reduced to the good guys wearing white hats & the bad guys wearing black ones then how can anyone understand it, eh?! Put sadists & vampires in white coats & it's all good, right?!

The subtitle of this bk is "The Expropriation of Health". As I understand Illich's position, & I hope I don't misrepresent it, that means that the less autonomy the individual has to deal w/ their own health in their own way b/c of the encroachment of domineering impersonal medical institutions, the more a person's health ceases to *even be **theirs** anymore*: it becomes the 'property' of 'care*takers*' rather than 'caregivers' - thusly undermining what's truly healthy for the individual. Keep in mind that this review is *my interpretation* of Illich's highly articulated points. It's possible that I slant this interpretation too much one way or another w/o intending to.

In the author's opening "*Acknowledgments*" he references what I think we cd all use more of: intelligent conversation between well-informed & thoughtful friends. I don't mean the regurgitation of what passes for such things amongst the lazy: viz. stock propaganda phrases from NPR or Fox News or whatever, I mean actual discourse between people who take their own opinions seriously enough to actually have them *be their own* instead of just dummy-speak provided by their ventriloquists.

"My thinking on medical institutions was shaped over several years in periodic conversations with Roslyn Lindheim and John McKnight. Mrs. Lindheim, Professor of Architecture at the University of California at Berkeley, is shortly to publish *The Hospitalization of Space*, and John McKnight, Director of Urban Studies at Northwestern University, is working on *The Serviced Society*. Without the challenge from these two friends, I would not have found the courage to develop my last conversations with Paul Goodman into this book." - p v

Lindheim's bk title struck my fancy but, alas, if it ever made it to publication I didn't find it online. Instead I found her <u>Changing Hospital Environments for Children</u> (1972) as her only available publication. Illich may very well have had a ms copy of *The Hospitalization of Space*.

"The medical establishment has become a major threat to health. The disabling impact of professional control over medicine has reached the proportions of an epidemic. *Iatrogenesis*, the name for this new epidemic, comes from *iatros*, the Greek word for "physician," and *genesis*, meaning "origin." Discussion of the disease of medical progress has moved up on the agendas of medical conferences, researchers concentrate on the sick-making powers of diagnosis and therapy, and reports on paradoxical damage caused by cures for sickness take up increasing space in medical dope-sheets." - p xi

"*Dorland's Illustrated Medical Dictionary*, 25th ed. (Philadelphia: Saunders, 1974): "Iatrogenic (*iatro*—Gr. physician, *gennan*—Gr. to produce). Resulting from the activity of physicians. Originally applied to disorders induced in the patient by autosuggestion based on the physician's examination, manner, or discussion, the term is now applied to any adverse condition in a patient occurring as the result of treatment by a physician or surgeon."" footnote 5, pp 4-5

I've made myself very unpopular during the time of the QUARANTYRANNY by expressing opinions similar to this. Think of deaths caused by putting people on ventilators to hypothetically save their lives, think of drug addiction enabled by prescriptions of pain-killers, think of mental states changed negatively by an excessive medical tendency to diagnose 'disorders' that might've previously just been moods or temporary states of mind, think of people slotted into bureaucracies who get oversimplified as a result.

"This book offers the lay reader a conceptual framework within which to assess the seamy side of progress against its more publicized benefits. It uses a model of social assessment of technological progress that I have spelled out elsewhere1 and applied previously to education2 and transportation,3 and that I now apply to the criticism of the professional monopoly and of the scientism in health care that prevail in all nations that have organized for high levels of industrialization.

******

1 *Tools for Conviviality* (New York: Harper & Row, 1973).

2 *Deschooling Society* (New York: Harper & Row, 1971).

3 *Energy and Equity* (New York: Harper & Row, 1974)." - pp xii-xiii

I include the footnotes in the above for 2 main reasons: 1. the footnotes in this bk are very important, Illich references bks, journal articles, manuscripts, & even mimeographs in at least 5 languages: English, Spanish, French, German, & Italian - making this one of the most exhaustively researched things I've ever read - there're pages where the footnotes cover almost the entire page: that will be undoubtedly tedious & distracting for many or most readers but I appreciate the opportunity to be referred elsewhere; 2. these are the titles of 3 of his other bks, I shd probably read them to see what such an extraordinary mind is capable of coming up w/ in relation to other subjects.

"The layman in medicine, for whom this book is written, will himself have to acquire the competence to evaluate the impact of medicine on health care. Among all our contemporary experts, physicians are those trained to the highest level of specialized incompetence for this urgently needed pursuit." - pp xiii-xiv

Illich has an amazing ability to cut thru the bullshit. Note that he refers to "the impact of medicine on health care" - to many or possibly even most people "medicine" has come to be synonymous w/ "health care": an aspect of Illich's importance is that he's capable of separating the 2.

"The recovery from society-wide iatrogenic disease is a political task, not a professional one. It must be based on a grassroots consensus about the balance between the civil liberty to heal and the civil right to equitable health care. During the last generations the medical monopoly over health care has expanded without checks and has encroached on our liberty with regard to our own bodies. Society has transferred to physicians the exclusive right to determine what constitutes sickness, who is or might become sick, and what shall be done to such people." - p xiv

&, as w/ so much in this bk, I find that to be extremely prescient of the present tension (2020-2021). Talk of forced vaccination & vaccine passports foisted upon the public against our will but ostensibly 'for our own good'. I'm reminded of forced sterilizations & the testing of dangerous birth-control devices on people in '3rd world' countries by drs in more industrialized nations. There's always some 'superior' person who 'knows what's best' for the people they pretend to care about but actually look down upon.

Belief in mass-media hysteria, the cowardice of conformity to peer-pressure, & the overmedication of people who simply don't have the mental resourcefulness or strength to cope w/ their own lives are things that've led to a phenomenal level of stupidity. This particular stupidity doesn't necessarily cut across all intellectual activities, only the ones dominated by fear. A lack of fear doesn't necessarily mean damage to the amygdala or even bravery, simply a common-sensical ability to see thru lies that contradict one's own experience. A friend of mine recently informed me that I'm 100 times more at risk for not being vaccinated - &, yet, he's still alive & I'm still alive. One might think that being 100 times more at risk wd've killed me off by now. IMO, a sensible person wd recognize the ridiculousness of such "100 times" nonsense simply by seeing how many people are still alive who defy such illogic - but, no, the fear is too compelling for logic to take hold.

""Health," after all, is simply an everyday word that is used to designate the intensity with which individuals cope with their internal states and their environmental conditions."

[..]

"Health levels can only decline when survival comes to depend beyond a certain point on the heteronomous (other-directed) regulation of the organism's homeostasis. Beyond a critical level of intensity, institutional health care—no matter if it takes the form of cure, prevention, or environmental engineering—is equivalent to systematic health denial." - p xv

& it's this basic premise of the bk that many people are likely to find challenging. Illich's position is that health is dependent on autonomy, the individual is healthiest when they're self-regulating - if self-regulation becomes undermined by the imposition of both toxic environmental conditions & an external health system that's essentially invasive then the individual becomes deprived of their ability to cope. 'Health' that's *imposed* is no longer health at all b/c it's not natural to itself, it's parasitic to the host organism.

"This study of pathogenic medicine was undertaken in order to illustrate in the health-care field the various aspects of counterproductivity that can be observed in all major sectors of industrial society in its present stage." - p xvi

People seem to have a tendency to respect power as having been earned. As such, domineering institutions can be taken for granted as somehow 'deserving' their place. If the propaganda shield around these institutions is omnipresent & effective enuf then they become even less subject to paradigm-shifting criticisms. Illich is a paradigm-shifter. It's about time that more people realized that the medical industry's self-promotion as a wholy benevolent phenomena is glossing over greed, sadism, reckless experimentation, & substantial harm.

"Almost everyone believes that at least one of his friends would not be alive and well except for the skill of a doctor, there is in fact no evidence of any direct relationship between this mutation of sickness and the so-called progress of medicine.3"

[..]

"3 René Dubos, *The Mirage of Health: Utopian Progress and Biological Change* (New York: Anchor Books, 1959), was the first to effectively expose the delusion of producing "better health" as a dangerous and infectious medically sponsored disease. Thomas McKeown and Gordon McLachlan, eds., *Medical History and Medical Care: A Symposium of Perspectives* (New York: Oxford Univ. Press, 1971), introduce the sociology of medical pseudo-progress. John Powles, "On the Limitations of Modern Medicine," in *Science, Medicine and Man* (London: Pergamon, 1973), 1:1-30, gives a critical selection of recent English-language literature on this subject. For the US situation consult Rick Carlson, *The End of Medicine* (New York: Wiley Interscience, 1975)." - pp 3-4

"The study of the evolution of disease patterns provides evidence that during the last century doctors have affected epidemics no more profoundly than did priests during earlier times. Epidemics came and went, imprecated by both but touched by neither. They are not modified any more decisively by the rituals performed in medical clinics than by those customary at religious shrines.8"

[..]

"8 On the clerical nature of medical practice, see "Cléricalisme de la fonction médicale? Médicine et politique. Le 'Sacerdoce' médical. La Relation thérapeutique. Psychanalyse et christianisme," *Le Semeur*, suppl. 2 (1966-67)." - p 5

"For more than a century, analysis of disease trends has shown that the environment is the primary determinant of the state of general health of any population." - p 7

That seems accurate enuf to me. If you live in an industrialized area where there's air & water pollution is it any surprise that you're sick? &, yet, if you went to a dr about a cough wd s/he advise shutting down the locally air polluting company? Or wd s/he give you some sort of medicine to make yr throat more insensitive to the damage being done to it?

There are paradoxes & ironies.

"27 J. E. Davies and W. F. Edmundson, *Epidemiology of DDT* (Mount Kisco, N.Y.: Future, 1972). A good example of paradoxical disease control from Borneo: Insecticides used in villages to control malaria vectors also accumulated in cockroaches, most of which are resistant. Geckoes fed on these, became lethargic, and fell prey to cats. The cats died, rats multiplied, and with rats came the threat of epidemic bubonic plague. The army had to parachute cats into the jungle village (*Conservation News*, July 1973)." - footnote, p 11

A similar example wd be the importation of cane toads to Australia to eat cane beetles. This backfired. "The long-term effects of toads on the Australian environment are difficult to determine, however some effects include "the depletion of native species that die eating cane toads; the poisoning of pets and humans; depletion of native fauna preyed on by cane toads; and reduced prey populations for native insectivores, such as skinks."" (https://en.wikipedia.org/wiki/Cane_toads_in_Australia) It seems to be more common for the natural environment to have its own checks & balances although some people, such as the author Michael Crichton (see my review of his State of Fear here: http://www.goodreads.com/book/show/15860.State_of_Fear), mock these. In recent times (2020-2021) there seems to be an anti-naturist mvmt that somehow believes itself to be rooted in science w/o, as far as I can tell, any real substance to that belief.

"28 A good example of medical persecution of innovators is given by G. Gortvay and J. Zoltan, *I. Semmelweis, His Life and Work* (Budapest: Akademiai Kiado, 1968), a critical biography of the first gynecologist to use antiseptic procedures in his wards. In 1848 he reduced mortality from puerperal fever by a factor of 15 and was thereupon dismissed and ostracized by his colleagues, who were offended at the idea that physicians could be carriers of death. Morton Thompson's novel *The Cry and the Covenant* (New York: New American Library, 1973) makes Semmelweis come alive." - footnote, p 11

This review isn't very saucy - am I getting too serious in my old age?

"31 Alain Letourmy and François Gibert. *Santé, environnement, consommations médicales: Un Modèle et sonestimation à partir des données de mortalité; Rapport principal* (Paris, CEREBE (Centre de Recherche sur le Bien être), June, 1974). Compares mortality rates in different regions of France: they are unrelated to medical density, highly related to the fat content of the sauces typical of each region, and somewhat less to alcohol consumption." - footnote, p 12

Even tho I prefer to not be vaccinated, I'm not entirely against vaccines. One of the reasons for this is that I was born in 1953 & was vaccinated against polio. I've had friends who were born slightly before me who *weren't* vaccinated b/c the vaccine didn't exist yet & who got polio. Some of them have disabled legs as a result. Illich addresses this.

"The reappearance of malaria is due to the development of pesticide-resistant mousquitoes and not any lack of new anti-malarial drugs.33 Immunization has almost wiped out paralytic poliomyelitis, a disease of developed countries, and vaccines have certainly contributed to the decline of whooping cough and measles,34 thus seeming to confirm the popular belief in "medical progress."35 But for most other infections, medicine can show no comparable results." - pp 13-14

The perceptive reader will've noted by now that I'm only up to page 14 but I've already quoted an enormous amt. That gives you an idea of how important I think every statement in this bk is.

Over the time of the PANDEMIC PANIC I've been ruminating over my position re disease. Given that I haven't gotten sick & that no-one I know has gotten sick-unto-death I'm probably a bit insensitive to the horror felt by others closer to their mortality & the mortality of their loved ones. Nonetheless, I will postulate something that's bound to be, ahem, controversial: *what if disease is just another part of nature that we have difficulty accepting b/c we think that humanity is **more important than nature** rather than just part of it?!* When strong winds happen tree limbs that're no longer as healthy as the rest of the tree tend to fall off - what if a person dying w/ the aid of a virus is like those tree limbs? Thinking of ourselves or the ones we love in that way might seem callous, & perhaps it is, but we all deteriorate & eventually die - it's simply a part of the natural process. I object less to a virus than I do to a bullet or a reckless driver.

I was diagnosed as having Diabetes Type II in September, 2019. I told the dr that I'd try to change this w/ diet & exercise. She told me that that was "impossible", she wanted me to start injecting insulin immediately & tried to terrorize me as much as possible into spending, spending. She outlined various ways in wch I was likely to die in the near future. I told her that I'd try my way 1st & if it didn't work I'd return to her in a mnth. I made a documentary about this that's on my onesownthoughts YouTube channel here: https://youtu.be/2GLu66dgpKl . W/in 5 days of following my own changes all my symptoms were gone, w/in 2 wks my blood sugar was normalized - so much for the 'impossibility' of it, it wasn't even hard. I'll never go to that dr again & wd certainly recommend that no-one else go to her either. W/ this in mind I'm always interested in positions regarding diabetes type II that go contrary to mainstream medical industry ones.

"[..]"Effects of Hypoglycemic Agents on Vascular Complications in Patients with Adult-Onset Diabetes," *Journal of the American Medical Association* 217 (1971): 777-84. Cochrane, *Effectiveness and Efficiency*, comments on the last two. They suggest that giving tolbutamide and phenformin is definitely disadvantageous to the treatment of mature diabetics and that there is no advantage in giving insulin rather than a diet." - footnote 39, p 15

That certainly jives w/ my own experience. One of the main differences is that the medical industry will profit enormously off the insulin & not at all off my diet change. Hence, the heavy promotion of the insulin as basically 'my only chance for survival'. Funny, it's been almost 2 yrs since the dr-predator tried to rope me in & I'm still alive & kicking.

"Unfortunately, futile but otherwise harmless medical care is the least important of the damages a proliferating medical enterprise inflicts on contemporary society. The pain, dysfunction, disability, and anguish resulting from technical medical intervention now rival the morbidity due to traffic and industrial accidents and even war-related activities, and make the impact of medicine one of the most rapidly spreading epidemics of our time. Among murderous institutional torts, only modern malnutrition injures more people than iatrogenic disease in its various manifestations.47"

[..]

"47 Literally "iatro-genic" means an action that produces physicians, i.e. something that only medical schools or parents of future doctors do. But for over 80 years the term has been commonly used for health-damage induced by doctors. For some of the standard textbooks see Robert H. Moser, *The Disease of Medical Progress: A Study of Iatrogenic Disease*, 3rd ed. (Springfield, Ill.: Thomas, 1969). David M. Spain, *The Complications of Modern Medical Practices* (New York: Grune & Stratton, 1963). H. P. Kümmerle and N. Goosens, *Klinik und Therapie der Nebenwirkungen* (Stuttgart: Thieme, 1973 [1st ed., 1960]). R. Heintz, *Erkrankungen durch Arzneimittel: Diagnostik, Klinik, Pathogenese, Therapie* (Stuttgart: Thieme, 1966). Guy Duchesnay, *Le Risque thérapeutique* (Paris: Doin, 1954). P. F. D'Arcy and J. P. Griffin, *Iatrogenic Disease* (New York: Oxford Univ. Press, 1972)." - p 17

As I explained to the dr-predator, I'm more concerned w/ maintaining the quality of my life than I am w/ just keeping myself going indefinitely. If I had followed her dr's pressure the quality of my life wd've gone way down & she wd've been vampiristically able to suck every last drop of funds that cd be siphoned from or thru me until I was little more than a braindead empty sack, an exhausted bank-bag. It wd've been hell for me & a cash-flow for her. It wd've been an iatrogenic tort, something I feel fortunate to've had my wits about me enuf to avoid.

"Medicines have always been potentially poisonous, but their unwanted side-effects have increased with their power51 and widespread use."

[..]

"51 L. Meyler, *Side Effects of Drugs* (Baltimore: Williams & Wilkins, 1972). *Adverse Reactions Titles*, a monthly bibliography of titles from approximately 3,400 biomedical journals published throughout the world; published in Amsterdam since 1966. *Allergy Information Bulletin*, Allergy Information Association, Weston, Ontario." - pp18-19

I'm particularly fascinated by there being an *Adverse Reactions Titles* bibliography! How many people even realize that such resources exist(ed)? It's to the medical industry's credit that they at least keep track of the iatrogenesis. How many people realize that the CDC (Center for Disease Control and Prevention in the USA) has The Vaccine Adverse Event Reporting System (VAERS) at this URL: https://wonder.cdc.gov/vaers.html ? In my own quick search it was revealed that there've been 6,374 deaths reported - 0.56% of 1,130,123 vaccinated as of July 16, 2021. This in people who probably got vaccinated 1. out of fear of dying (they died anyway), 2. under peer pressure (always a stupid 'reason').

Death is only one of maaaannnnnyyyyy adverse reactions reported. I shd interpolate here that the CDC VAERS website has an "I agree" button to click before using. They wdn't want you to think that there aren't potential inaccuracies w/ the data. 7,497 people experienced a "burning sensation"; 12,896 people (1.14%) experienced "cellulitis"; 2,403 experienced a "cerebrovascular disorder"; 13,237 (1.17%) experienced "chest discomfort"; 17,705 (1.57%) experienced "chest pain"; 94,280 (8.34%): "chills"; 22,894 (2.03%): cough; 8,720 (0.77%): COVID-19.. You get the idea. People get vaccinated to protect themselves against COVID-19 & then? get COVID-19 as a result. If 8.34% of a restaurant's clientele reported "chills" after eating there wd the restaurant be faced w/ possible closure?

"*Disabling nondiseases* result from the medical treatment of nonexistent diseases and are on the increase:59"

[..]

"59 Clifton Meador, "The Art and Science of Nondisease," *New England Journal of Medicine* 272 (1965); 92-5. For the physician accustomed to dealing only with pathogenic entities, terms such as "nondisease entity" or "nondisease" are foreign and difficult to comprehend. This paper presents, with tongue in cheek, a classification of nondisease and the important therapeutic principles based on this concept. Iatronegenic disease probably arises as often from treatment of nondisease as from treatment of disease." - p 20

"Nondisease": what about all the false positives that then get treated, generally w/ drugs that're harmful? It's so easy for the medical industry to gloss over such harmful mistakes w/ gobbledygook. The naive & trusting patient takes it for granted that the dr's acting in their best interests w/ a brilliance beyond that of mere humans - thusly having a hard time believing that mistakes might be made.. often.

"The United States Department of Health, Education, and Welfare calculates that 7 percent of all patients suffer compensable injuries while hospitalized, though few of them do anything about it. Moreover the frequency of reported accidents in hospitals is higher than in all industries but mines and high-rise construction. Accidents are the major cause of death in American children. In proportion to the time spent there, these accidents seem to occur more often in hospitals than in any other kind of place. One in fifty children admitted to a hospital suffers an accident which requires specific treatment.67"

[..]

"67 George H. Lowrey, "The Problem of Hospital Accidents to Children," *Pediatrics* 32 (December 1963): 1064-8." - p 23

I have a minor incident from my own childhood to relate: I was in a hospital, just for overnight observation as I recall, & my mother was leaving to go home. I was very young & I stood next to a window w/ venetian blinds to wave goodbye to her. Somehow I managed to cut my leg on the blinds. I still have the scar 60+

yrs later so it must've been a somewhat substantial cut. I wdn't necessarily blame the hospital or, more generally, the medical industry for that; I might blame instead the unfamiliar environment & my own youthful clumsiness.

"Technical and managerial measures taken on any level to avoid damaging the patient by his treatment tend to engender a self-reinforcing iatrogenic loop analogous to the escalating destruction generated by the polluting procedures used as antipollution devices.72

"I will designate this self-reinforcing loop of negative institutional feedback by its classical Greek equivalent and call it *medical nemesis*. The Greeks saw gods in the forces of nature. For them, nemesis represented divine revenge visited upon mortals who infringe on those prerogatives the gods enviously guard for themselves. Nemesis was the inevitable punishment for attempts to be a hero rather than a human being. Like most abstract Greek nouns, Nemesis took the shape of a divinity. She represented nature's response to *hubris*: to the individual's presumption in seeking to acquire the attributes of a god. Our contemporary hygenic hubris has led to the new syndrome of medical nemesis. 73"

[..]

"73 The term was used by Honoré Daumier (1810-79). See reproduction of his drawing "Némésis médical" in Werner Block, *Der Artzt und der Tod in ildern aus sechs Jahrhunderten* (Stuttgart: Enke, 1966)." - pp 25-26

In a sense, Illich 'loses me' here. If the medical industry's hubris is arrogance, then ok; if it's being ambitious in its endeavors then I don't really fault it for that. By framing the title of the bk in Greek mythology the overall premise becomes more problematic to me. The sense in wch I agree is that I think that there's a presumption to medicine & science that humans have a wisdom re nature that justifies interventions that're actually far from wise & that produce a backlash that signifies ignorance disguised as supreme knowledge. I reckon I prefer a more straightforward descriptive title like "Backlash to Arrogance" over "Medical Nemesis" b/c the latter takes too much of a risk of being turned into a metaphor - but I'm quibbling.

At any rate, I found this bk to be fantastically prescient in relation to the arrogance of the QUARANTYRANNY & the way it's been used politically for the advancement of forces pretending to be for our own good but transparently dictatorial to people who can keep their heads out of their assholes.

"Zola, Irving Kenneth, "Medicine as an institution of social control." in: *The Sociological Review*, vol. 20 No. 4 (new series) Nov. 1972. pp. 487-509. "The theme of this essay is that medicine is becoming a major institution of social control, nudging aside, if not incorporating, the more traditional institutions of religion and law. It is becoming the new repository of truth, the place where absolute and often final judgments are made by supposedly morally neutral and objective experts . . . in the name of health."" - footnote 6, pp 32-33

& what a big lie that is. Drs *are not* "morally neutral and objective", they're biased by all sorts of things, primary among them being a sense of class superiority. They might maintain a poker face b/c they don't want the people who're 'confessing' to them by describing intimate details of their life to realize just how biased they are. More than ever before, people have been suckered into a new submission based on this lie. The QUARANTYRANNY wd've never been pulled off w/o the sham of for-yr-own-good 'objectivity'. People don't recognize that the gradual deterioration of their trust in priests has been replaced by an equally senseless trust in the medical industry.

"Social iatrogenesis designates a category of etiology that encompasses many forms. It obtains when medical bureaucracy creates ill-health by increasing stress, by multiplying disabling dependence, by generating new painful needs, by lowering the levels of tolerance for discomfort or pain, by reducing the leeway that people are wont to concede to an individual when he suffers, and by abolishing even the right to self-care." - p 33

During the time of the PANDEMIC PANIC the push has been on to demonize anyone whose medical opinions contradict the mainstream narrative being heavily pushed by mass media, to demonize anyone who rejects wearing masks as both unhealthy & dictatorially imposed, to demonize anyone who rejects vaccines for the same reason. The new dictatorships are as much 'for our own good' as nazism was 'good for the German people', as capitalism is 'good for freedom', as the Soviet Republics were 'good for the proletariat'. Somehow, it never really works out that way does it? Nazism put everyone in uniform as a form of job security & then put them to work for a death machine, capitalism is good for the rich & their toadies (when they can stand their position in the hierarchy) but bad for everyone else - or how else wd there be profits?, the Soviet Republics had to keep narrowing down its conception of the good comrade until being human barely even fit anymore. Now we have a new creation of 2nd class citizens, the scapegoats that nazism, capitalism, & the soviets had to create to keep the larger populace in line - the new 2nd class citizen is just another artificially manufactured 'threat to the general well-being', in reality, one of its only hopes.

"14 P. M. Brunetti, "Health in Ecological Perspective," *Acta Psychiactrica Scandinavica* 49, fasc. 4 (1973): 393-404. Brunetti argues that the concentration of power and the dependence on extrametabolic energy can make the vital milieu uninhabitable for beings whose integration depends on the exercise of their autonomy. Medicine is used to rationalize this transfer." - footnote, p 36

Society as it exists now seems more & more insect-hive-like to me. No autonomous beings welcome. I'm a Lone Wolf Individualist Anarchist. Nonetheless, I support mutual aid & prefer having social options. The medical industry's dictatorial rationalizations in the name of a collective good not only strike me as completely false but damaging to the autonomous mindset most likely to be imaginative & productive of innovation. Pro-Capitalists make the claim that innovation comes from profit-motive: ie: competeing companies develop better products to capture more of the market. That's probably accurate enuf w/in that context but any innovations that've originated from me haven't originated w/ profit-motive but more from the sheer joy of creativity. Different types of innovation result from those different contexts.

"The Greeks' only word for "drug"—*pharmakon*—did not distinguish between the power to cure and the power to kill.21"

"21 On the double meaning of this term from archaic texts to the Hippocratic corpus, see Walter Artelt, *Studien zur Geschichte der Begriffe "Heilmittel" und "Gift": Urzeit-Homer-Corpus Hippocraticum* (Darmstadt: Wissenschaftliche Buchgesellschaft, 1968)." - p 37

In German, *heilmittel* means remedies & *gift* means poison. I had the misfortune of being lovers w/ a German-speaking paranoid schizophrenic who fixated on the idea that I was trying to poison her after I used the English word "gift" in conversation w/ her. Making matters worse, she then proceeded to tell other people that I was trying to poison her w/o referencing that "gift" means "poison" in German.

But what about health care costs you ask? My mom told me that it cost $25 to give birth to me in a hospital in 1953. In an article entitled "When Childbirth Cost $100" (https://www.cryo-cell.com/blog/april-2017/when-childbirth-cost-100-dollars) it's stated that childbirth cd cost $29.50 in the 1950s. A receipt is shown that itemizes that cost as follows: $10.50 for "Board & Nursing", $10.00 for "Delivery Room", $5.00 for "Anesthetic", $3.00 for "Laboratory Exam." & $1.00" for "Drugs". When I asked the internet what giving birth in a hospital costs now: "According to data collected by Fair Health, the average cost of having a vaginal delivery is **between $5,000 and $11,000** in most states." (https://www.parents.com/pregnancy/considering-baby/financing-family/what-to-expect-hospital-birth-costs/) According to the CPI Inflation Counter (https://www.in2013dollars.com/us/inflation/1953?amount=25) that 1953 $25 wd now be $254.40. That leaves $4,745.60 highly suspect as called for by GREED.

"In all industrial nations—Atlantic, Scandinavian, or East European—the growth rate of the health sector has advanced faster than that of the GNP.41"

[..]

"41 An excellent general introduction to the cost explosion in health care is R. Maxwell, *Health Care: The Proving Dilemma; Needs vs. Resources in Western Europe, the U.S. and the U.S.S.R.* (New York: McKinsey & Co., 1974). Ian Douglas-Wilson and Gordon McLachlin, eds., *Health Service Projects: An International Survey* (Boston: Little, Brown, 1973). This international comparison shows "the extreme heterogeneity in organization and ideology" of different systems. Everywhere "the rationalization is motivated, not by the politics of the left or the right, but by the sheer necessity to secure more effective use of scarce and expensive resources." No country can indefinitely sustain unchecked increases in funds allocated for the treatment of illness." - p 42

"Since 1950 the cost of keeping a patient for one day in a community hospital in the United States has risen by 500 percent.45 The bill for patient care in the major university hospitals has risen even faster, tripling in eight years. Administrative costs have exploded, multiplying since 1964 by a factor of 7; laboratory costs have risen by a factor of 5, medical salaries only by a factor of 2.46"

[..]

"46 John H. Knowles, "The Hospital," Scientific American 229 (September 1973): 128-137. Contains charts and graphs on the evolution of hospital expenditures." - p 43

Illich & all of his sources pointed these things out by 1976. It's 2021 as I write this & things seem even worse. I'm reminded of the early 1970s when friends of mine & I 1st became aware of preplanned obsolescence. It seemed so transparently egregious that we thought &/or hoped that the greed of corporations wd have to curtail themselves, that they were busted. Instead? It's far worse now than it was then. It helps, of course, if the marks, the suckers, are BELIEVERS, if they believe that the medical industry is entirely benign, if they're too passive & complacent to even bother to do a little simple research.

"In the past decade, while a few rich countries began to control the damage, waste, and exploitation caused by the licit drug-pushing of their doctors, physicians in Mexico, Venezuela, and even Paris had more difficulty than ever before in getting information on the side-effects of the drugs they prescribed.89"

[..]

"89 The American physician can easily gain access to this information from such sources as Medical Letter on Drugs and Therapeutics, Medical Library Association, 919 N. Michigan Avenue, Chicago, Ill. This is an unbiased source of drug information mailed fortnightly. Nothing comparable is available in French, German, or Spanish. Also see Richard Burack, The New Handbook of Prescription Drugs: Official Names, Prices, and Sources for Patient and Doctor, rev. ed. (New York, Pantheon, 1970) (See below, note 99, p. 61, for description of this book.)" - p 58

The Medical Letter still exists & still appears to have the same mission as when Illich wrote about it:

"**Medical Letter, Inc. (EIN: 13-1881832)** is a nonprofit organization* that relies solely on subscription fees and donations to support our mission of providing objective, practical, and timely information on drugs and therapeutics.

"Our work relies on support from people like you who value credible, unbiased drug information that is free of any commercial interest." - https://secure.medicalletter.org

I bought a 3 yr subscription.

"medical memories have proved particularly short.93"

"93 Memory is no guide to which drugs have been prescribed or consumed in the past. A search in the national registry of prescriptions in England and Wales show that 8 out of 10 women who had born a defective child after taking thalidomide on prescription denied that they had taken the drug, and that their physicians denied having ordered it. See A. L. Speirs, "Thalidomide and Congenital Abnormalities," *Lancet*, 1962, 1:303." - p 60

"Thalidomide was a widely used drug in the late 1950s and early 1960s for the treatment of nausea in pregnant women. It became apparent in the 1960s that thalidomide treatment resulted in severe birth defects in thousands of children. Though the use of thalidomide was banned in most countries at that time, thalidomide proved to be a useful treatment for leprosy and later, multiple myeloma. In rural areas of the world that lack extensive medical surveillance initiatives, thalidomide treatment of pregnant women with leprosy has continued to cause malformations." - https://pubmed.ncbi.nlm.nih.gov/21507989/

Did the *Medical Letter* warn people about thalidomide at any time during the 1960s? The previous issues available to me online as a subscriber only go back to 1988 so I can't check that way. A search for thalidomide on their website didn't yield any results. Asking the question online that opens this paragraph yielded this:

"In July 1962, the Food and Drug Administration sent an urgent message to its field offices with an assignment it said was "one of the most important we have had in a long time.""

"Overseas, thousands of babies in Germany, England and other countries were being born with severe defects tied to their mothers' use of thalidomide, a drug widely taken for insomnia, morning sickness and other ailments.

"Meanwhile, the federal government sought to figure out what had happened in the United States, and how many babies had been affected.

"The drug was not approved in the United States in the 1960s, but as many as 20,000 Americans were given thalidomide in the 1950s and 1960s as part of two clinical trials operated by the American drug makers Richardson-Merrell and Smith, Kline & French.

"Here is the story of the F.D.A.'s investigation, told through a sampling of the more than 1,300 pages of documents obtained through a Freedom of Information request.

"'Great public interest'

"On Aug. 1, 1962, President John F. Kennedy issued a warning: "Every woman in this country, I think, must be aware that it's most important that they check their medicine cabinet and that they do not take this drug."

"Just days earlier, the F.D.A. had instructed its inspectors to interview every doctor who had received thalidomide and to investigate if any babies were harmed."

- https://www.nytimes.com/2020/03/23/health/thalidomide-fda-documents.html

SO, while I don't know whether the *Medical Letter* was on top of this at the time it's obvious that powerful government forces were &, yet, pharmaceutical companies still managed to cause severe health harm to large numbers of people.

"Notably, within one week after the Chilean military junta took power on September 11, 1973, many of the most outspoken proponents of a Chilean medicine based on community rather than on drug imports and drug consumption had been murdered.104" - pp 62-63

"104 Albert Johnsen et al., "Doctors in Politics: A Lesson from Chile," *New England Journal of Medicine* 291 (1974): 471-2. Describes the particular violence with which physicians were persecuted by the junta." - footnote, p 63

"On November 5, 1766, the Empress Maria Theresa issued an edict requesting the court physician to certify fitness to undergo torture so as to ensure healthy, i.e. "accurate," testimony; it was one of the first laws to establish mandatory medical certification." - p 71

Are you getting the picture yet? There are political forces in the world that know that using the medical industry is key to social control. Some of those forces are people w/in the medical industry who're allied w/ other forces. Then there're people w/ medical philosophies that go contrary to the power. These are the ones murdered by the Chilean junta. These days, in countries where murdering is less viable, people w/ contrary opinions are 'discredited' by fact chokers &/or shadow-banned, etc..

"Once a society is so organized that medicine can transform people into patients because they are unborn, newborn, menopausal, or at some other "age of risk," the population inevitably loses some of its autonomy to its healers." - p 72

"At each stage of their lives people are age-specifically disabled. The old are the most obvious example: they are victims of treatments meted out for an incurable condition." - p 73

True dat. Instead of getting more & more embroiled in a system that's ultimately parasitic on you why not try to separate yourself from it as much as you can? My neighbor of 25 yrs ago was 98 yrs old, she didn't go to drs. She finally died at age 100 when she was taken to the hospital by her daughter after she hurt herself at home. Her daughter, who might've been as old as 80 at the time, had openly told me that she was sick of taking c/o her mom & her husband. I think she wanted to take c/o loose ends before she, herself, died. This daughter died soon thereafter.

""For the sick," Hippocrates said, "the least is best."" - p 74

"As more old people become dependent on professional services, more people are pushed into specialized institutions for the old, while the home neighborhood becomes increasingly inhospitable to those who hang on." - p 77

Old people often own their own homes & try to stay in them until they die. In the meantime the real estate vultures are hovering overhead hoping for the old to disappear, one way or another, as obstacles so that their houses can be flipped. As such, it's in the vultures's interests for the old person to be 'put away in medical care for their own good' - thusly 'freeing up the house'. It's not a pretty picture.

"Although physicians did pioneer antisepsis, immunization, and dietary supplements, they were also involved in the switch to the bottle that transformed the traditional suckling into a modern baby and provided industry with working mothers who are clients for a factory-made formula.

"The damage this switch does to natural immunity mechanisms fostered by human milk and the physical and emotional stress caused by bottle feeding are comparable to if not greater than the benefits that a population can derive from specific immunizations." - p 81

I include such quotes b/c I agree w/ them. It seems obvious to me that breast-fed children are likely to be receiving benefits both psychological & physical that bottle-fed babies aren't. The increasingly intervening methods of producing babies that seem progressive b/c they enable, e.g., lesbians to become pregnant w/o the insemination provided by a man's penis during sex, are an extension of bottle feeding, a further dislocation of reproduction & baby-raising from natural processes that I think it's the medical industry's hubris to think they can do better than.

"When a veterinarian diagnoses a cow's distemper, it doesn't usually affect the patient's behavior. When a doctor diagnoses a human being, it does." - p 84

How many children have been pronounced ADD (Attention Deficit Disorder) or ADHD (Attention Deficit Hyperactive Disorder) or Autistic, been prescribed Ritalin

& had their identity shaped negatively into adulthood?

"Diagnosis always intensifies stress, defines incapacity, imposes inactivity, and focuses apprehension on nonrecovery, on uncertainty, and on one's dependence upon future medical findings, all of which amounts to a loss of autonomy for self-definition." - p 90

I've always joked that my cure for cancer is for myself not to know I have it - hoping that if I do it just kills me w/o all the attendant stress of obsessing over it.

"180 Peter Schrag, Diane Divoky, *The Myth of the Hyperactive Child. And Other Means of Child Control.* Pantheon, 1975. The definitive repertory on an "entire generation slowly being conditioned to distrust its own instincts, to regard its deviation from the narrowing standards of approved norms as sickness and to rely on the institutions of the state and on technology to define and engineer its "health." The book also provides a guide to the U.S. literature on the subject. Peter Schrag in personal communications with the author has not only shaped his view on the medicalization of society but has also been of invaluable help in editing parts of the definitive edition of *Medical Nemesis.* Minimal brain damage in children is as often as not a creation of Ritalin; it is a diagnosis determined by the treatment. See Roger D. Freeman, "Review of Medicine in Special Education: Medical-Behavioral Pseudorelationships," *Journal of Special Education 5* (winter-spring 1971): 93-99." - footnote, p 89

"The medical label may protect the patient from punishment only to submit him to interminable instruction, treatment, and discrimination, which are inflicted on him for his professionally presumed benefit.165"

[..]

"165 Wilhelm Aubert and Sheldon Messinger, "The Criminal and the Sick," *Inquiry* 1 (1958): 137-60. Discusses the different forms social control can take, depending on the special way in which stigma impinges on moral identity." - pp 84-85

"Diagnosis may exclude a human being with bad genes from being born, another from promotion, and a third from political life." - p 85

Thought of that way, diagnosis IS a powerful tool for social control. What if the criteria for negative judgment have hidden biases other than what're on the surface? What if a person who feels fine about their self finds some of their characteristics 'discovered' to be a disease? I jokingly (but w/ critical intent) coined the term AMD (Administrator Meeting Disorder). (see my relevant movie here: https://youtu.be/IQgkC5OwPmw) What if showing characteristics of AMD were to become grounds for firing? Hhmm.. I might actually like that.

In today's day & age the socio-political push is on in a big way to get everyone on the same page, using the 'collective good' as an excuse, people were forced into wearing masks regardless of whether they believed they were healthy or not, people are being pushed to get a COVID-19 vaccination or face being banned from various aspects of social life - again regardless of what their own opinions are on the matter. Illich saw this coming from a mile away.

"The ultimate triumph of therapeutic culture183 turns the independence of the average healthy person into an intolerable form of deviance."

[..]

"The individual is subordinated to the greater "needs" of the whole, preventative procedures become compulsory" - p 91

"In the detection of sickness medicine does two things: it "discovers" new disorders, and it ascribes these disorders to concrete individuals. To discover a new category of disease is the pride of the medical scientist.170"

"170 *New England Journal of Medicine* 256 (1957): 253-8. Describes the costly discovery of an incurable "disease" that neither kills nor impairs and seems to be endemic wherever people come in contact with chickens, cattle, cats, or dogs." - p 86

What if instead of this being a "disease" it's an immunization?

"The medical-decision rule pushes him to seek safety by diagnosing illness rather than health.173 The classic demonstration of this bias came in an experiment conducted in 1934.174 In a survey of 1,000 eleven-year-old children from the public schools of New York, 61 percent were found to have had their tonsils removed. "The remaining 39 percent were subjected to examination by a group of physicians, who selected 45 percent of these for tonsillectomy and rejected the rest. The rejected children were reexamined by another group of physicians, who recommended tonsillectomy for 46 percent of those remaining after the first examination. When the rejected children were examined for a third time, a similar percentage was selected for tonsillectomy so that after three examinations only sixty-five children remained who had not been recommended for tonsillectomy. These subjects were not further examined because the supply of examining physicians ran out."175" - p 87

Thank the holy ceiling light they weren't testing for castration or lobotomy. I was given a tonsillectomy when I was 6. My adult perception is that the drs just wanted to practice their surgery skills on live subjects. I was put under w/ ether, something I still vividly remember & something I reckon was a bit too dangerous to ultimately justify the risk.

"more than 90 percent of all tonsillectomies performed in the Unites States are technically unnecessary, yet 20 to 30 percent of all children still undergo the operation. One in a thousand dies directly as a consequence of the operation and 16 in a thousand suffer from serious complications." - p 106

"The conjuring doctor perceives himself as a manager of crisis.192 In an insidious way he provides each citizen at the last hour with an encounter with society's deadening dream of infinite power."

[..]

"192 This spread of legitimacy for the institutional management of crisis has enormous political potential because it prepares for irreversible crisis government." - pp 92-93

Hence we have people like Governor Wolf of Pennsylvania prolonging his coronavirus emergency powers over & over again despite substantial legal moves against his doing so.

"Under the stress of crisis, the professional who is believed to be in command can easily presume immunity from the ordinary rules of justice and decency. He who is assigned control over death ceases to be an ordinary human. As with the director of a triage, his killing is covered by policy." - pp 93-94

Hence anyone who's peer-pressured into vaccination who then dies hasn't been murdered. It's so convenient.

"Death without medical presence becomes synonymous with romantic pigheadedness, privilege, or disaster. The cost of a citizen's last days has increased by an estimated 1,200 percent, much faster than over-all health-care." - p 95

The myth here, of course, being that if one dies in a hospital one at least dies w/ the 'experts doing everything they can for you'. More realistically, one might die as a result of simply being hospitalized, debilitated by being taken out of comfortable familiar surroundings & attacked by superbacteria resistant to the sterilization - or, even worse, deliberately killed off in a hospice by sleeping pills & starvation.

"People think that hospitalization will reduce their pain or that they will probably live longer in the hospital. Neither is likely to be true." - p 98

[..]

"But by staying at home, they avoid the exile, loneliness, and indignities which, in all but exceptional hospitals, awaits them." - pp 98-99

"Physical participation in a ritual is not a necessary condition for initiation into the myth which the ritual is organized to generate. Medical spectator sports cast powerful spells. I happened to be in Rio de Janeiro and in Lima when Dr. Christiaan Barnard was touring there. In both cities he was able to fill the major football stadium twice in one day with crowds who hysterically acclaimed his macabre ability to replace human hearts. Medical-miracle treatments of this kind have worldwide impact. Their alienating effect reaches people who have no access to a neighborhood clinic, much less to a hospital. It provides them with an abstract assurance that salvation through science is possible. The experience in the stadium at Rio prepared me for the evidence I was shown shortly afterwards which proved that the Brazilian police have so far been the first to use life-extending equipment in the torture of prisoners. Such extreme abuse of medical techniques seems grotesquely coherent with the dominant ideology of medicine." - p 107

Lest you think that torture is just a problem in countries 'more barbaric' than the US, let's 1st look at the subject as it's been presented in the last 2 decades:

"After the September 11, 2001 attacks on the United States, the US government authorized the use of so-called "enhanced interrogation techniques" on terrorism suspects in US custody. For years US officials, pointing to Department of Justice memorandums authorizing these techniques, denied that they constituted torture. But many clearly do: International bodies and US courts have repeatedly found that "waterboarding" and other forms of mock execution by asphyxiation constitute torture and are war crimes, Other authorized techniques, including stress positions, hooding during questioning, deprivation of light and auditory stimuli, and use of detainees' individual phobias (such as fear of dogs) to induce stress, violate the protections afforded all persons in custody – whether combatants or civilians – under the laws of armed conflict and international human rights law, and can amount to torture or "cruel, inhuman, or degrading treatment." Accordingly, the United Nations Committee against Torture and the UN Special Rapporteur on Torture have clearly stated that these techniques are torture." - https://www.hrw.org/news/2014/12/09/usa-and-torture-history-hypocrisy#

Torture IS illegal according to US law.

"Bill of Rights

"Torture as a punishment falls under the cruel and unusual punishment clause of the Eighth Amendment to the United States Constitution. The text of the Amendment states that:

"Excessive bail shall not be required, nor excessive fines imposed, nor cruel and unusual punishments inflicted.

"The U.S. Supreme Court has held since at least the 1890s that punishments which involved torture are forbidden under the Eighth Amendment.

"18 U.S.C. § 2340 (the "Torture Act")

"An act of torture committed outside the United States by a U.S. national or a non-U.S. national who is present in the United States is punishable under

18 U.S.C. § 2340. The definition of torture used is as follows:

"As used in this chapter—

"(1) "torture" means an act committed by a person acting under the color of law specifically intended to inflict severe physical or mental pain or suffering (other than pain or suffering incidental to lawful sanctions) upon another person within his custody or physical control;

"(2) "severe mental pain or suffering" means the prolonged mental harm caused by or resulting from—

"(A) the intentional infliction or threatened infliction of severe physical pain or suffering;

"(B) the administration or application, or threatened administration or application, of mind-altering substances or other procedures calculated to disrupt profoundly the senses or the personality;

"(C) the threat of imminent death; or

"(D) the threat that another person will imminently be subjected to death, severe physical pain or suffering, or the administration or application of mind-altering substances or other procedures calculated to disrupt profoundly the senses or personality; and

"(3) "United States" means the several states of the United States, the District of Columbia, and the commonwealths, territories, and possessions of the United States."

- https://en.wikipedia.org/wiki/Torture_and_the_United_States

Alas, as the Wikipedia entry goes on to explain,

"In October 2006, the United States enacted the Military Commissions Act of 2006, authorizing the President to conduct military tribunals of enemy combatants and to hold them indefinitely without judicial review under the terms of habeas corpus. Testimony coerced through humiliating or degrading treatment would be admissible in the tribunals. Amnesty International and others have criticized the Act for approving a system that uses torture, destroying the mechanisms for judicial review created by Hamdan v. Rumsfeld, and creating a parallel legal system below international standards. Part of the act was an amendment that retroactively rewrote the War Crimes Act, effectively making policymakers (i.e., politicians and military leaders), and those applying policy (i.e., Central Intelligence Agency interrogators and U.S. soldiers), no longer subject to legal prosecution under U.S. law for what, before the amendment, was defined as a war crime, such as torture. Because of that, critics describe the MCA as an amnesty law for crimes committed during the War on Terror."

The wikipedia article is fairly exacting in providing details to show that torture is part of US history. My own experience is that a neighbor, when I was young, returned from Vietnam, gathered a few of the young boys around, & told us in extremely graphic detail about how the US military tortured Viet Cong suspects. It was far too brutal for me to recount here. In the US, a friend of mine was hogtied by L.A. police & beaten on the soles of his feet. He had been protesting the incarceration of illegal immigrants in suspicious conditions. This wd severely injure him w/o causing easily detectable bruising. Another friend was dragged into an alley by Baltimore police & repeatedly beaten on the same part of his stomach to cause internal organ damage. He was known to the police as a junkie, an 'undesirable'. Neither victim stood a chance of prosecuting their tormenters.

Do I digress? I'm simply trying to drive Illich's points closer to home for many of the readers of this review.

"When he *assigns* sick-status to a client, the contemporary physician might indeed be acting in some ways similar to the sorceror or the elder; but in belonging also to a scientific profession that *invents* the categories it assigns while consulting, the modern physician is totally unlike the healer." - p 113

As I've already mentioned, I coined the term AMD (Administrator Meeting Disorder). What if I were in a position accepted as authoritative & I were to start diagnosing people as 'suffering' (really, causing other people to suffer) from this "disorder"?

"In the same way, it can be argued that medical testing becomes an increasingly powerful means for classification and discrimination, as the number of test results accumulate for which no significant treatment is feasible. Once the patient role becomes universal, medical labeling turns into a tool for total social control. E. Richard Brown. *The Rockefeller Medicine Men: Medicine & Capitalism in the Progressive Era*. Berkeley, U. of California Press, forthcoming 1977. Demonstrates how medical tests were used for social control in Guatemala." - footnote 258, p 115

"260 Franco Basaglia, *La maggioranza deviante: L'ideologia del controlla sociale totale*, Nuovo Politecnico no. 43 (Turin: Einaudi, 1971). Since the sixties a citizen without a medically recognized status has come to constitute an exception. A fundamental condition of contemporary political control is the conditioning of people to believe they need such a status for the sake not only of their own but of other people's health." - footnore, p 115

On the subject of social control using community health as a pretext, the French government has passed the Vaccine Passport bill as of today, July 23, 2021. From my perspective, this is full-blown-Fascism. Here's some info about that sent to me by a friend living in Paris:

"Last night, around 5 a.m., the Government passed the vaccine passport law

"Paradoxically, an amendment proposing to apply the health pass also to the National Assembly, a hotbed of brewing, was rejected

"Worse, we will also note the adoption by the National Assembly - subject to contrary amendments, of the examination of the text in the Senate, then by the Constitutional Council - of various serious amendments:

"Authorizing the Regional Health Agencies (ARS) to access the Covid Vaccine database to carry out checks for the "tens of thousands of liberal health professionals"

"Increasing, in the event of use of a false PCR test or a false "vaccination" certificate, the penalties incurred up to 5 years in prison and a € 75,000 fine

"Reducing to the consent of a single parent the authorization required to vaccinate a minor

"Introducing a sentence of up to one year in prison and a fine of € 15,000 for the person who does not scan the QR Code at the entrance of a restaurant or does not leave his contact details in a paper reminder book

"Authorizing middle and high school principals to consult the virological status of students as well as their vaccination status

"Creating an unpaid suspension of the employment contract for two months then a "real and serious" cause of dismissal for employees who do not have a health pass"

My Parisian friend ammended this info in a follow-up email w/ this:

"out of 550 deputies, only 160 or so (I forget the exact number) showed up for the vote"

"The Aesculapian power of conferring the sick-role has been dissolved by the pretensions of delivering totalitarian health care. Health has ceased to be a native endowment each human being is presumed to possess until proven ill, and has become an ever-receding goal to which one is entitled by virtue of social justice." - p 116

Hence we come to what I'll call the *Era of the Asymptomatic*: a person who's healthy becomes demonized as a probable carrier of something that will afflict others, no-one is free of the taint of illness & the propagandistic mutation of 'social justice' becomes the pretext for crushing all suspects who refuse their "sick-role".

"Medicine converges with education and law enforcement. The medicalization of all diagnosis denies the deviant the right of his own values: he who accepts the patient role implies by this submission that, once restored to health (which is just a different kind of patient role in our society), he will conform. The medicalization of his complaint results in the political castration of his suffering. For this see Jesse R. Pitts, "Social Control: The Concept," *International Encyclopedia of the Social Sciences* (1968), 14:391." - footnote 261, p 116

"As a priest," [the doctor] "becomes the patient's accomplice in creating the myth that he is an innocent victim of biological mechanisms rather than a lazy, greedy, or envious deserter of a social struggle for control over the tools of production. Social life becomes a giving and receiving of therapy: medical, psychiatric, pedagogic, or geriatric. Claiming access to treatment becomes a political duty, and medical certification a powerful device for social control." - p 117

"The medicalization of industrial society brings its imperialistic character to ultimate fruition." - p 118

Hence we're confronted w/ a global economy in wch small businesses are destroyed in favor of giant corporations & international measures such as imposed 'health care' become excuses for internationally rewriting laws.

"Progress in civilization became synonymous with the reduction of the sum total of suffering. From then on, politics was taken to be an activity not so much for maximizing happiness as for minimizing pain. The result is to see pain as essentially a passive happening inflicted on helpless victims because the toolbox of the medical corporation is not being used in their favor.

"In this context, it now seems rational to flee pain rather than to face it, even at the cost of giving up intense aliveness. It seems reasonable to eliminate pain, even at the cost of losing independence. It seems enlightened to deny legitimacy to all non-technical issues that pain raises, even if this means turning patients into pets." - p 148

How many people in industrialized countries are now dependent on SSRIs & other drugs to flee from their dismay w/ their life? How many are willing to admit that the very sources of their drugs are intimately tied in w/ the sources of their misery? Being unable to face, & therefore cope, w/ their problems enables the sadist-predators in their lives to profit from both their suffering & the pseudo-cure for it.

"By equating all personal participation in facing unavoidable pain with "masochism," they justify their passive life-style. Yet while rejecting the acceptance of suffering as a form of masochism, anesthesia consumers tend to seek a sense of reality in ever stronger sensations. They tend to seek meaning for their lives and power over others by enduring undiagnosable pains and unrelievable anxieties: the hectic life of business executives, the self-punishment of the rat-race, and the intense exposure to violence and sadism in films and on television." - p 149

During the time of the QUARANTYRANNY I started going to the drive-in movies b/c they were among the few entertainment venues open. As such, I've witnessed more mainstream movies in the last yr+ than in the decade before. One thing I've noticed is the intensity of the upping the ante of the thrill ride. Most recently, I saw "Escape Room - Tournament of Champions", "Black Widow", & "Fast & Furious 9". "Escape Room" probably outdid them all in the constant-adrenaline-rush category. "Fast & Furious 9" had a subtxt that the reason why the 'good guys' survived is b/c *god was on their side* - that wasn't explicitly stated but it was heavily implied. What I wondered is how much further these types of movie can push the excitement button? Personally, I found "Escape Room" to be exhausting w/o being particularly rewarding.

Illich surprised me w/ the following:

"The French Revolution gave birth to two great myths: one, that physicians could replace the clergy; the other, that with political change society would return to a state of original health."

[..]

"For several months in 1792, the National Assembly in Paris tried to decide how to replace those physicians who profited from care of the sick with a therapeutic bureaucracy designed to manage an evil that was destined to disappear with the advent of equality, freedom, and fraternity." - p 151

& now, 229 yrs later, the National Assembly in Paris is using a "therapeutic bureaucracy" to dispense w/ "equality, freedom, and fraternity" once & for all.

"The first clinical trials using statistics, which were performed in the United States in 1721 and published in London in 1722, provided hard data indicating that smallpox was threatening Massachusetts and that people who had been inoculated were protected against its attacks. They were conducted by Dr. Cotton Mather, who is better known for his inquisatorial fury at the time of the Salem witch trials than for his spirited defense of smallpox prevention." - p 158

Ok, I've been w/ Illich all along but here's where that gets somewhat derailed. The "United States" was founded in 1776, therefore it didn't exist in 1721 & was, therefore, the colonies - in this case British colonies. Mather's also reputed to've been a believer in the Hollow Earth.

"In Latin *norma* means "square," the carpenter's square. Until the 1830s the English word "normal" meant standing at a right angle to the ground. During the 1840s it came to designate conformity to a common type. In the 1880s, in America, it came to mean the usual state or condition not only of things but also of people." - p 161

Once humanity is merged into a superorganism by global medical government everyone shd be named Norma or Norman. Everything'll be so much simpler then.

"The age of hospital medicine, which from rise and fall lasted no more than a century and a half, is coming to an end." - p 162

Oh, really?! I live in Pittsburgh, the yr is 2021. "Pittsburgh has always been a city built on industry, and though the landscape of these industries has changed throughout history, Pittsburgh is still a leader. Within Pittsburgh's top five leading industries - advanced manufacturing, healthcare, energy, financial and business services, and information technology – some of its leading companies include, Alcoa, Inc., Kraft Heinz Co, and Highmark Health. Find more of Pittsburgh's leading companies." (https://www.visitpittsburgh.com/pcma2019/industries-and-corporations/) Notice that "healthcare" is in there as a leading industry. I find 38 hospitals listed as ones in or around Pittsburgh (http://www.ushospitalfinder.com/hospitals-by-city/hospitals-in-PITTSBURGH-PA). The 2021 metro population of Pittsburgh is sd to be 1,700,000 people. There're at least 4 hospitals w/in easy walking distance of my house.

"An advanced industrial society is sick-making because it disables people to cope with their environment and, when they break down, it substitutes a "clinical" prosthesis for the broken *relationships*. People would rebel against such an environment if medicine did not explain their biological disorientation as a defect in their health, rather than as a defect in the way of life which is imposed on them or which they impose on themselves." - p 165

A friend of mine had been stressed by a growth in her mouth wch she was afraid was cancerous. After mnths of back'n'forth w/ medical figures she was finally able to arrange an appt to have the growth removed by an oral surgeon. She asked me to pick her up after surgery & I offered to drop her off & pick her up. She then revealed to me that it was *required* by the institution that I stay there & wait for her while she was operated on. I told her that I rejected their 'right' to *require* me to do any such thing & told her that it was my intention to go elsewhere while she was being operated on & to return when they were done. She then told me that they sd I *must* stay there & talk to the surgeon about this before leaving. As a compromise to relieve my friend's stress I agreed to do this.

As it turned out my presence there was completely unnecessary & was only '*required*' b/c they wanted to make sure that someone wd be there to help her when she left in a very groggy state. What boggled my mind was the way the receptionist, a woman at least 20 yrs younger than me, felt entitled to boss me around - telling me that it was against the rules for me to leave & that I *must not leave before talking to the doctor.* What did she think gave her the authority to lord it over me? A person who was simply doing a favor for a friend? Obviously, in her hierarchy I was low-man-on-the-totem-pole in relation to the Doctor Gods. Needless to say, my position on this hierarchy that they attempted to impose on me, was never inquired about. I didn't stay. Unlike any dr I've ever dealt w/ I WAS very punctual about picking up my friend. As such, it seems more reasonable to me that if there's to be a hierarchy I must be considerably higher up in it than any unpunctual dr.

"The sickness that society produces is baptized by the doctor with names that bureaucrats cherish. "Learning disability," "hyperkinesis," or "minimal brain dysfunction" explains to parents why their children do not learn, serving as an alibi for the school's intolerance or incompetence" - p 166

Despite my scholarly excellence I almost failed high school b/c I was so dreadfully bored. When it came time to go to a community college that wd accept someone who'd graduated w/ a D average they thought that I might have to be put in a Remedial Reading course, assuming that I cd barely read. Instead, around this time I was reading William S. Burroughs, Alfred Jarry, & James Joyce's Finnegans Wake - all literature so far above the head of my teacher there that I cd've taught *her* far more about writing than she wd've ever been able to comprehend. As it turned out, I wasn't put in a remedial reading course so we studied The Great Gatsby, a bk I'd already read in high school when I was 16. During the class, the teacher wd ask questions about the bk & no-one but me ever answered the questions, my classmates sat there as if dead. When it came to test time, the same questions that I'd answered in class were now in written form. I wrote on the test that I refused to answer the questions a 2nd time. The teacher then failed me - this despite her knowing that I was the one who'd answered the questions already. Let that serve as an example of a school's incompetence.

"Language is taken over by the doctors: the sick person is deprived of meaningful words for his anguish which is then further increased by linguistic mystification." - p 166

When I was in my 20s I told a dr that I had blood in my sperm & asked him what that meant. He had me masturbate into a jar, hypothetically he studied the results, & then he informed me that I had "hematospermia". I think he took it for granted that I wd be mystified by the high-falutin' language. Instead I had enuf sense to know that he was telling me that I had blood in my sperm b/c I had blood in my sperm. Duh.

"Finally, increasing dependence of socially acceptable speech on the special language of an elite profession makes disease into an instrument of class domination. The university-trained and the bureaucrat thus become their doctor's colleague in the treatment he dispenses, while the worker is put in his place as a subject who does not speak the language of his master." - pp 167-168

Illich's chapter entitled "Death Against Death" was historically illuminating.

"From the fourth century onwards, the Church struggled against a pagan tradition in which crowds, naked, frenzied, and brandishing swords, danced on the tombs of the churchyard. Nevertheless, the frequency of ecclesiastical prohibitions testifies that they were of little avail, and for a thousand years Christian churches and cemeteries remained dance floors." - p 173

Centuries pass, &..

"The new image of death helped reduce the human body to an object. Up to this time, the corpse had been considered something quite unlike other things: it was treated almost like a person. The law recognized its standing: the dead could sue and be sued by the living, and criminal proceedings against the dead were common. Pope Urban VIII, who had been poisoned by his successor, was dug up, solemnly judged a simonist, had his right hand cut off, and was thrown into the Tiber. After being hanged as a thief, a man might still have his head cut off for being a traitor. The dead could also be called to witness." - pp 184-185

Interesting, eh? Calling on the dead to be witnesses must've been tricky at all sorts of levels. Did they appear in court? How did the court know that it wasn't the worms testifying? Was there a cut-off point before they were no longer viable? If there was only a fragment of them left cd that fragment testify? & what about Pope Urban VIII? Simony: "The buying or selling of ecclesiastical offices or of indulgences or other spiritual things." (https://www.thefreedictionary.com/ Simonists) Was he tried as a Simonist to justify his murder?

"During the late Middle Ages, the dictionary of "natural" death became one of the mainsprings of European lyric and drama." - p 196

""Natural death" now appeared in dictionaries. One major German encyclopedia published in 1909 defines it by means of contrast: "Abnormal death is opposed to natural death because it results from sickness, violence, or mechanical and chronic disturbances." A reputable dictionary of philosophical concepts states that "natural death comes without previous sickness, without definable specific cause." It was this macabre hallucinatory death concept that became intertwined with the concept of social progress. Legally valid claims to equality in clinical death spread the contradictions of bourgeois individualism among the working class. The right to a natural death was formulated as a claim to equal consumption of medical services, rather than as a freedom from the evils of industrial work or as a new liberty and power for self-care." - p 194

That last passage is a bit tricky for me. Is there such a thing as "natural death"? Illich seems to think not. My understanding of the idea is that a natural death is one that's simply the final wearing-out of the body, a death esp not imposed on one by obvious external forces, such as by murder. That aside, I'm w/ Illich that "a claim to equal consumption of medical services" is a poor compromise in relation to "freedom from the evils of industrial work or as a new liberty and power for self-care".

"Like all other major rituals of industrial society, medicine in practice takes the form of a game. The chief function of the physician becomes that of an umpire. He is the agent or representative of the social body, with the duty to make sure that everyone plays the game according to the rules.60 The rules, of course, forbid leaving the game and dying in any fashion that has not been specified by the umpire. Death no longer occurs except as the self-fulfilling prophecy of the medicine man.61

"Through the medicialization of death, health care has become a monolithic world religion" - p 202

Once again, I find Illich to be astonishingly prescient in relation to the NOW of the medical industry's use of 'our own good' as a pretext for establishing a New World Odor. Consider this April 2, 2020 quote from my bk Unconscious Suffocation:

"Most people seem to 'need' a monolithic narrative. E.G.: Rump is BAD or Rump is GOOD. People are desperately reaching for a monolithic narrative here in relation to the PANDEMIC PANIC. & anything that undermines the oversimplification of that, according to the desperate, is in complete denial of all facets of that monolith. Therefore, if I say that the PANIC serves economic interests I'm denying that there's disease or death. Voila! Therefore, I'm obviously wrong. It hardly even matters when I tell people that I've never denied that there's a virus & that I've been consistently talking about the sociopolitical conditions that have come to surround the pandemic that use the pandemic as an excuse." - p 103, Unconscious Suffocation - A Personal Journey through the PANDEMIC PANIC

In chapter 6, "Specific Counterproductivity", there's only one footnote at the very end. I reckon this is b/c Illich does a sort of philosophical summing-up instead of taking the reader thru any more background basis.

"Iatrogenesis will be controlled only if it is understood as but one aspect of the destructive dominance of industry over society, as but one instance of that paradoxical counterproductivity which is now surfacing in all major industrial sectors. Like time-consuming acceleration, stupefying education, self-destructive military defense, disorienting information, or unsettling housing projects, pathogenic medicine is the result of industrial overproduction that paralyzes autonomous action." - p 207

Autonomy, autonomy - it's important to me, what about you?

"Like school education and motor transportation, clinical care is the result of a capital-intensive commodity production; the services produced are designed for others, not with others nor for the producer.

"Owing to the industrialization of our world-view, it is often overlooked that each of these commodities still competes with a nonmarketable use-value that people freely produce, each on his own. People learn by seeing and doing, they move on their feet, they heal, they take care of their health, and they contribute to the health of others. These activities have use-values that resist marketing. Most valuable learning, body movement, and healing do not show up on the GNP. People learn their mother tongue, move around, produce their children and bring them up, recover the use of broken bones, and prepare the local diet, and do these things with more or less competence and enjoyment. These are all valuable activities which most of the time will not and cannot be undertaken for money, but which can be devalued if too much money is around." - pp 210-211

Indeed. The monetizing of most things, if not ALL things, devalues them for me. Monetized sex, e.g., makes sex not worth having. If sex isn't driven by a pheremone-enhanced pursuit of pleasure then it's been derailed. How many people take the opposite position, due "to the industrialization of our world-view"? In other words, how many people have become so out-of-touch w/ standards intrinsic to themselves that they can only value things that're monetarily expensive to them?

"Fifteen years ago" [1961] "it would have been impossible to get a hearing for the claim that medicine itself might be a danger to health." - p 217

Is it any different now? In the process of writing Unconscious Suffocation in 2020 it became obvious to me that any criticism of the medical industry was something that most people became very defensive about - presumably b/c *they deeply believed that it's their only ticket out of mortal woes*.

"It was not predicted that soon, in a regional screening, only sixty-seven out of one thousand people would be found completely fit and that 50 percent would be referred to a doctor, while according to another study, one in six people screened would be defined as suffering from one to nine serious illnesses.4 Nor had the health planners forecast that the threshold of tolerance for everyday reality would decline as fast as the competence for self-care was undermined, and that one-quarter of all visits to the doctor for free service would be for the untreatable common cold." - p 218

"untreatable common cold"? Some people wd take exception to that. Check out what Peter Simpson, an Australian biomedical scientist has to say about that in my movie Don't Walk Backwards: https://youtu.be/kODzM_2_bRM?t=60 . This section is only a little over one minute long so don't be intimidated by the overall 8:25:34 of the full movie.

It seems to me that for most of humanity's existence to be sick was to be in a minority. I wonder if now we've reached a point where everyone must have some sort of sickness to talk about or otherwise be somehow on the 'outside' of the herd - some sort of suspect extraordinary creature.

"In addition, nobody knows if the most advantageous form of health care is obtained from medical producers, from a travel agent, or by renouncing work on the night shift." - p 229

How often is dr care just glossing over the problem w/ drugs so that we can continue to bear something that we might be better off getting rid of? If a person is miserable about something that they can change why not do so & give up the "mood brighteners" in the process? I've worked the graveyard shift, I can attest that it negatively impacted my life, I wdn't do it again.

"Consumer-protection movements can translate information about medical ineffectiveness now buried in medical journals into the language of politics, but they can make substantive contributions only if they develop into defense leagues for civil liberties and move beyond the control of quality and cost into the defense of untutored freedom to take or leave the goods. Any kind of dependence soon turns into an obstacle to autonomous mutual care, coping, adapting, and healing, and what is worse, into a device by which people are stopped from transforming the conditions at work and at home that make them sick." - p 233

Once again, Illich hammers his point home: if you aren't autonomous, you're at the mercy of whoever you're dependent on - & that 'mercy' is inevitably predatory. Still, I'd settle for more people having the ability to concentrate long enuf to even *read* articles in medical journals w/o having to take Adderall to help them focus.

"In Venezuela, one day in a hospital costs ten times the average daily income; in Bolivia, about forty times the average daily income." - p 235

Is it better or worse in the United States than the 2 examples given? In one instance that I know of, the patient was charged twice his yrly income for 6 days in the hospital. That worked out to $4,500 a day in contrast to his $36 a day earned income. That's 125 times his average daily income. Much of that bill was simply padding by corrupt drs who'd come into his rm to gratuitously take his temperature so they cd charge a large visitation fee.

"Insofar as medicine is a public utility, however, no reform can be effective unless it gives priority to two sets of limits. The first relates to the volume of institutional treatment any individual can claim: no person is to receive services so extensive that his treatment deprives others of an opportunity for considerably less costly care per capita if, in their judgment (and not in the opinion of an expert), they make a request of comparable urgency for the same public resources. Conversely, no services are to be forcibly imposed on an individual against his will: no man, without his consent, shall be seized, imprisoned, hospitalized, treated or otherwise molested in the name of health." pp 239-240

That last limit particularly strikes home for me. The legislators in Paris who've enabled the passing of the Vaccine Passport legislation wd be well-advised to reverse their position w/ Illich's critique in mind. Of course, they have priorities that go contrary to Illich's or they wdn't've passed such laws in the 1st place. BEWARE.

"In general, people are much more the product of their environment than of their genetic endowment. This environment is being rapidly distorted by industrialization. Although man has so far shown an extraordinary capacity for adaption, he has survived with very high levels of sublethal breakdown. Dubos96 fears that mankind will be able to adapt to the stresses of the second industrial revolution and overpopulation just as it survived famines, plagues, and wars in the past. He speaks of this kind of survival with fear because adaptability, which is an asset for survival, is also a heavy handicap: the most common causes of disease are exacting adaptive demands." - p 256

I find that latter to be a very interesting point. It seems to me that our "genetic endowment" is this ability to adapt. The problem is that some adaptation is a reduction of human potential while others is an increase. Adapting to being forced to wear a mask is a reduction of one's ability to give & receive facial cues, to taste, to smell, etc.. UNCONSCIOUS SUFFOCATION results.

"Much suffering has been man-made. The history of man is one long catalogue of enslavement and expoitation, usually told in the epics of conquerors or sung in elegies of their victims. War is at the heart of this tale, war and the pillage, famine, and pestilence that came in its wake. But it was not until modern times that the unwanted physical, social, and psychological side-effects of so-called peaceful enterprises began to compete with war in destructive power.

"Man is the only animal whose evolution has been conditioned by adaptation on more than one front. If he did not succumb to predators and forces of nature, he had to cope with use and abuse by others of his own kind. In his struggle with the elements and with his neighbor, his character and culture were formed, his instinct withered, and his territory was turned into a *home*." - p 259

In my own life, humans have managed to keep challenging aspects of nature at bay: I wear clothes that protect my skin & keep me warm, I live in a house that keeps me from exposure to the elements, food is easily enuf come by (most of the time) as is water. My fellow humans, on the other hand, have been a major pain-in-the-ass. They've also been my greatest source of joy. Can't live w/ them, can't live w/o them. Humans are the most dangerous predator & keeping real estate vultures, e.g., at bay & medical industry vultures, to make a reference closer to the heart of this bk, at bay is very hard indeed. Let's not empower the most dangerous predator any more than they're already empowered.

"E. R. Dodds, *The Greeks and the Irrational* (Berkeley: Univ. of California Press, 1951), especially chap. 2. Irving Kenneth Zola. "In the Name of Health and Illness;" on some political consequences of medical influence. in: Social Science and Medicine, Feb. 1975, vol. 9 pp 83-87 . . . the medical area is the arena of the example par excellence of today's identity crisis, the area where the *banality of evil* is best masked as a technical, scientific, objective process engineered for our own good." - footnote 1, pp 260-161

The "*banality of evil*" presumably being a reference to Hannah Arendt's excellent Eichmann in Jerusalem - A Report on the Banality of Evil (see my review here:

http://www.goodreads.com/book/show/13367624-eichmann-in-jerusalem). I
quoted from that bk in my movie entitled Robopaths that also looks into
Eichmann & beyond: on my onesownthoughts YouTube channel here: https://
youtu.be/-PR7C8nFKtA - on the Internet Archive here: https://archive.org/details/
Robopaths . It's my opinion that present-day conditions, partially fostered by the
insufficiently questioned godlike status of the medical industry & its pushers, the
politicians, are generating a whole new breed of Adolf Eichmanns & if we're not
on our guard their new breed of atrocities will be done "for our own good".

"Better health care will depend, not on some new therapeutic standard, but on
the level of willingness and competence to engage in self-care. The recovery of
this power depends on the recognition of our present delusions." - p 268

It's amazing that Illich cd get this bk published at all (Thank You Bantam). I
wonder if he had friends in the publishing industry. He certainly wasn't preaching
to the converted b/c now, 45 yrs later, I think there're probably even fewer people
w/ minds open enuf to appreciate such a thorough critique enuf to even make it
thru it. Are you one of them? I can only hope.

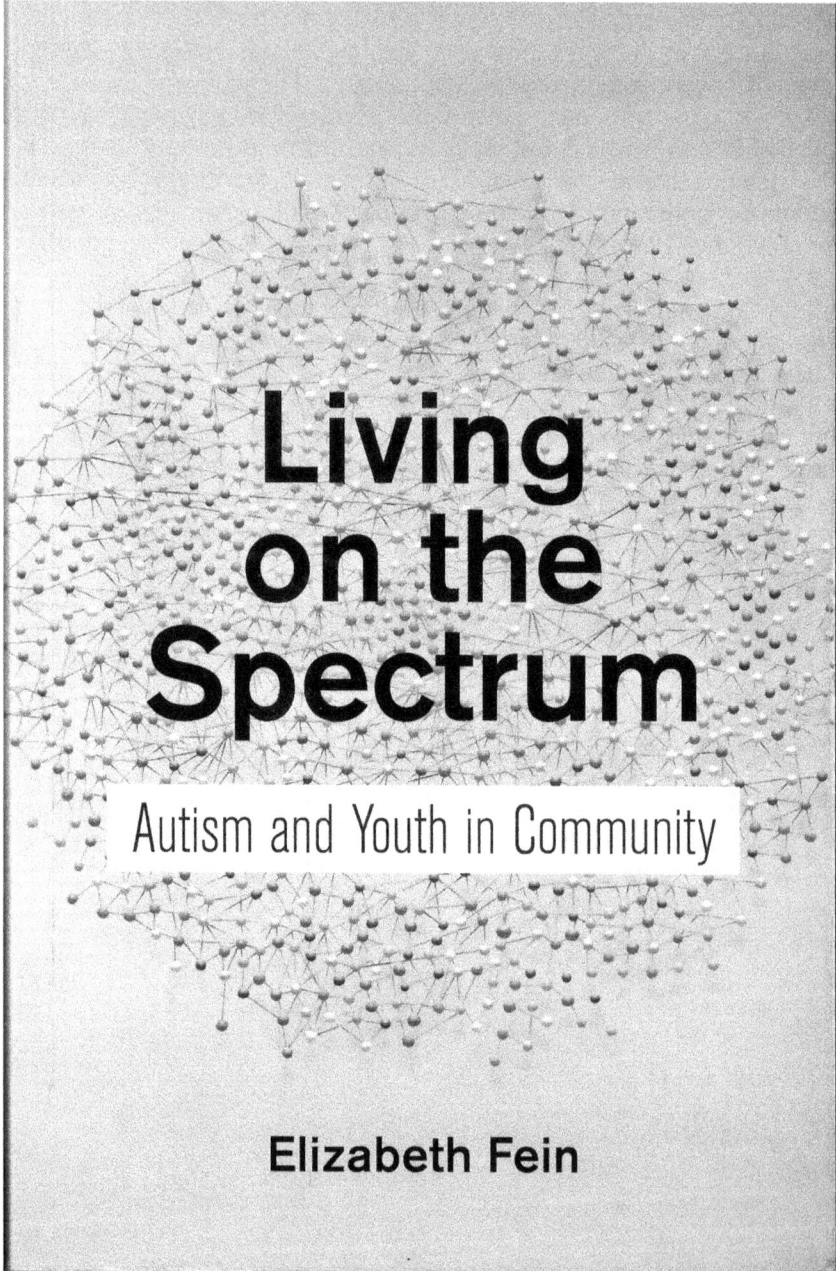

Living on the Spectrum

Autism and Youth in Community

Elizabeth Fein

review of
**Elizabeth Fein's <u>Living on the Spectrum - Autism and Youth in
Community</u>**
by tENTATIVELY, a cONVENIENCE - December 9-14, 2021

1st off, I may as well put it out there that this bk was written by a friend/ acquaintance of mine, someone who lives very close to me in my neighborhood. 2nd, I have a particular interest in autism/Asperger's (wch I still question putting on the "spectrum") b/c I see it as potentially a 'sign of our times' that bears closer scrutiny in relation to medicalization & the pernicious interests & influence of big business, in particular (but not limited to), the pharmaceutical industry. People who deny the extent to wch the greedy will cover their crimes in the interest of continually escalating profit are naive indeed. These factors make this review almost impossible for me to write. Fortunately, I actually like this bk very much so I don't have to risk alienating its author over points of quality. If I'm to alienate the author at all, wch isn't necessarily a given, it'll be more likely over points of socio-philosophy.

"Autism is a way of being. It is pervasive; it colors every experience, every sensation, perception, thought, emotion, and encounter, every aspect of existence. It is not possible to separate the autism from the person. . . .

"Therefore, when parents say, I wish my child did not have autism, what they're really saying is, I wish the autistic child I have did not exist, and I had a different (non-autistic) child instead." - p 1

Fair enuf.. & I tend to agree - & that more or less establishes the author's position on whether autism shd be characterized as a 'disease', something to be framed by pathology, or as a personality characteristic, something to be framed as intrinsic to the person, potentially viewable as flawed, but no more so than other personality characteristics that people may have who're not necessarily thought of as 'diseased' at all.

Again, I tend to agree.. but then is it also agreable to extend such logic to other states of being? By using the term "Autism Spectrum" a wide range of characteristics get lumped together. At one 'end' of the "spectrum" there might be people whose development is so severely impeded that it might come close to being deprived of a sense, of being blind, e.g.. Some people might consider giving a blind person sight to be something that wd make them no longer them. Consider this:

"**Martha's Vineyard Sign Language** (MVSL) [mre] Extinct. Formerly in Martha's Vineyard, Massachussetts. *Class* Deaf Sign Language. *Dialects:* The early sign language was based on a regional one in Weald, England, where the deaf persons' ancestors had lived. French Sign Language was introduced to Martha's Vineyard in 1817. MVSL was later combined with American Sign Language (ASL), but never became identical to ASL. *Lg Use:* From 1692 to 1910 nearly all hearers on Martha's Vineyard were bilingual in English and sign language. *Other:* The first deaf person arrived in 1692. From 1692 to 1950 there was a high rate of hereditary deafness. In the 19th century, 1 in 5700 Americans were deaf, 1 in 155 in Martha's Vineyard, 1 in 25 in one town, 1 in 4 in one neighborhood." - p 304, Ethnologue - Languages of the World

"Deaf MVSL users were not excluded by the rest of society at Martha's Vineyard, but they certainly faced challenges due to their deafness. Marriage between a Deaf person and a hearing person was extremely difficult to maintain, even though both could use MVSL. For this reason, the Deaf usually married the Deaf, raising the degree of inbreeding even beyond that of the general population of Martha's Vineyard. These Deaf-Deaf marriages contributed to the increase of the Deaf population within this community. The MVSL users often associated closely, helping and working with each other to overcome other issues caused by deafness. They entertained at community events, teaching hearing youngsters more MVSL. The sign language was spoken and taught to hearing children as early as their first years, in order to communicate with the many Deaf people they would encounter in school. Lip movement, hand gestures, mannerisms, and facial expressions were all studied. There were even separate schools specifically for learning MVSL. Hearing people sometimes signed even when there were no Deaf people present. For example, children signed behind a schoolteacher's back, adults signed to one another during church sermons, farmers signed to their children across a wide field, and fishermen signed to each other from their boats across the water where the spoken word would not carry."
- https://en.wikipedia.org/wiki/Martha%27s_Vineyard_Sign_Language

Is it so hard to imagine that in the deaf-deaf marriages there wd be no stigma attached to being deaf? - especially when seeing the language prove useful to hearing people? A language that was develped especially for the deaf to communicate w/?

W/o ever being declared pathologically impaired in any way, I was still raised as if many of my most defining characteristics were aberrations that shd be forced out of me in order to make me 'normal'. It's having been thru that experience that helps me be sympathetic to people who undergo similar pressures that're excused by pathology. Whose decision shd it be to decide what characteristics of a person are to be stigmatized & wch aren't? It gets very problematic, eh?!, b/c some people might embrace their difference while others might be too intellectually limited to even understand what that difference is.

One thing I hope to put forth is the thought experiment of imagining a medicalization *tabula rasa*, a world in wch indivduals are just perceived as individuals; a world in wch people are just identifiable as being who they are & not categorized otherwise. As it is now, who gets stigmatized & who doesn't partially depends on how adept they are at defending their self-identity. Stephen Hawking managed to be categorized as an English theoretical physicist, cosmologist, Lucasian Professor of Mathematics at the University of Cambridge, and author who, at the time of his death, was director of research at the Centre for Theoretical Cosmology at the University of Cambridge *rather than* exclusively as someone suffering from motor neurone disease (amyotrophic lateral sclerosis – ALS, for short).

As a sidenote, it's interesting to me that Hawking was born into a family of physicians. I'm always looking for signs of iatrogenesis. Did his physician family do anything to him that resulted in the ALS? My stepfather had ALS. He was awarded a regular military payment specifically b/c of his ALS b/c so many soldiers had contracted it that, apparently, it was cheaper for the government to pay off as if there 'might' be a connection rather than have closer investigation reveal using the soldiers as guinea pigs &/or exposing them to toxicities that the powers-that-be wd rather have remain hush-hush.

"Autism is a deeply contested condition. Sometimes, people talk about it as a public health crisis, an invading enemy to be identified and eliminated. Sometimes, people talk about it as a fundamental and valued aspect of personhood and identity, something to be protected rather than something protected against. As autism rose to its extraordinary state of prominence over the past several decades, these two ways of talking about autism—and the conflicts between them—dominated a great deal of the public conversation." - p 2

What if it's *neither* of the above? My interest in autism is in a very nascent stage. I'm keeping my mind open to the mainstream arguments & the controversial ones. As a 68 yr old who's witnessed autism grow from a very rare condition into something practically omnipresent (esp during the broadened diagnostic era of the 'spectrum'),

["The estimated prevalence rate of autism spectrum disorder among children in the United States at the time I conducted this interview was 1 in 150 (see Centers for Disease Control and Prevention, 2019, for a historical summary of rates over time)." - endnote, p 264]

I'm beginning to think of it as an aspect of environmental change. If the world were to flood wd human babies born w/ gills & fins be considered 'deformed'? I think not. If the overpopulation of the world by humans is recognized an an ecological threat might it be considered an 'adaptation' if more & more people are born infertile? Then again, if infertility were to be a side-effect of medical experimentation wd it be inconceivable that some people wd consider it to be a boon? & might want to amplify that side-effect?

What if humans are becoming more insect-like? That seems to be the case to me. Ants & humans seem to have a fair amt in common. What if humans really are just divided into worker-humans, army-humans, queen-attendant-humans? After all, it appears that musician-humans are now "non-essential" - what good are they to the hive?

"At the same time, a growing movement of individuals on the autism spectrum, often referred to as (part of) the *neurodiversity movement*, argue that autism is instead a natural and valuable aspect of human diversity, a cultural identity calling for accommodation rather than prevention. A cure for autism would thus be not a mercy but a genocidal suppression of difference, equivalent to "curing" someone's race, gender, or sexual orientation." - p 2

I find that valid enuf - but at the same time I return to my proposed thought experiment "of imagining a medicalization *tabula rasa*, a world in wch indivduals are just perceived as individuals; a world in wch people are just identifiable as being who they are & not categorized otherwise". What accommodations wd be made then? Of course, such a thought experiment involves the extra complication of imagining a world in wch there *isn't discrimination* against these people hypothetically perceived as individuals - rather than as membors of a particular race, sex, sexual orientation, or psychological state, etc. It's hard to imagine as a possibility, isn't it? Even in today's day & age a person considered to be a high-functioning Aspie still has to deal w/ whether their family is rich or poor, whether they can manage to access services or not, etc.. In other words, privilege doesn't just revolve around one factor - there's a whole hierarchical plethora - can we 'level' those? - as the English Civil War's *Levellers* might've proposed?

"The **Levellers** were a political movement during the English Civil War (1642–1651) committed to popular sovereignty, extended suffrage, equality before the law and religious tolerance. The hallmark of Leveller thought was its populism, as shown by its emphasis on equal natural rights, and their practice of reaching the public through pamphlets, petitions and vocal appeals to the crowd." - https://en.wikipedia.org/wiki/Levellers

A black lesbian woman on the dysfunctional end of the autism spectrum (if there can even be such a person) might not have the same idea of self-advocacy as a rich white male heterosexual diagnosed as high-functioning.

"Discourses rooted in autistic self-advocacy and the celebration of neurodiversity are still intertwined with the assumptions of the medical paradigms within which they originated, in intricate and subtle ways. And people living with autism spectrum diagnoses must find ways to live under both of these descriptions, integrating these seemingly irreconcilable discourses in the business of their daily lives." - p 3

What if the *neurodiversity movement* were to make being so-called 'autistic' accepted as just another human possibility? Wd it then be fair for both the diagnosis to go away, to be discredited, & for the autism-specific services to go away too? A sort of sink-or-swin Darwinism in wch autism is considered potentially valuable, thereby putting the autist in a position where they'd fend for themselves as much as other 'non-autistic' people? What if the 'autism spectrum' were to become so definitionally broadened that *everyone* wd be considered 'autistic' at any moment when they seemed developmentally challenged, prone to repetitive behavior, socially inept, or whatever else might seem relevant?

"This book takes a different approach: It is based on ethnographic fieldwork in communities where people on the autism spectrum come together." - p 4

As a person who's often found myself in social situations where the people around me are suicidal or otherwise troubled, I've consistently stated that it's the community that people have, the quantity & quality of their *friends* that gives them the strength to keep going against hostile forces - & *NOT* the Medical Industry, no matter how well-intentioned. As such, Fein's "ethnographic fieldwork in communities" is important to me as something that may or not bear out my assertion. Where were we?

"Through a process I call "divided medicalization," the former get misrepresented as the latter: Complex, multivalent neurodevelopmental conditions are produced and then reduced to fit within a preexisting, disease-oriented clinical paradigm. Through this process of expansion and constriction of diagnostic categories, phenomena that play out between the individual and the environment are *individualized*: Their social and sensory dimensions get mapped back onto the individual as the locus of explanation and site of intervention. Aspects of experience that transcend the bounds of the physical body—the connections between our selves and our social partners, between our senses and the things being sensed, between our attention and the material to which we attend, and so forth—are thus occluded, rendered invisible." - p 5

Given that everything is potentially relevant, how can we pay attn to it all w/o being exclusionary? We can't. The more we try to pin down the essentials, the more the *other essentials* escape us. Is it possible to establish a fluid state of mind that doesn't over-categorize but still remains sensitive to aspects of the subject under scrutiny at the times when they come to the fore w/o becoming fixated as things move on? Probably. This approach, then, also probably comes into conflict w/ trying to establish firm diagnostic criteria.

"Through attending to their shared social practices, organized around the care and maintenance of locally interconnected systems, we can gain some insight into how to mitigate the loneliness and isolation that too often accompanies autistic difference, by intervening at the troubled intersection between the individual and their social surroundings. That space of intersection, I argue, is where autism happens." - p 5

I still feel stuck w/ the problems of whether the 'autism spectrum' is ultimately 'viable' - at least for me as the free thinker thinking about this outside the box of current training on the subject. It still seems to me that having a 'spectrum' that has people w/ developmental limits at one end & high-functioning at the other end pretty much describes human possibility in general. If I were better informed this wd probably not puzzle me so much. Perhaps as I work my way thru this bk, the idea of the spectrum will be clearer to me. Then again, maybe I shd consult & quote the DSM-5 & get it over w/. *SO,* here it is, pp 50-51 of <u>Diagnostic and Statistical Manual of Mental Disorders 5th edition</u>. I'll intersperse commentary in [brackets]:

"Autism Spectrum Disorder
Diagnostic Criteria 299.00 (F84.0)

A. Persistent deficits in social communication and social interaction across multiple contexts, as manifested by the following, currently or by history (examples are illustrative, not exhaustive; see text):

["examples are illustrative, not exhaustive": I always appreciate qualifiers - in this case, it seems an appropriate caution]

1. Deficits in social-emotional reciprocity, ranging, for example, from abnormal social approach and failure of normal back-and-forth conversation; to reduced sharing of interests, emotions, or affect; to failure to initiate or respond to social interactions.

[Does this go both ways? In other words, when I have a conversation w/ someone in wch they don't express any interest in what I'm saying who's being the bad communicator? Example: I'd just posted a movie online of a friend of mine signing another friend of ours's poem ("Rhoda Mappo signs John M. Bennett's "The Spitter" May, 1986 Texas" [actually, the poem was called "Cup of Spit"]: online here: http://youtu.be/l7H8DJ0CYJE). Since the poem is unusual & wd require advanced sign-language skills to enunciate it in this way, I thought it was interesting. I approached 2 of my coworkers & asked them if they'd be interested in seeing it. They gave me a very firm NO & proceeded to talk about whether it was more in fashion to wear one's shirt-tail tucked in or left out. We obviously didn't have shared interests, despite our both working for a museum - but are either or both of us showing deficits "in social-emotional reciprocity"?]

2. Deficits in nonverbal communicative behaviors used for social interaction, ranging, for example, from poorly integrated verbal and nonverbal communication; to abnormalities in eye contact and body language or deficits in understanding and use of gestures; to a total lack of facial expressions and nonverbal communication.

[How often are such deficits brought on by medication? How many seemingly affectless people are people who are 'numbed-down' (a variation of 'dumbed-down')?]

3. Deficits in developing, maintaining, and understanding relationships, ranging, for example, from difficulties adjusting behavior to suit various social contexts; to difficulties in sharing imaginative play or in making friends; to absence of interest in peers.

[If the people in one's social environment are more interested in the issue of shirt-tails-in or shirt-tails-out & one is interested music, literature, art, anarchism, social planning, performance, psychology, wherein lies the problem in making friends? Is it 'normal' to numb oneself down to such an extreme that all intellectual interests disappear? Just for the sake of having profoundly vapid conversations?]

Specify current severity:

Severity is based on social communication impairments and restricted, repetitive patterns of behavior (seeTable 2).

B. Restricted, repetitive patterns of behavior, interests, or activities, as manifested by at least two of the following, currently or by history (examples are illustrative, not exhaustive; see text):

1. Stereotyped or repetitive motor movements, use of objects, or speech (e.g., simple motor stereotypies, lining up toys or flipping objects, echolalia, idiosyncratic phrases).

[Drummers use "repetitive motor movements", people seem to like such things very much. As for "idiosyncratic phrases"?: How is one to express oneself creatively if one only speaks in stock phrases? Does the objection to "idiosyncratic phrases" extend into objection to coining neologisms? It seems to me that "idiosyncratic phrases" is the opposite of "echolalia" - thusly creating a spectrum of possibilities that might be overly inclusive.]

2. Insistence on sameness, inflexible adherence to routines, or ritualized patterns of verbal or nonverbal behavior (e.g., extreme distress at small changes, difficulties with transitions, rigid thinking patterns, greeting rituals, need to take same route or eat same food every day).

[Does "insistence on sameness" include an objection to "idiosyncratic phrases"?]

3. Highly restricted, fixated interests that are abnormal in intensity or focus (e.g., strong attachment to or preoccupation with unusual objects, excessively circumscribed or perseverative interests).

[Isn't a scholar someone w/ "highly restricted, fixated interests that are abnormal in intensity or focus"? W/o people who're capable of abnormal focus how does anyone expect deep developments to occur? W/o abnormal focus how wd I be able to even write this review? No-one else I know wd be capable of doing so, at least in the way that I will. I'm sure I'm making my point. W/o abnormal focus how wd an incredible bk like the DSM ever be compiled?]

4. Hyper- or hyporeactivity to sensory input or unusual interest in sensory aspects of the environment (e.g., apparent indifference to pain/temperature, adverse response to specific sounds or textures, excessive smelling or touching of objects, visual fascination with lights or movement).

[I find all of the above to be somewhat objectionable. Is it autistic to be a sensualist? Just about any experimental filmmaker cd be sd to have a "visual fascination with lights or movement" - check out The Visual Music Village (my page is here: http://visualmusic.ning.com/profile/tENTATIVELYacONVENIENCE). When does it become "excessive" to smell or touch objects? Wd you rather have sex w/ someone who can focus on the sensuality of yr body for long periods or for a 'normal' length of time (whatever that might be). As for "adverse response to specific sounds or textures"?: isn't that what some people call "taste"? Ok, I exaggerate a little: most of my friends have strong negative reactions to my

(M)Usic: I find this to be a sensual/intellectual limitation on their part - I'm not sure that I can go along w/ diagnosing them pathologically over it, tho - even if they do themselves. "apparent indifference to pain/temperature"?: I reckon that depends on how far this goes: if a person just stands still while their environment catches fire around them, w/o bothering to try to escape burning alive, then, YES, I'd say that's a problem, a big problem - but are the authors of the DSM referring to this or to something less severe?]

Specify current severity:

Severity is based on social communication impairments and restricted, repetitive patterns of behavior [..]

C. Symptoms must be present in the early developmental period (but may not become fully manifest until social demands exceed limited capacities, or may be masked by learned strategies in later life).

[Aren't most social behaviors for people, autistic or not, "masked by learned strategies in later life"? If people are capable of learning such strategic masks, what does that mean?]

D. Symptoms cause clinically significant impairment in social, occupational, or other important areas of current functioning.

[Let's say the patient under scrutiny is incomprehensible to the doctor for reasons that're the result of the doctor's own intellectual deficiencies. I remember reading about a woman who was lobotomized b/c she spoke a language the examining doctor didn't understand, he thought she was speaking gibberish. Having the power to be able to commit such a crime w/o punishment is artificially created by standards of dubious value. Doctors need to be taken down quite a few notches.]

Autism Spectrum Disorder 51

E. These disturbances are not better explained by intellectual disability (intellectual developmental disorder) or global developmental delay. Intellectual disability and autism spectrum disorder frequently co-occur; to make comorbid diagnoses of autism spectrum disorder and intellectual disability, social communication should be below that expected for general developmental level.

[That's 'funny': I generally find most people I know & meet to be what I call *bad communicators* - but their communication seems to be inhibited as much by socially-induced anxiety as it is by anything else. Are we living in an autistic society?]

Note: Individuals with a well-established DSM-IV diagnosis of autistic disorder, Asperger's disorder, or pervasive developmental disorder not otherwise specified should be given the diagnosis of autism spectrum disorder. Individuals who have marked deficits in social communication, but whose symptoms do not otherwise meet criteria for autism spectrum disorder, should be evaluated for social (pragmatic) communication disorder.

Specify if:

With or without accompanying intellectual impairment

With or without accompanying language impairment

Associated with a known medical or genetic condition or environmental factor (Coding note: Use additional code to identify the associated medical or genetic condition.)

Associated with another neurodevelopmental, mental, or behavioral disorder (Coding note: Use additional code[s] to identify the associated neurodevelopmental, mental, or behavioral disorder[s].)

With catatonia (refer to the criteria for catatonia associated with another mental disorder, pp. 119–120, for definition) (Coding note: Use additional code 293.89 [F06.1] catatonia associated with autism spectrum disorder to indicate the presence of the comorbid catatonia.)"

I can't exactly fault the DSM for any lack of thoroughness. My general tendency as a creative thinker is to question whole systems - in this case, I think the Medical Industry is highly overrated & entirely too powerful, I'm suspecting more & more than there's much more iatrogenesis at work than generally meets the eye. Consider this example:

Eric Dolphy, I claim, is generally undisputed as one of the greatest jazz musicians to ever live. His talents as an alto saxophonist, clarinetist, bass clarinetist, & flute-player are simply phenomenal. There's no such thing as a mediocre Eric Dolphy recording. How did he die?:

"On June 27, 1964, Dolphy traveled to Berlin to play with a trio led by Karl Berger at the opening of a jazz club called The Tangent. He was apparently seriously ill when he arrived, and during the first concert was barely able to play. He was hospitalized that night, but his condition worsened. On June 29, Dolphy died after falling into a diabetic coma. While certain details of his death are still disputed, it is largely accepted that he fell into a coma caused by undiagnosed diabetes. The liner notes to the *Complete Prestige Recordings* box set say that Dolphy "collapsed in his hotel room in Berlin and when brought to the hospital he was diagnosed as being in a diabetic coma. After being administered a shot of insulin he lapsed into insulin shock and died". A later documentary and liner notes dispute this, saying Dolphy collapsed on stage in Berlin and was brought to a hospital. The attending hospital physicians did not know that Dolphy was a diabetic and decided on a stereotypical view of jazz musicians related to substance abuse, that he had overdosed on drugs. He was left in a hospital bed for the drugs to run their course. Unbeknownst to doctors, Dolphy was a teetotaler who didn't smoke cigarettes or take drugs.

"Ted Curson remembers, "That really broke me up. When Eric got sick on that date [in Berlin], and him being black and a jazz musician, they thought he was a junkie. Eric didn't use any drugs. He was a diabetic—all they had to do was take a blood test and they would have found that out. So he died for nothing. They gave him some detox stuff and he died, and nobody ever went into that club in Berlin again. That was the end of that club". Shortly after Dolphy's death, Curson recorded and released *Tears for Dolphy*, featuring a title track that served as an elegy for his friend." - https://en.wikipedia.org/wiki/Eric_Dolphy "

Is that bad enuf? For me, Dolphy was a genius, there's no doctor in the world that comes even close. If the doctor(s) &/or nurse(s) whose racist neglect & poor 'treatment' resulted in Dolphy's death were to've been summarily executed I might very well have applauded if I'd been there. Instead, doctors routinely kill people w/ impunity while the rest of us are at their mercy. Consider a related story:

I was in a band w/ Eric Mingus, Charles Mingus's son. Eric told me that he'd been named after Dolphy: Eric Dolphy Mingus. However, when the attending nurse at his birth saw the name "Dolphy" on the birth certificate she changed it to "Michael" b/c "there's no such name as Dolphy". No doubt she saw Mingus & his wife as ignorant people, when in reality it was the nurse who was ignorant, completely unaware that she was in the company of musical greatness. Such mediocre imbeciles are given power by a system whose hypothetical positivity is dubious at best & transparently egregious at worst.

& to think I'm only on p 5 of this bk. SHEESH.

"Through attending to their shared social practices, organized around the care and maintenance of locally interconnected systems, we gain some insight into how to mitigate the loneliness and isolation that too often accompanies autistic difference, by intervening at the troubled intersection between the individual and their social surroundings. That space of intersection, I argue, is where autism happens." - p 5

I find that last sentence to be particularly noteworthy. Is autism a thing that "happens"? I'm not really sure what to make of that language. Stating things in that way seems akin to saying: "The soccer field is where soccer happens" or "The boxing ring is where boxing happens". If autism is something that "happens" in "the troubled intersection between the individual and their social surroundings", is autism then like war or crime or bullying? I have great respect for this bk & I particularly respect the author's focus on the "ethnographic fieldwork in communities" but saying the autism "happens" in a social environment seems to take it away from the individuals in that social environment & to somehow disembody it. Ok, people can only drown if they're immersed in fluid - but that doesn't take away from their not having bodies that can breathe in water, e.g.. Perhaps the following helps qualify the preceding:

"As for restricted interests, nobody is born into this world obsessed with Thomas the Tank Engine; the interest emerges through exposure to the content." - p 6

"The philosopher Ian Hacking (1995) uses the phrase "the looping effect of human kinds" to describe the ways that living under a social scientific classification changes people—their self-concept, their behavior, even their biological makeup—in ways that then lead to changes to the classification itself, thus creating an ongoing feedback loop. This process applies to a wide variety of perceived human types, from "breast cancer survivor" to "child abuser." People react to the ways in which they are classified—sometimes living up to their labels, sometimes seeking to prove them wrong." - pp 7-8

The interesting &, perhaps, usually unfortunate thing about these labels is that the more dramatic they are the more they tend to dominate all other possible descriptions. Let's say a woman is being raped & she manages to kill her attacker: she's then found guilty of murder & sentenced to life-without-parole in prison. Yes, it happens. She then becomes a "murderer", no matter if she'd been a person w/ many other prominent characteristics prior to that, murderer is likely to trump them all. Does a person who's a "breast cancer survivor" really want to be *that* for the rest of their life?

"Now, a National Council on Severe Autism is being formed by parents increasingly concerned that it is the needs of the people with what is sometimes called "classic autism," on "the severe end of the spectrum," that are currently being left out of conversations about autism (Opar, 2019). (This, I am uncomfortably aware, is a problem to which this book, which began as a research project on "Asperger's syndrome" and has now become a book about "the autism spectrum," could be said to contribute.) The word *autism* describes something very different than it once did." - p 8

Yes, I continue to find this a conundrum, this whole idea of an autism spectrum covers such a variety of conditions & severity of conditions that it seems more than a little ridiculous to lump them all together to me - &, yet, it seems that the author is forced to do so *because that's the way it's done these days.* If she had just written about Asperger's she wd've seemed to be unprogressive, out of the loop of the latest lingo & attitudes.

"Listening to the patterns that get aggregated under autism diagnoses allows us instead to more fully recognize their social and moral implications.

"Clinical Ethnography

"I do so, here, through an approach know as *clinical ethnography*. Clinical ethnography works at the intersection of personhood and social world, illuminating the inextricability of domains associated with psychology and those associated with the anthropological study of culture." - p 10

I very much appreciate this approach but I continue to wonder whether it's really any more valuable than the friendliness of a sociable person who interfaces w/ a large variety of people & tries to make all of them feel comfortable, & good, & to draw them out w/ a sincere personal interest. Such people, IMO, may do more good than most psychologists ever will. The author seems to at least hint at a similar questioning attitude toward her profession in passages like the following:

"In my research and practice, I turn a critical eye toward the clinical traditions and tools of my own society while I also apply them in settings of care with people urgently seeking my help; the ongoing tension has produced in me a sort of uncomfortable, curious stance toward my own ways of knowing." - p 11

"I first met A when he came along to a movie with a group of my friends. I remember a tall, blond fellow hanging out at the fringes, aloof, speaking only to offer cryptic commentary on the proceedings. *Who the hell does he think he is,* I remember thinking crossly, unimpressed. The next time I saw him was at a New Year's Eve party, the last night of 1998. He was standing in the corner of the host's living room, wrapped up in a long black overcoat and black gloves, unmoving and silent. Someone had decorated him with a string of Christmas lights. Later that evening, I found him sitting under a table and brooding about the apocalyptic potential of the Y2K bug, due to hit a year later. I ducked under there to join him. We spent the rest of the night talking and woke up, the first day of the last year of the millennium, wrapped in each other's arms. It was the beginning of a long love affair and a deep friendship. It was also the beginning of my awareness of Asperger's syndrome, sometime later, when his mother sent him a copy of Tony Attwood's (1997) *Asperger's Syndrome: A Guide for Parents and Professionals* as a sort of gentle hint. Nonplussed, he read a few sections to me over the phone. I was fascinated at the refraction of what I immediately recognized as a well-known folk category—the nerd, socially awkward, physically clumsy, and obsessed with arcane topics—into a medical diagnosis. I started going to talks about Asperger's at the hospital where I was working as an administrative assistant as I prepared to go to grad school, and took it up as my topic when I got there." - p 12

That's very romantic, I like romantic stories. I, too, had a love affair w/ a probable Aspie. What ruined it wasn't any of her Aspie characteristics, I found them most endearing, it was her self-proclaimed selfishness. Is there a cure for selfishness? Is it recognized pathologically? Maybe psychologists are barking up the wrong tree.

"In 1994, Asperger's disorder was added to the *DSM-IV* (American Psychiatric Association, 1994) as one of the five "pervasive developmental disorders" that make up the autism spectrum. In 2013, with the *DSM-5*, the American Psychiatric Association removed it, as these individual disorders were rolled into one single "autism spectrum disorder."

"Over the past several decades, the number of people diagnosed with autism spectrum conditions has risen sharply, with prevalence estimates rising from fewer than 5 per 10,000 in studies published in the 1980s to 1 in 59 today (Baio et al., 2018; Burd, Fisher, & Kerbeshian, 1987; Matson & Kozlowski, 2011; Ritvo et al, 1989). This rise has led to windespread concern about the emerging "autism epidemic," with some attempting to identify some pathogenic influence— most notoriously, but not exclusively, vaccination—that can explain and potentially mitigate it (e.g., Blaxill, 2004; Herbert, 2008). Others, however, have argued that the rise in autism's prevalence can be explained by changes in diagnostic classification criteria and the broader infrastructures in which they are embedded (Gernsbacher, Dawson, & Hill Goldsmith, 2005; Grinker, 2007)." -p 14

NOW, as a layperson who's utterly cynical about the Medical Industry (to quote myself: "The Medical Industry compromises your Immune System & then gets rich renting you an Inferior Substitute"), I tend to believe that there's a serious illness about that consists of essentially diagnosing *everyone as sick* so that one's health can be turned into a product for sale for outrageous prices by the 'experts'. If no-one's healthy & no-one believes in the possibility that their health can be anything other than something provided for them by the DOCTOR GODS then the situation's great for greedy sadists but pretty bad for the rest of us.

In other words, it's easiest for me to think that the diagnostic criteria for "autism spectrum" are so broad that almost everyone can be put under its thumb. Nonetheless, I'm open to alternate explanations: it may just be that human nature is dramatically changing, most likely in sync w/ the dramatic overpopulation & explosion of technological change. 50 yrs ago, when I was a teenager, very few people wd've had any knowledge about or access to computers. Over the last 20 yrs, almost everyone, at least in the society I inhabit, has a computer that they carry w/ them at all times & that allows them access to a mind-boggling amt of sophisticated tools. This is no insignificant development & it's no surprise if humans change dramatically in tandem w/ it.

Fein refers obliquely to the allegations of Andrew J. Wakefield & his fellow doctors who noticed that some children given the MMR vaccine in the 1990s developed gastro-intestinal problems that were quickly followed by developmental regression. Wakefield has been 'discredited' but, esp in this day & age, I've noticed that just about anyone who refutes the agenda of Big Pharma gets somehow 'discredited' by what I call FACT CHOKERS, so I'm giving Wakefield a chance. I'll also give his critics & detractors a chance. I'll be reading bks by both, until I do so, I'll retain an open mind & won't let the propagandists dissuade me from at least researching a variety of opinions.

"In writing this book, I have taken real people—myself included—and made them into book characters, figures who are there to represent and convey particular elements of a story, and, in doing, so have frozen some of the dynamism of their actual lives. I have written in this way to preserve a sense of vividness and immediacy, inviting readers to feel today as the characters in this book have in the past to recognize a little of themselves. But as far as the actual people upon whom this book is based are concerned, know that this use of present tense is a literary convention, a piece of artistic license on my part; time has passed, and they may see things very differently now." - pp 20-21

An unfortunate tendency for some writers is to pretend to objectivity & to not even acknowledge that they're *writing the bk* - as if they're writing from an ominiscent position that brooks no contestation. I think it's to Fein's credit that she acknowledges that *she's a writer, taking a particular approach* & that this approach, like any other, will probably have its pluses & its minuses. That helps keep the pomposity & pretentiousness to a minimum & I'm grateful for it.

"The Journeyfolk, among whom I had found myself, are a close-knit collective of young people dedicated to transforming themselves and building community through the art of live-action role-playing games. These games combine elements of collaborative storytelling, improvisational theater, and tabletop games of the Dungeons and Dragons sort. Their busiest time is the summer, when they run a series of "Summer of Adventure" camps—one of which, this year, was the "Aspie Camp" developed in collaboration with the Unity Center." - p 26

If it works, good; if role-playing games help people feel part of a community that they then get positive reinforcement from, good. It wdn't work for everyone, of course, & that's one of the strange side-effects for me, as a reader of this bk w/ particular cultural interests: I realized how much the culture that helps the community under study *bond* is something that I find very much something for many people & *very much not for me personally*. That's nether here nor there as far as whether such role-playing is good for Aspies, at least, if not for everyone on the "autism spectrum"; it's just a strange thing that I didn't really expect to think about along the road of reading this: *the culture that works is a culture that I, personally, find unimaginative & highly limited.* Here's an example of something that works for me: "**HiTEC 051 - New Hazlett Theater**" (- the complete gig is on the Internet Archive here: https://archive.org/details/HiTEC051full - complete version also on my onesownthoughts YouTube channel here: https://youtu.be/lc20DsVW718). This is my own 'game', a collective histrionic thought experiment. There aren't any swords (but there cd be) or monsters (but there cd be).

"Several years earlier, the mother of a boy diagnosed with Asperger's syndrome, Mariel Mattheson, had become intrigued by the possibilities of autistic culture. A few years later, she herself would be diagnosed with Asperger's as well. Inspired by the nascent neurodiversity movement, and its roots within autistic self-advocacy, she wondered how her son's struggles in school might change if he were among other autistic students in an environment that treated his condition not as a disorder but as a valued difference." - p 28

The "nascent neurodiversity movement"? Perhaps there's been such a thing already from here-to-eternity, perhaps it's called "Free Thinking" - I mean it's not like there haven't always been people who accept & seek out difference! It's just a matter of what the difference is. As for "autistic self-advocacy"? Is that really what it is? Or is it really Asperger's Syndrome self-advocacy presented as if it represents the whole spectrum? How many people who don't speak at all & who're dependent on others for basic survival needs are involved in self-advocacy?

"Mark had observed that campers with Asperger's often seemed to respond to the game in a special way. "The kids with Asperger's tend to really get—as we called it—*into character*," Mark told me. "They really could embody a character. Because I think they really could imagine, identify—and *had* identified, through their reading or fascination with fantasy literature or video games. And they could understand how to be that character.["]" - p 29

Wch brings to my mind an issue I'm often preoccupied w/: what I continue to consider to be *over-diagnosing*. Some things referred to as "disorders" just seem point-blank idiotic. SAD, Seasonal Affective Disorder - can't we just acknowledge that people are bound to get depressed when it's cold & there's less sun? Why turn doing so into a "disorder"?! & what about HPD, Histrionic Personality Disorder - why not just recognize that such a person is a natural-born performer? This tendency that the Medical Establishment has to turn just about everything into a "disorder" is a way of stigmatizing people AND a way of Medicalizing society, of creating a hierarchy in wch doctors & pharmaceutical companies & all the rest are here to *regulate* us all - as if we need their fake superiority to rule over us. SO, let's hereby announce a new disorder: ODD, Over-Diagnosing Disorder. Any doctor found suffering from it must lose their license to practice. That shd help cut down on the idiocy somewhat.

"The most well-known theory of autism proposes that people diagnosed with autism are experiencing a deficit in *theory of mind*, or what Baron-Cohen (1995) has called "mindblindness." Numerous studies have documented that people with this cognitive profile have difficulty correctly attributing mental states (beliefs, desires, intentions, etc.) to others, especially when those mental states are different from their own" - pp 30-31

"Individuals on the autism spectrum often show difficulty planning and carrying out tasks in the absence of external structure, difficulty in flexibly shifting approaches when task demands change, and a vulnerability to extreme fluctuations in mood, all signs of executive dysfunction."

[..]

"Individuals on the autism spectrum are also often said to display what has alternatively been called *weak central coherence* (Happé & Firth, 2006). In other words, when taking in a scene, individuals with this perceptual style tend to focus on individual details and local, clearly observable relationships between those details, rather than displaying the more typical perceptual style of integrating details into a fuzzier "big picture" and making inferences" - p 31

"Moving a piece of furniture in a room, changing the order of tasks on a schedule, taking different turns on the drive home—these alternations that might seem insignificant to those with a more globally coherent style can produce a sense of deep ontological uncertainty, an existential sort of distress in those who are proceeding locally." - p 32

That's all very interesting & I can't really find fault in the observations. I'm not sure I've ever known or noticed anyone who fits those descriptions. Or have I? I have a friend who'd get up at 5 in the morning to try to give herself enuf time to decide wch clothes to wear to work, where she had to be by 9AM. She was always late anyway. Naturally, her friends found her exasperating. She eventually moved back to living w/ her parents. If she'd managed to get a more executive position at her job she wdn't've had to care at all whether she even showed up or not.

I had another friend who often made plans to do things w/ me. I joked that I cd be sure that the one place she wdn't be at at the time scheduled was where we'd agreed to meet. What do *you* do in response to friends like that? In the latter case I simply stopped making plans w/ her. Wch is worse?: not making plans or pathologizing her & putting her on 'punctuality drugs'. I put the latter expression in single quotes b/c I'm not saying that there is such a thing, I'm creating an imaginary 'cure' that might be negative - just b/c *I* prefer punctuality & reliability doesn't mean that I think it 'needs' to be a rule for all humanity. At any rate, my intuition tends to agree w/ the following:

"we need to engage instead with the ways that human cognition is inextricable from culture, and the possibility that the cognitive processes that get labelled "autistic," rather than being exceptions to this rule, are instead exemplars of it." - p 33

Alas, just b/c "human cognition is inextricable from culture" doesn't mean that the culture isn't harmful. The culture that I see the most people in the thrall of is dumbing them down & numbing them down. The LCD (Lowest Common Denominator) is borderline vegetative. The culture is OUR culture - wch means we have a responsibility to *make it* rather than just consume it. The following illustrates my point:

"children on the spectrum were more likely to tell "prepackaged" fictional narratives taken from movies or television, rather than relate events from their own life." - p 35

It's my opinion that as the "prepackaged" narratives invade & permeate us more & more the less capable we are of living our own lives at all. As such, I tend to wonder whether autistic people aren't an evolutionary adaptation to this encroachment, whether autistic people might not be an indicator species of sorts. This next quote tends to reinforce that idea:

"cultures themselves can be disabling—or enabling—for particular members, and the nature of a disability depends very much on the particular demands and opportunities of the given environment." - p 36

In other words, if autism really is becoming more & more common, rather than just overdiagnosed as such, I'm open to imagining it as intertwined w/ the ways in which the social environment has become increasingly mediated by forces not necessarily friendly to local connections.

"The Journeyfolk prided themselves on being improvisors and collaborators; the thrill of not knowing what your fellow players would add to the story was part of the fun. Each time a game is run, with the same basic setup manifesting through different players in a different place, a very different set of things will probably happen. The ex-lovers who are joyously reunited in one go-round might, in the next, both be murdered by assassins before they even meet." - p 43

I've been preoccupied w/ (m)usically (& otherwise) improvising since 1972 & w/ collaborating from about the same time (http://idioideo.pleintekst.nl/ Collaborations.html). However, I'm more or less ignorant when it comes to the types of role-playing games that the Journeyfolk play. Somehow, our respective ideas of collaboration & improvisation don't seem to be the same since they seem to be working w/in a pre-established framework, a "prepackaged" narrative. I did get more interested in what autistic musicians might be doing so when the author of this bk informed me of an event entitled "Divergent Musicalities" I considered going.

Unfortunately, from my POV, the event stipulated: "Proof of vaccination or recent negative test result required." Since it was claimed that the organizers were "really excited to share it with anyone who might be interested" I found that to be extremely contradictory since there were probably plenty of people who were interested who find "Proof of vaccination or recent negative test result required" to be an exclusionary Medical Police State tactic of a most reprehensible (& absolutely unnecessary health-wise) kind. As such, I didn't attend the event. Alas, the author is probably ultimately a BELIEVER in the Medical Establishment (&, by extension, its Police State) or she wdn't be employed as she is.

"Elise, battling to save the beleaguered Unity Center, worked hard to translate our adventures with the Journeyfolk into the kind of language that would be compelling to funding organizations, as well as to the public school district. These were important sources of potential financial viability, as the district could cover the tuition for a student's participation in Unity Center programming if it were considered a necessary component of their official education plan. After each day, we were supposed to go over our lesson plan, in which the Journeyfolk's practices had been translated into special education lingo, and evaluate whether we had met our goals and objectives for that day. I liked this ritual—it gave us a rare moment or two to catch our breath and reflect—but it never seemed to have all that much to do with what we were actually doing." - p 48

"By the end of the summer, Elise had confirmed what we had all begun to suspect: The Unity Center had not been able to secure the funds it needed to open the Unity School in the fall." - p 53

This bk certainly convinced me that the Journeyfolk & The Unity Center were a good thing & that, therefore, a lack of funding to the point of closure was very sad indeed. I often disagree w/ funding priorities. Somehow, the Military-Industrial Complex, despite decades of heavy criticism of it, still gets along just fine w/ billions of taxpayer dollars. In the meantime, spaces that support friendliness & creativity are constantly struggling. If one were cynical (who, me?) one might think that war-mongering imperialism is valued more highly than peace. Locally, a large number of the cultural centers that I've deemed most important have gone under in the last few yrs - partially b/c of understandable reasons, partially b/c of gentrification, & partially b/c of the disruptions caused by the QUARANTYRANNY. See my movie about Pittsburgh Filmmakers as an example: "**FILMMAKERS don't let it be *Dead at 47***" (- uploaded to my onesownthoughts YouTube channel August 23, 2018: https://youtu.be/6ZEkIgWa5vQ - uploaded to the Internet Archive August 24, 2018: https://archive.org/details/DeadAt47). Funny, how the eagerness of the participants in the summer activities to be able to do it again strikes me as a positive thing. Am I one of the those naive idealists I've heard so much about my whole life?

"Once the Summer of Adventure was over, I stayed in Brookfield to continue my dissertation research and stayed in touch with many of the campers as I did so. I had the chance to interview many of them, a few of them multiple times. They were very polite, and did their best to give me good answers to my interview questions about Asperger's syndrome and how they felt about their diagnoses and so on. But what they really wanted to know, what they asked me about over and over again, was, *When are we going to do the camp again?*" - p 50

The camp apparently made them happy. Aspies or not, why shd any kid have to give up something that makes them happy? Is our society really so sadistic & uncaring?

"I asked her what the school district had recommended. "They want to put him in an ILP with kids with multiple disabilities, with no other autistic kids. They have a padded time-out room. I'm trying to fight that."

""What's an ILP?" I asked her.

""Intensive Learning Program. It's a code name for kids with behavior disorders, kids who are in and out of psychiatric hospitalizations. And those kids do need programs—but I think those padded time-out rooms should be made illegal. We've been finding out that people have been dying because of those interventions." Indeed, just that morning I had read an article, circulated on a local autism-related e-mail list, about people on the autism spectrum who had died after being restrained by insufficiently trained personnel." - p 54

Naughty, naughty! Be careful about what you say, the Fact Chokers will be all over that shit. As for "insufficiently trained personnel" being the cause of the deaths? Gosh, if only we'd trained them how to put down those troubled kids *properly* - why is it that nagging in the back of my brain there's a little subversive thought that maybe even the 'sufficiently trained personnel' might cause a death now & then too, eh? & what about chemical strait-jackets? They're BBBBBBIIIIIGGGGGGGGG BUSINESS. It's my opinion that most of the worst damage done *to* humans is done *by* humans - there's war, economic injustice, abuse of power, greed, nuclear disasters - how many victims get treated to extreme prejudice under the guise of health care?

"Zander, whom I haven't met before, is on the computer making greeting cards. "Sorry I Sent the Zombies to Your House! Happy Easter!" reads one of them, against the backdrop of a giant Christmas tree."

[Zander has been diagnosed w/ SHD]

"This kind of creative bricolage characterizes the classroom culture. Linguistic anthropologists Elinor Ochs and Olga Solomon (2008; see also Solomon, 2008) have identified a conversational technique often employed by people on the spectrum that they call "proximal relevance": participating in interactions by making a comment that is relevant to what has just been said based on structural algorithms grounded in shared objective knowledge rather than based on its relevance to the broader topic under consideration. At ASPEN, proximal relevance is elevated to a collective art form, and perhaps even a form of resistance to the teacher determined concerns of the classroom context. We spend a lot of time disassembling discursive systems and tacking them back together in new and funny arrangements," - pp 69-70

[SHD = Sense of Humor Disorder]

"Like ASPIRE, the ASPEN curriculum encourages students to consider Asperger's syndrome as a complex condition bringing gifts and challenges. The day I arrived in Alice's classroom for older teens, the students were reading Daniel Tammet's (2006) *Born on a Blue Day: Inside the Mind of an Autistic Savant*. The book is an autobiography of a man on the autism spectrum whose ability to perceive numbers as beautiful colors assists him in performing astounding calculations." - p 71

I'm reminded of A.R. Luria's The Mind of a Mnemonist - A Little Book About A Vast Memory about Solomon Veniaminovich Shereshevsky, a journalist & performer who, essentially, cd remember *anything* & who turned this ability into a stage act.

"On the basis of his studies, Luria diagnosed in Shereshevsky an extremely strong version of synaesthesia, fivefold synaesthesia, in which the stimulation of one of his senses produced a reaction in every other. For example, if Shereshevsky heard a musical tone played he would immediately see a colour, touch would trigger a taste sensation, and so on for each of the senses. The images that his synaesthesia produced usually aided him in memorizing. For example, when thinking about numbers he reported:

"Take the number 1. This is a proud, well-built man; 2 is a high-spirited woman; 3 a gloomy person; 6 a man with a swollen foot; 7 a man with a moustache; 8 a very stout woman—a sack within a sack. As for the number 87, what I see is a fat woman and a man twirling his moustache.

"The above list of images for digits is consistent with a form of synesthesia (or ideasthesia) known as ordinal linguistic personification but is also related to a well-known mnemonic technique called the number shape system where the mnemonist creates images that physically resemble the digits. Luria did not clearly distinguish between whatever natural ability Shereshevsky might have had and mnemonic techniques like the method of loci and number shapes that "S" described." - https://en.wikipedia.org/wiki/Solomon_Shereshevsky

In The Mind of a Mnemonist, "S" is presented as discussing how he memorized a particularly difficult string of nonsense syllables:

"This is the way I worked it out in my mind. My landlady (*Mava*), whose house on Slizkaya Street I stayed at while I was in Warsaw, was leaning out of a window that opened onto a courtyard, With her left hand she was pointing inside, toward the room (NASA) [Russian: *nasha*, "our"]: while with her right she was making some negative gesture (NAVA) [Yiddish expression of negation]", etc, etc.. - p 53 (Basic Books, 1968)

In other words, he translates the long string of repetitive nonsense syllabes into a narrative & by remembering the narrative remembers the syllables. This isn't a technique exclusive to him:

"**Akira Haraguchi** (原口 證, Haraguchi Akira) (born 1946, Miyagi Prefecture), a retired Japanese engineer, is known for memorizing and reciting digits of pi.

"Memorization of pi

"He holds the current unofficial world record (100,000 digits) in 16 hours, starting at 9 a.m. (16:28 GMT) on October 3, 2006. He equaled his previous record of 83,500 digits by nightfall and then continued until stopping with digit number 100,000 at 1:28 a.m. on October 4, 2006. The event was filmed in a public hall in Kisarazu, east of Tokyo, where he had five-minute breaks every two hours to eat onigiri to keep up his energy levels. Even his trips to the toilet were filmed to prove that the exercise was legitimate.

"His previous world record of 83,431 was performed from July 1, 2005, to July 2, 2005.

"On Pi Day, 2015, he claimed to be able to recite 111,701 digits.

"Despite Haraguchi's efforts and detailed documentation, the Guinness World Records have not yet accepted any of his records set.

"Haraguchi views the memorization of pi as "the religion of the universe", and as an expression of his lifelong quest for eternal truth.

"**Haraguchi's mnemonic system**

"Haraguchi uses a system he developed, which assigns kana symbols to numbers, allowing for the memorization of pi as a collection of stories. The same system was developed by Lewis Carroll to assign letters from the alphabet to numbers, and creating stories to memorize numbers. This system preceded the system above which Haraguchi developed."

- https://en.wikipedia.org/wiki/Akira_Haraguchi

In Haraguchi's defense, it's worth noting that Guinness is basically a business that wants to be hired by other businesses to perform publicity stunts & isn't necessarily interested in extraordinary people. See documentation of my own unpleasant experience w/ them near the bottom of this webpage: http://idioideo.pleintekst.nl/tENTFeatures.html .

"A well-earned suspiciousness toward an often unpredictable and often unwelcoming social world interacts with emotional volatility and a tendency toward mimetic repetition to create what I came to think of as a sort of echolalia chamber, resulting in the deep entrenchment of certain stances within autistic communities." - p 77

I usually appreciate new coinages, hence "echolalia chamber" appeals to me. Beyond that, I'm not sure what to make of the above. When I think of "mimetic repetition" it's mostly as a positive thing - but in the case of humans, if the repeater is fixed in that mode it cd be very limiting. Nonetheless, I think of the mimetic birds: Lyre Bird, Mockingbird, African Gray Parrots, Gray Catbirds, Thrashers, etc.. These, for me, are the birds w/ the most advanced (& hence interesting) vocal abilities. A good human mimic might be a good actor. As for "the deep entrenchment of certain stances" thanks to people imitating each other rather than thinking for themselves? Alas, that's far from limited to "autistic communities" & applies across the board to more or less *every subculture* - producing the phenomenal stupidity of BELIEVERS.

"In his early ethnographic study of early brain visualization technologies, anthropologist of science Joseph Dumit (2004) observes how much of the appeal of these technologies rests on the "taboo nature of subjectivity in science" in which "every possibility of subjectivity must be eliminated in order to produce something reliable—that is, something real, something known" and a subsequent romanticizing of "automation, which stands as the opposite of interactivity" (p

122)." - p 91

Personally, I've always found "subjectivity" & "objectivity" to be dubious concepts. If subjectivity is a perception limited by the idiosyncracies of the perceiver then how is objectivity achievable given that the same limitations apply to any perceiver? Conversely, is it possible for anyone to be separate from a larger whole? In other words, an individual perceiver is nonetheless infinitely interconnected w/ a larger whole by virtue of breathing, eating, existing w/in an environment, etc, etc, ad infinitum.. Therefore, if objectivity is a perception from a larger vantage point than that of the individual that removes idiosyncratic interference w/ a more applicable generalization then everyone can make the claim of not simply being a detached individual incapable of thinking outside their own box. 45+ yrs ago I proposed the term "ogjectivity" to suggest a 3rd possibility, a sort of idiosyncratic objectivity, an acknowledgment that a completely detached omniscient perception is impossible. Objectivity, as it stands now, is most 'proveable' when it describes relations that can be manipulated to predictable ends. Person A has an 'objective' perception of the nature of a transaction when they can manipulate Person B to give them money - that doesn't mean that the money given isn't toward an imaginary end &, hence, a subjective one.

"Within the two groups of students in the ACER building (ACER is the acronym for Brookfield's special education services), the ontological assumptions and traditions of knowledge production underlying each group's diagnosis organize the interpretations of each group's behaviors into a neat and compelling parallel. Brains, as understood by a psychiatric science driven by replacable experimentation on fixed physical systems, are material, predictable, and morally innocent—just like the Asperger's students that best enact and exemplify them. The minds and behaviors of the "psychiatric" kids, on the other hand, are seen as unpredictable, fluid, socially embedded, morally culpable, and messily human—reflecting the psychiatric traditions out of which their classification developed." - p 92

This sussing-out process can go both ways & it's probably in the best interests of the people who find themselves in interaction w/ psychologists w/ power over their lives to get good at it ASAP. When I was about 20, I was applying to be a research volunteer for simulated space stn living. I was sincerely interested in the experience &, as always, desperate for money. Applicants had to pass a psychological test. The questions were generally somewhat obvious: one asked whether the applicant thought they were being poisoned, an obvious way of weeding out paranoids.

Figuring that I'd better give them something to question me about, something that made me seem 'human', I confessed to post-high-school loneliness. This wasn't actually true, I had plenty of friends & a girlfriend. Sure enuf, after I completed the application the *one thing they asked me about was my 'loneliness'*. These people were completely predictable & manipulatable. One of my coapplicants sd he was being poisoned. He was, inevitably, questioned about that & he explained that he was just kidding b/c he thought it was such an absurd question. All of us were allowed in.

"she observed an affinity between students with Asperger's and technology so intense and intimate as to be almost symbiotic" - p 92

"She also proposes that there is some fundamental *simpatico* between people with Asperger's and this technology, imagining that they are in some way made to go with computers, born into a biologically determined symbiosis of human and technology." - p 93

As I wrote earlier: "it may just be that human nature is dramatically changing, most likely in sync w/ the dramatic overpopulation & explosion of technological change. 50 yrs ago, when I was a teenager, very few people wd've had any knowledge about or access to computers. Over the last 20 yrs, almost everyone, at least in the society I inhabit, has a computer that they carry w/ them at all times & that allows them access to a mind-boggling amt of sophisticated tools. This is no insignificant development & it's no surprise if humans change dramatically in tandem w/ it."

In my bk entiled Unconscious Suffocation - A Personal Journey through the PANDEMIC PANIC (written as "Amir-ul Kafirs"), I discuss this phenomena in relation to the current oligarchy-induced dystopia:

"Parents of some of her friends refuse to let their kids to socialize at all,

"Talk about BAD PARENTING! Imagine being a child & being not allowed to socialize, being forced to sit at a computer screen most of the day for 'education' & being deprived of human cues. Imagine being forced to wear a mask & to be around other people wearing masks. I mean the old adage of "Children should be seen and not heard" has reached a new level of viciousness against natural ebullience.

"Most of what I see about kids in quarantine is just parents complaining about how
annoying it is to be around their own offspring.

"It can't possibly be good for them. It's already been claimed that there's a correlation between the preponderence of computer use & the rise of Asperger's Syndrome. This is going to create a whole new class of automatons, getting their primary social contact via computers AND their primary rewards from the same. It's easy to imagine being sociable, in the old physical sense, becoming criminalized. In fact, we're already there." - p 757, Unconscious Suffocation

"Re: TR: Free virtual summer camps

"what a hellish idea

"Absolutely hellish & against everything that makes summer camps valuable & important. HOWEVER, it does serve Bill Gates's apparent interests perfectly: reroute everyone's socialization away from actual human contact to mediated contact & turn them into zombies for big business: the perfect Asperger's Syndrome Manchurian Candidiates!" - pp 795-796, Unconscious Suffocation

I suspect that only open-minded people will manage to get past the last part of this review w/o scoffing. Some will be offended at my implying, or outright stating, that Asperger's is a calculated *product* of sadistic big business interests - or has at least been turned into that by opportunism. So be it. Bill Gates makes appearances in this bk.

""Computers," the cheerful gentleman in the video told us, "were designed by and for people with Asperger's syndrome. Why? Because you don't have to socialize with it. It does not get moody. Unlike teachers, it does not get premenstrual tension, or hangovers . . . . The major way of learning is visual and logical. Not emotional. Emotion and logic are contradictory. . . . Computer science may be a very appropriate career choice for many of these individuals who find the logic of computers is just so natural to them" (Attwood, 2002). This association of Asperger's students with logical technology over emotion-laden sociality resonated throughout the Brookfield district, with tangible consequences for the students." - p 94

I've got news for you, folks, there's only so far away you can get from PMS & testosterone *unless human biology is fundamentally changed* wch there are people foolish enuf to try doing at this very moment. What about Asperger's women w/ PMS or hot flashes or other emotional fluxes determined by menstrual cycles? What about the *urge to reproduce or to simply satisfy one's desire for the intense pleasure provided by satisfying one's DNA drives?!* Well, there are those who'd like to get rid of those too.

"Women with Asperger profiles are less likely to be diagnosed and more likely to be misdiagnosed for a number of reasons. Additionally, many professionals have been trained to recognize typical Asperger/autism spectrum expression more easily in males than in females." - https://www.aane.org/women-asperger-profiles/

If you're a woman sitting at a computer that doesn't get PMS b/c it's a machine but *you're experiencing PMS* it's still there w/ you isn't it? If you're a woman or a man & you're hot-to-trot being at yr computer won't exactly solve that will it?

"However, this conception of Asperger's students as innocent machines can also be deeply alienating."

[..]

"Consequently, the semiotic equations across autism, technology/science, social deficiencies and lack of personal reflexivity contain the potential to dehumanize autism and the people associated with this label. (p. 131)" - p 100

"incapable of change, of self-control and thus belonging to a "they" fundamentally different from the capable, agentic "we" to which he & I belong. In such moments, the potential for change, for the capability to "do the things he should be doing," is foreclosed.

"In *Recovery's Edge: An Ethnography of Mental Health Care and Moral Agency*, psychiatric anthropologist Neely Myers (2015) argues that a key component of mental health is *moral agency*: "the ability to be recognized as a 'good' person in a way that makes possible intimate connections to others" (p. 13). Only through the relationships of mutual accountability, she argues convincingly, can we take our place as members of a community and thus live full lives. When students with Asperger's are conceptualized as fundamentally estranged from such networks of accountability, the meaningfulness of their communications become obscured." - p 102

That might be a bad thing for the Aspies but it's not necessarily a bad thing for their prospective employers. The 'perfect employee' doesn't necessarily have to have any emotion that might interfere w/ their obedience, they're much more 'perfect' if they just do their repetitive job w/ an obsessive devotion uncomplicated by such factors as "intimate connections to others".

"I first met Kevin when he was 11. Curious and astute, he was at the time enthusiastic about his Asperger's diagnosis because he associated it with a community of so-called nerds that he'd found on the Internet. Over the years, however, he became more skeptical about the diagnosis. By the time of this interview, when he was 17, he had become quite wary of the way the Asperger's diagnosis was used to ascribe a lack of agency to those living under its description. He emphasized, instead, the importance of interpreting behavior as a deliberate choice. "This is how it is with a lot of things with Asperger's," he told me.

"When I was younger it was really hard for me to shower. Just because it's such a monotonous thing to do. It still is, but I get it done anyway. People think like: *oh, this person doesn't understand that it's important to shower if you have Asperger's.* But no: it's a personal decision that they don't think it's that important. They might fully understand how it impacts them and be like, *well, if you're going to judge me based on whether I'm showering, I don't want to be friends with you.* That's fair." - p 104

I can relate. Friendships shd be based on something more substantial than the degree to wch one conforms to social conventions.

"The topic of today's discussion is medications, and the overall consensus is that they don't work very well.

""I don't want to take something to fix me. I'd rather work on things myself. This is who I am and I need to learn to live with myself the way I am," says Eduard, a young man in his mid-20s who is a regular attendee at these meetings." - p 108

Thank the Holy Ceiling Light for people like Eduard. IMO, it's the responsibility 1st & foremost of the individual to decide *whether they even have a problem* & to then develop their own solution to it. The drug companies have a vested interest in having *everyone* diagnosed as needing their product, the profit to be made from such a process is just too enormous to not pursue ruthlessly - assuming that the CEOs of the companies are greedy, wch I think we can take for granted.

Naturally, not every problem is solvable by the individual experiencing it. Society is full of every imaginable obstacle & prejudice & disadvantage that can be   ⁃ imposed on those on its bottom. & the privileged snobs at the top feel every bit entitled & justified to demean anyone whose intelligence might get in the way of their own megalomania. THAT's a problem, a problem bigger than Asperger's will ever be.

"I asked him what ASPNET is trying to do. "Building community, advocating for ourselves to get the things we need, speaking for ourselves until we get what we need to handle living in this neurotypical world, which we find difficult to do." He told me the story of a teenager on the autism spectrum who had committed suicide. Would she have done so, he wondered, if she'd had access to a group like this?" - p 116

"Neurotypical" isn't that different from "normal" or "New Normal" wch some of us have been objecting to the dictatorship of for most or all of our lives. "Neuroatypical" people aren't just people on the "autism spectrum". "Neurotypical" might be more aptly named "Neurovacuii", nobody home upstairs. Perhaps I suffer from "Horror Neurovacuii", the fear of people who just don't think, the Puppet People, the Sheeple, the people who go along w/ the program w/o questioning it.

I've had at least 7 friends commit suicide during the QUARANTYRANNY. That's alot. If more people wd've gone against the grain (of mass media morphine) those 7 people might still be alive - so it's not just a matter of providing support groups, wch I value, but of also not becoming Neurovacuii yourself. Living in a world of Pod People is no fun for the Neuroatypical, it's a form of hell.

"Participants shared strategies for getting along in an often mis-fitting, ever-changing, and confusing world while also striving to develop a shared sense of their condition and its implications—a sense developed out of their own experiences and on their own terms rather than one handed to them by medical professionals." - pp 116-117

& *THAT'S CRUCIAL.* Recognizing & holding dear one's own identity *rather than shopping for a "prepackaged" one.* You are YOU, you aren't this, that, 'n' the other subcultural member. Medical Professionals have been allowed to hand us shit-on-a-stick called by one euphemism or another for far too long. RESIST

MEDICALIZATION & IATROGENESIS. & that includes resisting comparing the human brain to a computer as if it's an analogy.

""I would say it's a neurobiological difference in our minds. It's a different kind of wiring that makes it difficult for us to understand social cues and social interactions, and it also creates some sensory issues. I tend to react to loud noises and bright lights, some textures of food, clothing textures—things like that. But it's not a disease, and it's not an illness. It's a part of our personality and part of who we are."" - p 119

People use the expression "wiring" entirely too casually, IMO. Perhaps people just say that metaphorically, to use an expression that's common enuf for people to 'understand'. I object to it b/c people seem to think that comparing the brain to a computer is a compliment. I quote from Hubert Dreyfuss:

"Adherents of the psychological and epistemological assumptions that human behavior must be formalizable in terms of a heuristic program for a digital computer are forced to develop a theory of intelligent behavior which makes no appeal to the fact that a man has a body, since at this stage at least the computer clearly hasn't one. In thinking that the body can be dispensed with, these thinkers again follow the tradition, which from Plato to Descartes has thought of the body as getting in the way of intelligence and reason, rather than being in any way indispensible for it. If the body turns out to be indispensible for intelligent behavior, then we shall have to ask whether the body can be simulated on a heuristically programmed digital computer. If not, then the project of artificial intelligence is doomed from the start." - p 235, <u>What Computers Can't Do (revised edition) The Limits of Artificial Intelligence</u>, Harper Colophon Books, 1972

Perhaps it seems anachronistic to quote a critique from almost 50 yrs ago. I think its prescience still applies.

"Autism/other forms of neurodiversity are the equivalent of a Mac or Linux in a world of PCs. PCs think you're defective because you appear to be a PC with a virus when in reality you're a WHOLE OTHER OPERATING SYSTEM with all the strengths and weaknesses of any other, that just happens to be different from the majority of computers. But being a Mac is not a problem in the same way that being a PC with a virus can be a problem." - p 121

I shd explain that I'm quoting Fein quoting a "post on Tumblr on "Mental Illness vs. Autism/Neurodiversity" by autistic blogger alice-royal". The point of the blogger is clear enuf but I find it problematic for various reasons, one being that humans don't have operating systems & aren't brand-manufactured. Even if we accept the idea of the brain's ways of doing things as being determined by a metaphorical OS I don't think that there's a "WHOLE OTHER OPERATING SYSTEM" between any people b/c if the 'OS' were that different it might be entirely lacking respiratory or circulatory functioning, e.g.. In other words, regardless of how different people might be from each other there's still far more that we have in common.

"The metaphor of autism, and the group of similar conditions that it occupies, as the operating system of a computer defines the neurostructural self as a processing system, evoking the sort of neurocognitive domains associated with developmental disability and casting them as inherent and definitional elements of that self." - p 122

Operating Systems are developed as something that can be upgraded & up/down-loaded. Can whatever drives a human being be detached in that way? I think not.

"Recently, a number of scholars have claimed that the dominant mode of neural-selfhood is no longer neurochemistry but *neural plasticity*, characterized by a greater openness to the environment and potential for deeper change."

[..]

"anthropologist and historian of science Tobias Rees describes the shift as follows:

"What was at stake was a shift from a conceptualization of the brain—the human—as a neurochemical machine largely determined in its basic design, fixed in its being, toward a conceptualization of the brain—the human—as a ceaselessly emerging cellular organ, an organ that never stands still, that is defined by an irreducible openness. At stake was a shift from neurochemistry to plasticity. (p 8)" - p 128

It's that *organic fluidity* that's a primary distinguishing characteristic of the biomorphic vs the mechanical.

"He cautioned me in our interview against "medicalizing autism, and treating autistic people as broken nonautistic people with something wrong with them." - p 122

I generally find the process of medicalization to be very problematic. However, to across the board reject it in the case of the "autism spectrum" strikes me as equally problematic. What if a young child were 'neurotypical' until a physical problem arrested or caused the regression of their development? Wdn't they then have been 'broken' by specific physical causes? Once again, it seems that the broadness of the "autism spectrum" generalization leads to people objecting to medicalization applied to people diagnosed as manifesting Asperger's syndrome as if the objection applies to everyone of the spectrum. After awhile, my sympathy for Aspies started turning into feeling that there's a bit too much conceit involved.

"We wouldn't have at least one president, you know, if it wasn't for Asperger's. We wouldn't have a lot of the great scientists. We wouldn't have a lot of people running the country. Some of the lucky ones who have the condition, and just happen to be—I'm thinking of Bill Gates, even though we don't really know in fact if he has it or not. But let's just take somebody like that. Let's say the mother had an abortion. You'd be depriving the world of Bill Gates possibly." - p 125

It seems that at least this one Aspie being quoted seeks to redefine the ability to focus as only an Aspie characteristic. I can focus far more than anyone else I know, at least on certain things, & I don't consider myself to be an Aspie. Maybe the "great scientists" were just great scientists, maybe they were more people like me: people who pursued their particular interests w/ concentrated passion & don't need to be labeled "Asperger's syndrome" to please one special interest group or another. As for "a lot of people running the country"? Given that I'm an anarchist I find it more than a bit difficult to think of such people positively - instead, I see greed & abuse of power. & Bill Gates? Gimme a break! Bill Gates is pushing thru an agenda that he's trying to force on everyone in the world, there's nothing positive, for me, about Gates. "Let's say the mother had an abortion. You'd be depriving the world of Bill Gates possibly." That cd be sd about any abortion, or miscarriage, or death of a child. Perhaps that fetus, that baby, that child wd've been [insert yr most respected type of human here] if they'd only had a chance to grow up.

"His deep and often thwarted desire to feel accepted informs his attitudes toward autism treatment. "I don't believe in treatment for autism and Asperger's," he told me, "because we are how we are—if the world can't live with it, it's too bad. So don't try to treat us, we are who we are, you've gotta live with it."

"I asked him what he meant by "treatment"—what does he include under that term?

"I don't know. It could be anything. It could be pills, could be vaccines, I don't know. I just know that if people can't live with who we are, then the problem is not us, it's them." - p 127

In general, I'm sympathetic to the above-quoted person's opinion - but, at the same time, it seems myopic. Once again, there's the currently widely accepted lumping together of all possibilities on the spectrum. To be irritated by other people's inability to accept &/or appreciate one's personality is one thing; to equate that as the same as being a parent trying to help a dysfunctional child develop the skills that wd enable them to leave the nest is quite another. No doubt, parents can get the 2 confused at times & think that a child's personality shd be 'normalized' w/o that's really being necessary for them to survive. But back to "neuroplasticity":

"Neuro-philosopher Catherine Malabou":

[..]

"The brain's capacity for plasticity, she argues, is potentially revolutionary; many of the existing discourses around plasticity, on the other hand, merely work to naturalize models of capitalist society. "*What should we do so that the consciusness of the brain does not purely and simply coincide with the spirit of capitalism?*" she asks in *What Should We Do With Our Brains?* (p. 12, emphasis in the original). "When does openness to the future simply become pliability and compliance—and how can one resist such an appropriation?" Nima Bassiri (2016, para. 13) similarly asks at the end of his review of *Plastic Reason.*" - p 129

I don't interpret "plasticity" as synonymous w/ "pliability" & "compliance", I think of it as more synonymous w/ "fluidity" & "flexibility". To be flexible doesn't necessarily mean that one allows other people to lord it over you, it can mean that one can quickly adapt one's self-defenses.

The author *does* address autism spectrum differences:

"For the parents of a severely disabled autistic adult who cannot speak, dress themselves, or go to the toilet unassisted, it is often hard to make sense of this articulate, college-educated young woman, and others like her, arguing passionately against a cure for autism, marshalling rhetoric about eugenics and the need for genetic diversity. *You're not autistic*, she's been told, even though she has an in-depth diagnostic evaluation that says she is." - p 139

But, obviously, there're drastic differences in degress of functionality between a "disabled autistic adult who cannot speak, dress themselves, or go to the toilet unassisted" & an "articulate, college-educated young woman" so, once again, lumping them together in the same category is ridiculous. Furthermore, the "articulate, college-educated young woman" shd not be so pretentious that she speaks for the "disabled autistic adult". The "autism spectrum" really needs to go, it's counterproductive. If the "disabled autistic adult" were simply left to fend for themselves, like non-disabled people are, they'd probably die fairly quickly.

If one philosophically supports the notion of "suvival of the fittest" then maybe that's ok - but most of us probably have a little more sympathy for people less capable of surviving than we are. The spectrum that we're all on is the survival spectrum. The people above us in that spectrum are usually born privileged & don't have to do much of anything to lead an easy life. The people below us may've been born w/ even fewer advantages than we may have. If we can feel the unfairness of the privileged above us we might also be able to feel the unfairness of our own privilege.

"In 2013, the fifth edition of the *Diagnostic and Statistical Manual of Mental Disorders* (*DSM-5*; American Psychiatric Association, 2013) radically changed the diagnostic nomenclature for autism, combining all the separate categories of "pervasive developmental disorder" (including Asperger's syndrome) into a single diagnostic term: "autism spectrum disorder." This change had been in the works since the beginning of the 21st century, part of a broader move away from the categorical, symptom-based diagnoses and toward a reconceptualization of disorders as continuous spectrums dictated by neurobiology." - pp 139-140

"This new approach to diagnosis is producing a different sort of diagnostic construct from the disease entities produced under the previous system. They are less like pathogens and more like—as Lana says—"a whole package." Rather than producing conditions that are exclusively negative, these new diagnostic constructs are *multivalent*: They contain elements that are wanted, elements that are unwanted, and elements that are neither, or both, depending on the context." - p 142

Obviously, I'm arguing against the "autism spectrum" as something that counterproductively lumps together people of dramatically different degrees of problems. I'm *not* arguing against a perception of the "whole package". It seems to me that *everyone* has personality or circumstantial "elements that are wanted, elements that are unwanted, and elements that are neither, or both, depending on the context." So, if we're going to have a "spectrum" why not broaden it to include *everyone*?

""From each according to his ability, to each according to his needs" (German: Jeder nach seinen Fähigkeiten, jedem nach seinen Bedürfnissen) is a slogan popularised by Karl Marx in his 1875 Critique of the Gotha Programme. The principle refers to free access to and distribution of goods, capital and services." - https://en.wikipedia.org/wiki/From_each_according_to_his_ability,_to_each_according_to_his_needs

Mack Reynolds is a good science fiction writer for exploring the possibilities of this concept. See, e.g., my review of his Lagrange Five: https://www.goodreads.com/story/show/1380533-review-of-mack-reynolds-s

"The government, such as it is, consists of a unpd council.

"""Then, what *do* they get out of it?"

"""The honor of serving. The honor and respect granted them by their fellow La-grangists."

"""To each according to his needs and from each according to his abilities, eh?"

"""Why, I suppose so," Susie said, not recognizing the quotation.

"""Syndicalism,"" - p 87"

"The speaker was the chief executive officer of a clinic that had helped to spon-sor the conference, one of a number of alternative medical centers that strive to identify signs of contamination through various techniques, including laboratory testing for heavy metals in hair and urine samples, and then attempt to pull these pathogens out of children's bodies, using interventions that range from infrared saunas to chelation, a treatment for heavy metal poisoning. Many of these trea-ments are scientifically unproven and come with their own set of risks. But symptoms of physical distress can also be read as signs of progress, ways of making disorder newly visible in the body." - p 144

There're plenty of treatments that're scientifically proven that have such long lists of possible harmful side-effects that such 'scientific proof' shdn't necessarily be trusted. One of my many medically-related slogans of late is "Industrial Society creates Industrial Toxins. The Medical Industry creates Industrial Strength harmful Side-Effects." I think it's sensible to view any industrial-strength treatment w/ caution. The side-effects just may compound the initial problem. "laboratory testing for heavy metals in hair and urine samples" seems like a good idea for me - as long as the tests aren't invasive.

If the presence of heavy metals is detected then the obvious next question is: How did they get there? Presumably the answer, if it can be found, will be in the child's environment. The next obvious step, if at all possible, is removing the child from exposure to the toxins. After that? If there are any *non-industrial-strength measures to take to reduce the deleterious effects then by all means take them.* Alas, not everyone's in a position to follow even such rudimentary steps, the poverty trap being the most common impediment. People seem to want industrial-strength solutions as if they're miracle cures when they're most likely going to produce additional problems. It's a gamble not worth taking.

The author generally manages to maintain the even keel that's expected of academics but her disapproval of the following seeps thru in the concluding parenthetical statement:

"The "Ransom Notes" campaign of advertisements placed by the New York University (NYU) Child Study Center in 2007 depicted conditions such as autism alongside ADHD, bulimia, and depression as kidnappers holding kids for ransom (payment to be delivered, presumably, to the NYC Child Study Center)." - p 146

"Medical historian Charles Rosenberg (2002) chronicles how *disease specificity,* "the notion that diseases can and should be thought of as entities existing outside the unique manifestations of illness in particular men and women" has become an increasingly prevalent organizing force in an "increasingly technical, specialized, and bureaucratized" (p 237) context of care." - p 150

Ah, well.. what if we think of bodies as smaller ecosystems w/in larger ecosystems - not exactly autonomous, not exactly NOT autonomous - ogautonomous if you will. What if we then accept that the dis-ease of the body is part & parcel of both ecosystems? Assuming the dis-ease is *not acceptable* then what if we change the ecosystems as much as we can so that they no longer accomodate the dis-ease's presence? Easier sd than done, right? That's what we go to doctors for, to determine what's to be done to rid our semi-autonomous ecosystem from an element that's generally harmful - but what if the basic systemic approach of the doctors is based on the idea that the dis-ease is an *invader* to be expelled at all cost? Imagine a flame-thrower used on a tree to get rid of a certain bug that's eating the tree, the flame-thrower might kill the bug but it's likely to do more damage to the tree than the bug was. The "increasingly technical, spe-cialized, and bureaucratized" health care system is largely incapable of thinking holistically - despite this having been pointed out for at least the last 50 yrs.

Then again, I recently learned about the idea of "suicide genes": "A **suicide gene**, in genetics, will cause a cell to kill itself through apoptosis. Activation of these genes can be due to many processes, but the main cellular "switch" to in-duce apoptosis is the p53 protein." What if any condition destroying its host body is a sort of "suicide gene" that's responding to its environment w/ an impulse to reduce it to raw material from wch something new may develop? That's not a very pleasant thought for the sufferer to have & few wd respond to it 'philosophi-cally'. Nonetheless, from an 'amoral' natural perspective it might be true.

"Darrel Regier (2007), the vice-chair of the task force planning the *DSM-5* (Amer-ican Psychiatric Association, 2013), observed that the expectation that "highly specific phenomenological descriptions of disorders . . . would result in phenome-nological subtypes that would eventually correlate with etiological and pathophys-iological factors needed to validate these more precise clinical syndromes" has not been supported by the findings emerging out of genomics and neuroscience." - pp 150-151

Barking up the wrong tree, it was probably worth a try & the results were proba-bly very discouraging to researchers who were putting their faith in their tech-niques. Maybe it's time for the research's metaphorical "suicide genes" to do their job & for the research to take a new direction.

"this revision" [from *DSM-IV* to *DSM-5*] "was seen as an opportunity for a com-prehensive overhaul in the basic assumptions around which the classification system itself was designed, addressing pervasive dissatisfactions with the cur-rent system." - p 153

The *DSM*, presumably no matter wch edition, is an astounding accomplishment made possible by an almost superhuman effort to comprehend & comment on states of human consciousness. Far be it from me to knock it as an overall achievement. The question is: *What wd happen if it & the whole monumental ed-ifice that it represents were to disappear?* Ok, that's a ridiculous question, that's even less likely to happen than for weapons of mass destruction to disappear, still, I like to imagine it as a sort of thought experiment. I'd definitely miss psy-chology, b/c I find it very interesting, but I'm not absolutely sure that human psy-chological suffering wd really increase that much as a result, if at all. That's an 'extreme' thing for me to say & I reckon any person who's dedicated their life to a psychology profession wd have deep disagreement w/ it.

Now, on the other hand, if I imagine, as a thought experiment, the complete dis-appearance of, say, avant garde classical music, I practically shudder w/ horror at the loss - that's how important such music is to me. That loss wd be like having part of my brain cut out, a very nasty thought. Between the 2, the music is much more likely to disappear.

"clinical research structures are delineating an entity with different contours, not limited to impairment and more focused on fundamental human capacities deeply intertwined with surrounding meaning-worlds. Such approaches produce diagnostic entities that are more package-like than pathogen-like—and that thus feel closer to a description of not merely what we suffer from but also who we are." - p 157

& I'm fine w/ that - but what we're still dealing w/ here is *medicalization* more than it is *philosophy* wch means that there's still an underlying attitude that there's a right-to-intervention in another person's life 'for their own good'. I have a profound distrust of that.

"Conditions once defined as sets of impairments are increasingly being reconceptualized as comprehensive forms of neurogenetic organization, and as they do, they are increasingly recognized to involve a wide range of attributes. For example, while dyslexia had traditionally been defined as a learning disability that produces difficulty in reading, a recent spate of research takes dyslexia as a brain-wide difference in information processing patterns; looking at it in this way reveals advantages, in some contexts, as well as difficulties in others." - p 157

Missing in the above is what is, for me, the real crux, often ignored or underacknowledged by the people who benefit: viz. that advantages & disadvantages are often most deeply based in economic privilege. E.G.: a former close associate was dyslexic, he was also from a very wealthy family. It didn't matter whether his dyslexia got in the way of certain accomplishments, he simply hired people to do the work for him, took most of the money & most of the credit. Dyslexia wasn't a problem for him b/c he never really had to face it. Therefore, it doesn't matter whether dyslexia is a "brain-wide difference" or not; what matters is what financial protection one has, there's no meritocracy involved.

"As the lens through which we view dyslexia widens through a big-picture of whole-brain organizational patterns, the phenomena that emerge are no longer limited to functional impairments. Instead, they expand to include phenomena such as entrepreneurship (and, as in Charlton's case, a knack for creative writing —he's written a series of fantasy novels featuring a young wizard cursed to misspell his spells)." - p 158

I had a conversation recently w/ someone about innovations in tool design. He attributed these innovations to capitalism, as if people wdn't make them if they weren't motivated by profit. I didn't necessarily disagree that there are people so motivated but I've been an innovator my whole life & what's motivated me is the joy of innovating. I can see entrepreneurship as just a way of getting rich, wch I don't find interesting; & writing novels as a sign of imagination, wch I DO find interesting. Therefore, I question the 'need' to factor in dyslexia.

"Although the need to include some participants who are not presenting with clinically significant distress is briefly acknowledged, the focus is on recruiting participants through clinical settings that aggregate those who are suffering and fore-ground that suffering over other elements of experience. This approach pro-duces a distorted image of a condition, in which problematic parts stand in for the entirety of the whole." - p 160

Indeed. But isn't that the problem w/ *every description*?! Take the above in-stance: presumably most, if not all, of the participants, including researchers & clients, are BELIEVERS. What wd happen to their data if an unbeliever, such as myself, were to provide observations & opinions?

I remember reading a long time ago that one cdn't see anything thru an electron microscope unless it was still. As such, no observation cd apply beyond a fairly limited scope. Things have changed since then insofar as now motion imagery can be taken, but, still, there's a limitation.

"Electron microscopes have been helping us see what the things around us are made of for decades. These microscopes use a beam of electrons to illuminate extremely small structures, but they can't properly capture something while it's moving." - https://www.inverse.com/innovation/a-new-device-turns-electron-mi-croscopes-into-high-speed-atom-scale-cameras

The point is that even the most fabulous means are still inadequate. That doesn't mean we shdn't try, it does mean that there must always be *qualifiers*, statements explaining the limits & what these limits mean to a more overall picture. This springboard review is an attempt to add qualifiers to this bk. But what are my own personal limits? For one thing, my experience w/ people who accept an autism spectrum diagnosis is limited to a few friends.

"Meanwhile, the rolling together of a wide variety of life experiences into a single *autism spectrum* supports attention to their common problems over their multiva-lent variations in manifestation." - p 162

Indeed. It seems that the author & I are on roughly the same page w/ this one. In fact, I generally find the author to be paying deep attn.

"The kinds of linguistic slippage posed by divided medicalization, in which one kind of autism (autism as a pathogen) comes to represent another (autism as a whole package), presents a particular kind of risk. When conditions such as autism are targetted for intervention and change as if they consisted only of their unwanted attributes, we risk suppressing a wider range of human qualities *de-spite*—and, ironically, in many ways *because of*—medical infrastructures that de-liberately exclude such attributes from their domain of intervention." - p 164

"fear that the war against mental illness is now being waged against the sympho-ny of the self." - p 165

"At the Asperger Center, laughter signaled moments when the roles and respon-sibilities of these health care providers were shifting in unexpected directions; it acknowledged discrepancies between their expectations and their reality. Such laughter indexed the paradox that the Group Group faced: They were working at a medical center to treat a medical condition, but the best ways to treat it didn't feel very medical at all." - p 168

Ha ha! Well, there, the author's POV & mine practically merge!

"I recognized that making noise, and even in art, making a mess was not something that could happen. Having fun wasn't [something that could happen] in the therapeutic world." - p 172

I direct the reader of this review to "Paper Dolls in Dava's Class": https://youtu.be/x0fkIlvVAl4 .

"Having fun, or at least creating situations in which other people had fun, was, in fact, a general goal of many of the Group Group's interventions. However, this cultivation of pleasure and play often felt out of place in a space whose main work was remediating life's tragedies and misfortunes. On a very practical level, insurance companies frequently refused to pay for interventions that appeared to be "recreational" rather than medical, even though they took place in a medical center and were arguably the best means to remedy a medically diagnosable set of problems." - p 172

How's this for "remediating life's tragedies and misfortunes" by "having fun"?: "Neoist Guide Dog": https://youtu.be/vM92UGWzMPM . I might as well mention that I have the utmost contempt for insurance companies.

"Such perpetual management of potential contagion served as an ongoing reminder that we were in a space of containment and control, a place whose practices are dictated by the logics of diseases in bodies. Interventions to contain transmission conflicted with interventions to encourage generalization and cross-contextual spread." - p 175

Gee, somehow I'm reminded of "Proof of vaccination or recent negative test result required." But, of course, questioning the righteousness of this particular act of medicalization is taboo - punishable by complete shunning & exile.

"By emphasizing [the] medical model, we're putting Band-Aids on it by medicating them, and there's no medication for this. . . . "Oh, Ritalin didn't work." Well, of course Ritalin didn't work. They're inattentive because they're so anxious they can't get through their day, not because they have ADHD. . . . *We've tried everything.* No you haven't. You've tried medical model interventions. . . . The interventions for autism are not medical, they're educational. You are creating different cognitions, and different brain patterns, within them by the teaching." - p 184

"Well, of course Ritalin didn't work." - but what difference does that make? It made profits for the pharmaceutical companies didn't it? That's the bottom line, that's what counts. The rest is all pretense.

"But last week I had a potluck for my clients on the spectrum and their families. And it was at my house.

"I know the people who trained me wouldn't approve. You're not supposed to let patients know this or that about you. But with this population, with people on the spectrum, it's so important to build community. These parents need to be able to go to parties. And the children! One of my teenage patients said to me: are you going to have another party?" - p 186

& this is where it's at, something as basic as a potluck might very well be more important than any medical intervention - but the drug & insurance companies can't profit from it - wch just shows how useless they are.

"At first, Mr. Trent took this as a reflection on Trevor's character. But then the diagnosis, he told me, "gave me an answer to *why doesn't he get it*, especially with things regarding interactions with other people and certain accepted norms, things you do, things you don't do in interacting with people. And so realizing that there was something more behind it, other than just a conscious decision on his part, made it a little bit easier."" - p 190

In other words, an Aspie kid is being an obnoxious brat & his stepdad's bending over backwards to excuse it b/c, you know, it's not the kid's fault, he's sick n'at. Then again, maybe he's just another insensitive jerk & his being an Aspie is no excuse.

"Can he live independently? "He's going to be 18 in October. And the long-standing joke was that when you're 18, you're out of here." Mrs Trent told me. "And I don't see that happening." She worries about how he will fare on his own when he still needs so many reminders to get up in the morning, to go to bed at night, to complete and turn in his work. She worries about how he will get and keep a job in a tight employment market that prioritizes soft skills—communication skills, time management, teamwork—when he doesn't pay attention to things that aren't interesting to him and doesn't seem to get that he shouldn't insult his boss like he insults his peers." - p 191

Oh, don't worry, we'll just socially isolate him & have him work from home. We won't even give him a chance to interact w/ others at all. That'll solve it. All 'for his own good', of course.

But there's still the problem of overvaluing Asperger's syndrome as if it's the only form of intelligence:

"They're very sharp people. They talk about Bill Gates [as having Asperger's]. Edison. Just different people who sit and think. They come up with different ways to deal with things." - p 201

Right, w/o Aspies we might not have sharp people.. or inventors.. or thinkers.. or people who think differently. Gimme a fucking break.

"I have read the political or the philosophical notion that this is wonderful. I mean, this is how you get the Bill Gates and the Einsteins of the world." - p 204

Some might say that we get Bill Gates by having him be the son of an attorney who was the founder of a law firm - in other words, born into a wealthy family & having all the attendant privileges. As for Einstein? Is he being rehistorified to fit the Aspie narrative too? 'Oh, yeah, all the geniuses were Aspies & I'M AN AS-PIE so I'm a genius too (by implication).' 'My little boy isn't a spoiled brat, he's a *genius*!' Whatever.

"Outside of their families, Andrew and Trevor and Thomas will need to earn themselves a place and a social role—a job, friendships, a place in the social community—by meeting high and specific demands for which their characteristics and inclinations might not be a good fit." - p 211

Welcome to the real world. Then again, if you don't fit there make it fit you instead. It's a bit of an up-dam battle but, HEY!, you might not have much of a choice, eh?!

"Many associated autism with the traits most central to their own self-concept. For example, Trevor's social presentation relied heavily on his quirky, random sense of humor. "Everything has to be humorous with me," he told me. "If someone dies, I make it funny."" - p 219

I can relate.. but others might find it hebephrenic.

"Darren, on the other hand, thought of himself as being highly logical. "I look for efficiency in things, you know. How they'll work best. Logical is the sense of, the science of seeing how things will work best." What he remembers the most clearly about Asperger's is "the logical side—you know, more logical than social? I think that was something about it. I mean, that's one thing I've noticed and kind of relate to a bit."" - pp 219-220

Now what wd happen to the above quotes of Aspies if one were to just remove the Aspie references? It seems to me that what we'd have is maybe some teenagers talking about how their personality is developing, they don't really seem any different from anyone else.

"He was already planning his first book: the autobiography of a Revolutionary War-era pirate whose anti-social behaviors, both notorious and thrilling, led Thomas to suspect he may have had autism. When he was speaking about history or his own development as a historian, he was more likely to mention the ways in which autism made people smarter and more capable" - p 223

So smart, in fact, that they're capable of writing the autobiographies of dead people to serve their own self-glorifying agendas.

"In their art, their games, and their informal creative practices, they often imagined themselves as half-human and half-demon, pierced by shards of evil swords, or possessed by powerful ancestors—reenvisioning themselves as mutant, hybrid creatures who contained a force they struggled to manage and control. In doing so, they drew on a shared folk mythology from fantasy literature, comic books, anime, role-playing games, and other such forms of speculative culture." - p 232

I find this mythified inner struggle w/ one's 'demons' to be interesting. I can respect that. Otherwise, I prefer analysis based on socio-economic conditions to fantasies using dragons & swords & the like. As for "speculative culture"? For me, the word "speculative" is likely to be paired w/ "fiction" as an alternative to "science fiction" & it's something very far away from the fantasy alluded to above.

"I never heard these two teenagers talk together about their *symptoms* or their *coping strategies* or their *behavior plans*. When they talked to me about such things, they were shy, a bit abashed—aware of the stigmatizing baggage of these terms and practices. However, the shared folk mythology of the mutant antihero allowed them to reframe what had been a source of shame into a condition whose coexisting extremes were a source of motivation, pleasure, and meaning —both potentially generative and a little bit romantic." - pp 234-235

But wd it be healthy to live in such a fantasy world when they become adults? It'll be fine if other people are taking care of them & they don't have to be realistic. There're plenty of stoners out there in a similar mindset.

"But I saw something very different at Jake's bowling league. Although it was a league by and for adults on the autism spectrum, Jake was very clear to me that the point of the group was not autism, it was *bowling*." - p 247

Redirecting their attn away from introspective pathologizing. I can respect that too.

"Glass Man does not bend but shatters, getting upset over problems others think are unimportant. He gets especially upset, the worksheets tell us, over things that seem "unfair" (scare quotes in original)." - p 249

The thing is, there *are* many things in life that're unfair - say you're highly qualified for a job at work but you see the boss's drug dealer or husband or whatever hired for it instead. That's unfair. People who deny this unfairness tend to be people in denial of their own privilege, it's easier for them to look the other way instead of looking in the mirror.

"The end goal is to create a brain/society that is free from problematically idiosyncratic obduracy." - p 250

Oh, really? Anyone who embraces such a goal might as well openly admit to being a nazi b/c it's pretty fucking transparent. People w/ such goals have this strange notion that they're entilted to control other people's opinions & lives, they're an intellectual elite & they *know what's best for you* - regardless of whether you agree or not. To be opposed to anything idiosyncratic is to be opposed to individualism.

"As I try to figure out what it means for me to be a psychologist, the Society for Qualitative Inquiry in Psychology, a section of the American Psychological Association's Division of Quantitative and Qualitative Methods, has provided an intellectual home both cozy and revolutionary. (Want to join? Find us at sqip.org.) That project has also been nurtured by my fellow members of the Pizza Illuminati" - p 257

Naturally, I find the idea of the "Pizza Illuminati" entertaining. As for what it means for the author to be a psychologist? I hope it means that Fein is interested in observing & studying & caring about human consciousness & *not* about telling other people how to be or think.

All in all, I found this bk to be very stimulating. I only write & quote at such length when I find substantial stimulus to respond to. I'm glad I've started my projected readings in autism here. Of all the bks I plan to read, this is probably the one I'll be able to relate to the most. I truly thank Fein for writing it & acknowledge that I found it sincerely & deeply motivated.

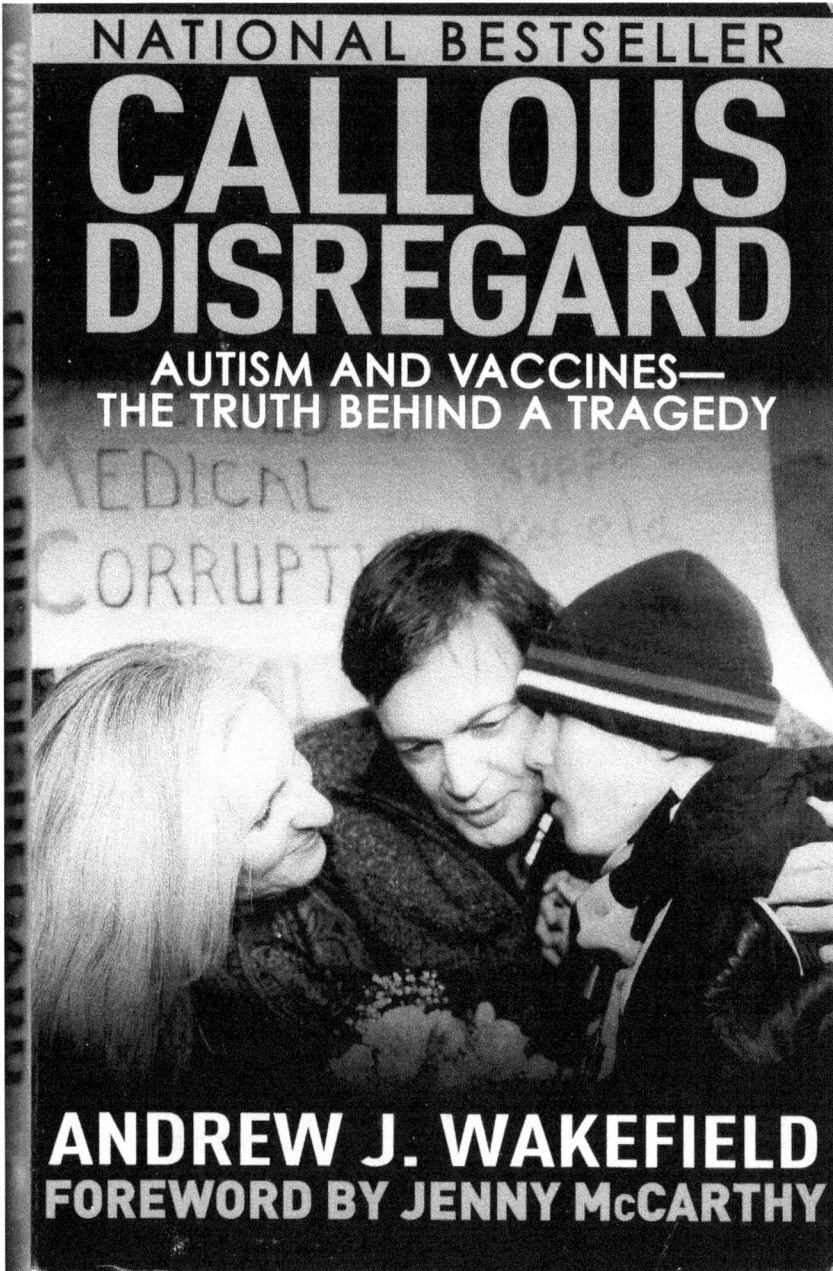

review of
Andrew J. Wakefield's **Callous Disregard—Autism and Vaccines —The Truth Behind A Tragedy**
by tENTATIVELY, a cONVENIENCE - January 12-18, 2022

"Everything's Peachy-Keen: Don't Look, Behind the Curtain"

As far as I can recall, I'd never heard of Wakefield until I started researching in connection w/ the newly announced 'pandemic' in mid-March, 2020. I immediately recognized that the monolithic narrative we were being fed by the mass media was propaganda of a most insidious nature but the details of that took some time to accrue for me. Wakefield was someone who was referred to disparagingly in the mass media as espousing the notion that vaccination causes autism. Ha ha, what a fool. Given that I cd see that a stupendous amt of spin doctoring was going on about everything medical I decided to learn more about what Wakefield himself was saying so that I wdn't just be another fool accepting the word of the FACT CHOKERS & not thinking for myself. Hence I bought 2 bks of his, this being the 1st one I've read. It ISN'T the 1st bk I've read on autism, I didn't want to start w/ him, I decided to start w/ a bk by a friend that I know & trust (w/o necessarily agreeing w/ her 100%) so I read Elizabeth Fein's Living on the Spectrum - Autism and Youth in Community wch I found to be excellent & wch gave me a good introduction to some aspects of the subject.

I really wasn't sure what to expect of Wakefield. On Wikipedia, wch I have both respect for & criticisms of, I find Wakefield disparaged to a degree that shows that the authors of the article feel that they're above a retaliatory lawsuit. The entry begins:

"**Andrew Jeremy Wakefield** (born 1956) is a British anti-vaccine activist, former physician, and discredited academic who was struck off the medical register for his involvement in *The Lancet* MMR autism fraud, a 1998 study that falsely claimed a link between the measles, mumps, and rubella (MMR) vaccine and autism." - https://en.wikipedia.org/wiki/Andrew_Wakefield

Callous Disregard is Wakefield's defense against these allegations, wch are presented as 'fact' in the above entry. In anticipation of reading Wakefield's bk I wasn't really sure of what to expect. It seemed possible to me that I might actually agree w/ the allegations against him & that I might just find him to be a bad doctor, a shallow sensationalism seeker, a fraud. However, I found Callous Disregard to be very convincing &, at the very least, I'm inclined to consider Wakefield as a sincere & knowledgable doctor who's been scapegoated. I also found that Wakefield's position on the matters he's associated w/ is far from what he's presented as having sd. I find it particularly egregious that on the Wikipedia page sidebar he's summarized as "**Known for** Lancet MMR autism fraud".

It's interesting to note that in *The Lancet*'s entry sidebar there's nothing discrediting the journal - this despite their having published at least 3 papers that've been retracted, the most recent of wch was something far more clearly fraudulent than what Wakefield's been accused of:

"In May 2020, *The Lancet* published a metastudy by Dr. Mandeep R. Mehra of the Harvard Medical School and Dr. Sapan S. Desai of Surgisphere Corporation, which concluded that the malaria drugs hydroxychloroquine and chloroquine did not improve the condition of COVID-19 patients, and may have harmed some of them.

"In response to concerns raised by members of the scientific community and the media about the veracity of the data and analyses, *The Lancet* decided to launch an independent third party investigation of Surgisphere and the metastudy. Specifically, *The Lancet* editors wanted to "evaluate the origination of the database elements, to confirm the completeness of the database, and to replicate the analyses presented in the paper." The independent peer reviewers in charge of the investigation notified *The Lancet* that Surgisphere wouldn't provide the requested data and documentation. The authors of the metastudy then asked *The Lancet* to retract the article, which was done on June 3rd 2020."

- https://en.wikipedia.org/wiki/
The_Lancet#Metastudy_on_the_use_of_hydroxychloroquine_and_chloroquine_(
2020)

Now what's particularly noteworthy about this latter scandal is that it seems to've largely slipped past the notice of the general public. Hydroxychloroquine still seems to be 'discredited' even though it's a cheap drug that's been long-since readily available all over the world. It doesn't seem to be far-fetched that Wakefield is so 'discredited' b/c he questions issues of vaccine safety while the fake discrediting of hydroxychloroquine goes largely unnoticed b/c hydroxychloroquine has been presented as being an alternative to COVID-19 vaccination. In other words, in both cases the degree to wch the subject is demonized or swept under the rug depends on the degree to wch it supports Big Pharma's vaccine narrative.

Now, before I go any further w/ this review, I want to give the reader what's my most important take-away from this bk. I make no claim to be entirely accurate in my representation of what Wakefield says, I may be misunderstanding or unintentionally slightly misrepresenting him. Curious readers shd read his words for themselves to get a better understanding. I don't want to add to what appears to be the distortion of Wakefield's public statements. *SO*, w/ that disclaimer of sorts out of the way, here goes:

Wakefield & his fellow doctors got young patients referred to them who were experiencing gastrointestinal distress AND developmental regression that may've been autism. Some of the parents claimed that these problems started shortly after they rc'vd their MMR (Measles Mumps Rubella) vaccinations. Clinical study of the patients suggested that the vaccination in a minority of recipients might possibly be causing gastrointestinal problems & that these problems could be so severe that the children were unable to develop any further in a normal way but were, instead, regressing under the influence of extreme pain & inadequate digestion of nutrients. It was never claimed by Wakefield & co that the MMR vaccine *directly caused autism*. It was, instead, suggested that the connection between the MMR vaccine & the gastrointestinal problems in *some children*, not all, merited further investigation - as did the relationship between the gastrointestinal distress & arrested & regressed development. The patients were mostly very young. Wakefield took the position that single-disease vaccinations were safer b/c they were less of an assault on the body.

The mere suggestion of this possibility was taken as threatening by the medical establishment b/c the possibility of a lawsuit seemed likely. Wakefield was singled out to be scapegoated b/c he had been hired by a lawfirm to do research relevant to side-effects of the MMR vaccine. This hiring did not fund Wakefield's work w/ the patients who were reported on in the eventually retracted *Lancet* paper in wch the observations alluded to above were set forth. Wakefield had several enemies who wanted to discredit him. In one case, a doctor who eventually testified against him claimed that he hated Wakefield b/c Wakefield was trying to get a Nobel prize & was, therefore, skewing the case studies from the very beginning to support a sensational outcome. In other words, Wakefield's enemies had various ulterior motives for scapegoating him - some of wch might've been related to their own financial relationship to the purveyors of the MMR vaccine. B/c the persecution of Wakefield was so intense, many people saved their own ass by turning against him or, at least, distancing themselves.

That's my summary. I find it reasonable. I find it reasonable that some children might suffer side-effects from a vaccination of 3 viruses at once & that these side-effects cd be so severe that development in, say, an 18 mnth old might be so affected by the side-effects that the child's development wd also be severely impeded &/or regressed. I find it predictable that people in the medical industry wd rush to protect themselves against disgrace & lawsuit by scapegoating & disgracing any doctor who might put forth such a hypothesis.

This bk was published in 2011. I don't know what Wakefield is saying now, I'm mainly addressing myself to what I've found in this bk & how it relates to things in my own life. Callous Disregard seems to be basically Wakefield's defense against his so-called 'discrediting'. I still intend to read at least one bk that takes a different position from Wakefield's but I admit that I find Callous Disregard fairly convincing AND I admit that I have an extreme bias against the medical industry & an increasing bias against vaccination in general - mainly as a side-effect of what I consider to be the completely unjustified social pressure in favor of being injected w/ a GMO to ostensibly ward off COVID-19 (something wch, according to personal observations, isn't even working anyway given that my vaccinated friends are still getting more sick than I am).

In the "Acknowledgements" section at the beginning of the bk, there's a website listed: www.callous-disregard.com . I took the precaution to determine if this website is still valid & discovered that it's apparently been domain-grabbed & is now used as a website to promote CBD - this even though the "Callous Disregard" heading is till there. Too bad. Doing a search for "Andrew J Wakefield website" yielded results w/ headings like "the discredited doctor hailed by the anti-vaccine movement". It took my reaching the 5th page of the search results to find a website connected to Wakefield that presents his statements: https://vaxxedthemovie.com/dr-andrew-wakefield-deals-with-allegations/ . Also appearing on the same page of search results is a link to an article entitled "The Lancet Should Reinstate the Andrew Wakefield Paper" by Martin Hewitt (https://

static1.squarespace.com/static/57698667d2b8574e2db7710a/t/
590ef6f7f5e231410337f5ec/1494152953115/
The%2BLancet%2Bshould%2BReinstate%2Bthe%2BAndrew%2BWakefield%2B
Paper+%281%29.pdf).

SO, I'll get to some details:

The front cover of the bk claims that it's a "National Bestseller". I don't know whether to take that as pure hype or if there really is such a big readership for a bk written by a doctor defending himself against charges brought about b/c his case studies showed a possible connection between MMR vaccinations & severe deleterious reactions. It seems like the type of bk parents of autistic children might read in an attempt to get an explanation for their child's condition; it also seems like a bk that doctors might read to determine if they find Wakefield's arguments convincing or bogus.

From the "PREFACE":

"*Letter from Dr. Peter Fletcher, Ex-Principal Medical Officer with responsibility for the UK's Committee on Safety of Medicines and later Senior Principal Medical Officer and Chief Scientific Officer*" - p i

[..]

"My fourth comment relates to the safety evaluation of medicinal products intended for healthy people. The two biggest examples are hormonal contraceptives and vaccines. The differences between the two are mind-boggling. The contraceptives have been evaluated more intensively than any other group of medicinal products both in humans and animals. In contrast, vaccines have been minimally investigated and there seems to be no hope of an improvement in the future." - p ii

From the "FOREWORD":

"As the parent of a child who regressed into autism after his vaccinations, I have always considered Andy Wakefield to represent the kind of doctor and scientist who will ultimately help us end the epidemic of children with autism."

[..]

"I really wish the primary trigger for autism was something everyone could dislike like cigarettes or rat poison. It would make ending this epidemic so much easier. Unfortunately, it appears that a product intended for good—vaccines—also has a dark side, which is the ability to do harm in certain children. This ability to do harm has unfortunately increased quite a bit in the last few decades because children today receive so many more shots than when most parents were kids." - p iii

This latter was written "April 22, 2010". Keep in mind that this was roughly 11 yrs before the FDA gave the EUA (Emergency Use Authorization) for the Pfizer-BioNTech COVID-19 'vaccine' (I prefer putting 'vaccine' in single quotes to question whether this even qualifies as a vaccine given that it's a GMO). The point being that questions & criticisms regarding vaccine safety significantly predated the current raging debate.

Wakefield is usually the one who takes the hit for being the author of *The Lancet* paper but the coauthors included these other Royal Free Hospital School of Medicine doctors: Department of Pediatric Gastroenterology Professor John Walker-Smith, Department of Histopathology Dr. Amar Paul Dhillon, Department of Child and Adolescent Psychiatry Dr. Mark Berelowitz, Department of Neurology Dr. Peter Harvey, Dr. Simon Murch Senior Lecturer, Dr. Andrew Anthony Research Pathologist, Dr. Mike Thompson Consultant, & some junior doctors. (list taken from diagram on p v)

The title is explained in an endnote:

"At the GMC [General Medical Council] I was accused and found guilty of "callous disregard" for the suffering of children." - endnote 1, p 7

Obviously, the title of the bk is intended to show that the "callous disregard" applies more accurately to the medical system that thwarted Wakefield & co's investigations into health problems possibly caused by the MMR vaccine.

"This book is composed of a series of essays that deal with the now infamous paper – a humble case series – written by doctors at the Royal Free Hospital and published in *The Lancet* in February 1998. The essays were originally intended to stand alone and some repetition is inevitable." - p 3

"It is the story of how the powerful deal with threats to their interests. It was recently suggested to me in an interview with a major US network that this was really just a conspiracy theory. As it happened, earlier that week, internal memos from the pharmaceutical giant Merck were disclosed to the Australian court in the Vioxx litigation. They talked about how Merck had to "neutralize" dissent from those doctors who questioned the safety of this drug. In relation to these concerned doctors, one of the emails read:

"*We may need to seek them out and destroy them where they live.*3

"It would seem that rather than being conspiracy theory, this can sometimes be corporate policy." - p 4

"3 Rout M. (2009, April 1). Vioxx maker Merck and Co drew up doctor hit list. Retrieved from http://aftermathnews.wordpress.com/2009/04/27/vioxx-maker-merck-and-co-drew-up-doctor-hit-list/" - endnote 3, p 7

I double-checked to see if the link still works & I'm happy to say it does. Here's the beginning of the article:

"Vioxx maker Merck and Co drew up doctor hit list

"Posted on April 27, 2009
The Australian | Apr 1, 2009
By Milanda Rout

"AN international drug company made a hit list of doctors who had to be "neutralised" or discredited because they criticised the anti-arthritis drug the pharmaceutical giant produced.

"Staff at US company Merck &Co emailed each other about the list of doctors – mainly researchers and academics – who had been negative about the drug Vioxx or Merck and a recommended course of action.

"The email, which came out in the Federal Court in Melbourne yesterday as part of a class action against the drug company, included the words "neutralise", "neutralised" or "discredit" against some of the doctors' names.

"It is also alleged the company used intimidation tactics against critical researchers, including dropping hints it would stop funding to institutions and claims it interfered with academic appointments.

""We may need to seek them out and destroy them where they live," a Merck employee wrote, according to an email excerpt read to the court by Julian Burnside QC, acting for the plaintiff.

"Merck & Co and its Australian subsidiary, Merck, Sharpe and Dohme, are being sued for compensation by more than 1000 Australians, who claim they suffered heart attacks or strokes as a result of Vioxx.

"The drug was launched in 1999 and at its height of popularity was used by 80 million people worldwide because it did not cause stomach problems as did traditional anti-inflammatory drugs.

"It was voluntarily withdrawn from sale in 2004 after concerns were raised that it caused heart attacks and strokes and a clinical trial testing these potential side affects was aborted for safety reasons.

"Lead plaintiff Graeme Peterson, 58, claims the drug caused him to have a heart attack in 2003 after he took it for back pain and arthritis every day from May 2001.

"Merck last year settled thousands of lawsuits in the US over the effects of Vioxx for $US4.85billion ($7.14 billion) but made no admission of guilt.

"The company is fighting the class action in Australia."

Some of us take it for granted that this sort of thing is business-as-usual & find plenty of evidence to support it. Other, considerably more naive people, dismiss such things as 'conspiracy theory'. It's my opinion that such naive people are in denial about the lengths that many people will go to to get to be rich & powerful.

"To the vaccine industry, the regulators, public health officials and doctors, pediatricians, and Bill and Melinda Gates,7 I would say this: the success of vaccination programs requires the willing participation of consumers. Key to any success, therefore, is public confidence in the scientists, doctors, and policy makers (including industry) that shape these programs. In turn, the key to that confidence is a safety first vaccine agenda. Those whose priority is *safety first* are not anti-vaccine. By analogy, those who ordered the recall of multiple Toyota brands for sticking gas pedals are not anti-car." - p 5

"7 Bill and Melinda Gates Foundation (2010, January 29). Bill and Melinda Gates Pledge $10 Billion in Call for Decade of Vaccines. Retrieved from: http://www.gatesfoundation.org/press-releases/Pages/decade-of-vaccines-wec-announcement-100129,aspx." - endnote 7, p 7

That URL did take me to the Bill & Melinda Gates Foundation website but I got this error message: "We're sorry - the page you requested was not found." *SO*, I searched their site for "decade of vaccines". That led to this: https://www.gatesfoundation.org/ideas/media-center/press-releases/2010/12/global-health-leaders-launch-decade-of-vaccines-collaboration :

"NEW YORK -- The World Health Organization (WHO), UNICEF, the National Institute of Allergy and Infectious Diseases (NIAID) and the Bill & Melinda Gates Foundation have announced a collaboration to increase coordination across the international vaccine community and create a Global Vaccine Action Plan. This plan will build on the successes of current work to achieve key milestones in the discovery, development and delivery of lifesaving vaccines to the most vulnerable populations in the poorest countries over the next decade."

Isn't it *touching* how the "poorest countries" are to be the beneficiaries of this 'benevolence'? Now, imagine substituting the words "most defenseless" for "poorest" & make a few other changes & maybe we'll get a taste of reality: 'This plan will build on the successes of current work to achieve new markets for the delivery of experimental vaccines to the most vulnerable populations in the most defenseless countries over the next decade.' That just about brings us to the present, doesn't it?

Alas, I'd have to put Wakefield's prefatory statement that "the success of vaccination programs requires the willing participation of consumers. Key to any success, therefore, is public confidence in the scientists, doctors, and policy makers (including industry) that shape these programs." I think it's been proven fairly solidy that what's required is NOT the "willing participation" & "public confidence" but total intimidation thru mass-media-induced peer pressure & fear of being cut out of social life & access to services & travel. If this fear is induced strongly enuf, & it certainly has been since March, 2020, then even statistics about death from vaccination AND the common evidence of the ineffectual nature of the vaccines for their purported purposes are no longer strong enuf to overcome the terror of being demonized as an 'anti-vaxxer'. B/c of this intense fear-mongering, even pointing out that the CDC's own VAERS (Vaccine Adverse Effects Report System) lists 21,382 deaths from the COVID-19 vaccines as of December 31, 2021 & 36,758 permanent disabilities (https://medalerts.org/vaersdb/findfield.php?TABLE=ON&GROUP1=CAT&EVENTS=ON&VAX=COVID19) isn't enuf to generate from many people anything other than a shrug & some glib comment to the effect that 'people always die from vaccinations' *as if the very commonness of it makes it somehow 'OK'.*

Is it unrealistic of me to think that at one time if a 'cure' was known to not work people might've denounced it as quack medicine? Now, it seems that when people get 'vaccinated' against COVID-19 & get sick *anyway* they go along w/ the unbelievably gullible 'solution' that *they just need to be vaccinated some more.*

"*As CBC News has reported, the government has been settling vaccine injuries that resulted in autism and/or autistic symptoms since at least the early 1990's, while at the same time telling the public there is no cause for concern. Not all of the cases are published, but some of them are and can be found by searching legal case databases.9*" - p 6

"9 Attkisson S. (2008, June 19). Vaccine Watch. CBS News Investigates: Primary Source. retrieved from http://www.cbsnews.com/8301-501263_162-4194102-501263.html." - endnote 9, p 7

Clicking on that link resulted in this error msg: "**The page cannot be found** The page may have been removed, had its name changed, or is just temporarily unavailable." *SO* I tried doing a search on "Vaccine Watch" wch yielded 462 results - the ones immediately visible being about the COVID-19 'vaccines'. I tried specifically looking for anything from June 19, 2008 but it seems that the available material is only from the last 3 yrs max. Too bad.

Chapter One, "That Paper", begins w/ a key explanation.

"On February 28, 1998, twelve colleagues and I published a case series paper in *The Lancet*, a respected medical journal, as an "Early Report." The paper described the clinical findings in 12 children with an autistic spectrum disorder (ASD) occurring in association with a mild-to-moderate inflammation of the large intestine (colitis). This was accompanied by swelling of the lymph glands in the intestinal lining (lymphoid nodular hyperplasia), predominantly in the last part of the small intestine (terminal ileum). Contemporaneously, parents of 9 children associated onset of symptoms with measles, mumps, and rubella (MMR) vaccine exposure, 8 of whom were reported on in the original paper (see also Child PH's story on following page). The significance of these findings has been overshadowed by misunderstanding, misrepresentation, and a concerted, systematic effort to discredit the work. This effort, and specifically the complaint of a freelance journalist and an intense political desire to subvert inquiry into issues of vaccine safety and legal redress for vaccine damage, culminated in the longest running and most expensive fitness to practice case ever to come before the United Kingdom's medical regulator, the General Medical Council. At this point, the guilty verdict is in." - p 9

Here's an excerpt from an article from *The Guardian* about the outcome of the case:

"At the end of the GMC's longest case, lasting 217 days, a disciplinary panel found Wakefield guilty of serious professional misconduct on a number of charges relating to a paper published in the Lancet medical journal in February 1998, some of them related to research ethics and others to financial conflicts of interest and failing to put patients first.

"Young and vulnerable children were subjected, in the interests of research, to invasive medical procedures such as colonoscopies and lumbar punctures which they did not need and the ethics committee of the Royal Free hospital in north London, where Wakefield was based, had not approved. "The panel is profoundly concerned that Dr Wakefield repeatedly breached fundamental principles of research medicine," the GMC said.

"The GMC also struck off Professor John Walker-Smith, 73, who retired 10 years ago as head of the Royal Free's department of paediatric gastroenterology, where the panel conceded he was held in high esteem. The panel accepted he posed no danger to patients and "was and remains a well-respected doctor whose contribution to paediatric medicine has been exemplary". Nonetheless, the GMC said he had failed in the care of vulnerable children and was guilty of "irresponsible and misleading reporting of research findings potentially having such major implications for public health".

"A third doctor, Simon Murch, at the time a junior consultant but now professor of paediatrics and child health at Warwick medical school, was cleared of serious professional misconduct. He had raised concerns about the research project and acted in good faith under the instructions of his superior, Walker-Smith."

- https://www.theguardian.com/society/2010/may/24/andrew-wakefield-struck-off-gmc

I admit to being sympathetic to Wakefield & Walker-Smith. I also admit to finding this statement: "Young and vulnerable children were subjected, in the interests of research, to invasive medical procedures" to be utterly hypocritical & ridiculous since the problem postulated is the result of what I, at least, wd call an "invasive medical procedure": viz: an injection of the MMR vaccine. How does that *not* qualify as an "invasive medical procedure"?!

Wakefield puts a great deal of effort into explaining that he *didn't* make strong unsupported claims against vaccination in *The Lancet* paper but simply pointed out clinical observations made by his fellow doctors & organized by himself that he felt deserved further investigation. Some might see this as back-pedalling & nit-picking but it seems reasonable to me - esp given the incredible ordeal that he & Walker-Smith & Murch were subjected to. Finer points are usually lost on most people but that doesn't make them any less important.

"Hennekens and Buring make the crucial point that the purpose of a case series is to **generate new hypotheses** about potential causation. It is **not** designed to investigate possible causality. *The Lancet* paper was hypothesis generating; it stimulated a series of subsequent papers – rarely if ever acknowledged by critics – that confirmed and characterized the bowel disease as novel, relatively frequent, and potentially treatable and tested ideas about causation." - pp 11-12

Somehow, these fine points, wch Wakefield is capable of pointing out, are overlooked in the apparent frenzy to discredit case studies that might point to Big Pharma & government being culpable.

"Was the research hypothesis clearly stated?
She observes, "The paper does not state a research hypothesis at all." This is quite true. Case series studies are neither required nor expected to do so. Having established that there was no hypothesis, Professor Greenhalgh goes on to pose the ridiculous question:

"Was this design an appropriate way to test the research hypothesis?
She concludes that the study design was not an appropriate way to test "the research hypothesis." However, since she has already identified the fact that no hypothesis was stated, she rather begs the question as to which hypothesis the study was designed to test." - p 12

Much of the reason why I find it so easy to find Wakefield's acct truthful is b/c I've been thru similar experiences myself. In 1992 I gave a presentation to a group of students who clearly found my creative activities repulsive b/c they were, in some cases, sexually explicit, but also politically critical in an anarchist way as well as mocking of bulemics. Not to mention actually creative instead of an unresonable facsimile thereof. I explained to these students that I didn't consider myself to be an artist b/c I find the art context to be an uncreative one that inhibits originality, etc.. SO, what was the student take-away? At least one student sd something to this effect: 'He claims to be an artist but I think he fails at that.' The point is that the student was too incapable of thinking outside the art box that he cd only 'evaluate' my activities in terms that I had explicitly rejected. Wakefield makes the believable claim that he & his colleagues made a "case study" & is then criticized for his poor "research hypothesis". That's like *giving* someone a bicycle & then being prosecuted for *selling* them a car w/o a motor.

Something that I quoted from Wikipedia above is worth requoting:

"**Andrew Jeremy Wakefield** (born 1956) is a British anti-vaccine activist, former physician, and discredited academic who was struck off the medical register for his involvement in *The Lancet* MMR autism fraud, a 1998 study that falsely claimed a link between the measles, mumps, and rubella (MMR) vaccine and autism." - https://en.wikipedia.org/wiki/Andrew_Wakefield

NOW, note what Wakefield has to say:

"It is notable that despite 5 years of investigation by the GMC, no charge of scientific fraud has been made against any of the defendants. The allegation of fraud was made by the same freelance journalist who had actually also initiated the GMC inquiry, continuing his litany of false allegations. There is no evidence at all that the data had been "fixed" as was alleged, and the newspaper in question has failed to produce any, despite a request to do so from the Press Complaints Commission. Paradoxically, the price paid for diligent science has been a headline proclaiming fraud. In my opinion, the intended goal – to reinforce the false belief that the work is discredited – has been achieved." - p 16

Despite my sympathy for Wakefield's claim, my devil's advocate question is: *was there a charge of non-scientific fraud?* I don't mean to say that there was, I'm just trying to point out how tricky language can be.

"I have been excluded from presenting at meetings on the instructions of the sponsoring pharmaceutical company." - p 17

For those of you who are so incredibly naive as to think that drug companies don't exert powerful pressure to insure that their profits are ever on the increase, I highly recommend reading Marcia Angell, M.D.'s The Truth About the Drug Companies - How They Deceive Us and What to do About It. I quote from this bk in my review of it:

""I witnessed firsthand the influence of the industry on medical research during my two decades at *The New England Journal of Medicine*. The staple of the journal is research about causes of and treatments for disease. Increasingly, this work is sponsored by drug companies. I saw companies begin to exercise a level of control over the way research is done that was unheard of when I first came to the journal, and the aim was clearly to load the dice to make sure their drugs looked good. As an example, companies would require researchers to compare a new drug with a placebo (sugar pill) instead of with an older drug. That way the new drug would look good even though it might actually be worse than the older one." - p xviii" - https://www.goodreads.com/story/show/1319242-big-pharma

In this 1st chapter, Wakefield debunks what he calls the "Myths". All of these debunkings are important to Wakefield's defense but I'll just present one:

"*Findings have not been independently replicated*

"**False** – The key findings of lymphoid nodular hyperplasia (LNH) and colitis in ASD children have been independently confirmed in five different countries.25" - p 18

I don't quote the entire endnote 25, just a small sample to give you the idea that Wakefield's claim is substantiated.

"Krigsman A, et al. http//www.cevs.ucdavis.edu/Cofred/Public/Aca/WebSec.cfm?confid=238&webid=1245 (last accessed June 2007) [no longer available; full paper now published below as:

"Galiatsatos P. Gologan A, Lamoureux E. Autistic enterocolitis: fact or fiction. *Canadian Journal of Gastroenterology.* 2009;23:95-98." - excerpt from endnote 25, p 22

Then, under the heading "The legacy of *The Lancet* paper":

First study to seek evidence of a mitochondrial disorder by measurement of lactate:pyruvate in cerebrospinal fluid
"Mito" disorders appear to be common in ASD children and may be acquired. The US government conceded that vaccines triggered autism in Hannah Poling, a child with "mito" disorder.29" - p 19

Wakefield supplies an endnote list of 5 articles about the Poling case. I only quote the 1st one.

"29 Poling J, Poling T. (2008, April 5). Vaccines, autism and our daughter Hannah. *The New York Times.*" - p 22

Here's a fairly typical description of one of the case study children:

"Child 1 had developed normally to 18 months of age and regressed soon after MMR with a clearly delineated onset with loss of words, comprehension, and social interaction plus secondary fecal and urinary incontinence. In his history, the passage of blood and undigested food in his feces provided more than enough indication for ileocolonoscopy." - p 26

These observations didn't only originate w/ the doctors who were eventually persecuted for them:

"In her referral letter to Walker-Smith, Child 3's general practioner (GP) was not so dismissive:

"[Child 3] *developed behavioural problems of autistic nature, severe constipation and learning difficulties after MMR vaccination. The batch incriminated was D1433, incidentally, which was the discontinued batch following adverse reactions.*" - p 29

When a witch-hunt is going on, it's always easy to convert cowards to being testifiers against the 'witch'.

"The early concerns about Child 8's development were reported accurately in *The Lancet*. In spite of this, however, in her evidence to the GMC, Child 8's GP felt able, without having read the paper in detail (as she admitted), to voice her concern to the prosecution that we had reported this little girl's devlopment as "normal." We had not. Despite having no basis in fact, this effectively amounted to an allegation of scientific fraud." - pp 35-36

Endnotes continue to cite reinforcing information:

"In a report of 12 cases in India seen between 1989 and 1998, Malhotra and Gupta note onset in 4 cases following infectious/vaccine exposures, including fever with seizures, acute gastroenteritis and vaccination. The type of vaccine is not stated. See: Malhotra S and Gupta N. Childhood disintegrative disorder. Re-examination of the current concept. *Eur J Child and Adolescent Psych.* 2002:11:108-114" - endnote 18, p 47

"There are also those idosyncratic features of the children's behavior that turned out to be related to gastrointestinal distress. They included posturing, a behavior that often involved leaning for hours at a time over the edge of a piece of furniture that was, as it turned out, done in order to apply pressure to the abdomen and relieve their pain. Such behaviors continue to be misinterpreted as "Oh, that's just his autism," when, in fact, they are entirely appropriate for a child with abdominal pain who can find no other form of expression or relief." - pp 41-42

Once again, this seems entirely reasonable to me. Is it any surprise if an insensitive/stupid/unimaginative adult interprets unusual behavior as some sort of twisted psychological deviance instead of recognizing it as a symptom of pain? Not in my world.

"Cerebellar ataxia was first reported as a possible complication of MMR by Dr. Anne-Marie Plesner in Denmark. This association had not been detected with any other vaccine administered to children of the same age, including the single measles vaccine, indicating that a novel adverse reaction might be associated with the combined MMR vaccine. In a more recent follow-up of the mandatory passive reporting system operated in Denmark, Plesner not only confirmed this association, but also indicated that the more severe ataxias following MMR were associated with residual cognitive deficits in some children. This sounds suspiciously like our own experience." - pp 43-44

Again, this seems reasonable to me. A vaccine that combines THREE illnesses in one might just increase the likelihood of bad side-effects.

"The thing about pattern recognition is that, for the process to be enabled, one has to *allow* the line of enquiry; that is, one has to explore a symptom or pursue an aspect of the past medical history through the narrative maze to a final, considered determination of its significance. There are several constraints on this process, none of which make for good medicine. The first is the view that the "doctor knows best," even for a disease like autism about which so much remains to be discovered. The doctor seems deaf, even hostile, to anything outside his or her specific realm of interest or belief system. The second constraint is the sheer, unmitigated fear of calling MMR vaccine safety into question." - p 44

Not to belabor what I hope is, by now, painfully obvious, but a similar statement applies to COVID-19 'vaccines'. How can "doctor know best" when these 'vaccines' are really GMOs (Genetically Modified Organisms) produced under so-called 'emergency' conditions that allow bypassing ordinary safety precautions? GMOs are still relatively new, the evolution that produced biology as we know it, biology *not interfered w/ by arrogant humans*, has been developed over an unknown number of yrs that can be safely estimated at far beyond our historical accts. As such, the GMOs may have unpredictable effects far more extreme than the combining of 3 viruses into one vaccine - & these effects might show *over generations* rather than just w/in wks of the vaccine's administration. Even from a conventional experimental standpoint, having *controls*, i.e.: unvaccinated people, to compare the guinea pigs to is sensible - so *why so much emphasis on getting everyone in the world vaccinated?!* It's not for our 'safety', that's for fucking sure. PROFIT is a number one suspect. As for "the sheer, unmitigated fear of calling" any COVID-19 'vaccine' unsafe: anyone who's done so has encountered enuf ignorant name-calling to last them more than a lifetime.

"Clinical trials of vaccines and drugs are funded by the manufacturers for the principle purpose of profit. This is not a judgment or a criticism, but an economic reality for an industry that is answerable first and foremost to its stockholders. It appeared to me that, unspoken, Zuckerman was proferring the ethical paradox of medical academia endorsing – indeed embracing – the conduct of clinical trials funded by the pharmaceutical industry, but denouncing as something distasteful and prohibitively conflicted17 the investigation of children whose lives may have been irreparably damaged by an inadequately tested vaccine." - p 51

"17 In 1997 while expressing his concern about such conflict with regard to LAB funding for a scientific study, Zuckerman wrote an editorial for the *New England Journal of Medicine (NEJM)* strongly supporting universal use of hepatitis B vaccination of the world's infants. At that time, he was a named inventor on a patent related to the hepatitis B vaccine and, as such, might have stood to benefit from the policy endorsed in his editorial. At the time, in 1996, strict rules existed for the *NEJM* barring patent holders from writing editorials." - endnote 17, p 64

BUST*eeeddddd*! But such inconsistencies didn't prevent Zuckerman from being one of Wakefield's persecutors. Now might be a good time to bring up the idea of *corruption*.

"I wrote to Zuckerman advising him that because of my concerns about a possible MMR-autism connection, I had proactively arranged a meeting with representatives of the JCVI for the purpose of communicating these concerns. The JCVI is a committee in the UK charged with offering independent advice to the DoH on vaccines and their safety. In recent years, it has been revealed that – far from being independent – many of the committee members have links to various pharmaceutical companies in the forms of grants and consultancies." - p 55

"One question that had taxed Zuckerman – the funding of research where there was a "clear financial interest in the outcome" – was dealt with succinctly: Armstrong wrote:

"...*funding of research by special interest groups is commonplace and as long as the findings, or uses to which the data is put, are not influenced by the wishes of the funder, this should not be problematic.*

"As a final slap in the face to Zuckerman, Armstrong concluded that

"...*to delay or decline to conduct research which appears to be in the public interest on the grounds that it may embarrass the government or a particular health facility does not appear to be a sound moral argument.*" - p 57

Now everything I quote here is obviously taken out of context in the interest of not just retyping the entire bk. Zuckerman has written to Armstrong hoping to get back-up from him in his manoeuvering to thwart Wakefield. Armstrong thwarted Zuckerman instead. Wakefield explains elsewhere that funding that he was getting *did not represent a conflict-of-interest*. Nonetheless, Armstrong's saying that the "uses to which the data is put, are not influenced by the funder" is a little strange for Wakefield to quote given that earlier he'd written that "Zuckerman was proferring the ethical paradox of medical academia endorsing – indeed embracing – the conduct of clinical trials funded by the pharmaceutical industry" - in other words, he's calling out Zuckerman's hypocrisy (fine) but also trying to use a similar attitude to defend his own taking of funding in a potential conflict-of-interest situation (even tho he demonstrates that the conflict-of-interest didn't actually take place). It's these gray areas that sometimes make this bk tricky for me.

An important part of Wakefield's presentation is that the MMR vaccine used in the UK was known to be dangerous & was still pushed thru under Big Pharma pressure. A whistleblower then provides the substance for this claim.

"George had moved to the UK from Canada, where he had worked as a principal program immunization advisor in Ontario and was closely involved with the MMR vaccination program. He had been actively recruited to join the Scottish Office as a senior medical officer, particularly to advise on evolving UK MMR vaccination policy.

"At the time, a brand of the MMR vaccine had been withdrawn in Canada because it was unsafe; Canada had introduced an MMR vaccine containing the Urabe AM-9 mumps virus (*Trivirix*) in 1986. It soon became apparent that this vaccine caused unaccceptably high rates of meningitis. George was invited to sit on the UK's Joint Committee on Vaccination and Immunization (JCVI) as its Scottish representative. In Canada, he had collated extensive information about the MMR hazard; therefore, he was in an excellent position to advise the DoH and, particularly, Dr. David Salisbury, the UK's chief MMR strategist, on the introduction of the MMR. George claimed:

"*...he was very cautious not to openly release the damning evidence that he had retained from Canada regarding the safety of the MMR vaccine...*" - pp 65-66

I suggest that people in positions of high authority tend to reduce the humans that they're responsible for to statistics & to, therefore, have a high acceptance of collateral damage b/c it becomes largely dehumanized for them - but what's an *acceptable* rate of meningitis to someone who gets it as a result of a vaccine that's supposed to *protect their health* but, instead, has the opposite effect? Obviously, there is no *acceptable* rate of meningitis to its victim. Think of the military: the lower the rank of the soldier, the more their deaths are considered to be 'acceptable' collateral damage. Obviously, this is class warfare. These same 'expendable' people are also used in scientific experiments by exposure to Agent Orange, Depleted Uranium, Chemical & Biological Warfare Agents, etc.. Do you really think that the high & mighty decision makers, who aren't, personally, taking risks themselves, are that much different in the Medical Industry from their counterparts in the Military?! I don't. ESPECIALLY when they stand to gain from it financially - & let's not forget to throw in potential sadism.

"Ominously, having withdrawn *Trivirix* in Canada, SmithKline Beecham (SKB) simply repackaged the identical vaccine ingredients as *Pluserix* for use in the UK, apparently irrespective of the risk to children,

""*...there was a determination to sell a* [cheaper] *MMR vaccine from SmithKline French Beecham* [*sic*]," a British company.

"George stressed:

""*...at the time they announced the MMR vaccination campaign, Pluserix was not actually licensed to be used in* [the UK]. *Nonetheless, the Government had certainly internally announced that it was going to be used and its licensing procedure was rushed through on a 'fast track' basis... safety trials were circumvented to allow the [Pluserix] vaccine to be licensed and widely used.*" - p 67

Do you think I'm just being cynical to find the above easy to believe? That a drug withdrawn in Canada for safety reasons wd be renamed & sold in the UK so that SKB can recoup their expenses? That UK government officials wd assist SKB in doing this b/c they're a British company & b/c they're offering a 'deal'? That safety concerns wd be brushed aside b/c it's only 'insignificant collateral damage'? & let's not forget the possibility of kickbacks, the possibility of backroom deals between the good ole boys & girls. Alas, it's all-too-easy for me to believe, esp these days, that the Medical Industry's concern for patient safety often runs only 'skin deep'.

"George's expert concerns were ignored. In fact, the high risk *Pluserix* vaccine was to take the lion's share of the UK MMR market. In light of its known dangers, one would have expected that vigilant surveillance of adverse events would have been put in place. George and the rest of the Scottish Committee strongly advocated for such surveillance and

"...*pressed for an increase in funding to allow more active surveillance.*

"Active surveillance (as opposed to passive surveillance, which is awaiting spontaneous reports of adverse reactions from doctors) involves the prospective ascertainment of adverse reactions through active canvassing of data from primary care doctors. George was involved in trying to set up some sort of surveillance system, but no money was given him to do this. According to George, the money had to come through a senior member of the medical secretariat who controlled millions of pounds to implement the vaccination program but was very resistant to spending any money on monitoring safety." - p 70

"Clearly he felt somewhat guilty that "*instead of going to the top and exposing all this, he did not do that at the time, but stayed quiet because he had children to support at University.*" - p 71

The high cost of going to university, at least in many 'Western" countries has become yet-another manipulative racket. In a recent trial in Pittsburgh, a person found guilty of stealing some rare things in his care got a lighter sentence than the person he sold them to b/c he had daughters to put thru university. Apparently that pathetic excuse tugged at the judge's bourgeois heartstrings. It's my observation that when one pays $40,000 or so a yr to attend university, esp if one is funding it w/ school loans, one is putting oneself in a trap from wch one then has to have a certain type of job to even have the illusion of possible escape. Sadly, one becomes so entrenched in the system, esp if one is paying off those school loans, that even after one has 'advanced' one is trapped by habit in a system one might've preferred to stay out of.

"It was of greater concern to George that he was a signatory to the Official Secrets Act, which forbade him from taking home copies of official documents. He was worried that if it were found out that he had been in breach of this Act, he would likely go to prison." - p 72

Look at the torment & persecution that the heroic Julian Assange is being put thru. Even tho he's not even a US citizen, our 'leftist' (**haHAhaHaaaahhhhhaaaaa, ha!!!**) president, Bidentity Crisis, is still prosecuting the guy. Once upon a time, 'Leftism' supposedly stood for Free Speech & respected Whistle-Blowers. No more, now it's all about the Police State, 'for our own good', of course.

"The vaccine manufacturers and JCVI members had reasonable fears that they might be liable, and SKB, for their part, appear to have been given a "Get Out of Jail Free" card by Her Majesty's Government. Confirmation of this was later to appear in the JCVI minutes of May 7, 1993, where it states:

"*...SKB continue to sell the Urabe MMR without liability.*10

"The vaccine proved to be unsafe for UK children, just as it had for other children around the world. Permanent damage to children was denied and this denial continues. Years later, Salisbury was to deny any knowledge of an indemnity (initially, at least) in spite of the fact that the JCVI minutes are unambiguous and entirely consistent with George's firsthand experience." - pp 73-74

"10 JCVI minutes of May 7, 1993." - p 76

& what about Informed Consent? It's nice in theory but practically nonexistent in practice. I had a friend whose aunt was arrested for something trivial or another. She was put in a mental hospital. The doctor there wanted to give her a lobotomy but needed the consent of a family member so he called the aunt's mother to get it. The problem was that the mother was a Polish immigrant & barely spoke English. The doctor presumably became aware of this in his conversation w/ her but he had *deniability* & he wanted to do his little mad scientist experiment on the hapless captee so he got his permission & ruined the woman for life. & got away w/ it, of course. Informed Consent isn't gotten by actually informing people, it's gotten by conning them into thinking they understand what's going on when they don't.

"*Informed consent is a crucial element of the foundation upon which ethical medical practice rests. Providing patients, parents, or guardians with an honest assessment of the risks and benefits of any medical procedure requires the physician to be, to the best of his or her ability, "informed."*"

[..]

"an experimental vaccine combination – measles and rubella (MR) – that was administered to approximately 8 million UK school children over a 1-month period in November 1994. The justification for the campaign was a mathematically-predicted measles epidemic. The principal architects of the campaign were Salisbury and his boss, Dr. Kenneth Calman, the UK's Chief Medical Officer. Through an intense and frightening advertising campaign,3 parents were motivated to get their children revaccinated by the threat of up to 50 deaths from measles." - p 77

"3 Minutes of JCVI meeting May 5, 1995; 6.5 para 2. *HEA measles/Rubella campaign report: The HEA* [Health Education Authority] *did acknowledge the view that the TV advert used had been a little frightening, and also that not enough information on the possible side-effects of the vaccine had been provided for some people.*" - endnote 3, p 82

& when they can't con the public into thinking they're "informed" there's always fear, that works even more reliably. & that's where we're at today w/ COVID-19 'vaccines'. How many people who've gotten vaccinated &/or, worse yet, forced their vulnerable children to get vaccinated, understand this?:

"How mRNA Vaccines Work

"To trigger an immune response, many vaccines put a weakened or inactivated germ into our bodies. Not mRNA vaccines. Instead, mRNA vaccines use mRNA created in a laboratory to teach our cells how to make a protein—or even just a piece of a protein—that triggers an immune response inside our bodies. That immune response, which produces antibodies, is what protects us from getting infected if the real virus enters our bodies.

"1   First, COVID-19 mRNA vaccines are given in the upper arm muscle. The mRNA will enter the muscle cells and instruct the cells' machinery to produce **a harmless piece** of what is called the spike protein. The spike protein is found on the surface of the virus that causes COVID-19. After the protein piece is made, our cells break down the mRNA and remove it.
"2   Next, our cells display the spike protein piece on their surface. Our immune system recognizes that the protein doesn't belong there. This triggers our immune system to produce antibodies and activate other immune cells to fight off what it thinks is an infection. This is what your body might do to fight off the infection if you got sick with COVID-19.
"3   At the end of the process, our bodies have learned how to protect against future infection from the virus that causes COVID-19. The benefit of COVID-19 mRNA vaccines, like all vaccines, is that those vaccinated gain this protection without ever having to risk the potentially serious consequences of getting sick with COVID-19. Any temporary discomfort experienced after getting the vaccine is a natural part of the process and an indication that the vaccine is working."

- https://www.cdc.gov/coronavirus/2019-ncov/vaccines/different-vaccines/mrna.html

Why that's so simple that even a child might understand it, right? Well.. maybe not, maybe it's too simple & maybe some of its claims bear examination. E.G.: It's stated that "mRNA vaccines use mRNA created in a laboratory to teach our cells how to make a protein": I ask *IS it possible that this 'teaching' isn't 100% effective? Or, to put it a different way, IS it possible that this 'teaching' isn't really teaching at all but is, instead, an attempt to generate a reflex reaction that ISN'T universal in effect?* It seems to me that the word "teaching" is, in fact, a metaphor - if we were to take it literally we might imagine a sentence like this: 'Today I taught my class how to write the word "stupid".' Does that mean that *everyone* in this class actually learned that? I think not, everyone is different & has different circumstances effecting their ability to learn at any given point in time. Even if they're not 'stupid' they might be tired or distracted or restless or whatever. What if there're similar viccissitudes in every BODY? What if not every cell being 'taught' has exactly the same reaction?

One of the 'facts' that appears later in this CDC article is that "mRNA vaccines do not use the live virus that causes COVID-19 and cannot cause infection with the virus that causes COVID-19 or other viruses." What they somehow neglect to mention is that the subject can still have COVID-19 itself enter their bodies, after all the purpose of the vaccines is to combat it once it's there, so I, at least, wonder why can't the virus then leave their body in the ordinary way & spread in the ordinary ways?

Another claim that's made in the CDC article is that "mRNA vaccines are <u>safe</u> and <u>effective</u>." Oh, really? If that's true why does the CDC's very own VAERS website (Vaccine Adverse Effects Report System) list *so many adverse reactions*?! Cd it be b/c there's no way the Medical Industry can know what effect something will have on *everyone's* health w/o actually trying it out on *everyone* - wch just reduces us all to guinea pigs? So much for safety. As for "effective"? Gee, do you know *anyone* who's gotten sick even tho they're vaccinated? I sure do. So I think we can scratch off "effective" too.

NOW, apply such questioning to ALL medical interventions. To me, they're ALL suspect.

"To make matters worse, the same information leaflet claimed that reactions were expected to be less likely with the booster dose. I am not aware of any evidence that anaphylaxis is less likely or less severe on re-exposure. The data from New York with MCV boosters would suggest quite the opposite." - p 79

People seem to be generally reluctant to question the Medical Industry. There seems to be an inclination to believe that anything that's been developed over 100s of yrs must be on 'the right track' if only b/c it's persevered for so long. Ever the HERETIC I ask whether it's possible that basic premises of medical practice might be based in a questionable philosophy. Premises that, if demonstrated to be faulty, wd be undermined to the extent that the whole edifice collapses. As I understand the way our immune system evolves it's a constant process that has built on experience over a vast amt of time in every being. Scientific research is paltry in contrast - even on millions of people over hundreds of yrs. To me, the faulty premise is the completely arrogant conceit that a small number of humans can adequately understand & *generalize* from this evolution enuf to 'improve' on it. I'm not opposed to people *trying* as much as I'm opposed to the egomania that gloats over a false success. If one doubts or outright rejects this conceit, as I do, then all medicine that follows from it is suspect.

"Trying to persuade parents of the merits of an MR campaign on the basis of up to 50 possible measles deaths while ethically warning them of the possibility of up to 14,337 anaphylaxis deaths from the MR vaccine would have doomed the campaign to failure. In my opinion, parents were deliberately frightened by a powerful advertising campaign to get their children revaccinated with an experimental vaccine in an untested mass revaccination strategy. The outstanding question is whether or not a deliberate decision was taken not to warn of the risks of anaphylaxis." - p 80

&, once again, that's where we are today. When one judges the merits of something based on whether or not it 'wins' scruples tend to fall by the wayside. Advertising psychologists who are given the task of making as many people as possible terrified of not getting vaccinated are likely to lose sight of whether that's really such a healthy thing to succeed at. I had a friend who was a propagandanist for the Shah of Iran. He used the classic justification: 'If I didn't do it somebody else would have.' &, gee, I'm sure the cushy living he made off of it didn't exactly rub him the wrong way either. Might as well ignore SAVAK, the secret police who were torturing dissidents.

"I approached the idea of a press briefing with some anxieties; by this stage, I had examined the issue of MMR vaccine in great detail. I had reviewed all of the published scientific literature about measles and MMR vaccine safety studies on the basis that, as part of investigating parental concerns and before calling into question MMR vaccine safety, it was essential to have done so. On a personal level, I was dismayed that I hadn't done this research before vaccinating my two older children. On a global level, it was clear that the safety studies had been wholly inadequate. "George" the whistleblower had major concerns about the attitude of many of his public health colleagues toward concerns over vaccine safety. The forced withdrawal of MMR vaccines that had been "spun" as being *completely safe* was testament to their failings." - p 83

"*In view of this, if my opinion is sought, I cannot support the continued use of the **polyvalent** MMR vaccine. I have no doubt of the continued use of the **monovalent** vaccines, and will continue to support their use until the case has been proven one way or another of the measles link to chronic inflammatory bowel disease.*" - p 84

Again & again & again, I find Wakefield's position to be completely reasonable. A "polyvalent" vaccine, one that combines 2 or more diseases, might be too strong for its recipient's body to reflexively react to in the targetted way. Zuckerman, a figure in authority over Wakefield in the hospital hierarchy informs him that:

"*You support the continued use of the **monovalent** vaccines and you write that you have no doubt of their value. To my knowledge this has not been repeated by the media... It is vital, in your own interest and that of children, that you clearly state your support for **monovalent** vaccination.*" - p 86

Alas, as Wakefield's position became more under fire, Zuckerman protects his ass w/ denialism.

"*Coonan: Professor Zuckerman*"

[..]

"*you told them that you did not know that Dr Wakefield would suggest the use of monovalent vaccines in place of the MMR vaccine.*

Zuckerman: That is correct.

Coonan: That is correct, is it? You did not know that?

Zuckerman: I knew that he held that view. I knew that the press briefing was to be restricted to the pathological changes in the gut. The issue of vaccines was not relevant. I was reassured on this by Professor Pounder, I was reassured on this by Dr Wakefield" - p 92

Well, if Wakefield is assumed to be telling the truth (after all, this is *his bk* & he is likely, therefore, to present himself in a good light) then Zuckerman is lying. I'm more inclined to believe Wakefield b/c his presentation of the matter fits better into my own experiences w/ human nature. Otherwise, I have no firm basis for believing *anybody* that I don't know personally.

"*Zuckerman:*" [..] "*Let me just qualify this. It was an exchange of correspondence between Dr Wakefield and I where he wrote to me, assuring me that he had confidence in **polyvalent** vaccines.*" - p 93

Now, it's always possible that Zuckerman's mind was a bit fuddled, perhaps his memory was bad, maybe he wasn't just lying. Maybe it was time for him to retire. Then again, maybe it was time for him to take a stand w/ some integrity to it, maybe he failed to do so.

"Zuckerman was a man of conflicting agendas: on the one hand his role as dean had been to support academic freedom. On the other, he was put under considerable political pressure from the Department of Health and World Health Organization (WHO) in respect to my legitimate research into vaccine safety (see endnotes 2, 3, 5 and 7 of Chapter 3, "The Dean's Dilemma" and endnote 8 of this chapter). While initially using the autism work to promote the medical school in the media, it rapidly became a decaying albatross about his neck. In a rather clumsy *volte face* before the GMC hearing, Zuckerman's imperfect memory was exposed." - p 98

Wakefield was accused of conflict-of-interest & he makes a good case there were other conflicts-of-interest that were behind his persecution.

"Crispin Davis, Chairman of Reed-Elsevier, *The Lancet's* proprietor and Horton's boss, which took place at the UK goverment's Licence and Technology

Committee on March 1, 2004. Referring to the Rouse/Wakefield exchange, Davis told Harris:

"*You can imagine that it is virtually impossible for every editor to research every single author in terms of conflict of interest, and in this one Dr Wakefield said there was no conflict of interest, and in fact three months later in written form repeated that there was no conflict of interest. In all fairness, I do not hold our editor to blame.*

"As an aside, readers may be interested to know that on July 1, 2003, Crispin Davis was made a non-executive director of Glaxo SmithKline, one of the largest pharmaceutical companies in the world, manufacturers of MMR, and one of the codefendents in the MMR litigation with which I was assisting the LAB. In the summer of 2004, Davis was knighted for his services to the information industry." - p 111

Imagine how much it's worth to be a "non-executive director" of a major pharmaceutical company. As of 2021, the average salary of a director of GSK is $180,747 per year according to this job-seekers website: https://www.indeed.com/cmp/Glaxosmithkline/salaries/Director . I, at least, wonder what Davis had to do in his role as director to earn this salary.

Wakefield further establishes that the editor of *The Lancet* lied about his knowledge or lack thereof regarding Wakefield's connection to lawyers seeking redress for autistic children hypothetically damaged by the MMR vaccine.

"In a nutshell, therefore, it is clear that, as a matter of fact, Horton *was* aware of the law firm Dawbarns, *was* aware of Mr. Barr the lawyer and of his central role in the MMR litigation on behalf of Dawburns, *was* made aware of my relationship with Barr and my involvement in the litigation – and *all* of this happened one year before the paper's publication. He was reminded of these matters in the Rouse letter and in my response, and was provided with full references to these facts – facts that were never secret – just one working day after the paper was published in February 1998. He was reminded once again by Laurence of *The Independent* newspaper in 2004. Despite all of this, Horton has claimed repeatedly in print, on radio and television, through the law firm Olswang, and under oath upon the witness stand at the GMC, that until 2004 he knew nothing of these matters, claiming instead that he took my response to Rouse to mean that the agreement to work with Barr had started *following* publication of *The Lancet* paper." - p 124

Now some of you might be asking: 'If Wakefield established his innocence & the dishonesty of those testifying against him so well why was he eventually decided against by the GMC?' My reply to such a hypothetical question is: 'When the fix is in, it's in.' I've had personal experience seeing photographic proof presented of something contrary to the prosecution's completely unproven assertion - the judge just ignored the evidence & in his concluding statement acted like it wasn't even there. 'Guilt' was a foregone conclusion.

We come to CDD (Childhood Disintegrative Disorder) & the issue of its relationship to autism.

"The combination of these atypical features along with the fact that, for the majority, there was onset followed by an infectious (vaccine) exposure, led our colleagues in the Department of Child Psychiatry at the Royal Free Hospital to suggest that what we were dealing with was not Kanner's autism, but childhood disintegrative disorder".

[..]

"CDD is a pervasive developmental disorder that fulfills behavioral criteria for childhood autism/autistic disorder, but where the pattern of onset is different. CDD requires documented normal or near-normal development up to 24 months of age with subsequent regression and loss of skills in at least two of the following: expressive/receptive language, play, social/adaptive skills, continence, and motor skills". - p 134

Wakefield & co decided that CDD & autism are not easily distinguished from each other & have the same core symptoms. Rutter, an expert witness against them, expresses a contrary opinion.

"For Rutter, as a key prosecution witness at the GMC hearing, however, the matter was black and white. When asked whether "in embarking on a study of children with behavioral disorder, would [he] expect a distinction between CDD and autism to be made," he replied, "Yes." He continued, "and the literature would support drawing a clear distinction at the time [1996]." It is somewhat surprising, therefore, to find that he had earlier written that "The clinical picture [in CDD] after the phase of regression is often similar to autism and the differentiation may be difficult, **if not impossible**, in cases with an onset before 30 months." It is notable that regression and onset before 30 months applies to virtually all of *The Lancet 12*." - pp 135-136

The **emboldening** in the above is Wakefield's. Now, obviously, Rutter's testimony conflicts w/ his earlier written statement. There are many possible explanations for this disparity: he might've changed his opinion, he might've forgotten that he ever had the earlier opinion, he might've sd whatever he felt fit the agenda of the hearing regardless of whether it also fit legitimate medical findings. Another way of looking at such things is that people who're called on to testify are expected to really be on top of their game, they have to be CONVINCINGLY EXPERT - regardless of whether that's true or not - their reputation depends on it. I think that people in such positions become accustomed to bluffing, to acting, they have their 'dignity', their reputation, to protect. The more esoteric the subject, the less likelihood of contradiction. That makes it easier to be a bullshit artist. To me, doctors are 2nd only to lawyers in bluffing - although I reckon politicians & career artists are high up there too.

""If autism is a consequence of vaccination it should have been a consequence of natural infection"

"Paul Offit, in interview with Melanie Howard, *Babytalk* magazine" - p 137

I think that's a fair enuf statement up to a point - but Wakefield's primary complaint is that a *polyvalent* vaccine is what's the cause of the trouble. THAT's not a "natural infection". Wakefield goes on to counterargue that autism *has been* connected to natural infections.

"In support of a causal role for prenatal measles in autism, Ring et al., used sophisticated statistical modeling of the number of autism births in Israel compared with epidemics of measles, rubella, poliomyelitis, viral meningitis (inflammation of the lining of the brain) and viral encephalitis (inflammation of the brain) and found that peaks in the number of births of children with autism followed peaks of epidemics of measles and viral meningitis.

"The authors concluded that "Autistic birth patterns are partially explained by the rates of measles and viral meningitis [incidentally a frequent feature of measles] in the general population. There is a statistically significant environmental association between autism and both viral meningitis and measles that should be further investigated."" - p 139

So there. The obsessive amt of work & research that Wakefield must've gone thru in order to create the arguments in this bk signifies to me that he's not likely to be a fraud. IMO few frauds wd ever go to so much trouble, they'd be moving on to a simpler game.

Wakefield concludes chapter 9 w/ this:

"It is proposed that autism and CDD are on the same continuum of clinical disease. Measles virus exposure has been linked to both CDD and autism. The timing of this exposure – i.e., early (*in utero*) or later, in childhood – may determine the clinical presentation, including the presence and extent of regression. Infantile autism without regression may be linked to early exposure, whereas CDD with regression may be linked to later exposure. It is entirely plausible that measles, in combination with two other viruses which have themselves been linked independently to autism – as MMR – may increase the risk for this condition in certain children. Whether or not MMR is guilty as charged remains to be determined." - p 140

Yes, "Whether or not MMR is guilty as charged remains to be determined" but will the manufacturers of the MMR vaccines under scrutiny even allow their investment to be put at risk?! Not if they have any say-so - & their money talks BIG.

Chapter 10 begins:

"I have often wondered where autism might be today had it not fallen into the hands of child psychiatrists. Would things have been very different if, for example, the first child with autism presented before Drs. Gilles de la Tourette,2 Joseph Babinski,3 and Pierre Marie4 at one of the Tuesday lectures of the great French neurologist Professeur Jean-Martin Charcot?5 I think so. Charcot, with his supreme diagnostic skills and clinical intuition, would, I believe, have deferred to his medical training rather than being influenced by the emergent psychoanalysts in his audience." - p 143

"2 Georges Albert Édouard Gilles de la Tourette (1857-1904) was a French neurologist who described what became known as *Tourette syndrome* in nine patients in 1884 as "*maladie des tics.*" Charcot gave the syndrome the eponymous title "*Gilles de la Tourette's illness.*"

"3 Joseph Jules François Félix Babinski (1857-1932) was a Polish neurologist. In 1896, he described what became known as Babinski's sign, a pathological plantar reflex (upward movement of the toes elicited by stroking the sole of the foot ("*phenomène des orteils*") indicative of central nervous system (*corticospinal tract*) damage.

"4 Pierre Marie (1853-1940) was a French neurologist who, among other things, described a disorder of the pituitary gland known as acromegaly, an overproduction of growth hormone leading to "giantism."

"5 Jean-Martin Charcot (1825-1893), "the Napoleon of the neuroses," was a French neurologist and professor of anatomical pathology. He is one of the pioneers of neurology, and his name is associated with at least 15 medical eponyms, including joint manifestations of neurosyphilis, Charcot-Marie-Tooth disease, and amyotrophic lateral sclerosis (Lou Gehrig's disease)." - p 163

I particularly like that Wakefield refers to "clinical intuition" since I think that intuition is an underappreciated, but highly vital quality these days.

"Had autism existed in late 19th century Paris, it would doubtless have been described. But it seems these men were unaware of autism as were other equally eminent European and American physicians of the time. Notwithstanding Theodore Heller's report of CDD in 1908, it was not until 1943 that child psychiatrists first laid claim to autism." - p 143

This is my question too: *Did autism start during a particular time in the 20th century?!* Given that I think that many of the things that plague humans the most are human in origin, but not always recognized as such, I tend to look for those human origins. The 20th century was a time of unprecedented change. In the arts, I've loved those changes. Still, the 'improvements' in warfare are something I detest. It's easy to find problems created by things like the horrific & unforgiveable atomic bombings of Hiroshima & Nagasaki - but what about more subtle changes brought about by other human acitivities? Some people claim that the so-called Spanish Flu pandemic of 1918-1919 was primarily something that affected *vaccinated* soldiers, soldiers being used as guinea pigs, as usual. This wd've been further complicated by the devastation of the environment by trench warfare.

"Immunization with three live, modified viruses given together by injection at a much younger age than is typical for natural infection is most certainly an *unusual circumstance.*" - p 145

As usual, that seems reasonable to me, so reasonable that it seems like common sense. In other words, while it might be exciting for medical researchers to try out a triple whammy it might not be as safe as they want you to believe.

I reckon there's an ongoing debate in many areas about what's more important: genetics or environment. I'm sure both are important & that dismissing one in favor of the other is foolish. However, the argument goes beyond foolish if something happens like a person gets their arm cut off on their job - no-one's going to claim that that was exclusively caused by genetics. The problem is, that in some cases the cause isn't so easy to pinpoint. Let's say that for some people there's a genetic disposition to being harmed by a particular food - then that person shd avoid that food. Alas, w/ vaccines there seems to be a tendency to act like they're equally helpful to almost everyone (some medical exemptions are acknowledged). What if there're more people who have negative reactions to vaccines than the vaccine makers & endorsers are willing to admit?

One of the areas where I'm probably the most sharply divided from Wakefield is the appropriateness of using spinal taps, Lumbar Punctures (LP), to get data from a patient. Wakefield &/or his colleagues performed some LP on some patients in order to get data. I had a friend who had 8 LPs performed on her as a child, probably in the 1970s. She sd it was the most painful thing she ever experienced, that she cdn't move for a long time after the LP was performed b/c moving was agonizing. Her description of it made it seem absolutely horrible. As I understand it, Wakefield & co's use of LP on their patients was one of the things they were criticized for. As a person who generally rejects invasive procedures I'm inclined to agree.

"The role of LP in autism is more contentious than for CDD, and expert opinion is sharply divided. In fact, debate over the merits of this procedure reflects, in some ways, the larger debate over the priority of genetics or environment in this disorder." - p 150

"Christopher Gillberg, a professor of child and adolescent psychiatry and autism expert from Sweden, advocates LP in the routine clinical investigation of children with autism, and in his hands, when specific hypothesis-testing studies have been performed on CSF, these have consistently identified differences between children with autism and non-autism controls that support the likelihood of an underlying organic pathology." - p 151

Wakefield provides numerous endnotes directing the reader to studies using LP that revealed important things about autism. These, hypothetically, justify the procedure. Truthfully, I'm not buying it. Admittedly, I'm highly biased by my friend's description of what getting a spinal tap was like but it seems like such an extreme procedure that, to me, it seems like another potential cause for iatrogenesis, for harm done *by* physicians. My attitude *against* vaccines is largely rooted in a larger rejection of *injections* in general. The older I get the more suspicious I get of these industrial strength interventions in the body.

"In this study, measles virus genetic material was present in CSF from 19 of 28 (68%) cases and in one of 37 (3%) non-autism controls. Further tests confirmed that where there was sufficient amount of sample available, the genetic material was consistent with having come from the vaccine virus.

"The draft paper concludes by saying,

"*The data indicate that virological analysis of CSF is indicated in children undergoing autistic regression following **exposure** to live vaccine viruses.*

"The Paper's conclusions stop well short of any claim that the MMR vaccine causes autism. The most one can say from the findings of measles viral genetic material in CSF is that there is a strong statistical association between the presence of this virus and the autism group." - p 158

As the reader has probably deduced by now, "CSF" = Cerebrospinal Fluid. As the reader has probably also deduced, the emboldening of **exposure** is the author's.

On to the 1st paragraph of chapter 11, "Disclosure":

"I have been accused and ultimately found guilty of professional misconduct for not disclosing in *The Lancet* paper that I was a medical expert involved in assessing the merits of litigation against the manufacturers of MMR on behalf of plaintiff children possibly damaged by the vaccine. Notwithstanding the fact that – long before publication – details of my involvement as an expert in the litigation had been provided to my senior coauthors, the dean of the medical school, and the editor of *The Lancet*, it is a matter of fact that it was not disclosed in the publisher paper." - p 169

Some, or much, of this bk borders on legalese - in other words, technical fine points that are potentially valid but still have the stink of lawyer-manipulation-speech about them. This essentially applies to both the charges against Wakefield, wch are all-too-easy to write off as an attempt to discredit something that might lead to exposure of both Big Pharma & the UK's medical establishment as not living up to their responsibilities, AND to Wakefield's defense of himself. My inclination is to favor Wakefield even when I find some of the logic tricky. Wakefield argues that the disclosure rules of *The Lancet* did not require the particular type of disclosure that he's accused of not presenting at the time of publication. These rules then changed to be stricter in the yrs that followed.

"What have been the practical consequences for this move to stricter disclosure requirements from 1998 to 2007 for *The Lancet*?"

[..]

"With the stricter rules in place, between 1998 and 2007 the rate of disclosures per *Lancet* article went from one in two hundred to more than one in two articles." - p 170

OK, that's obviously a huge & very significant difference.

Now, one of the main people testifying as an expert witness in Wakefield's case is a guy named Rutter. Wakefield demonstrates that Rutter has his own conflict-of-interest.

"As referred to above, Rutter was a paid expert in at least three separate litigation projects on behalf of vaccine manufacturers and US government; in these projects, he was to offer an expert opinion that thimerosal-containing vaccines and MMR do not cause autism"

[..]

"Rutter served as a defendants' expert in US litigation where the plaintiffs alleged that mercury (thimerosal) in vaccines caused autism." - p 172

Well, well.. I admit that I have a bad attitude toward so-called 'expert witnesses' & I just consider them to be people for sale to the highest bidder . My opinion is reinforced by the above. I'm one of those people who also has a bad attitude toward mercury - viz b/c I remember when mercury contaminated fish caused harm to humans in Japan in what was probably the 1960s or 1970s. Here's the beginning of an abstract from an article explaining that mercury is harmful:

"Acute or chronic mercury exposure can cause adverse effects during any period of development. Mercury is a highly toxic element; there is no known safe level of exposure. Ideally, neither children nor adults should have any mercury in their bodies because it provides no physiological benefit. Prenatal and postnatal mercury exposures occur frequently in many different ways. Pediatricians, nurses, and other health care providers should understand the scope of mercury exposures and health problems among children and be prepared to handle mercury exposures in medical practice. Prevention is the key to reducing mercury poisoning. Mercury exists in different chemical forms: elemental (or metallic), inorganic, and organic (methylmercury and ethyl mercury). Mercury exposure can cause acute and chronic intoxication at low levels of exposure. Mercury is neuro-, nephro-, and immunotoxic. The development of the child in utero and early in life is at particular risk. Mercury is ubiquitous and persistent. Mercury is a global pollutant, bio-accumulating, mainly through the aquatic food

chain, resulting in a serious health hazard for children." - https://
www.ncbi.nlm.nih.gov/pmc/articles/PMC3096006/

Note that this article states that "Mercury is a highly toxic element; there is no
known safe level of exposure." W/ that in mind, one has to wonder about
Rutter's fluffing off any claims about any harmful effects of mercury in vaccines.

"The GMC's ethical guidance for doctors in this case, which is posted as "Acting
As An Expert Witness – Guidance for Doctors," requires that an expert witness
give honest testimony:

"*If you are asked to give evidence or act as a witness in litigation or formal
inquiries, you must be honest in all your spoken and written statements. You
must make clear the limits of your knowledge or competence.*

[..]

"James Moody, Esq., an attorney acting pro bono on behalf of several autism
organizations, has written to Rutter, bringing to his attention the fact that these
instances of lack of disclosure are to be put in the public domain and offering him
the right of reply. The e-mail was sent on March 6, 2010, and a response was
requested by March 12, 2010. The respective journal editors have also been
contacted. At the time of going to press – April 28, 2010 – no reply has been
received." - p 176

In the interest of fairness, it seems to me that 6 days response time might not be
enuf for many busy people.

"this essay is about what amounts to, in my opinion, hypocrisy, double-standards,
and professional retribution dressed in sanctimonious piety." - p 177

That seems about right, at least from Wakefield's side of the story.

The main person to cause all this trouble for Wakefield is a journalist named
Brian Deer. The next chapter is dedicated to refuting his claims. This must've
been particularly painful & difficult for Wakefield to get into since Deer obviously
really has it in for him.

"*The dynamite in The Lancet was the claim that their conditions could be linked
to the MMR vaccine, which had been given to all 12 children.*

"False: *The Lancet* paper did not "claim that their conditions could be linked to
the MMR vaccine." No such claim was ever made in the paper; on the contrary,
it was explicitly stated in that paper that *no association* – let alone a causal
association – had been proved between MMR and the syndrome described. It
reported only that the parents said onset of symptoms started after MMR
vaccination in 8 of the 12 cases." - p 186

Endnote 80 on p 221 includes an open letter from 8 of the 12 parents with this concluding paragraph:

"*We have been following the GMC hearings with distress as we, the parents, have had no opportunity to refute the allegations. For the most part we have been excluded from giving evidence to support these doctors whom we all hold in very high regard. It is for this reason we are writing to the GMC and to all concerned to be absolutely clear that the complaint that is being brought against these three caring and compassionate physicians does not in any way reflect our perception of the treatment offered to our sick children at the Royal Free. We are appalled that these doctors have been the subject of this protracted enquiry in the absence of any complaint from any parent about any of the children who were reported in the Lancet paper.*"

That seems particularly significant. The parents of the children who were under Wakefield & co's care respect the doctors. The complaints originate w/ a journalist. Journalists frequently seek sensational 'scandals' to boost their careers. Was that Deer's case? What cd really justify this persecution?

"*Vaccination is designed to protect the majority, and it does so at the expense of a minority of individuals who suffer adverse consequences. Although the case against MMR is far from proven, it is one that we are obliged to investigate in view of the consistent history given by those patients' parents and by the observations made in the United States. If this disease is caused by the MMR vaccination, then these children are the few unfortunates that have been sacrificed to protect the majority of children in this country. If this is the case, our society has an absolute obligation to compensate and care for those who have been damaged by the vaccine for the greater good. This is an inescapable moral imperative and is the principle reason that I have decided to become involved in helping these children pursue their claims.*" - pp 206-207

We're back to the idea of collateral damage. To my mind it's not acceptable to even put a minority at risk in order to, ostensibly, protect the majority. If that's what vaccines do then they're unacceptable. Doctors & pharmaceutical companies, legally, are rarely at risk for the negative consequences of their vaccines. Imagine such an exemption applied in any other circumstance where death or disabling harm are a possible result: 'Welcome to our restaurant. Fewer than 1% of 1% of our customers die or are permanently brain-damaged from food poisoning. We all have to eat. This establishment is not legally responsible for any negative consequences. By the by, we're also vvveeerrrryyyyy expensive - so be prepared to pony up big time.'

Another person working against Wakefield was Professor Tom MacDonald, a scientist studying bowel disease.

"The GMC vetted MacDonald as a potential witness against me and his erstwhile colleagues. The attendance note of his meeting with GMC lawyers in 2005 reads:

"*He* [MacDonald] *believes Wakefield is a charlatan, who has been pursuing his own agenda since 1995, this being to win the Nobel Prize. He believes Wakefield's alleged link between measles vaccine and Crohn's was entirely fabricated in order to obtain publicity for this reason.*" - p 210

"He may also have failed to disclose conflicting agendas, one scientific (as above) and one personal; as related to me by John Walker-Smith, when MacDonald declined the invitation to transfer to the Royal Free with Walker-Smith, he had reportedly vowed to his boss to destroy my career. Deer has been useful to him in that respect." - p 211

It seems to me that something is missing from this depiction of Deer & MacDonald - viz: Why do they hate Wakefield so much? What this hatred reminds me of the most is moments from my own life when people have attacked me, ostensibly for one reason, not b/c of the ostensible reason but b/c they were jealous that I'd had sex w/ someone they wanted to have sex w/ but cdn't. Does that seem petty? Sure it is.. but it's all too human. Maybe Wakefield was popular & Deer & MacDonald weren't. It just seems to me that there's more going on here than the bk addresses.

"Consider the recent revelation during the course of Vioxx class action hearings: the publishing house Elsevier (owner of *The Lancet* and over 500 other medical and scientific titles) created six fake journals that were dressed up to look like scientific journals, funded by Merck without any disclosure, and strongly favorable to Merck in their content.8" - p 228

"8 The allegations, reported in theScientist.com (http://www.the-scientist.com/blog/display/55679/) involve the *Australasian Journal of Bone and Joint Medicine*, a publication paid for by pharmaceutical company Merck that amounted to a compendium of reprinted scientific articles and one-source reviews, most of which presented data favorable to Merck's products. *The Scientist* obtained two 2003 issues of the journal – which bore the imprint of Elsevier's *Excerpt Medica* – neither of which carried a statement obviating Merck's sponsorship of the publication." - endnote 8, pp 230-231

Here's something I found online:

"Scientists from the pharmaceutical giant Merck skewed the results of clinical trials in favor of the arthritis drug, Vioxx, to hide evidence that the drug increased patients' risk of heart attack.

"To increase the likelihood of FDA approval for its anti-inflammatory and arthritis drug Vioxx, the pharmaceutical giant Merck used flawed methodologies biased toward predetermined results to exaggerate the drug's positive effects. Internal documents made public in litigation revealed that a Merck marketing team had developed a strategy called ADVANTAGE (Assessment of Differences between Vioxx And Naproxen To Ascertain Gastrointestinal tolerability and Effectiveness) to skew the results of clinical trials in the drug's favor. As part of the strategy, scientists manipulated the trial design by comparing the drug to naproxen, a pain reliever sold under brand names such as Aleve, rather than to a placebo.

"The scientists highlighted the results that naproxen decreased the risk of heart attack by 80 percent, and downplayed results showing that Vioxx increased the risk of heart attack by 400 percent. This misleading presentation of the evidence made it look like naproxen was protecting patients from heart attacks, and that Vioxx only looked risky by comparison. In fact, Vioxx has since been found to significantly *increase* cardiovascular risk, leading Merck to withdraw the product from the market in 2004." - https://www.ucsusa.org/resources/disinformation-playbook

Now that's an article credited as published by the "Union of Concerned Scientists". The article's entitled "Merck Manipulated the Science about the Drug Vioxx" & it's from Oct 12, 2017. Superficially, it looks genuine. But do I really know that it is? I just did a search for "Merck Vioxx" & this is what I picked. Now I've quoted from it here - but for all I know it cd be fake. *SO*, I do a search for "Union of Concerned Scientists" & enuf info about them appears to reassure me that they're for real.

HOWEVER, the idea that there were "six fake journals" created to favor a big pharmaceutical company doesn't strike me as preposterous as it might to some. In the process of writing my bk <u>Unconscious Suffocation - A Personal Journey through the PANDEMIC PANIC</u> I came across multiple articles written by one "Eric Dolan". An example is one entitled "Psychopathic traits linked to noncompliance with social distancing guidelines amid the coronavirus pandemic" & it contains this paragraph:

""I knew that traits from the so-called Dark Triad (narcissism, Machiavellianism, and psychopathy) as well as the traits subsumed within psychopathy are linked to health risk behavior and health problems, and I expected them to be implicated in health behaviors during the pandemic. There is also prior research suggesting that people high on the Dark Triad traits may knowingly and even deliberately put other people's health at risk, e.g., by engaging in risky sexual behavior and not telling their partner about having HIV or STIs," Blagov told PsyPost."

Now to someone looking for confirmation that anyone who resists mask wearing & social distancing is a monster, this article's perfect & it *appears* to be in an online psychology magazine - thusly potentially validating it. Well, a little research revealed this:

""Eric is the founder, publisher, and editor of PsyPost." (https://www.psypost.org/author/edolan). A superficial search revealed that he's basically a gun-for-hire:

"Eric W. Dolan - clearvoice.com

"Feed your marketed channels the content they need with talented writers from ClearVoice. Hire vetted writers in more than 200 business categories. Power all your content marketing. All in One Place. Easy to Use. DIY or Fully Managed. Articles & Blogs, Ebooks.

"However, clicking on the link didn't work. He's also listed on "Muck Rack""

- p 723, Unconscious Suffocation - A Personal Journey through the PANDEMIC PANIC

Now, whether *PsyPost* has any claim to legitimacy or not I don't currently know. It still has web presence but neither Dolan nor *PsyPost* have Wikipedia entries. For that matter, neither do I. The point is that I find Dolan's motives suspicious & I'm not convinced that his articles that conveniently appeared at a time when the propaganda push for 100% belief in the quarantine was on big time were anything but motivated-by-profit gun-for-hire snipes at resistance to this propaganda. Where did the funding for these articles come from? Just as Merck wd fund fake journals to make them look good, cdn't someone have funded Dolan's articles for a similar purpose? Pure speculation might posit the Democratic Party. I cdn't, personally, stand Trump – but that doesn't mean that I'd endorse dirty politics used to get rid of him. But, I digress..

Wakefield & co were, if I understand correctly, under suspicion by the GMC of using the 12 children studied for *The Lancet* paper as *research subjects* rather than as patients. A part of the argument against them was that they'd used the wrong permissions. Wakefield demonstrates that this isn't true:

"The fact that the files of *The Lancet* children all contained the 162-96 EC consent for research biopsies and neither the EC-approved information sheet for the 172-96 research study nor the associated consent form makes confusion impossible. The GMC allegation and finding that *The Lancet* children were enrolled in EC project 172-95 is objectively impossible as shown by the consent forms in each child's record." - p 238

Making matters even more confusing is that, as far as I can tell, what's meant by "162-96" in the above is *actually "162-95"* & what's meant by "172-95" is *actually "172-96"*. Brazil, here we come!

"The GMC did not find Professor Walker-Smith, the senior clinician, guilty of dishonesty. The panel must have accepted the integrity of his position on the clinical merits of these tests both in 1996 and now. The documents confirm not only this doctor's position on the clinical need for investigation and the inadequacy of previous investigations on these children, but make it clear that the Royal Free EC, the dean, the CEO, and even Sir David Hull knew all along that autistic children with bowel disease were being seen and investigated acording to clinical need. In 1996-98 senior medical staff knew what was being done and what approval was in place." - p 242

"Perhaps this whole GMC case has not been an honest effort to protect patients but politically motivated scapegoating after all?" - p 243

"Perhaps this is just one part of an ongoing campaign to stop research into the safety of MMR and vaccines on the one hand, and on the other to conceal the appalling refusal of the NHS to provide proper care for autistic children with severe GI problems, which is itself an egregious violation of basic medical ethics." - p 244

"In the US, increasingly coercive vaccine mandates and fear-mongering advertising campaigns are a measure of your failure" - pp 247-248

That was written in 2011. Wakefield's point being that public confidence in vaccines has become an increasingly forced issue. The fear campaign that's on in relation to getting the COVID-19 'vaccines' (over & over) demonstrates his point. W/o massive fear campaigns I suspect that most people wdn't've become so sheeplike in lining up for injections of dubious value & of a highly experimental nature.

"With the issue of vaccine safety in mind, in 2001 I got together with Dr. Laura Hewitson, an outstanding researcher, at that time at the University of Pittsburgh. With colleagues, we designed a study that should have been done many years before. We set out to examine the safety of the US vaccine schedule — starting with the hepatitis B vaccine (HBV) given on day one of life through to pre-school boosters. The first paper – the first of many – reported delayed acquisition of survival reflexes (e.g., feeding) in infant monkeys after the day one HBV shot (containing mercury preservative). Following rigorous peer-review and online publication in *Neurotoxicology*, the paper was withdrawn, not apparently, on the instructions of the scientific editor, but by the publishing company Elsevier. The links between this company and the pharmaceutical industry have been reported by Mark Blaxill in one of his excellent pieces for *Age of Autism*. Science, it would seem, is available to the highest bidder." - p 248

&, alas, it's my observation that more or less *everything* is available to the highest bidder. I, on the other hand, *have not been so far in my life & I've made it to 68 enduring some extremes of poverty*. As such, I trust MY opinions more than most.

In the "Postcript", its author, James Moody, Esq, has this to say:

"With a history including the clandestine meeting between US regulators and industry at the Simpsonwood Retreat Center in Norcross, GA. in 2000;8 highly biased and ineffectual Institute of Medicine "reviews" of vaccine safety in 2001 and 2004; and a deeply flawed string of "studies" reminiscent of the junk epidemiology once used to defend tobacco safety, all as water under the bridge, the CDC eventually admitted that its epidemiology was flawed9 and in a leaked media strategy document, that it did not have the science to dispel safety questions of "anti-vaccine" challengers.10" - p 268

"8 See http://www.safeminds.org/government-affairs/foia/simpsonwood.html

"9 Kirby D. (2009, August 5). CDC: Vaccine Study Used Flawed Methods. *Huffington Post.* retrieved from http://www.huffingtonpost.com/david kirby/cdc vaccine study used fl_b_108462.html.

"10 Moody J. (2009, August 5) CDC Media Plan Shocker – We Don't Have the Science – Some claims aganst vaccine cannot be disproved. *Age of Autism.* Retrieved from http://www.ageofautism.com/2009/08/cdc media plan shocker we don't have the science some claims against vaccine cannot be disproved .html" - endnotes 8-10, p 270

Now, you'll notice that the links provided in the last 2 endnotes have spaces that aren't ordinarily in URLs. I assume this is a mistake. I'll check all 3 links to see if any of them are still any good.

http://www.safeminds.org/government-affairs/foia/simpsonwood.html : The "Safe Minds" website exists but the link didn't lead to the searched-for article. *SO,* I tried a search on their site for "Simpsonwood Retreat Center". No luck. "Norcross, GA". Nada. "clandestine meeting between US regulators and industry". Also nada. Did the article ever exist? I don't know.

http://www.huffingtonpost.com/david kirby/cdc vaccine study used fl_b_108462.html : This URL worked, contrary to my expectations, but led to this:

"**Editor's Note**

"A previous blog post published on this site has been removed in the interest of public health. The article expressed the sole opinion of its author, who retains the rights to publish it elsewhere. Multiple studies have demonstrated that vaccines are safe and effective. Our letter from the editor has more on this decision."

I clicked on the "letter from the editor" link:

"**Letter From The Editor: Why We Are Removing Anti-Vaccine Blogs**

By Lydia Polgreen
06/21/2019 11:09am EDT I **Updated** June 21, 2019

"Open platforms have an extraordinary power to allow communities to come together in new and exciting ways. But in an age of misinformation, we have learned that they also allow falsehoods to spread unchecked and inaccuracies to jeopardize the common good.

"This is one of the reasons HuffPost decided at the end of 2017 to shut down our contributor platform. I noted at the time that "open platforms that once seemed radically democratizing now threaten, with the tsunami of false information we all face daily, to undermine democracy." As a news organization, we needed to shift from a model where anyone could access and distribute content on a platform with our name on it, to a model in which content was commissioned by our editorial staff."

- https://www.huffpost.com/entry/letter-from-the-editor-vaccines_n_5d0cefd6e4b0aa375f4ba1c6?9ok

SORRY (NOT), but I just don't buy it. We're NOT "in an age of misinformation" any more than we've ever been. Lying has been w/ us for a very long time & it's always been a primary tool for far more people than wd ever be admitted to. Think of the concept of "Yellow Journalism":

"**Yellow journalism** and **yellow press** are American terms for journalism and associated newspapers that present little or no legitimate, well-researched news

while instead using eye-catching headlines for increased sales. Techniques may include exaggerations of news events, scandal-mongering, or sensationalism. By extension, the term *yellow journalism* is used today as a pejorative to decry any journalism that treats news in an unprofessional or unethical fashion."

[..]

"The term was coined in the mid-1890s to characterize the sensational journalism in the circulation war between Joseph Pulitzer's *New York World* and William Randolph Hearst's *New York Journal*. The battle peaked from 1895 to about 1898, and historical usage often refers specifically to this period. Both papers were accused by critics of sensationalizing the news in order to drive up circulation, although the newspapers did serious reporting as well."

http://www.ageofautism.com/2009/08/cdc media plan shocker we don't have the science some claims against vaccine cannot be disproved .html : That link worked to take me to the initial page of the "Age of Autism" website. *SO*, I did a search for just "cdc media plan shocker" & this led to the correct article (here's the correct link: https://www.ageofautism.com/2009/08/cdc-media-plan-shocker-we-dont-have-the-science-some-claims-against-vaccine-cannot-be-disproved-.html - Note that it has the correct hyphens that one wd expect).

Moody's conclusion is that:

"Dr. Wakefield is the "Ralph Nadar of vaccine safety." He is the latest in a long line of scientists who use the power of the scientific method to challenge establishment orthodoxy. Thanks to Galileo, Semmelweiss, Needleman, and McBride, we now take it for granted that the Earth is not the center of the Universe, that doctors must wash their hands to prevent the spread of infection, that lead impairs neurological development in children, and that thalidomide causes birth defects. And now, thanks to Dr. Wakefield, we will one day have safer vaccines and better treatments for autism." - p 269

I tend to agree - but don't take my word for it, read something by Wakefield & something that refutes Wakefield - that's what I'm doing. For the moment, I'm on Wakefield's side.

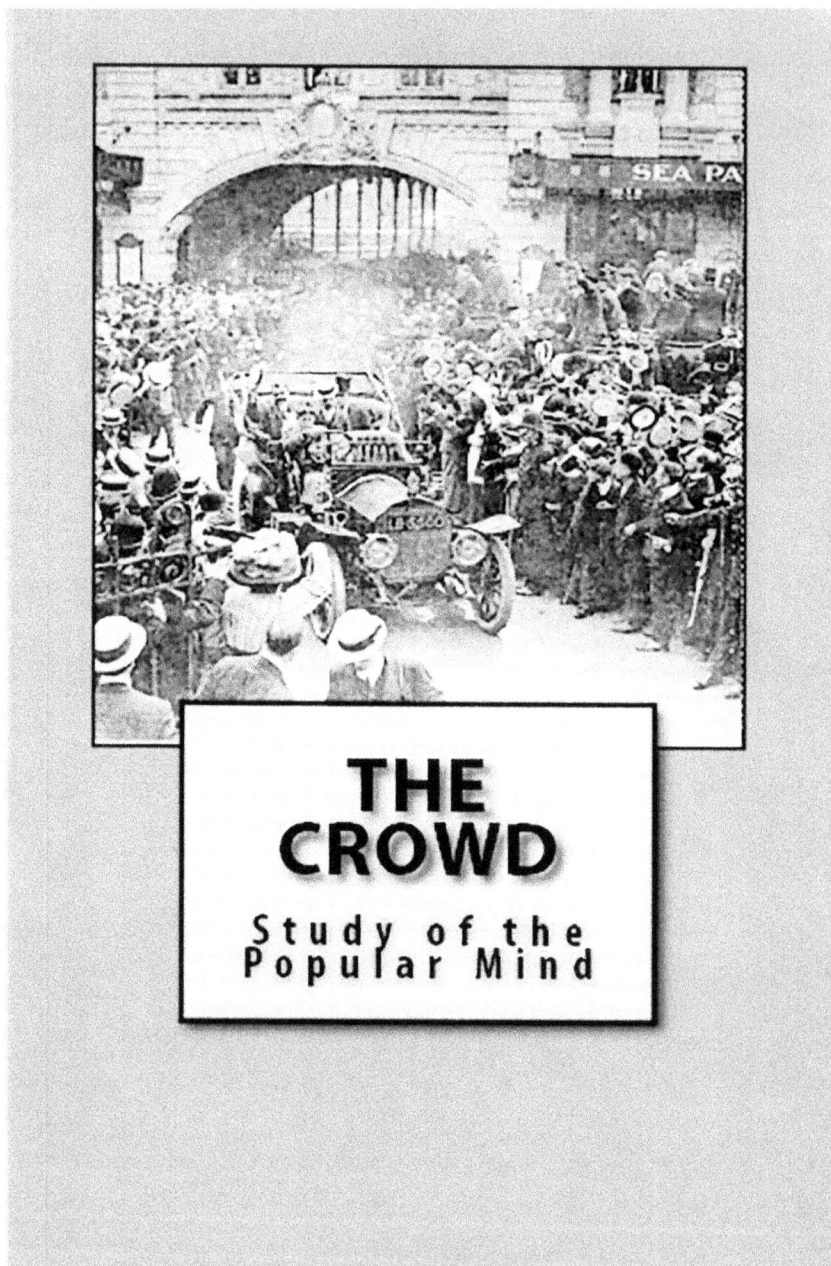

review of
Gustave Le Bon's **The Crowd - Study of the Popular Mind**
by tENTATIVELY, a cONVENIENCE - February 21-27, 2022

I got a copy of this bk b/c it was apparently influential on the idea of "mass formation". Since I now consider most people to be hopelessly conformist sheeple I'm interested in what crowd psychology observations have been made that predate this current hellish age. As such, I was predisposed to find any study of crowds as of interest. Still, it was w/ some trepidation that I read the description of Le Bon on the back cover:

"Gustave Le Bon (7 May 1841 - 13 December 1931) was a French social psychologist, sociologist, and amateur physicist. He was the author of several works in which he expounded theories of national traits, racial superiority, herd behavior and crowd psychology."

"racial superiority"?! What was I getting myself in for?! As it turned out, Le Bon's meaning of "race" seems to be more what we'd call "nationality" these days & doesn't seem to be based on skin color. Still, how does the "superiority" factor in? Well, in this bk, it doesn't seem to factor in at all so I'm relieved by that. IMO any notion of "superiority" except for, perhaps, the technical superiority of functioning of a machine designed for a specific purpose, is bound to cause more trouble than it's worth - esp when applied to people & peoples.

As w/ many publishers in this day & age of taking things from public domain & putting them out using POD printers, this particular edition is rather, ahem, bizarre since it claims "Copyright © 2018 All rights reserved." & "Your support to the author's rights is appreciated. First Edition". Ahem, the original was written in 1895, the author is long since dead. The publisher might be copyrighting this particular translation, there's no translator credited, MAYBE it's a "First Edition" of that. Whatever the case may be, the publisher's claims on this bk as represented by what I quoted above are very tenuous. The publisher is so pathetic that they also state on the same page:

"Except in United States of America, this book is sold subject to the condition that it shell no, by the way of trade or otherwise, be lent, resold, hired out, or otherwise circulated without publisher's prior consent in any form or binding or cover other then that in which it is published and without similar condition including this condition being imposed on the subsequent purchaser."

Wow, I won't even say 'Nice try' sarcastically. This copyright page appears to be the ONLY page that the publisher actually had to go to the trouble of writing - &, yet, they cdn't even get this tiny bit right. They're so clearly not literate people, they're so clearly just trying to make money off of someone else's labor & trying to own it that it's disgusting. "that it shell no"?! I do believe they were intending to write 'shall not' but that wd've taken being able to think past their bank acct. Maybe they cd've really strained their intellect & gone for 'Except in **the** United States of America'. & how about "other then that"? Perhaps they cd've striven to learn a little more about the English language, to not just rely on their spellcheck. THEN they might've written 'other than that' - but, no, they're just in it for the money, apparently. Of course, there's no publisher name on the bk. This seems to be the common way of things w/ these scammers, it makes it harder to pinpoint who, exactly, they are. I wonder if they ever even read the bk that they 'publish'. It seems just as likely, if not more so, that they just learned about the bk thru a college course &/or thru some online search for bks in the public domain. Oh, well.

The "Preface" begins w/ this paragraph:

"The following work is devoted to an account of the characteristics of crowds. The whole of the common characteristics with which heredity endows the individuals of a race constitute the genius of the race. When, however, a certain number of these individuals are gathered together in a crowd for purposes of action, observation proves that, from the mere fact of their being assembled, there result certain new psychological characteristics, which are added to the racial characteristics and differ from them at times to a very considerable degree." - p 5

There's that race thing again. Here're further relevant quotes to bear out my assertion that race = nationality in Le Bon's usage:

"We showed in a previous volume, what an historical race is, and how, its character once formed, it possesses, as the result of the laws of heredity such power that its beliefs, institutions, and arts — in a word, all the elements of its civilization — are merely the outward expression of its genius. We showed that the power of the race is such that no element can pass from one people to another without undergoing the most profound transformations." - p 41

"The assemblage of dissimilar units begins to blend into a whole, to form a race; that is, an aggregate possessing common characteristics and sentiments to which heredity will give greater and greater fixity. The crowd has become a people" - p 100

In Le Bon's preface he establishes himself as a subculture of one. I respect that, it's a better way than most to exercise impartiality.

"In a recent publication an eminent thinker, M. Goblet d'Alviela, made the remark that, belonging to none of the contemporary schools, I am occasionally found in opposition of sundry of the conclusions of all of them. I hope this new work will merit a similar observation. To belong to a school is necessarily to espouse its prejudices and preconceived opinions." - p 5

Moving on to the "Introduction":

"The age we are about to enter will in truth be the *Era of Crowds*.

"Scarcely a century ago the traditional policy of European states and the rivalries of sovereigns were the principle factors that shaped events. The opinion of the masses scarcely counted, and most frequently indeed did not count at all. To-day it is the traditions which used to obtain in politics, and the individual tendencies and rivalries of rulers which do not count; while, on the contrary, the voice of the masses has become preponderant." - p 8

It seems that he's referring to "the dictatorship of the proletariat", but maybe not quite.

"In Marxist philosophy, the **dictatorship of the proletariat** is a state of affairs in which the proletariat holds political power. The dictatorship of the proletariat is the intermediate stage between a capitalist economy and a communist economy, whereby the post-revolutionary state seizes the means of production, compels the implementation of direct elections on behalf of and within the confines of the ruling proletarian state party, and instituting elected delegates into representative workers' councils that nationalise ownership of the means of production from private to collective ownership. During this phase, the administrative organizational structure of the party is to be largely determined by the need for it to govern firmly and wield state power to prevent counterrevolution and to facilitate the transition to a lasting communist society." - https://en.wikipedia.org/wiki/Dictatorship_of_the_proletariat

Either way, it seems that the proletariat ultimately doesn't have much of a say b/c a 'higher authority' is allowed to run the show. The times when the crowd &/or proletariat seems to have the most say is when they run amok &, temporarily, override state authority. It all seems like a mess to me. I'm in favor of the individual, 1st & foremost, by wch I mean a true individual, a person who thinks for theirself. The problem, for me, w/ crowds, is that the mob mentality takes over - wch means the LCD (Lowest Common Denominator), wch is rarely any more than brutishness. But let's see what Le Bon has to say about the matter, eh?:

"History tells us, that from the moment when the moral forces on which a civilisation rested have lost their strength, its final dissolution is brought about by those unconscious and brutal crowds known, justifiably enough, as barbarians. Civilisations as yet have only been created and directed by a small intellectual aristocracy, never by crowds. Crowds are only powerful for destruction. Their rule is always tantamount to a barbarian phase. A civilisation involves fixed rules, discipline, a passing from the instinctive to the rational state, forethought for the future, an elevated degree of culture — all of them conditions that crowds, left to themselves, have invariably shown themselves incapable of realizing. In consequence of the purely destructive nature of their power crowds act like those microbes which hasten the dissolution of enfeebled or dead bodies." - p 10

I agree w/ much of the above. I don't think I agree w/ "Civilisations as yet have only been created and directed by a small intellectual aristocracy". It seems to me that a civilisation is like a brain: there may be top-down processes that seem to be 'directing' it but w/o the other autonomous processes the brain will be neither able to function or to create. In other words, while I consider myself, e.g., to be a creative intellectual, I think that my creativity & my intellect are only part of a much more complex organism that I'm interdependent w/ - even when the "inter" aspect of it may seem antagonistic. It's more likely the arrogance of the "small intellectual aristocracy", the arrogance of someone like Noam Chomsky or William F. Buckley Jr., that gives them the delusion that they've created & are directing the civilisation that they're hypothetically a part of.

"It is only by obtaining some sort of insight into the psychology of crowds that it can be understood how slight is the action upon them of laws and institutions, how powerless they are to hold any opinions other than those which are imposed upon them, and that it is not with rules based on theories of pure equity that they are to be led, but by seeking what produces an impression on them and what seduces them." - p 11

I basically agree w/ all of that while at the same time finding it somewhat self-contradictory & disturbing. I say "contradictory" b/c if "the action upon" [crowds] "of laws and institutions" is "slight" than by what means is it that opinions "are imposed upon them"? The answer given by Le Bon is whatever "produces an impression on them and what seduces them." Fair enuf - but it seems to me that the "laws and institutions" are a part of what produces this impression.

My disturbance is partially rooted in my opinion that crowd psychologists have come a long way since Le Bon's writing of this bk in 1895. It's known that CIA recruiters look for potential new agents at universities, probably in history departments but probably also in psychology ones. Surely an expert crowd psychologist is valuable to institutions other than the CIA too. Where have the top crowd psychologists of the last 30 yrs found gainful employment?

Wherever they're located, they seem to be at least partially behind the current era of extreme obedience to almost completely nonsensical mind control under the guise of 'for your own good'. From my POV the current SHEEPLE behavior is preferable to a murderous mob but it's far from preferable to free & critical thinking - something that's been entirely too close to being obliterated by the brainwashing.

An example: brainwashed people get vaccinated out of fear of dying otherwise & out of fear of being ostracized by their 'peers'. Then, even tho they believe that the vaccination has been gotten to protect them, they still consider unvaccinated people to be a threat. This, alone, makes no sense - either the vaccination protects or it doesn't. As it turns out, it doesn't so the SHEEPLE get vaccinated AGAIN b/c, this time, hypothetically, it'll work. Their faith in the vaccine is unwavering even tho the 'need' for a 2nd vaccination is predicated on the 1st one not working. It doesn't stop there, a 3rd & 4th vaccination are still somehow believed necessary. In the meantime, there're unvaccinated people in excellent health. Despite the evidence of what little of their senses are by now unblindered by propaganda, the unvaccinated people are still considered dangerous - even tho they, themselves, are healthy in some cases & not healthy in others. In the meantime, influential propagandists such as Bill Maher present outdated statistics from a time before many people were vaccinated to claim that it's 99% people who're unvaccinated who're dying from COVID-19.

In other words, I think that "some sort of insight into the psychology of crowds" *has* been obtained & that it's being applied in greater force than ever before. This recent force of application seems more 'gentle' than such predecessors as the nazis & the Khmer Rouge but it's similarly as devastating to free thinking. IMO it's all-too-easy to get into conversations w/ zombies who speak in prefabricated sentences. It's b/c of concerns w/ this state of affairs that I read this bk.

"Under certain given circumstances, and only under those circumstances, an agglomeration of men presents new characteristics very different from those of the individuals composing it. The sentiments and ideas of all the persons in the gathering take one and the same direction, and their conscious personality vanishes. A collective mind is formed, doubtless transitory, but presenting very clearly defined characteristics." - p 13

I think it might be more appropriate to say 'a collective lack of mind' than a "collective mind". It's my observation that there's an ongoing constant campaign originating w/ mass media to implant what're tantamount to post-hypnotic suggestions. Then all that needs to happen is for the trigger words to be massively applied in order to get the hypnotized to fall in line - oblivious to the process the whole time but still completely obedient. Simultaneously, they're given some outlets for trapped energies as if they're achieving liberation instead of being co-opted. I, personally, resist the "collective mind" as do a very few others. It's hard on all of us but it's better than not having a mind at all wch is what being part of the 'collective non-mind' wd entail.

"The most striking peculiarity presented by a psychological crowd is the following: Whoever be the individuals that compose it, however like or unlike be their mode of life, their occupations, their character, or their intelligence, the fact that they have been transformed into a crowd puts them in possession of a sort of collective mind which makes them feel, think, and act in a manner quite different from that in which each individual of them would feel, think, and act were he in a state of isolation. There are certain ideas and feelings which do not come into being, or do not transform themselves into acts except in the case of individuals forming a crowd. The psychological crowd is a provisional being formed of heterogeneous elements, which for a moment are combined, exactly as the cells which constitute a living body form by their reunion a new being which displays characteristics very different from those possessed by each of the cells singly." - pp 14-15

In other words, a superorganism of Manchurian Candidates. Given the latency of the new behavior I wonder whether the people who join a crowd were ever deserving of being considered individuals. If we're all just cells in a superorganism awaiting orders in a top-down fashion then our individual abilities are only 'important' for enabling us to carry out orders. Personally, I prefer to be as free an agent as possible. The thoughts I have are far more interesting to me that the prepackaged excuses for thoughts that people are force-fed thru near-omnipresence every day thru electronic gadgets.

"In crowds it is stupidity and not mother-wit that is accumulated." - p 16

Indeed. Nonetheless, there're people out there who counterbalance this. Vermin Supreme is an excellent example:

666. "**VerminSuprememeaversarry, day one**"
- a collaboration between Vermin Supreme; tENTATIVELY, a cONVENIENCE; & Tyler Lindsey on December 12, 2021
- tENTATIVELY, a cONVENIENCE edit finished December 15, 2021
- 2K, 60p, Stereo
- 2:11:35
- on my onesownthoughts YouTube channel here: https://youtu.be/sqdNKZuW8AI
- on the Internet Archive here: https://archive.org/details/vermin-suprememeaversarry

"The most careful observations seem to prove that an individual immerged for some length of time in a crowd in action soon finds himself — either in consequence of the magnetic influence given out by the crowd, or from some other cause of which we are ignorant — in a special state, which much resembles the state of fascination in which the hypnotised individual finds himself in the hands of the hypnotiser." - p 17

I've already commented on this here. I've also commented on it in my bk Unconscious Suffocation - A Personal Journey through the PANDEMIC PANIC (http://idioideo.pleintekst.nl/Book2020.09PANDEMIC.html). Here's a sample from that:

"people easily have their chains jerked. In fact, their chains are so easily jerked that doing so is like flipping a posthypnotic suggestion switch on Pavlovian Zombies. Most of these people are Manchurian Candidates programmed by mediated non-experience into being suppressors of Free Thinkers. This, unfortunately, is what I've been commenting about all along: viz. that people are being manipulated by mass media signals into having completely predictable responses." - p 903

That was written before I read The Crowd. The Crowd just reinforces it. As long as Le Bon sticks to observations about crowds I find what he writes to be accurate. Alas there're times when he strays into what I might call 'patriarchical biases' that I find.. beyond annoying.

"It will be remarked that among the special characteristics of crowds that there are several — such as impulsiveness, irritability, incapacity to reason, the absence of judgment and of the critical spirit, the exaggeration of the sentiments, and others besides — which are almost always observed in beings belonging to inferior forms of evolution — in women, savages, and children, for instance." - p

18

"inferior forms of evolution": "women, savages, and children". Ahem. Was Le Bon an inferior form of evolution when he was a child? Did his mom's DNA have anything to do w/ producing his apparently 'superior' form of evolution as an adult? Is it possible that 'savages' just have a non-European way of being for wch Le Bon's standards of 'superior' evolutionary form are irrelevant? I'm having a hard time even being sarcastic here.

""Nevertheless, the neighbors, the brother-in-law, the schoolmaster, and the mother were mistaken." [about the identity of a corpse of a child] "Six weeks later the identity of the child was established. The boy, belonging to Bordeaux, had been murdered there and brought by a carrying company to Paris."

"It will be remarked that these recognitions are most often made by women and children — that is to say, by precisely the most impressionable persons. They show us at the same time what is the worth in law courts of such witnesses. As far as children, more especially, are concerned, their statements ought never to be invoked." - p 24

Who exactly CAN be trusted in a courtroom? Almost everyone will lie to try to produce the outcome they want. Police will almost certainly lie - esp if they consider the person on trial to be a criminal, regardless of whether they're guilty of what they're being charged w/. The lawyers will try to manipulate the judge &/ or jury, that's what they're there for. The person on trial, esp if they're 'guilty', will probably lie in an attempt to protect theirself. People's witnessing is completely determined by their biases.

I was in an airport shortly after 9/11. There was much emphasis at the time on noticing abandoned luggage that might contain a bomb. I noted some people leave their voluminous luggage & not come back. I went to the counter & pointed this out to the woman there. She didn't seem to care in the least. I sat down again. Almost immediately, another passenger, a man, went to the counter &, as far as I cd tell, informed the counter-person that the luggage was actually mine. I deduced that this person had a bias against 'my kind of person' but how cd I know under those conditions?

"The certainty of impunity, a certainty the stronger as the crowd is more numerous, and the notion of a considerable momentary force due to number, make possible in the case of crowds sentiments and acts impossible for the isolated individual. In crowds the foolish, ignorant, and envious persons are freed from the sense of their insignificance and powerlessness, and are possessed instead by the notion of brutal and temporary but immense strength." - p 26

& there you have it - why I'm more than wary of crowds. I've participated in many protests, w/ the intention of adding myself to the body politic to show that there's a serious quantity of people trying to make a point & showing our potential strength - &, in my experience, it's just as likely, if not more so, for the police to riot than it is for the protestors. In fact, it's always seemed to me that when the police use violence & blame it on the protestors it helps serve their cause. As such, my sympathy is for the protestors - but *not* for a **mob**, *any mob*. Mobs are driven by their "brutal and temporary but immense strength" &, as far as I'm concerned, we need *less of that* & not more. Of course, a part of the problem is that the police often operate using that mob mentality as a built-in aspect of their function so protestors just turn into what they feel is the only available countering force. The whole dynamic is dangerous but it doesn't necessarily originate w/ the people who get blamed for it.

"An orator wishing to move a crowd must make an abusive use of violent affirmations. To exaggerate, to affirm, to resort to repetitions, and never to attempt to prove anything by reasoning are methods of argument well known to speakers at public meetings." - p 26

Having recently finished reading an autobiography by a public speaker & having reviewed it (Bonar Thompson's Hyde Park Orator) I'm inclined to think that Le Bon's generally accurate. But does it really have to be that way? As effectively manipulative as anger-mongering is, humor is also powerful & has, IMO, a much more desirable outcome. Check out just about any of Vermin Supreme's movies online - or those of JP Sears.

For that matter, I haven't done so bad in that area myself. Since I'm not trying to make money off my political speech-making & since I'm trying to encourage people to be thoughtful & strong individualists, instead of members of a crowd, I take a completely different tack from that of anger-mongering. My world wd be worse if more people were simply *angry* in preference to imaginatively & creatively in resistance. But not yielding to anger is one of the hardest paths to take & often seems close to impossible under oppressive &/or brutal conditions. In that sense, having a sense of humor can be more of a luxury than it is a solution. Here's a link to some of my offerings online: http://idioideo.pleintekst.nl/tENTActivism.html . Money Against Capitalism, created & enacted in collaboration w/ etta cetera, is one of my best ventures along the desired lines of what I call IMP ACTIVISM:

246. "**Money Against Capitalism & May Day 2000 in Melbourne, Australia**"
- incorporating in its entirety the Access News / SKA TV May Day 2000 piece
- VHS
- 13:21
- may '00 / january '04
- broken into 2 parts for YouTube:
- May Day
- on my onesownthoughts YouTube channel here: https://youtu.be/5GrIEM3uvMg
- Anarchist World this Week
- on my onesownthoughts YouTube channel here: https://youtu.be/-yi9PTR99xE
- see the May 1 & 17, 2000 entries here: http://idioideo.pleintekst.nl/MereOutline2000.html

SO, yes, I think there's hope of there being more & more people in the world who choose humor as a way of addressing crowds & I think the outcome will be a much happier one.

That sd, I hope that the reader of this review can recognize the type of oration that Le Bon refers to & counterbalance it by retaining their 'common sense'. Terms like "anti-vax Trumper" are best 'appreciated' as propaganda manipulators w/ more than a grain of salt, but, rather, a whole beach of sand. A recent movie of mine, conveniently only 5.5 seconds long for people w/ short attn spans, mocks this propaganda:

672. "**Breaking News!**"
- 5.5 seconds
- 1920X1440 - 30fps, stereo
- on my onesownthoughts YouTube channel here: https://youtu.be/NEFKZ1n_muw
- on the Internet Archive here: https://archive.org/details/breaking-news_202202

It's a conflicted sensation, reading Le Bon, for me, b/c I think he's very astute about crowds & mind-numbingly unperceptive about other things. Here's his take on the Luddites:

"Their fetish like respect for all traditions is absolute; their unconscious horror of all novelty capable of changing the essential conditions of their existence is very deeply rooted. Had democracies possessed the power they wield to-day at the time of the invention of mechanical looms or of the introduction of steam-power and of railways, the realization of these inventions would have been impossible, or would have been achieved at the cost of revolutions and repeated massacres. It is fortunate for the progress of civilisation that the power of crowds only began to exist when the great discoveries of science and industry had already been effected." - pp 28-29

Le Bon's position here is firmly rooted in the propaganda of ruling elites. The Luddites weren't objecting to 'progress', they were objecting to the forced reduction of their wages &, subsequently, to the degradation of their standard of living. The manufacturers introducing the Spinning Jenny knew exactly what they were doing & were calculatedly destroying the quality of life for the many to increase the quality of life for the few, for themselves.

Interestingly, in September of 2020, I was going for a walk at my favorite park in Pittsburgh when I encountered a former friend of mine who works as a union organizer. She's from a wealthy family & has lived a life of privilege as a result but is sincerely dedicated to what one might call progressive politics. For most of the 20 yrs we'd known each other we'd been pretty good friends. Since our birthdays are around the same time it was a tradition fo sorts for us to get together in September for a meal.

HOWEVER, all that changed w/ the coming of the QUARANTYRANNY. She believes that the Medical-Industrial Complex & its political puppets are working for our own good. IMO, people like my friend have been *astroturfed*, fooled into thinking that an oppression created from 'on high' is somehow 'grassroots'. Orders from a governor, followed by fear of losing their businesses by business owners, are somehow 'grassroots' b/c they're 'protecting us' from an exaggerated natural phenomenon. It's my opinion that my former friend believes these things b/c she's from wealth & has an in-bred respect for global planners, as if their self-appointed lording-it-over is somehow their birthright. That explains for me why, even though she's been a very well-pd union organizer for around a decade now, *she's never once succeeded in organizing a union* - she can never possibly understand the actual situations of the workers that she addresses. (See my recent relevant critique of Noam Chomsky: https://medium.com/@idioideo/putting-chomsky-in-his-place-da6a6f90202f)

ANYWAY, I saw her & her boyfriend at the park & we talked for awhile. Her boyfriend, who teaches researching, commented to the effect that *Everyone's a Luddite these days* in reference to people being against the quarantine (if I understood correctly). This struck me as just a continuation of ruling elite party line. I continued on my way & encountered another friend who told me that she was on her way to the former friend's *birthday picnic in the park*. I had been *not invited* by the 'union organizer' b/c I have opinions about the QUARANTYRANNY that go contrary to those of the ruling elites. As such, I suppose that makes me a 'Luddite' in the actual sense rather than the propaganda-rehistorified sense. In other words, during these times of the so-called "Fourth Industrial Revolution", the quality of life is once again being debased for smaller business people in the interests of further solidifying the corporate oligarchy's profiteering - & I'm *not going to be conned by the astroturfed justification for this.*

"To know the art of impressing the imagination of crowds is to know at the same time the art of governing them." - p 36

Then again, to not belong to a crowd may be a way to avoid being governed.

"We have shown that crowds do not reason, that they accept or reject ideas as a whole, that they tolerate neither discussion nor contradiction, and that the suggestions brought to bear on them invade the entire field of their understanding and tend at once to transform themselves into acts. We have shown that crowds suitably influenced are ready to sacrifice themselves for the ideal with which they have been inspired. We have also seen that they only entertain violent and extreme sentiments, that in their case sympathy quickly becomes adoration, and antipathy almost as soon as it is aroused is transformed into hatred." - p 36

That seems accurate to me.. but at the same time I'm usually distrustful of language like "We have shown" b/c it seems to me that the users of such language haven't necessarily 'shown' anything but have merely put forth a series of statements meant to serve as propositions & that these statements are then asserted to be somehow axiomatic.

"These general indications furnish us already with a presentiment of the nature of the convictions of crowds.

"When these convictions are closely examined, whether at epochs marked by fervent religious faith, or by great political upheavals such as those of the last century, it is apparent that they always assume a peculiar form which I cannot better define than by giving it the name of a religious sentiment.

"This sentiment has very simply characteristics, such as worship of a being supposed superior, fear of the power with which the being is credited, blind submission to its commands, inability to discuss its dogmas, the desire to spread them, and a tendency to consider as enemies all by whom they are not accepted."

[..]

"The Jacobins of the Reign of Terror were at bottom as religious as the Catholics of the Inquisition, and their cruel ardour proceeded from the same source." - p 36

&, NOW, we have social media, in particular Facebook - wch exemplifies the dynamic of "inability to discuss its dogmas, the desire to spread them, and a tendency to consider as enemies all by whom they are not accepted." During the time of the QUARANTYRANNY I've never seen such high levels of kneejerk intolerance of even the simplest debate over the merits of things such as mask-wearing, social distancing, & vaccination.

"Should a people allow its customs to become too firmly rooted, it can no longer change, and becomes, like China, incapable of improvement. Violent revolutions are in this case of no avail; for what happens is that either the broken fragments of the chain are pieced together again and the past resumes its empire without change, or the fragments remain apart and decadence soon succeeds anarchy." - p 42

That was written in 1895, decades before the Chinese Revolution. It seems to me that violent revolution WAS to some avail & that neither of the outcomes that Le Bon describes w/ such certainty was what happened. Whether that's been for better or for worse I don't know. I imagine that being on the low end of the imperial Chinese hierarchy must've been pretty horrible & that being on the low end of the new Chinese communo-capitalism is probably pretty horrible too. Otherwise, I haven't mentioned until now that "anarchy" is one of Le Bon's favorite boogaboos.

"The conclusion to be drawn from what precedes is, that it is not in institutions that the means is to be sought of profoundly influencing the genius of the masses. When we see certain countries, such as the United States, reach a high degree of prosperity under democratic institutions, while others, such as the

Spanish-American Republics, are found existing in a pitiable state of anarchy under absolutely similar institutions, we should admit that these institutions are as foreign to the greatness of one as to the decadence of the others. People are governed by their character, and all institutions which are not intimately modelled on that character merely represent a borrowed garment, a transitory disguise." - p 45

That assertion interests me. It seems borderline *anarchist* insofar as it essentially says that no system of governance can possibly represent those governed. What makes it NOT anarchist is that Le Bon's probably falling back on his preference for a small elite of guiding intellectuals whose only need for governance is to use the effective tools for manipulation.

"Statisticians have brought confirmation of these views by telling us that criminality increases with the generalization of instruction, or at any rate of a certain kind of instruction, and that the worst enemies of society, the anarchists, are recruited among the prize-winners of schools" - p 46

Ha ha! It seems that anarchists, even tho we're implied to be well-educated, are NOT the type of elite he has in mind. After all, we're an *anti-ruling elite* rather than a ruling one. I reckon I can find some 'humor' in his choice of the word "recruited" since it implies a *recruiting agent*, something that I find entirely absent, for good reason, among anarchists (except for, perhaps, among undercover cops pretending to be anarchists to provoke & entrap people).

"I have myself shown, in a work published some time ago, that the French system of education transforms the majority of those who have undergone it into enemies of society, and recruits numerous disciples for the worst forms of socialism." - p 46

There he goes again, 'showing' us something. I find this assertion highly problematic. What exactly is an 'enemy of society'? If the majority of people receiving a French education become "enemies of society" then what happened to France? It still seems to be there & it still seems to have a society. These days, under the rule of President Macron it strikes me that the plans of a ruling elite are being served more than ever. I'd love to have Le Bon's opinion on France's forced embracing of a Globalization agenda. Macron has transcended Socialism by finding nourishment in International Capitalism that is, yes, simultaneously one of "the worst forms of socialism" insofar as its police state tactics go.

"Instead of preparing men for life French schools soley prepare them to occupy public functions, in which success can be attained without any necessity for self-direction or the least glimmer of personal initiative. At the bottom of the social ladder the system creates an army of proletariats discontented with their lot and always ready to revolt, while at the summit it brings into being a frivolous bourgeoisie, at once sceptical and credulous, having a superstitious confidence in the State, whom it regards as a sort of Providence" - p 47

I enjoy his critique of the late 19th century French educational system but I suspect that he's overstating his case. From my point of view the best way to not have a "bottom" & a "summit" of a "social ladder" is to simply not have a social ladder at all. That may seem impossible to some people but, to me, if I can imagine it, it's possible. But, perhaps, maybe it's true that social ladders form naturally no matter what social system (or absence thereof) they exist in. After all, there are sadists & masochists who probably long for such things. Contrarily, in the founding texts of Sadism & Masochism, the relevant authors make it seem as if these sexual inclinations are an outgrowth of hierarchical society.

Anarchists, anarchists, yadda, yadda.

"M. Paul Ourget in his excellent book, "Outre-Mer." He, too, after having noted that our education merely produces narrow-minded bourgeois, lacking in initiative and will-power, or anarchists — "those two equally harmful types of civilised man, who degenerates into impotent platitude or insane destructiveness"" - footnote, p 50

It's at least fun to see "narrow-minded bourgeois" & "anarchists" lumped together. Maybe we shd form a "Narrow-Minded Bourgeois Anarchist Party". I might even go back to drinking in its honor as long as someone else foots the bill for some good booze.

"Have we digressed in what precedes from the psychology of crowds? Assuredly not. If we desire to understand the ideas and beliefs that are germinating to-day in the masses, and will spring up to-morrow, it is necessary to know how the ground has been prepared." - p 50

In other words, the psychological crowd in France has been prepared for its GroupThink by its education. I have to wonder, tho, whether Le Bon isn't at times somewhat inconsistent. In the preface he wrote "The whole of the common characteristics with which heredity endows the individuals of a race constitute the genius of the race. When, however, a certain number of these individuals are gathered together in a crowd for purposes of action, observation proves that, from the mere fact of their being assembled, there result certain new psychological characteristics, which are added to the racial characteristics and differ from them at times to a very considerable degree." In that description, heredity is the key factor & education isn't mentioned. THEN this "genius of the race" develops different characteristics when it becomes a crowd "which are added to the racial characteristics and differ from them at times to a very considerable degree." Again, no mention of education.

My point is that if education is such an important determinant then that's neither heredity or "new psychological characteristics" of the crowd - wch, to me, at least, implies the importance of institutions - institutions having been previously denied any role. Some of us DO pay attn when we read, y'know?!

"Crowds are somewhat like the sphinx of ancient fable: it is necessary to arrive at a solution of the problems offered by their psychology or to resign ourselves to being devoured by them." - p 51

OR to just avoid them altogether - the more people that do that, the smaller the crowd is, the less dangerous it is.

"All words and all formulas do not possess the power of evoking images, while there are some which have once had this power, but did lose it in the course of use, and cease to waken any response in the mind. They then become vain sounds, whose principle utility is to relieve the person who uses them of the obligation of thinking. Armed with a small stock of formulas and commonplaces learnt while we are young, we possess all that is needed to traverse life without the tiring necessity of having to reflect on anything whatever." - p 52

& that's what I call 'speaking in prefabricated sentences' or 'speaking in talking points provided by the puppet masters'. I've heard very little else since the campaign started in March, 2020, to forcefit the people of the world into the New World Odor thru Global Medicalization. One fill-in-the-blank phrase that appears to still retain some usefullness for keeping people braindead is "The War on _____". As I've pointed out in other writing of mine, this formula has been completed thusly: The War on Communism, The War on Drugs, The War on Terrorism, & The War on COVID-19. Strangely, people opposed this observation of mine seeing no connection between these phrases & the function that they served to homogenize behavior under the same flag of fear.

"For the Latin peoples the word "democracy" signifies more especially the subordination of the will and the initiative of the individual to the will and the initiative of the community represented by the State. It is the State that is charged, to a greater and greater degree, with the direction of everything, the centralisation, the monopolisation, and the manufacture of everything. To the State it is that all parties without exception, radicals, socialists, or monarchists, constantly appeal. Among the Anglo-Saxons and notably in America this same word "democracy" signifies, on the contrary, the intense devlopment of the will of the individual, and as complete a subordination as possible of the State" - p 54

I find that an interesting observation. I've noted something similar about the word "Libertarian". In the United States, in 1982, I lived w/ the local president of the Libertarian Party. At that time, Libertarianism was a type of Capitalism in which 'Free Trade' was unihibited by laws - esp laws stopping pot. In other words, it was a rich person's fantasy of not having to have much of a conscience about how spoiled they were. Contrarily, when I was in Spain in 2004, Libertarians were considered to be synonymous w/ Anarchists. The American Libertarianism had been mainly the 'liberation' of the rich while the Spanish version was the liberation of everyone. It seems that in the 21st century in the US Libertarianism is generally just considered to be a 'right wing' movement now to wch even the pro-marihuana/anti-police-state aspects are of little importance.

"In every social sphere, from the highest to the lowest, as soon as a man ceases to be isolated he speedily falls under the influence of a leader. The majority of men, especially among the masses, do not possess clear and reasoned ideas on any subject whatever outside their own speciality. The leader serves them as a guide. It is just possible that he may be replaced, though very inefficiently, by the periodical publications which manufacture opinions for their readers and supply them with ready-made phrases which dispense them of the trouble of reasoning."
- p 60

That was 1895. In those days there were just "the periodical publications which manufacture opinions for their readers" & not the plethora of such things we have 127 yrs later. This plethora is hypothetically more open to variety b/c it's harder to centralize the opinions offered. Nonetheless, the exaggerated pandemic has provided the perfect pretext for declaring any oppositional opinions to be 'misinformation', etc. In both eras, there's a fullsome supply of "ready-made phrases which dispense" [the populace] "of the trouble of reasoning". All the mental weaklings & cowards & conformists of this era have to do is evoke 'The Science' to pseudo-substantiate their rc'vd parroted pseudo-opinions - despite their having little or no understanding of what actual science might have to say about it.

"It is the need not of liberty but of servitude that is always predominant in the soul of crowds. They are so bent on obedience that they instinctively submit to whoever declares himself their master." - p 60

I think it's more a matter of people wanting to be free of responsibility for their actions so that they can unleash whatever frustrated brutality they feel compelled to as a reaction to their general impotence. By relinquishing their responsibility to a 'higher authority' they can be 'free' to be utter assholes as long as their assholism is directed against the people who actually have the courage to be free *w/o relinquishing their responsibility*, the people who're the ultimate impediment to the power-schemes of the 'higher authorities'.

"The influence of repetition on crowds is comprehensible when the power is seen which it exercises on the most enlightened minds. This power is due to the fact that the repeated statement is embedded in the long run in those profound regions of our unconscious selves in which the motives of our actions are forged. At the end of a certain time we have forgotten who is the author of the repeated assertion, and we finish by believing it. To this circumstance is due the astonishing power of advertisements." - p 62

Here's another relevant excerpt from Unconscious Suffocation:

"The trip was too long for me to make in one day so I stopped at a hotel on the way back. Hotels are the only place where I willingly watch TV. I was watching a Jackie Chan movie, whose work I love, but it was heavily separated by commercials, which I muted. The ads were so intensely PANDEMIC PANIC oriented that I can easily see how TV watchers would find the mind control pressure inescapable. There was the usual prescription drug mania; a justification of how pregnant women are being treated in terms of equating COVID-19 with the Spanish Flu; plenty about what to do in isolation (all of it involving spending loads of money) including buying stuff from Etsy no less!; a diabetes type 2 medication ad in which, as usual, the emphasis was on a wonder drug rather than on changing one's diet & getting more exercise; but this one wins the prize!: it was for people who "just can't wait to get a new car!": they were reassured that they could buy one ONLINE & HAVE IT DELIVERED TO THEIR HOUSE where they could take proud ownership of it wearing a mask." - pp 268-269

The Jackie Chan movie was the bait but the main purpose it served was to get the TV viewers exposed to an extraordinary amt of ads, all of wch were pushing the QUARANTYRANNY from one angle or another. I remember counting the ads & finding them to number something like 12 ads to every 5 minutes of the movie.

"Great power is given to ideas propagated by affirmation, repetition, and contagion by the circumstance that they acquire in time that mysterious force known as prestige." - p 64

Prestige is not a word I'd choose to use but I can imagine it's being more important in 1895 Paris. Basically, repetition seems to work by making something so *familiar* that it seems *real*, a commonplace, axiomatic. The constant hammering into the psyche by repetitive propaganda has done its work exceptionally well during the exaggerated pandemic b/c people talk about things as 'reality' that they have little or no experience of that's not just mediated hearsay - & any person who refers to actual observable reality is considered dangerous when they point out the disparity between the propaganda & what they see in their own environment.

"Crowds always, and individuals as a rule, stand in need of ready-made opinions on all subjects. The popularity of these opinions is independent of the truth or error they contain, and is solely regulated by their prestige." - p 66

This prestige being the rank they occupy in subcultures. For liberals, referring to something on NPR is like referring to something that's *absolutely known to be truth* & offering any counter-opinion, no matter how well-researched its basis might be, is like blaspheming against an 'irrefutable God'. No doubt there's a correlary for conservatives, perhaps Fox TV 'news', but I don't pay enuf attn to them to know. *The New York Times* is another 'unassailable fortress' of opinion - despite its having defamed Martin Luther King when King came out against the Vietnam War. Things like that can be shuffled under the rug easily when the readership can't remember for more than a few days back.

"Thus it is that, thanks to general beliefs, the men of every age are enveloped in a network of traditions, opinions, and customs which render them all alike, and from whose yoke they cannot extricate themselves. Men are guided in their conduct above all by their beliefs and by the customs that are the consequence of those beliefs. These beliefs and customs regulate the smallest acts of our existence, and the most independent spirit cannot escape their influence." - p 72

Really?! I don't think I can agree w/ that - but I mainly wonder whether Le Bon felt that this applied to himself as well.

"In literature, art, and philosophy the successive evolutions of opinion are more rapid still. Romanticism, naturalism, mysticism, &c., spring up and die out in turn. The artist and the writer applauded yesterday are treated on the morrow with profound contempt." - p 74

1st, It seems likely to me that Le Bon is exaggerating the social climate in late 19th century Paris. 2nd, It seems to me that he's really talking about the people who're trend-followers rather than sincere appreciators of the arts. In other words, trend-followers are by definition superficial, they *have* to depreciate what they appeared to appreciate in the recent past in order to 'show' that they're 'current w/ the times'. It's *their* pseudo-opinions that're meaningless, not the work they pretend to respond to.

"given the power possessed at present by crowds, were a single opinion to acquire sufficient prestige to enforce its general acceptance, it would soon be endowed with so tyrannical a strength that everything would have to bend before it, and the era of free discussion would be closed for a long time." - p 76

Alas, that's where I feel we're at now. All of the crowd manipulations have been applied w/ maximum propagandistic force: "To exaggerate, to affirm, to resort to repetitions, and never to attempt to prove anything by reasoning" & the result is that assertions that might've seemed highly questionable a mere few yrs ago such as the 'necessity' for masks, social distancing, & vaccination are now dictums closed to debate under threat of abuse from the majority.

Much of the history that Le Bon refers to is that of the French Revolution, certainly one of the most violent societal upheavels.

"It is not known exactly who gave the order or made the suggestion to empty the prisons by massacring the prisoners. Whether it was Danton, as is probable, or another does not matter; the one interesting fact for us is the powerful suggestion received by the crowd charged with the massacre.

"The crowd of murderers numbered some three hundred persons, and was a perfectly typical heterogeneous crowd. With the exception of a very small number of professional scoundrels, it was composed in the main of shopkeepers and artisans of every trade: bootmakers, locksmiths, hairdressers, masons, clerks, messengers, &c. Under the influence of the suggestion received they are perfectly convinced, as was the cook referred to above, that they are accomplishing a patriotic duty. They fill a double office, being at once judge and executioner, but they do not for a moment regard themselves as criminals." - p 80

So 300 people work themselves up into a frenzy & murder an unspecified number of prisoners - who, after all, may've been guilty of a wide variety of crimes ranging from stealing bread to murder - or guilty of nothing at all other than being poor. This is exactly the type of thing that my own dystopic Self-Negating Prophecies wd like to prevent. As I've stated before, I see the people who'd constitute this crowd to be robopaths, a term that wdn't've existed in Le Bon's day. The people reliquish their own responsibility & melt into a crowd not likely to face any consequences so that they can unleash their usually hidden brutality.

This brings me back, once again, to the present day - the more people are brainwashed into going along w/ whatever manipulative narrative is drilled into their heads by the puppet-masters the more likely it becomes that they, too, will become conscience & responsibility -free committers of the most atrocious crimes. Hence my encouragement for people to maintain their personal identity & to not be swept up into any mass identity - no matter how appealing & 'safe' it might seem to be.

"At various periods, and in particular previous to 1848, the French administration instituted a careful choice among the persons summoned to form a jury, picking the jurors from among the enlightened classes; choosing professors, functionaries, men of letters, &c. At the present day jurors are recruited for the most part from among small tradesmen, petty capitalists, and employés. Yet to the great astonishment of specialist writers, whatever the composition of the jury has been, its decisions have been identical." - p 82

Treating juries as a crowd &, therefore susceptible to all the weaknesses heretofore described, interests me. Whether Le Bon's assertion is verifiable or not is beyond me. I've never served on a jury & hope to never do so. I don't think I'm qualified to pass judgment on another person in a way that wd result in their imprisonment or death. Nor do I think that anyone else is qualified to do the same w/ me as the defendant. It seems to me that the composition of the jury IS important but that it cd never be ideal enuf to really be fair.

"The most important reforms may be fearlessly promised. At the moment they are made these exaggerations produce a great effect, and they are not binding for the future, it being a matter of constant observation that the elector never troubles himself to know how far the candidate he has returned has followed out the electoral programme he applauded, and in virtue of which the election was supposed to have been secured." - p 87

Well, there you have it: proof that people have been observing since at least 1895 the say-anything-to-get-elected procedure that politicians use &, **yet** people seem to be taken in by it again & again. It's as if wishful thinking were all anyone ever aspires to.

"It is terrible at times to think of the power that strong conviction combined with extreme narrowness of mind gives a man possessing prestige. It is none the less necessary that these conditions should be satisfied for a man to ignore obstacles and display strength of will in a high measure.

"Crowds instinctively recognize in men of energy and conviction the masters they are always in need of." - p 95

Here, I call into question the concept of "need" in this instance. To reiterate: I think that people relinquish personal responsibility by following a leader so that they can commit atrocities w/o fear of taking the blame. But, as for 'needing' leaders? I think people 'need' leaders like a hole-in-the-head. The hole in their head isn't necessarily a good idea in a healthy sense but it might relieve pressure like trepanation is reputed to do.

"This incessant creation of restrictive laws and regulations, surrounding the pettiest actions of existence with the most complicated formalities, invariably has for its result the confining within narrower and narrower limits of the sphere in which the citizen may move freely. Victims of the delusion that equality and liberty are the better assured by the multiplication of laws, nations daily consent to put up with trammels increasingly burdensome. They do not accept this legislation with impunity. Accustomed to put up with every yoke, they soon end by desiring servitude, and lose all spontaneousness and energy. They are then no more than vain shadows, passive, unresisting and powerless automata." pp 99-100

There he goes again, talking like his worst enemy, an anarchist! As for the desire for "servitude"? Why not just call it by name?: masochism.

All in all, this is an excellent bk. If I were to buy a copy for someone else I probably wdn't choose this edition. I've pointed out some of my differences w/ Le Bon. I see The Crowd as most useful as a warning. I also see it as something that cd be used by manipulative people to further their own agendas.

The
Poisoned
Needle

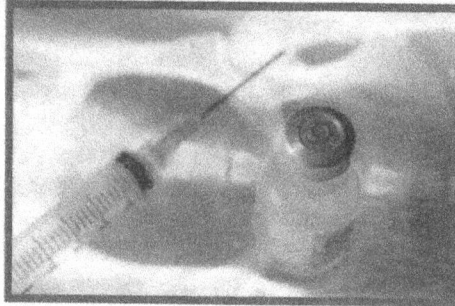

By
Eleanor McBean

review of
Eleanor McBean's **The Poisoned Needle - Sup[p]ressed Facts about Vaccination**
[ISBN 9780787305949 edition]
by tENTATIVELY, a cONVENIENCE - May 22, 2022

A HERETIC informed me about this bk. After making the unfortunate decision to buy a cheaper edition that turned out to be a disgrace to publishing in terms of how poorly done it was (deliberately or not) I was fortunate enuf to get this edition.

The bad cheaper edition (not this one) [ISBN 9798501688650 edition] is missing the following important quote from the beginning of the bk:

""It is exceedingly difficult to secure an honest hearing for any criticism of authority. Established beliefs are well nigh invulnerable because they are accorded infallibility by the masses who are educated to believe that they will be damned for thinking, and because of this, few will tolerate opposition of any nature to anything they have been educated to believe. People who have their thinking done for them are always intolerant."

"— *J. H. Tilden, M.D.*" - backside of title p

I'm an anarchist. For probably my entire life anarchism has been generally considered to be the extreme left of the left - w/ the left being the extreme left of liberalism. I've long-since rejected this idea. To me, being an anarchist means that I'm independent of such ideological simplicities. Some people might call such a positioning "Post-Left Anarchism". I consider myself to be a free thinker.

During the COVID-19 QUARANTYRANNY I've watched the so-called left, including anarchists who consider themselves to be on the leftist spectrum, turn into what I call neo-illiberals. Basic values of the left have been inverted. The left now supports Big Brother, the Police State, Censorship, & Globalization - all things that the left had previously traditionally stood against.

Why? IMO it's b/c the paradigm that's being used to get all the SHEEPLE marching in lockstep is one that was traditionally accepted as 'good' by the left - viz: the Medical Paradigm. Its axioms are: Medicine = good, Doctors = good. I find this paradigm to be completely stupid. However, as Dr. Tilden so astutely observed above, "few will tolerate opposition of any nature to anything they have been educated to believe." This, to me, is all too true. People who hypothetically rejected pro-military brainwashing accept pro-medical brainwashing w/o question. It's a pathetic state of affairs but it's also a dangerous one, dangerous to human intelligence.

This bk is an antidote to the unquestioning acceptance of the medical paradigm. It's also something that the SHEEPLE will never read. An aspect of their brainwashing is that they absolutely will not study *anything* that goes contrary to their programming. It's intrinsic to their programming that they not do so. They have FACT CHOKERS to steer them away. This review defies the Pavlovian conditioning.

The bad cheaper edition is also missing the dedication. That begins w/ this:

"THIS BOOK IS LOVINGLY DEDICATED —

"TO THE PATIENT SCHOOL CHILDREN who have too long been the innocent victims of medical exploitation.

"TO THE TRUSTING PARENTS who have searched hopelessly in the wrong direction for dependable health instruction." - p 2

The introduction is different in the bad cheap edition. It's missing the 1st paragraph. Perhaps it's a reprint of an earlier edition. I'll just quote the 2nd & 3rd paragraphs here:

"The purpose of this volume is to provide a working reference of various statements regarding vaccination and not to offer a "cure" of any specific disease in any specific individual.

"Preconceived opinions regarding vaccination will not be changed overnight nor by the reading of a single book. Herein are the opinions of many great scientists and doctors (of many schools). Many have spent a life-time of research—of trial and error before making their statements. The reader has the privilege to accept or reject the statements after due deliberation." - p 3

I find many things interesting about the above. The bk doesn't, obviously, offer "cure"s. That, equally obviously, requires an examination of the specific patient - something usually done by a dr. As such, the bk is addressing generalities. Nonetheless, *it's the generalities that're the 'axioms', the 'building blocks' that the specifics are based on.* If the building blocks, the foundation, on wch anything is based, are faulty then the whole edifice is faulty.

As such, if the belief that *viruses are the cause of certain illnesses & that these viruses can be fought by injecting sickening substances into the body to stimulate the production of anti-bodies* turns out to be false then all medicine based on it will also be false. People mock the more ancient beliefs that evil spirits cause disease - but is there really that much of a difference between belief in evil spirits & belief in viruses? Both are things invisible under ordinary circumstances to the layperson, both are 'treated' by experts who specialize in the invisible. To quote from an email I sent a friend this morning:

"Instead of priests speaking authoritatively for a higher power, God, we have doctors & researchers & scientists speaking authoritatively for a higher power, science. There's really entirely too little difference precisely b/c as long as the general public puts their faith in something that's essentially unknowable for them they're victims of induced thoughtlessness."

The Introduction goes on to state: "Preconceived opinions regarding vaccination will not be changed overnight nor by the reading of a single book." Note the emphasis on **book**. The Poisoned Needle was written in the mid-1950s, bks were still considered to be the preferred references, I still prefer them. These days, how many people do you know who actually read bks from front to back, who read bks for research purposes? I know entirely too few & most of them read simple bks, bks hypothetically targetted at an age group that the readers haven't been in for decades.

The Introduction portion quoted then concludes w/ "The reader has the privilege to accept or reject the statements after due deliberation." How often does such a qualification appear? I don't trust bks that make claims to ultimate objectivity - &, yet, isn't that what we're usually subjected to? The 'experts' appear on TV, in the 'news'papers, on the radio - don't they usually act as the ultimate purveyors of 'the truth'?! & yet no person is all-knowing - hence people's belief in 'God', but I don't believe in 'God' & I don't believe in the 'all-knowingness' of the 'experts' who speak as oracles for their science-god either. There're always lines to be read between, subtexts to ferret-out. If medical science really understood disease we wdn't have tortures to 'cure' cancer that don't work but that cost a fortune.

The Acknowledgement is also missing from the cheaper bad edition. Here's its 2nd paragraph:

"Truth is universal and no one had a monopoly on words or knowledge. But the statements of specific facts and findings, by recognized authorities, are essential in a research book composed of collected data from authentic sources." - p 4

Keeping in mind Tilden's quoted warning, one person's 'authorities' are another person's quacks or another person's cover-up for ulterior motives. Still, McBean's point is clearly stated: this is a "*research book composed of collected data from authentic sources.*" It's important to keep in mind that McBean wd've had to go to an enormous amt of trouble to compile all the quotes & reports that're in The Poisoned Needle. She probably went to libraries & studiously went thru newspapers, perhaps in microfiche form. She read, read, & read some more. Working thru such a massive amt of data in those days required a much more serious commitment than online research does these days. As a person who's researched in a wide variety of ways *I'm seriously impressed.*

McBean makes a bold statement in her Preface that many might consider to be insupportable but mainly I perceive her as challenging the dominant paradigm & I perceive The Poisoned Needle as attempting to back-up this challenge as thoroughly as the author can:

"Our worst epidemics **now** are **epidemics of vaccination** in which more people are killed every year by "vaccinal diseases" than by the diseases that the vaccinations were supposed to combat.

"Complete freedom from disease can scarcely be found except in remote islands or isolated communities that have not been invaded by medical mischief and commercialized products. Is our proud civilization to be degraded, corrupted, and destroyed by its own inventions designed for selfish profit and exploitation of the race?" - p 5

My opinion, based on observation, is that one is less likely to get sick in a natural environment that one is regularly existing in. My natural environment is the temperate middle east coast of the US. I think that simply by going for walks in the woods I'm gently in harmony w/ whatever else lives in those woods, microscopic or otherwise. However, if I were to go to the Amazon I'd be very much out-of-place, very much out-of-harmony &, therefore, more likely to get sick. OR, if my natural environment were to become polluted by human factors I'd be much more likely to get sick. Obviously, there was a time when a person walking thru the woods might drink from a stream - very few people wd risk that now given the prevalence of industrial pollution. It's this human-introduced toxicity that I find most concerning.

"INCREASE IN KILLER DISEASES DURING THE PAST 70 YEARS

"Insanity increased	400%
Cancer increased	308%
Anemia increased	300%
Epilepsy increased	397%
Bright's Disease increased	65%
Heart Disease increased	179%
Diabetes increased	1800% (In spite of or because of insulin)
Polio increased	680%" - p 7

What do I make of such statistics? Not much. I don't believe them, I don't disbelieve them. A very detailed treatise on how they were acquired wd have to be written to go into every bit of how they were arrived at & to what extent they shd be trusted. The author's paragraph that follows these statistics is as follows:

"Never in the history of this country have preventable diseases flourished with such wild abandon, continuously being fed by the very drugs and commercialized irritants that set them into operation in the first place." - p 7

I don't feel qualified to distinguish between whether the diseases themselves have increased or just the diagnosis of them or neither. I do tend to think that insanity increases under social conditions increasingly unfavorable to individuality - whether that's now or not is debateable. I also think that cancer increases w/ the increase of pollution, of human-introduced environmental toxins - that seems pretty obviously the case to me. Heart Disease & Diabetes seem intricately related to being overweight & consuming too much sugar & alcohol. I can imagine those are probably side-effects of an increasingly consumerist society. As for the rest? I don't have much of an opinion. Later parts of this bk make a pretty convincing case for me that Polio is a byproduct of accumulated toxins & that these toxins are often voluntarily introduced by drs as 'cures' that backfire. I tend to agree w/ this too. Before I read this bk, I became of the opinion that injecting anything into the body w/ a needle is a bad idea. The Poisoned Needle reinforced this for me.

"**R. T. Trall, M.D.,** was another doctor who gave the medical method a long and thorough trial before abandoning it as a failure. He turned to nature and its simplicity and upon its sound basic laws of healing he brought about miracles of recovery from all diseases. After trying to justify the inconsistencies of medical theory he had this to say:

""Why has not success in treating disease kept pace with the extraordinary progress of knowledge in the other sciences? The answer is ready: A successful practice of the healing art must be based upon the laws of life, the economy of vitality. **The only foundation, therefore, of true healing is correct physiological principles; and here is precisely where the whole orthodox medical system of the present day fails — utterly and totally fails. It has no physiological and biological science upon which to truly practice the healing art.**"" - p 8

One of the problems w/ reviewing this bk is that it's almost as if I have to quote the entire thing *& then search for additional souces that substantiate its positions*. Even writing it in the long-winded but still inadequate way that I'm planning to will fall far short of my ideal. E.G.: providing a bio of R. T. Trall wd be helpful - but I'm not going there b/c my plans for this review are already too ambitious.

"**Dr. Shelton** writes: "The whole of the modern medical practice of vaccine, serum and antitoxin therapy is based upon the **supposition** that the body manufactures substances called anti-toxins, anti-bodies, antigens, etc., which are capable of meeting and destroying toxins that get into the body. The idea **seems** to be sound, although it is possible that the work of destroying such toxins is the detoxification carried on by the liver, and lymph glands, etc. Anti-toxins, anti-bodies, antigens, etc., have never been isolated. They have only been **assumed**" - p 9

Dr. Shelton's <u>The Hygienic System</u> from wch the above quote comes, dates from 1939 so my 1st question is: Have anti-bodies been isolated? According to "Trends in Immunology" (2016): "A major advance in bNAb isolation and characterization was the development of single antigen-specific B cell cloning methods that allowed the rapid isolation of monoclonal antibodies (MAbs) [19,20,24,33–40]." (https://www.sciencedirect.com/topics/immunology-and-microbiology/antibody-isolation) That's 77 yrs later than Shelton's bk. Another recent (2013) source says this:

"Recent advances in the isolation, culture and expansion of human B cells and the recovery of genes encoding immunoglobulin (Ig) are enabling the isolation of large numbers of antibodies to be used for probing the humoral immune response and developing diagnostics and therapeutics. The history of these advances and the use of these techniques were recently described in a comprehensive review[1]."

[..]

"1. Wilson PC, Andrews SF. Tools to therapeutically harness the human antibody response. *Nat Rev Immunol.* 2012;12:709–719. [PMC free article] [PubMed] [Google Scholar]"

- https://www.ncbi.nlm.nih.gov/pmc/articles/PMC4844175/

So it seems that Shelton's assertion probably held true at the time of his writing but no longer does so. I imagine that the belief in antibodies might be so deeply entrenched in medical science that finding them might be ascribing a pre-existing explanation for something the finding of wch is wishful thinking. Still, having spent most of my life thinking that antibodies are a probability I won't abandon the idea just yet.

""The best, indeed the only, method of promoting individual and public health is to teach people the laws of nature and thus teach them how to preserve their health. Immunization programs are futile and are based on the delusions that the law of cause and effect can be annulled. Vaccines and serums are employed as substitutes for right living; they are intended to supplant obedience to the laws of life. Such programs are slaps in the face of law and order. Belief in immunization is a form of delusional insanity." (PRINCIPLES OF NATURAL HEALING — p. 478 — Shelton)" - p 10

I both agree & disagree w/ the above. I'm uncomfortable w/ expressions like "the laws of nature", I perceive nature as a very fluid state, something that adapts w/ an almost endless flexibility; I don't see it as something that has "laws", at least not in the sense that "laws" is usually used to mean fixed always-definable conditions. As such, Shelton's way of expressing his point is too far off from my feelings for me to easily relate or agree. Where I identify the most is when he writes "Vaccines and serums are employed as substitutes for right living" - even then, the language bothers me. I think that the medical attitude, at least in the case of 'Western medicine', is the 'quick fix' attitude: *Oh, you're an obese alcoholic? Here, take these pills.*

In other words, there're things that most people know they can do to improve their health: maintain a weight that doesn't put too much of a strain on their muscles & their organs - a person's body cannot handle 350 pounds; don't use or overuse things that're obviously toxic: sugar, alcohol, cigarettes, heroin, crack, speed, caffeine, etc.. Despite the common knowledge of such things, people go to drs instead & shoot insulin rather than cutting back on their sugar consumption, etc, etc.. Such an approach is a way of avoiding taking a hard look at one's failings & trying to improve oneself. It's the lazy way, the way that got them into their dilemma in the 1st place.

"**Dr. J. W. Hodge** had considerable experience with vaccination before he denounced it and wrote a book on his collected data. In his book THE VACCINATION SUPERSTITION (p 41) he states:

""After a thorough investigation of the most authentic records and facts in harmony with the physician's daily observations and experiences, the conclusion is drawn that instead of protecting its subjects from contagion of smallpox, vaccination actually renders them more susceptible to it. Vaccination is the implantation of disease . . . Every pathogenic disturbance in the infected organism wastes and lowers the vital powers, and this diminishes its natural resisting capacity.["]" - p 10

That seems common-sensical to me. As such, I reckon it approaches what I can accept as axiomatic. Nothing living & growing & moving is impervious to all harm. Instead, each living thing has barriers that protect its life. Trees have bark, animals have skins. To me, it seems simple & obvious enuf that our breathing has filtration to protect us. We sneeze if our nose takes in something too big or otherwise harmful. Our skin prevents the penetration of lesser attacks on it. A needle is desgined to penetrate that defense. As such, how can its use be not harmful? Everyone knows that once a drug user gets to the point of injecting they've gone to a new level of dangerous abuse. Injecting heroin is worse than snorting it. I've reached a point in my life where injecting *anything* seems like a very bad idea.

Hodges continues w/ a summarization list.

"20 — That compulsory vaccination ranks with human slavery and religious persecution as one of the most flagrant outrages upon the rights of the human race."" - p 11

Again, that seems common-sensical to me. Forcing someone to have a needle put in them that bypasses their defenses & injects them w/ clearly harmful matter is rape.

The full title of Dr. Hodges bk is <u>The Vaccination Superstition: Prophylaxis to Be Realized Through the Attainment of Health, Not by the Propagation of Disease; Can Vaccination Produce Syphilis?</u> It's from 1901 & is widely available thru the internet.

Chapter II is called "**SMALLPOX DECLINED BEFORE VACCINATION WAS ENFORCED**" & begins w/ a quote I like: ""**To mistake inferences or axioms for facts has been a curse of science.**" **—Sir Clifford Albutt, in "Nature."**" - p 12

Axiom:

"*noun*
1. a self-evident truth that requires no proof.
2. a universally accepted principle or rule.
3. *Logic, Mathematics.* a proposition that is assumed without proof for the sake of studying the consequences that follow from it." - https://www.dictionary.com/browse/axiom

How cd anything in science require no proof? It's my understanding that in math it's not just that axioms are accepted "for the sake of studying the consequences that follow from" them but also b/c no-one's devised a *way of proving them*. That's pretty significant given that pi, e.g., has been proven to be **infinitely** non-repeating. A discipline that can prove *that* has some powerful reasoning tools involved. IMO, *nothing* is a "self-evident truth". I quote from my own bk entitled Paradigm Shift Knuckle Sandwich & other examples of P.N.T. (Perverse Number Theory) (pp 16-17):

"Let's backtrack to AXIOMS & then segue to .. (but I fast-forward). If an axiom is: "1. a self-evident truth, 2. a universally accepted principle or rule, 3. *Logic, Math.* a proposition that is assumed without proof for the sake of studying the consequences that follow from it." (The Random House College Dictionary, 1975) then we're immediately presented w/ some obvious philosophical problems, eh?

"How is anything ever "self-evident"? (Anything is [Self-Evident]) P.N.T. accepts *no* "universally accepted principle or rule". If logic & math are dependent upon axioms as their bottom line then that's a demonstration of logic & math's most self-conflicted dilemma insofar as they require absolute irrefutable proof *w/o assumption*.

"One of the axioms of arithmetic is the *Commutative Law of Addition*:

"For any numbers m and n,
m+n=n+m

"That probably seems to most people to be a "self-evident truth". But, 'pataphysically speaking (perhaps), there must be an exception to this. Finding it & describing it precisely is the problem. Therefore, in P.N.T. we have:

"(m + n * n + m) -> x
((m + n does not equal n + m) is isomorphic to x)

"The problem is, therefore, what is x?

"What's being postulated here (w/o any substantiation) is a thought experiment providing an alternative to a "self-evident truth" that then, hypothetically, calls into question *any* "self-evident truth"s. This is, of course, certainly not unprecedented. Let's take the example of Euclidean Geometry.

"Euclid was a Greek mathematician from 300BC now known as the "Father of Geometry". The 4 postulates of his "absolute geometry" were followed by a 5th one, more problematic, commonly called the "parallel postulate". In this, it's stated that for every line w/ a point not on that line there is only *one line* running thru that point on the same plane that doesn't intersect the 1st line. Over 2,000 yrs later, Non-Euclidean Geometry was conceived of - in which this postulate no longer holds true. Hyperbolic Spaces are modeled on the possibility of at least 2 lines passing thru the point that don't intersect the 1st line."

ANYWAY, now that I've managed to promote one of my favorite bks that It seems no more than 5 people will probably read in my lifetime, let's get back to issues regarding science. My point is that science is reputed to provide a provable alternative to faith-based religion. Science's hypothetical strongpoint is that its method can be used to observe something in such a way that leads to ways of effecting what's observed in a repeatable manner.

Hence, a disease can be observed, the problem that it presents for the diseased can be identified & characterized & can then be treated in a repeatable way that enables a solution to a problem. Religious people claim that prayer does this. I've never observed that to be the case.

Alas, as soon as axioms become involved things become considerably less pragmatic & more faith-based. Thusly, science turns into religion, its hypothetical opposite. It's become glaringly obvious to me in recent yrs that there're subtexts, unspoken assumptions, basic errors that're axiomatic about medical thinking that ultimately make its whole approach untenable from my POV.

Let's take the example of cancer:

"Cancer is a disease in which some of the body's cells grow uncontrollably and spread to other parts of the body.

"Cancer can start almost anywhere in the human body, which is made up of trillions of cells. Normally, human cells grow and multiply (through a process called cell division) to form new cells as the body needs them. When cells grow old or become damaged, they die, and new cells take their place.

"Sometimes this orderly process breaks down, and abnormal or damaged cells grow and multiply when they shouldn't. These cells may form tumors, which are lumps of tissue. Tumors can be cancerous or not cancerous (benign).

"Cancerous tumors spread into, or invade, nearby tissues and can travel to distant places in the body to form new tumors (a process called metastasis). Cancerous tumors may also be called malignant tumors. Many cancers form solid tumors, but cancers of the blood, such as leukemias, generally do not.

"Benign tumors do not spread into, or invade, nearby tissues. When removed, benign tumors usually don't grow back, whereas cancerous tumors sometimes do. Benign tumors can sometimes be quite large, however. Some can cause serious symptoms or be life threatening, such as benign tumors in the brain."

- https://www.cancer.gov/about-cancer/understanding/what-is-cancer

The above is a definition of cancer. It has an unquestioned basis: viz: "abnormal or damaged cells grow and multiply when they shouldn't" - what exactly does that mean? I think that it's the arrogance of the medical world that's decided that it's understood what 'should' or 'shouldn't' be going on in the body. As a thought experiment, imagine that the production of cancer cells is just another natural process that results from specific stimulus in the body. That doesn't mean that the cancer cells aren't a harbinger of the death of the body, it might mean that whatever stimulus is its cause might be possibly counterbalanced &/or reduced. Instead, the medical approach is to attack the body in such a severe way that the cancerous cells are killed off b/c the body they're in is brought even closer to the brink of death. This technique is Draconian, to say the least, & isn't that successful. To me, it's also obviously an extremely ass-backwards approach.

Back to The Poisoned Needle:

"The most noticeable decrease in smallpox and other zymotic diseases began with the sanitation reforms just prior to 1800 and the improvement in nutrition brought about by such health crusaders as Trall, Graham and Jennings around 1840.

"THIS SANITATION AND HEALTH IMPROVEMENT PROGRAM included: (1) Sewage disposal, (2) cleaning of streets, back yards, stables, etc., (3) improvement of roads so that fresh vegetables, milk and other vital foods could be transported rapidly to the cities and distributed while still fresh. (4) The water supply was improved and protected from contamination. (5) Housing projects were built out in the suburbs to relieve congestion of population in the cities.

"THE NUTRITIONAL TEACHINGS stressed natural whole grain bread instead of white bread, fresh fruits and vegetables free from salt, sugar, chemicals and other harmful preservatives and the rejection of coffee, tea, alcohol, tobacco, drugs and other drastic poisons. Meat and other low grade proteins were denounced in favor of nuts, beans and other such proteins with more health value and less toxic effects." - p 12

Zymotic disease: "An old name for a contagious disease, which was formerly thought to develop within the body following infection in a process similar to the fermentation and growth of yeast." - https://www.oxfordreference.com/view/10.1093/oi/authority.20110803133614167

So, there you have it: some of the philosophical basis that runs thru The Poisoned Needle: What's considered to be healthy is elucidated & proposed as being more valuable than medical interventions considered to be doing more harm than good. I tend to agree w/ this but have to acknowledge that it too has an axiomatic subtext open to debate.

Nonetheless, I provide this example to explain why I'm sympathetic to this position: In the mid 19th century there was a cholera epidemic in London that was originally explained as having been caused by "Miasma", the transmission of harmful particles by air. A physician named John Snow hypothesized that the cause was the transmission of germs thru water mainly originating from a pump on Broad Street. As I remember it, it was eventually realized that human waste matter from a cesspool had been leaking into the well that the pump drew from. Improved sanitation, therefore, meant ensuring that the waste didn't leak into the drinking water.

This subject has particular interest for me b/c one of the earliest medical studies that I was a research volunteer for was a cholera one. Fortunately for me, I was one of the control subjects & wasn't, therefore, given cholera.

It seems to me that separating the waste from the drinking water is a sensible idea. It also seems to me that the vaccination approach is to inject the waste directly into the body - thusly introducing the potentially harmful substance in a way that bypasses the body's filters. Given that the original idea behind smallpox vaccination was to take pus from cowpox pustules & to smear it into cuts put on the patient's arm my description of vaccination as injecting waste directly into the body seems accurate. After all, pustules are produced by the body to eliminate harmful waste. Imagine the harmful effect of sealing yr anus shut so that you can no longer defecate.

"Before the passage of England's compulsory vaccination law in 1853, the highest authentic smallpox death-rate was only 2,000 for any two year period, even during their most serious smallpox epidemics; whereas, after almost 20 years of compulsory vaccination there occurred the most devastating scourge of smallpox in 1870 to 1871 that the world has ever known. It took 23,062 lives in England and Wales and spread over Europe in all the countries where vaccination and inoculation had been practiced on a large scale. After that the vaccination laws were enforced even more rigidly until the people began to notice that smallpox was not decreasing by this practice but continued to ravage the homes of the vaccinated. During the same epidemic in Germany 124,948 people died of smallpox. All had been vaccinated (according to their carefully kept records.) "In Berlin alone no less than 17,038 persons had smallpox after vaccination, and 2,884 of them died."" - p 13

It seems somewhat 'inevitable' to me that proponents of vaccination will claim that the above statistics are untrue or just twisted to serve the purposes of people who're anti-vaccination. It seems fairly obvious to me that if anti-vaxxers can be guilty of such distortion then so can pro-vaxxers. Here's a sample from Wikipedia:

"The Franco-Prussian War triggered a smallpox pandemic of 1870–1875 that claimed 500,000 lives; while vaccination was mandatory in the Prussian army, many French soldiers were not vaccinated. Smallpox outbreaks among French prisoners of war spread to the German civilian population and other parts of Europe. Ultimately, this public health disaster inspired stricter legislation in Germany and England, though not in France." - https://en.wikipedia.org/wiki/History_of_smallpox

Note that the Franco-Prussian War is sd to've "triggered" the pandemic. That seems fair enuf to me: wars are environmental disasters in wch normal bodily protection is usurped. Note next that it's the unvaccinated French soldiers who're ultimately blamed for spreading the disease. Note also that there's no mention of the German civilian population being vaccinated. Now, who are we to believe? Well, as usual, I tend to believe what's most supported by my personal experience - that means, for me, that the people most likely to be making money & to be causing the misery are also the most likely to lie to support their position. That means that the governments & the wealthy people behind them causing the war & making the money off the vaccinations are the least likely to be trustworthy.

""In 1942 one case of smallpox at Seindon (Britain) resulted in the vaccination of a large number of people. Only three cases of smallpox occurred and these all recovered, but 12 vaccinated individuals died from inflammation of the brain. (This is a common after effect of vaccination.) In the same year near Edinburg, Scotland, eight people died of smallpox (six of which had been vaccinated) while **ten died** from the effects of the vaccination . . .[""]" - p 12

"(Vaccine and Serum Evils, p. 23, by Dr. H. M. Shelton)" - p 13

"Meningitis is an inflammation (swelling) of the protective membranes covering the brain and spinal cord. A bacterial or viral infection of the fluid surrounding the brain and spinal cord usually causes the swelling. However, injuries, cancer, certain drugs, and other types of infections also can cause meningitis." - https://www.cdc.gov/meningitis/index.html

I quote from a relevant passage of my review of Andrew J. Wakefield's <u>Callous Disregard—Autism and Vaccines—The Truth Behind A Tragedy</u>:

""At the time, a brand of the MMR vaccine had been withdrawn in Canada because it was unsafe; Canada had introduced an MMR vaccine containing the Urabe AM-9 mumps virus (*Trivirix*) in 1986. It soon became apparent that this vaccine caused unaccceptably high rates of meningitis. George was invited to sit on the UK's Joint Committee on Vaccination and Immunization (JCVI) as its Scottish representative. In Canada, he had collated extensive information about the MMR hazard; therefore, he was in an excellent position to advise the DoH and, particularly, Dr. David Salisbury, the UK's chief MMR strategist, on the introduction of the MMR. George claimed:

""...*he was very cautious not to openly release the damning evidence that he had retained from Canada regarding the safety of the MMR vaccine...*" - pp 65-66

"I suggest that people in positions of high authority tend to reduce the humans that they're responsible for to statistics & to, therefore, have a high acceptance of collateral damage b/c it becomes largely dehumanized for them - but what's an *acceptable* rate of meningitis to someone who gets it as a result of a vaccine that's supposed to *protect their health* but, instead, has the opposite effect? Obviously, there is no *acceptable* rate of meningitis to its victim. Think of the military: the lower the rank of the soldier, the more their deaths are considered to be 'acceptable' collateral damage. Obviously, this is class warfare. These same

'expendable' people are also used in scientific experiments by exposure to Agent Orange, Depleted Uranium, Chemical & Biological Warfare Agents, etc.. Do you really think that the high & mighty decision makers, who aren't, personally, taking risks themselves, are that much different in the Medical Industry from their counterparts in the Military?! I don't. ESPECIALLY when they stand to gain from it financially - & let's not forget to throw in potential sadism."

More regarding meningitis from The Poisoned Needle:

"**Dr. R. C. Carter** of London reports:

""In looking over the history of vaccination for smallpox, I am amazed to learn of the terrible deaths from vaccination, amputations of arms and legs, foot and mouth disease, tetanus (lockjaw), septicemia (blood poisoning), cerebro spinal meningitis."" - p 21

"In spite of the 5,500,000 vaccinations in 1920 and the 14,600,000 vaccination[s] in 1926 and all the vaccinations in the intervening years, including the compulsory vaccinations of all infants, there occurred another larger and more terrible smallpox epidemic in 1932 (only 6 years after the alleged complete control of smallpox in Egypt). This sweeping epidemic continued for two years, aggravated, of course, by the meddling vaccinators."

[..]

""These figures speak for themselves and eloquently proclaim the utter futility of vaccination as a preventative or mitigant of smallpox." (Swan — "The Vaccination Problem" — p.291.)" - p 18

"At the beginning of the Second World War immunization was made compulsory in Germany and the diptheria rate soared up to 150,000 cases (1939) while in unvaccinated Norway there were only 50 cases." - p 19

The relationship between Nazi ideology & pretenses of good public heath as an excuse for genocide have been deeply pointed out. Here's an excerpt from an article entitled "Hitler and the fight against viruses" that was translated for & reproduced in my bk Unconscious Suffocation:

"[O]ne of Hitler's first steps after coming to power in 1933 was a major tuberculosis control program. He mobilized a fleet of trucks equipped with X-ray machines to detect the disease by X-ray of the lungs and isolate the sick.

"You had no choice whether to get tested or not. The motto of the Nazis, it should be remembered, was "common interest before individual interest" ("Gemeinnutz vor Eigennutz"). It was engraved on 1 Mark coins from 1933[.]

"After tuberculosis, Hitler launched a major program to disinfect factories and eliminate lice and rodents, which cause epidemics.

"Zyklon B gas, later used in gas chambers, was used for this purpose."

- pp 308-309, <u>Unconscious Suffocation - A Personal Journey through the</u>
<u>PANDEMIC PANIC</u>

Chapter III begins w/ a great epigraph:

""It often happens that the universal belief of one age — a belief from which no
one was, not without an extraordinary effort of genius and courage could, at that
time be free — becomes to a subsequent age so palpable an absurdity that the
only difficulty, then, is to imagine how such a thing can ever have appeared
credible." - John Stuart Mill" - p 21

Who are all these people that McBean quotes? Many people will recognize the
name of John Stuart Mill.

"John Stuart Mill (1806–73) was the most influential English language
philosopher of the nineteenth century. He was a naturalist, a utilitarian, and a
liberal, whose work explores the consequences of a thoroughgoing empiricist
outlook. In doing so, he sought to combine the best of eighteenth-century
Enlightenment thinking with newly emerging currents of nineteenth-century
Romantic and historical philosophy. His most important works include *System of
Logic* (1843), *On Liberty* (1859), *Utilitarianism* (1861) and *An Examination of Sir
William Hamilton's Philosophy* (1865)."

[..]

"We have seen, then, that Mill holds that "happiness is the sole end of human
action, and the promotion of it the test by which to judge of all human conduct"
(*Utilitarianism*, X: 237). The content of this claim, however, clearly depends to a
great extent upon what is *meant* by happiness. Mill gives what seems to be a
clear and unambiguous statement of his meaning. "By happiness is intended
pleasure, and the absence of pain; by unhappiness, pain, and the privation of
pleasure" (*Utilitarianism*, X: 210). That statement has seemed to many to commit
Mill, at a basic level, to *hedonism* as an account of happiness and a theory of
value—that it is pleasurable *sensations* that are the ultimately valuable thing.

"Mill departs from the Benthamite account, however, which holds that if two
experiences contain equal quantities of pleasure, then they are thereby equally
valuable. In contrast, Mill argues that

"[i]t would be absurd that while, in estimating all other things, quality is
considered as well as quantity, the estimation of pleasures should be supposed
to depend on quantity alone. (*Utilitarianism*, X: 211)

"Some pleasures are, by their nature, of a higher quality than others—and as
such are to be valued more.

"If I am asked, what I mean by difference of quality in pleasures, or what makes one pleasure more valuable than another, merely as a pleasure, except its being greater in amount, there is but one possible answer. Of two pleasures, if there be one to which all or almost all who have experience of both give a decided preference [...] that is the more desirable pleasure. (*Utilitarianism*, X: 211)"

- https://plato.stanford.edu/entries/mill/#MillConcHapp

I have yet to read Mill, although I'm sure I have at least one bk of his in my personal library. I'm sure that his emphasis on happiness can be twisted by malevolent parties to produce utmost misery. Nonetheless, Mill seems like my kinda guy.

"**Professor Adolf Vogt,** who held the chair of **Vital Statistics** and **Hygiene** in Berne University for 17 years said:

"After collecting the particulars of 400,000 cases of small pox I am compelled to admit that my belief in vaccination is absolutely destroyed."" - p 22

It seems that Edward Jenner is generally credited w/ instigating mass vaccinations & profiting from them. However, it's The Poisoned Needle's claim that: "Jenner and the modern experimenters in vaccination were not scientists and did not test and prove their theories by any **scientific** criteria before launching the hazardous practice on the public." - p 26

McBean does detail the claim against Jenner as a scientist later but it seems that the orthodoxy is predictably against her:

"Jenner went to school in Wotton-under-Edge and Cirencester. During this time he was inoculated for smallpox, which had a lifelong effect upon his general health. At the age of 14 he was apprenticed for seven years to Mr Daniel Ludlow, a surgeon of Chipping Sodbury, where he gained most of the experience needed to become a surgeon himself.

"In 1770 he moved to St. George's Hospital in London, to complete his medical training under the great surgeon and experimentalist John Hunter. Hunter quickly recognised Edward's abilities at dissection and investigation, as well as his understanding of plant and animal anatomy. The two men were to remain lifelong friends and correspondents.

"In 1772 at the age of 23, Edward Jenner returned to Berkeley and established himself as the local practitioner and surgeon. Although in later years he established medical practices in London and Cheltenham, Jenner remained essentially a resident of Berkeley for the rest of his life."

- https://www.jenner.ac.uk/about/edward-jenner

"INCREASED DEATH-RATE AFTER VACCINATION WAS ESTABLISHED

"The report of **Dr. William Farr, Compiler of Statistics** of the Registrar-General, London, states:

""**Smallpox attained its maximum mortality after vaccination was introduced** . . . The mean annual (smallpox) mortality to 10,000 population from 1850 to 1869 was at the rate of (only) 2.04, whereas (after compulsory vaccination) in 1871 the death rate was 10.24 and in 1872 the death rate was 8.33, and this after the most laudable efforts to extend vaccination by legislative enactments."" - p 27

One of the most thought-provoking statements of McBean's is probably the most heretical. I find it quite appealing.

"As has been stated before, disease is a cleansing effort of the body in its attempt to rid itself of excess poisons, waste matter, obstructions, and incompatible food. DISEASE IS NOT SOMETHING TO BE CURED; IT IS A CURE.

"Germs do not attack from without; they develop within the cells themselves when the need for them occurs. The whole framework of vaccination is based on the misconception that germs cause disease and must be counteracted with vaccines. But this procedure can bring no other result than harm." - p 28

Again, heretical - that's a profound paradigm shifter - or cd be if it were allowed to breathe. As usual, I approach considering things as thought experiments, I don't have to believe them one way or another. Imagine that a cancer tumor is an attempt by the body to call attention to an excess of toxicity that has no ordinary release b/c it's *too much of an overload*. What if a sensible person were to respond to it in a way similar to their response to pain? Imagine that you step on something sharp: it hurts, you recoil, the pain lessens, you know you've taken the right action. If we were smarter about such things we might be able to recognize cancer as the result of an action or series of actions that we can stop & reverse. Let's take an obvious example: you get lung cancer b/c you're constantly injuring yr lungs w/ cigarette smoke. You stop smoking. Will the cancer then go away? Maybe not - but stopping smoking is still a good idea & it may help prevent the cancer from getting worse.

More about Jenner:

"**Edward Jenner** inoculated his 18 months old son with swine-pox, on November 1791, and again in April, 1798 with cow-pox. The boy was never well after that and died of tuberculosis at the age of 21.

"In **Baron's Life of Jenner,** (Vol. II, p. 304) we learn that, "On the 14th of May, 1796 . . . Jenner vaccinated James Phipps, a boy about eight years old, with the matter taken from the hand of a dairymaid infected with **casual cowpox.**"

[..]

"The inoculation didn't "take" so on the strength of this one experiment and its questionable interpretation, Jenner based his claim that **one** vaccination would **"forever** secure a person from smallpox." No time had elapsed to prove whether it would last a lifetime or a month or not at all; but without any proof or any scientific basis or evidence for its practice, the doctors and the government adopted it and made it compulsory, no doubt, seeing the gold mine in profit it would yield.

"James Phipps was declared immune to smallpox but he too, died of tuberculosis at the age of 20." - p 29

I started wondering what difference there is, if any, between pimples & pox - other than severity of possible outcomes. After all, they're both eruptions on the skin that bring rejected matter to the surface. A brief search led to this, a *Science News* article entitled "Pox To Pimples; Study Says Acne-like Outbreak Not Vaccinia":

"The national smallpox vaccine program proved last year to be a bumpy road; it was also especially spotty for about 10 percent Vanderbilt's vaccinated volunteers who broke out in a curious acne-like rash.

""Several of the volunteers said their acne worsened, and we weren't sure it was the vaccine," said Dr. Tom Talbot, an Infectious Diseases fellow and co-investigator in Vanderbilt's smallpox vaccine clinical trials. (The national trials also were coordinated here.)

"He set the claims aside -- this type of reaction hadn't been documented in previous vaccinations -- but the investigative side of his public health brain kept it simmering.

""Then in late October (2002, during the clinical trial), we saw two volunteers with generalized rashes on their backs and legs and we became more attuned to a possible association with the vaccine," Talbot said. "The second light bulb to go off was in a news story in which one volunteer noted new "acne" on his back after vaccination. He said he 'felt like a teen-ager again'.""

[..]

"From October to the first of December last year, his study reports, 148 volunteers participated in the most recent of Vanderbilt's smallpox vaccine trial. All were between 18 and 32 years old, had never received the vaccine and passed a health battery. Talbot and Dr. Kathryn Edwards, professor of Pediatrics and vice chair for Pediatric Research, were testing the efficacy of the Aventis Pasteur smallpox vaccine at full strength and in dilutions of one-to-five and one-to-10.

"Between eight and 10 days post-vaccination, the time of maximal viral replication, four people had a generalized rash of pustules on their arms, legs, face, back and trunk; 11 had a similar, but more focal, reaction.

"Historical accounts of smallpox vaccine side effects listed "non-descript rashes or general vaccinia," a spreading of the virus throughout the body, a sometimes fatal side effect, Talbot said. But, he added, "they were never well described."

"Samples of the lesions seen at Vanderbilt tested negative for the virus, ruling out vaccinia and possibly establishing a new, less serious, reaction to the vaccine: folliculitis, benign eruptions like pimples."

[..]

"Three other smallpox vaccine studies, in civilian and military populations, ran in the same JAMA issue that carried Talbot's. An editorial by Dr. Anthony Fauci, director of the National Institute of Allergy and Infectious Diseases, said Talbot's study "illustrates the value of reevaluating seemingly established phenomena when new (diagnostic and technological) tools become available."

""If these cases had occurred 40 years ago, when cultures were not routinely performed on lesions occurring after vaccination, these individuals might have been diagnosed as having generalized vaccinia," Fauci and co-author Dr. Mary E. Wright, also of the NIAID, wrote."

- https://www.sciencedaily.com/releases/2003/06/030626074234.htm

So, let me get this straight, the vaccinated people broke out in folliculitis after being vaccinated but that doesn't qualify as vaccinia. Are you convinced? I'm not.

"**1** *or* **vaccinia virus** : a poxvirus (species *Vaccinia virus* of the genus *Orthopoxvirus*) that differs from but is closely related to the viruses causing smallpox and cowpox and that includes a strain used in making vaccines against smallpox
2 : a reaction to smallpox vaccine prepared from live vaccinia virus that may involve a rash, fever, headache, and body pain"

- https://www.merriam-webster.com/dictionary/vaccinia

"Dr. Clements in his pamphlet, "**A Superstitious Custom**" traces the inoculation practices through the various modern countries previous to Jenner's day."

[..]

""In 1754, Peverani introduced inoculation for smallpox into Rome, but smallpox soon began to spread and opposition rose to such a pitch that the practice was discontinued, until the medical profession, after years of labor, persuaded the people to again submit to it.

""One hundred forty-two years later, Carlo Ruta, Professor of Materia Medica at the University of Perugia, **Italy,** protested against the deadly custom in these scathing words:

"" '**Vaccination is a monstrosity, a misbegotten offspring of error and ignorance; and, being such, it should have no place in either hygiene or medicine . . . Believe not in vaccination, it is a world-wide delusion, an unscientific practice, a fatal superstition with consequences measured today by tears and sorrow without end.**' "" - p 30

"Regarding the increased death-rate due to vaccination, Herbert Spencer states in his **Facts and Comments:**

""Jenner and his disciples have assumed that when the vaccine has passed through the patient's system he is safe against smallpox, and there the matter ends . . . I propose to show that there the matter does not end. The interference with the order of Nature has various sequences than those counted upon. Some have been made known.

""A Parliamentary return issued in 1880 (No. 392) shows . . . that there was a decrease of 6,600 (per million births) in the deaths of infants from all causes; while the deaths caused by eight specified diseases, either directly communicable or exacerbated by the effects of vaccination, increased from 20,524 to 41,353 per million births per annum — more than double. It is clear that far more were killed by these other diseases than were saved from smallpox."" - p 33

In other words, it seems to be a common belief that vaccination against a specific disease only effects that particular disease. Why? It seems to me that if one introduces an illness into the body, no matter to what small degree this is presented as being, that the body's defense will be undermined - thusly increasing the likelihood of other negative conditions taking hold.

"According to **Thomas Morgan** in his **Medical Delusions** (p. 48-49) "Jenner soon discovered that vaccination did not give immunity from smallpox, including some who had been vaccinated by himself and had died from it. Not wishing to bring vaccination into disrepute, he endeavored to suppress reports, and in writing to a friend said, '**I wish my professional brethren to be slow to publish fatal reports after vaccination.**' and in 1810 he wrote: 'When I found Dr. Woodworth about to publish his pamphlet relative to the eruption (smallpox) cases at the Smallpox Hospital, I entreated him in the strongest terms, both by letter and conversation, not to do a thing that would so disturb the progress of vaccination.' (Barron's **Life of Jenner**).

""The foregoing plainly proves that Jenner himself was aware of the utter uselessness of vaccination; but, having received the bounty from the government . . . he preferred to resort to all kinds of schemes rather than acknowledge its failure.

""From its inception until the present day, the vaccination scheme has been an endless record of lies, deception, fraud, juggling statistics, and falsifying death certificates in order to preserve vaccination from reproach and to secure its continuation . . . and all this after more than a century of terrible experience, which has demonstrated that vaccination has killed more than smallpox, besides crippling and disfiguring millions more."" - p 35

The Poisoned Needle is full of testimonials of this kind & I, for one, am convinced by them - but that's at least partially b/c over the yrs I've seen entirely too much evidence of greed & sadism & general lack of concern among people whose PR image is to the contrary.

"In 1860 the Encyclopedia Britannica (Eighth Edition) stated:

""Nothing is more likely to prove hurtful to the cause of vaccination and render the public careless of securing to themselves its benefits, than the belief that they would require to submit to re-vaccination every 10 to 15 years."

"Years later, the Eleventh Edition of the Encyclopedia shows the change of policy when it states:

""It is desirable that the operation (vaccination) should be repeated at the age of from 7 to 10 years, and thereafter, if it be possible, at intervals during life." (The **more often** the vaccinations, the **greater the revenue** to the doctors and vaccine promoters. This is a prime consideration in this business.)" - p 39

As usual, differing viewpoints will provide differing explanations of the Britannica change-of-POV: pro-vaxxers will say that w/ the advancing yrs a more informed perspective was found, anti-vaxxers will say that the commercial interests became more powerful. It's interesting to note that when the COVID-19 'vaccines' were put forth they were presented as a one-shot deal. This progressively changed until 3 or 4 became the norm - & all this over a yr or so. To me, this is just marketing psychology at play - the same marketing psychology that's been at play all along in the history of vaccinations - if the 'need' for multiple vaccinations were presented from the beginning the likelihood of greater resistance wd exist. The suckers have to be slowly roped into the spider's web.

"Japan is one of the most glaring examples of the failure of vaccination and re-vaccination. Dr. Ruta, in his review of world statistics on smallpox reported:

""Between 1886 and 1892 there were 25,474,370 vaccinations and re-vaccinations performed in Japan, which meant that about two-thirds of the entire Japanese population, already vaccinated by the law of 1872, were re-vaccinated. **During that 7-year period (1886-1892) of thorough re-vaccination, there were reported 165,774 cases of smallpox with 28,979 deaths.**"" - p 41

""The chief, if not the sole cause of the monstrous increase in cancer has been vaccination." — Dr. Robert Bell, the famous cancer specialist of the British Cancer Hospital." - p 41

"**Robert Bell** FRFPSGlas (1845 – 1926) was a British physician, naturopath and medical writer, who advocated for alternative cancer treatments and vegetarianism." - https://en.wikipedia.org/wiki/Robert_Bell_(physician)

"In the **New York Press,** January 26, 1909, W.B. Clark stated:

""Cancer was practically unknown until cowpox vaccination began to be introduced. **I have had to do with 200 cases of cancer and I NEVER SAW A CASE OF CANCER IN AN UNVACCINATED PERSON.**" - p 42

Regarding the latter, see this article: https://www.sphir.io/cards/5dcaf04569702d4c63c72000

The "PHOTOGRAPHS OF FATAL CANCER CASES CAUSED BY VACCINATION" start on p 43. They're horrific since the people shown look otherwise healthy but're covered w/ truly huge tumors. Explanations accompany the photos. Naysayers against any criticisms of vaccination will easily dismiss such sensational photos as manipulatively over-the-top, & maybe they are - but as a person who's looked at medical textbooks I'd say they fit right in - in other words they're blunt presentations of a serious problem.

Much of what's stated in this bk will seem like common sense to some of us & like blasphemy to BELIEVERS in the medical system. One former friend of mine sd something to the effect that drs have 350 yrs of experience to validate their opinions. My retort is that the Inquisition went on for over 400 yrs & was justified as being for the good of the individual & for society but that doesn't justify its brutality & its obvious greedy & power-hungry ulterior motives. As far as I'm concerned, the same goes for the medical industry.

"One of the functions of the lymph glands is to filter the poisons from circulation before they reach the cells. When an excess of poison is generated in the body or is introduced from the outside as in **vaccination,** the lymph glands become enlarged in an effort to collect and hold in abeyance, the accumulation of poison. In the case of blood poisoning or a bite of a poison insect a lump usually forms under the arm or in the groin, to meet the emergency, but disappears again when the condition has healed. In the case of the continued intake of poison such as tobacco, coffee, drugs, etc., the glands most effected keep enlarging to accomodate the inflow of poison. A protective coating of tissue is formed around the gland to protect the body against collected poisons. It is this wonderful defense mechanism that the body provides against the ignorance and willfulness of the individual, that the doctors attack as an enemy. They call it **a wild growth of tissue** — a cancer or tumor. To cut or burn (as with radium, X-ray or surgery) this bulwark of protection is to add insult to injury and allow the **self-generated** poisons to flow unchecked through the body and do serious damage to the vital organs, tissues and glands." - p 48

That makes sense to me, let's see what the conventional position is on the function of the lymph glands.

"A small bean-shaped structure that is part of the body's immune system. Lymph glands filter substances that travel through the lymphatic fluid, and they contain lymphocytes (white blood cells) that help the body fight infection and disease. There are hundreds of lymph glands found throughout the body. They are connected to one another by lymph vessels. Clusters of lymph glands are found in the neck, axilla (underarm), chest, abdomen, and groin. For example, there are about 20-40 lymph glands in the axilla. Also called lymph node." - https://www.cancer.gov/publications/dictionaries/cancer-terms/def/lymph-gland

Well, well.. that seems to jive w/ McBean's position. Therefore, the conflict seems to enter when it comes to what to do about swollen lymph glands. McBean's position seems reasonable to me. There're reasons why the lymph glands wd get swollen & it seems sensible to me to stop doing whatever it is that one is doing to cause that swelling - thusly giving the glands time to work the poisons out & return to an unswollen condition. Surgery & all those other usual Draconian measures just, IMO, to quote McBean, "add insult to injury".

"The American Cancer Society, in a circular entitled FACTS ABD FALLACIES ABOUT CANCER says: "Cancer is not caused or cured by any known diet." Since most of the statements in that leaflet are false, this one is not to be taken seriously either. Abundant evidence to the contrary proves this claim to be without foundation in fact.

"**Dr. Willard Parker,** who for 30 years **held the chair of surgery** at **Columbia University Medical School** was quoted as saying:

""Luxorious living, and particularly excess in animal food, increases the waste products of the body, which if retained in the system have a tendency to produce abnormal growths . . . **Cancer is to a great degree one of the final results of a long continued course of error in diet,** and a strict dietetic regimen is, therefore, a chief factor in the treatment—preventative and curative."" - p 49

How the American Cancer Society can disassociate diet from *any* heath condition is beyond me. We are what we eat, what happens in our body is determined by what we put into it to create it. Obviously, if a person only drinks alcohol & doesn't eat their body's going to be in pretty sad shape & they won't last long. I keep referring back to when I went to a dr almost 3 yrs ago now worried that I was diabetic. She confirmed that I was & made dire predictions about my imminent death - saying that I had to immediately shoot insulin to save myself. I told her that I'd try changing my diet & getting more exercise 1st. She told me that was "impossible". I tried it anyway. W/in 5 days my symptoms were gone, w/in 2 wks my blood sugar was normalized. It wasn't even that hard. So much for the "impossible". That dr was grossly overpaid & close to useless. (See my relevant documentary here: https://youtu.be/2GLu66dgpKI)

"Many cases of external and internal cancer have been cured as a result of a fast followed by a 30-day diet of fresh (unsprayed) grapes. Some used crushed grape leaf poultices on the external cancer." - p 61

Of course, reports such as the above will be dismissed as dangerous nonsense by BELIEVERS in the medical system. After all, they're so brainwashed by capitalism that the only thing that makes sense to them is a sort of D-Day, a full-on assault on the body - & it's got to be expensive or how else can it be any good? Something that anyone can do w/o a dr & w/o money is impossible to them. I've fasted many times in the last 40 yrs & it certainly works for me. But it takes discipline.. it's so much 'easier' to passively submit to all the machines, & the surgery, & the chemicals.

"The newspapers quoted the doctors as saying that Bernarr MacFadden died of a three-day fast. They also quoted his lawyer as saying the same thing. The lawyer, Harry Gilgulin, was shocked when he read the statement he was supposed to have made. Excerpts from a published letter (*) refuting this statement are as follows: "I could never have made such a statement since I myself fasted some eight or nine times for a total of 175 days and benefitted enormously from each of these fasts . . . Three of my fasts were of 34 and 30-day duration . . . As a staunch supporter and enthusiastic advocate of hygienic living and an ardent believer in fasting, it must be obvious to you that I could never have made the statement attributed to me by the newspapers.

""What I did tell the reporters was that whenever Mr. MacFadden didn't feel well, he would go on a fast of several days duration . . . He had fasted three days but broke the fast improperly by one small glass of fruit juice in the morning and a **meat meal** with **eight** or more **pats of butter in the evening** and that the improper breaking of the fast is not only dangerous but may lead to dire consequences and even death."

"Mr. MacFadden was taken to the hospital and what they did to him in the way of shots and medicine was not stated, but he died soon after being taken to the hospital. People do not die of **three day fasts,** especially athletic men like MacFadden who are used to fasting." - p 51

The thought of dying from a 3 day fast is preposterous to me. Until I started intermittent fasting the shortest fast I ever did was one week - & one week isn't that big a deal. I've fasted for 4 wks twice now. That's more of a challenge. In general I agree that one shd be careful when breaking a fast but I've discovered that how I break a fast has never mattered much for me. After my 1st 28 day fast I went to a Japanese restaurant w/ the intent of cautiously only eating a bowl of miso soup. Feeling fine after the soup I ate some sushi. I eventually had a fairly large meal w/o the slightest negative consequences. I wdn't, however, recommend such recklessness to others.

The problem w/ the general public's attitude to such things is brainwashing. A person goes into a hospital or a dr's office feeling poorly & then dies & the medical staff are credited w/ 'doing all they could to save their life'. It seems to me that Occam's Razor suggests that the medical staff is more likely to've killed the person w/ incompetence & Draconian 'remedies' that're of the D-Day variety. To me, going into a hospital after having fasted is suicidal. The body can ingest food but *injections*?! No way.

As for the 'news'papers creating false quotes?! There must be many people out there who still believe in *Pravda* but I'm certainly not one of them. I have footage of a TV 'news' anchor quoting me as saying something that I never sd & never wd've sd. I have a friend who was interviewed about 2 different subjects. The 'news' took what he sd about one subject & attributed it as a quote about the other subject. This was no mistake, it was calculated propaganda .

Witness my movie entitled "TV 'News' Commits Suicide": https://youtu.be/hU-_aL7kKBl . This is a parody critiqueing the dishonesty of the 'news'. It's also got an obstacle course for you to navigate past on YouTube. Clicking on the link will take you to a warning screen that says: "The following content may contain suicide or self-harm topics. Viewer discretion is advised. Learn more." Then there's a button to click that reads: "I UNDERSTAND AND WISH TO PROCEED". Below this is a text that says:

"Talk to someone today

National Suicide Prevention Lifeline
Free Confidential Hours: Available 24 hours"

& gives a chat link & an 800 number.

Do not trust these people. I remember one time I called a poison control line b/c I'd eaten some mushrooms I shdn't have & got severe stomach cramps. The person who answered the phone offered *no help whatsoever* & only wanted to know who I'd gotten the mushrooms from. I didn't tell them. Later, I got a phone call from a hospital where 2 other people who'd taken the same mushrooms had been taken. I was told I'd either die in 3 days or I'd be fine. I foolishly decided to go to the hospital even though they told me there was nothing they cd do for me, as it turned out I was fine. I'll never make any of those mistakes again.

When I was 6, in 1959 or 1960, I had my tonsils removed. I'm sure my mom was scared into getting this done. It seems unlikely to me that it was in any way necessary. In fact, I think I've had a 'frog in my throat' ever since as a result. W/ that in mind it's interesting to read McBean's take on the tonsilectomy frenzy of the time.

"The teachers are instructed to examine the children every day for obvious symptoms of disease and report it to the school nurse. The nurse examines them, herself, each week and reports her findings to the doctors and the doctors and dentists come in for their check once or twice a year to round up considerable business for themselves by frightening the parents into needless treatments and operations for their children. The tonsils are the most vulnerable targets of these "mixed up" doctors. Tonsils are very necessary lymphatic glands of filtration that enlarge when the body is overloaded with incompatible foods and drugs." - p 52

As usual, that makes sense to me. It's my opinion that my body was damaged by drs 1st w/ circumcision, 2nd w/ tonsilectomy. The medical system is like a rigged deck: as long as they're dealing the cards you can't win. If their system does something damaging to the child & then the parent refuses to allow any further damage done by having the drs 'fix' the problem then the parents are deemed negligent, not the drs. It's a win-win situation for them - & always profitable.

"In the Encyclopedia Britannica (9th edition) under the heading **"Smallpox— Cowpox," Dr. Charles Creighton** says

"The real **affinity** of **cowpox** is not to **smallpox,** but to the **great pox (syphillis).** The vaccinal roseola is not only very like the syphlitic roseola, but it means the same sort of thing. The vaccinal ulcer of everyday practice is, to all intents and purposes, a **chancre** (syphlitic ulcer)."" - p 55

The same page features photos to back up this assertion. These photos are missing in the botched editions.

"Smallpox, to a large extent, went out when improved sanitation and nutrition came in, except in countries where vaccination is compulsory. There the smallpox epidemics continue to ravage the population at frequent intervals."

[..]

"**Dr. Chauven,** in an address before the French Academy of medicine, October, 1891, reviewed the results of his years of detailed and carefully tested experiments. In the face of undeniable facts he was forced to the conclusion that:

"(1) "Vaccine virus never gives smallpox to anyone." (Although it produces many diseases, some of which resemble smallpox and are diagnosed as smallpox and are more damaging than smallpox.)

"(2) "Vaccinia (disease produced by vaccination) is not even attenuated smallpox.

""Vaccinia is, in all probability, a modified form of syphilis, as has been clearly pointed out by Doctors Charles Creighton and E. M. Crookshank, Professor of Pathology and Bacteriology in King's College, London, two of the highest authorities on these subjects."" - p 59

The stakes are high in this rigged game, the pharmaceutical companies & the drs & medical institutions that front for them are making huge profits off of denying anything negative sd about vaccination - & the BELIEVERS are too afraid to even consider opinions that go contrary to the dogma.

"Even now, in what is sometimes called **the enlightened age,** we still hear some people (not so enlightened) claim or repeat what they have been taught to say, that "vaccination has reduced smallpox and certain other diseases, and has been a blessing to mankind." Those who make this claim have not bothered to consult authentic records or become acquainted with the most elementary facts on the subject.

"When traced down, these statements are found to originate from the fabricated propaganda of those who profit from the sale of serums and drugs. Unfortunately, many of these people are in places of authority where they influence the thinking of students, writers, public officials, teachers and others who mold the thinking of the masses." - p 59

I wdn't agree w/ that so much if I hadn't seen the same process heavily at work during the time of the QUARANTYRANNY. There's so much propaganda relentlessly at work & so many FACT CHOKERS telling people *what not to read* so that they won't even be exposed to the counter-propaganda that it's literally *sickening* - as in causing mental illness.

"The **Report of the Surgeon General of the Army,** (1919) vol. 1, page 37, gives the number of admissions to hospitals during the year 1918, on account of **vaccinia** (vaccination disease) as 10,830.

"The **Report of the Surgeon General of the Army,** (1918) gives me the number of admissions to hospitals during the year 1917, on account of **vaccinia** and **vaccinia-typhoid** combined as 19,608." - p 60

There're people who claim that the people who made up the statistics for the so-called 'Spanish Flu' epidemic casualties of 1918-1919 were primarily vaccinated soldiers. Is that so unlikely? Soldiers are always used as guinea pigs for chemical & biological experimentation & then the government exercises its usual denial. Soliders whose health was destroyed by exposure to the defoliant Agent Orange during the Vietnam War had to endure a 13 yr lawsuit before the government finally admitted culpability. It's still a mystery what American soldiers were exposed to during the Korean War but someone I was close to was apparently either effected by radiation exposure or by chemical weapons. Depleted Uranium took its toll during the Gulf War.

"**Dr, William Hitchman, Consulting Surgeon to the Cancer Hospital of Leeds** and formerly public vaccinator to the City of Liverpool, expressly stated that, **"Syphilis, abdominal phtisis, scrofula, cancer, erysipelas,** and almost all diseases of the skin, have been either conveyed, or **intensified by**

vaccination."

"**Dr. Peebles,** (world renowned researcher and authority on vaccination) stated in his book on Compulsory Vaccination:

""**We shall never stamp out smallpox, cancer, consumption, or leprosy, so long as we continue to STAMP THEM IN through the** idiotic and **vicious practice of vaccination.** The Germans endeavored to stamp out syphilis **by stamping it in** with syphilized vaccine. They have abandoned that practice now and in time they will abandon vaccination altogether."" - p 62

""It was found that nearly **all the cases of scarlet fever, measles and diphtheria had been recently vaccinated.** What is true of this city is true everywhere, and any close observer can satisfy himself in this direction."

"After reading this I decided to **"satisfy myself in this direction"** as suggested; so I obtained a LOS ANGELES COUNTY HEALTH INDEX, which gives the **weekly record** for diseases for the entire year. (This was 1954)

"I was amazed to find that the common diseases mentioned by Morgan, more than doubled after the annual June vaccination campaign. The figures are given below and will speak for themselves." - p 63

I won't quote the entire chart. It ends w/ "Total of the 48 diseases recorded" followed by "Before vaccinations" wch features the figure of "19,997" & "After vaccination campaign" wch features the figure of "47,070". That's pretty impressive. The Poisoned Needle provides plenty of evidence to support this. Naysayers will refute such evidence w/o any contradicting evidence of their own. Need I say? How many subject the basis for their pro-vax position to any such scrutiny?

"["]There is every reason . . . to believe the contrary, and to think that a system which has taken up a morbific principle, even though it should acquire thereby, immunity from cognate disease, must have its power of resistance to perturbing influences in general, diminished. Mr. Spencer (Herbert) finds evidence of this general debility in the greatly increased severity and **enlarged incidence of many diseases** such as measles, chickenpox, and influenza, **which, before vaccination became general, were of relatively rare** occurence and trifling in their effect. By a comparison of the infant mortality returns in quinquennial (five year) periods before and since compulsory vaccination, he finds that 'the mortality from 8 specified diseases, either directly communicable or exacerbated (aggravated) by vaccination, increased from 20,524 to 41,353 per million births per annum.' " (from — Vaccination a Disastrous Delusion, by McCormick - page 13.)" - pp 64-65

"BRIGHT'S DISEASE FROM VACCINATION

"**Earl Rohsbecker,** (Denver) was vaccinated for smallpox after which he

developed Bright's disease (an advanced form of degeneration of the kidneys).
The doctors do not hold out much hope for his recovery.

"**PARALYSIS FROM VACCINATION**

"**Raymond Nelson** (Denver) was given diphtheria toxin-anti-toxin after which he
developed paralysis. His son was given the antitoxin also, and died of it.

"**Mrs. Black** (Colorado Springs) was vaccinated for smallpox and soon
developed **spinal meningitis.** The case was so severe that the attending
physician had to call in assistance from Denver." - p 66

"According to the figures of the Office of Vital Statistics at Washington, D. C.,
there were "**33 deaths from vaccination** in the three years, 1949, 1950,
1951 . . . Post-Vaccinal Encephalitis 22; Generalized Vaccinia, 2; other
complications, 9." During this same 3 year period there were only 4 deaths from
smallpox. If vaccination kills people at the rate of 33 to every 4 protected (?) the
risk is greater than any possible advantage. (It was not stated how many of the
smallpox cases had been vaccinated also.)" - p 69

Now, I'm generally sympathetic to & in agreement w/ the observations &
conclusions of this bk. Nonetheless, in my attempt to be fair I feel that it's only
right for me to point out things that strike me as off. In the above paragraph the
statistics list 33 deaths from vaccination & 4 deaths from smallpox - but then
these 4 smallpox deaths are converted into their opposite by saying "If
vaccination kills people at the rate of 33 to every 4 protected" - "4 protected"?,
not '4 killed'? Then there's a parenthetical question mark. It makes no sense.

"In the years 1941 to 1948 there were (In U.S.) 107 deaths from "sequaclae of
preventative immunization, inoculation or vaccination". During this time there
were only 78 deaths from smallpox." - p 69

Now the apparent intended meaning of the above is that more people died from
the vaccination than people did from the disease vaccinated against. However,
it's all-too-easy for pro-vaxxers to say something to the effect of *'That's because
the vaccination worked & prevented more smallpox deaths!'* Nonetheless, the
overall statistical presentation of <u>The Poisoned Needle</u> gives more detail to
establish its assertion, some of wch I've already quoted here (such as the one
from p 13).

Since there's a spectrum of inter-related "pox" diseases that potentially puts
cowpox & syphilis together, it's asserted that smallpox vaccination, instead of
preventing smallpox caused syphilis. In cases where vaccinated babies died
from syphilis the drs asserted that this was b/c of prolifigate character of the
mothers.

"Some of the doctors at the inquest persisted in their morbid **character
defamation charges** against the mother and refused to believe the medical

report that gave her a **clean bill of health.** On the strength of their own opinions they insisted that she was syphlitic and would not be able to give birth to another normal baby.

"To prove that the accustations of the doctors were unjustified she had the following photograph taken after the birth of her next child, Frederick John, who was 10 weeks old at the time his picture was taken. It is apparent that the child is well and healthy —and unvaccinated. The fathers (in England) who refused to have their children vaccinated had to serve prison terms under compulsory vaccination but many of them accepted this punishment in order to save their children from the deadly effects of vaccination." - pp 75-76

That seems like class warfare to me - or, rather, the same old same old imposition of destructive force against the classes intended to be subjugated by the classes that consider themselves 'superior' - in this case the drs & lawyers, the usual suspects, *the real criminals*, the people who go to any extreme to increase their power whilst justifying it w/ a completely fictitious moral superiority.

"**Dr. Benjamin Rush** and **Dr. Josiah Bartlett** (both congressmen and signers of the Declaration of Independence) foresaw the medical tyranny that was, even then, laying its groundwork for domination of the healing field. In an effort to block this threatening menace they tried to introduce legislative measures that would insure freedom from medical tyranny as well as the other freedoms necessary to democracy. Their efforts were thwarted, however, by the preponderance of **orthodox** medical men in congress at that time, whose personal interest stood in the way of the public good. Some of these men even refused to sign the Declaration.

"An excerpt from one of the valiant speeches of Dr. Rush is as follows:

""**The Constitution of the Republic should make provision for medical freedom as well as for religious freedom. To restrict the art of healing to one class of men (such as medical doctors) and deny equal privileges to others (also trained in healing) will constitute the Bastile of medical science. All such laws are un-American and despotic. They are fragments of monarchy and have no place in a Republic.**"

"The failure of congress to establish freedom from medical domination has cost the nation countless billions of dollars and needless loss of life and health of our citizens.

"The medical dictatorship has extended its tentacles, since that time, until it now has control over a much wider area than the church ever did. Not only is our Public Health Service completely manned by the narrow and limited ability of this one branch of business, but it also wields a dictatorial power over the schools, colleges, charities, churches, public libraries, press, radio, women's clubs, P.T.A., life insurance companies and many industries and businesses, even the Civil Service, which of all things, **should respect the Constitution** in its law against

trusts." - pp 85-86

These days, more than ever before in my life, there's a propaganda campaign to dismiss critiques of ulterior motives as "conspiracy theories" as if there's no such thing as a conspiracy, as if all powerful transactions are conducted in completely above-board ways w/ total honesty & transparency. Such dismissal is a conspiracy in & of itself insofar as it serves the purposes that character-assassinated critics & researchers are trying to expose. People naive enuf to believe that everything done by any power elite is done 'for the good of the people' instead of for the gain of the elite are extremely foolish.

"A PLANNED MEDICAL CONSPIRACY

"This medical strangle hold did not just happen with the passing of time; it was a planned conspiracy against the American people, with an "eye" to the financial gains of the doctors. A sample of this strategy may be seen in the following speech of Dr. W. A. Evans, one of the top medical "bosses," and Health Commissioner for the city of Chicago, who gave these instructions to the doctors in their **annual convention of the American Medical Association** in 1911:

""The thing for the medical profession to do, is to get right into, and man every important health movement; man health departments, tuberculosis societies, housing societies, child care and infant societies, etc. The future of the profession depends on it . . . **The profession cannot afford to have these places occupied by other medical men."**

"This pronouncement was published in the journal of the A. M. A., September 16, 1911. Just how whole-heartedly this decree was carried out is clearly shown by how completely all the non-medical schools of healing such as chiropractic, naturopathic, religious science, hygienics, etc., have been excluded from such tax supported institutions as health boards, public hospitals, army camps, state prisons, workman's compensation bureaus, asylums, etc." - p 86

Then there're the 'expert witnesses': I've been saying for decades that so-called expert witnesses are generally just people w/ degrees who're willing to whore themselves out to the highest bidder. An example I often refer back to is this: When the government of Philadelphia commited mass murder by bombing the MOVE house & essentially destroying the entire block they lived on an 'expert witness' appeared in radio interviews to give his opinion that the MOVE people had deliberately blown themselves up by having a tank of gasoline on top of their house. Of course, the tank was really for collection of rainwater since the MOVE folks were back-to-nature types trying to live off-the-grid. Unfortunately for the 'expert witness' a photographer had gotten a picture of a helicopter dropping C4 military-grade explosive on the MOVE house - thusly proving that a deliberate massacre had taken place - & the 'expert witness' probably largely disappeared from the public memory.

""Twenty-four of these medical societies **furnish expert witnesses to testify on**

behalf of the accused (the doctor). Eighteen of these societies pay all expenses incurred by such witnesses, as well as **special remuneration for testifying favorably,** some of it running as high as $50.00 a day."

"These "expert witnesses" may live 1,000 miles from the scene of the crime and know nothing about it except what they are paid to say. Is this any different from the methods of the professed criminals of the underworld?" - p 87

The "$50.00" referred to is from a bk written in 1935 & may even harken back to an earlier time. $50 in 1935 is $1,055.14 dollars today (https://www.in2013dollars.com/us/inflation/1935?amount=50).

"Supreme Court Justice Cardozo said, in rendering a decision:

""Every human being of adult years and sound mind has a right to determine what shall be done to his own body, and a surgeon who performs an operation (vaccination is technically an operation) without his patient's consent, commits an assault for which he is liable for damages.""
- p 92

It's nice to see that there's some Supreme Court back-up for my opinion that forced vaccination is rape. For those of you who seek further confirmation:

"This was the judgement of Benjamin Cardozo in the case of Mary Schloendorff v the Society of New York Hospital in 1914.

"The plaintiff, Mary Schloendorff, was admitted to New York Hospital and consented to being examined under ether to determine if a diagnosed fibroid tumour was malignant, but withheld consent for removal of the tumour. The physician examined the tumour, found it malignant, and then disregarded Schloendorff's wishes and removed the tumour.

"The court found that the operation to which the plaintiff did not consent constituted medical battery. Unfortunately for the plaintiff, the judge also ruled that the Society of New York Hospital (in effect the management board) was not responsible for the actions of the doctor and the case failed."

- https://wchh.onlinelibrary.wiley.com/doi/pdf/10.1002/tre.288

Interesting, isn't it? When you're rich & powerful there's always a loophole. Then there's the whole problem for the ruling elites of how to censor competing info. Since censorship has gotten somewhat of a bad name, it's more 'clever' to block the flow of information in other ways. One way, very popular these days, is to declare any alternative to the dominant (& dominating) narrative "misinformation". Back in the day when this bk was written the Post Office cd ban bks w/ alternative health opinions from being shipped.

""e. That the Publication entitled "Health for ALL" contains information and instructions the following of which will effectively treat and cure such conditions and diseases as **sinusitus, bronchial asthma, goiter, colitis, peptic ulcer, diabetes mellitus, arthritis, rheumatism, high blood pressure and infantile paralysis."**

"The complaint goes on telling what the books promise in the way of healing. The fact that hundreds of people have been healed of these diseases by following these instructions doesn't concern the medically dominated Post Office Fraud Department. The **medical profession has no cure for any of these diseases** and, like the **dog in the manger** they intend to prevent anyone else from curing them either, or even from having the right to choose their reading matter that will teach them how to do it themselves." - p 94

Naysayers will make the claim that these are quack cures. They may be. Then again the mainstream medical approaches may also be quack cures. If I were to choose between the mainstream, very expensive, 'cures' & the alternative, cheap or free, 'cures', I'd choose the cheap path. At least that way I wdn't feel ripped off. Having been told by a ridiculously expensive 'legit' dr that it was "impossible" to cure my diabetes w/ diet & exercise & having quickly discovered just how wrong she was & having had a plethora of similar experiences w/ the medical industry I have no reason to think they're anything *other than quacks*.

"We like to believe that **crime doesn't pay.** But we see the constant multiplication of deaths from unneccessary operations and medically poisoned patients where the doctor "buries his mistakes" and then collects a small fortune from the survivors. We see the multi-million dollar "take" from the vaccination campaigns conducted at least twice a year in all American cities." - p 96

Much of The Poisoned Needle is concerned w/ the polio vaccine campaigns of the mid-1950s, the dominant vaccination push of the time of the writing of the bk. **"THE HIDDEN DANGERS IN POLIO VACCINE** starts on page 97 & essentially goes to the end of the bk.

""**Salk vaccine is hard to make and no batch can ever be proved safe before it is given to children."** This is an admission that was made by Dr. Scheele, (Surgeon General) before the Atlantic City Convention of the American Association in 1955. (Reported in *New York Times*, June 8, 1955.)

"Being fully aware of the hazardous aspects of the polio vaccine, Dr. Scheele announced it was the intention of the U.S. Government to inoculate 57 million people before August 1955. (Report from the *Lancet*, June 4, 1955)"

[..]

"$9,000,000 of public funds were gambled on this financially promising but highly questionable venture before it was even declared *safe* and usable. Later, when the over dramatized announcement *was* made that it was "safe" there was *still* no scientific evidence or factual proof of its safety. **The promoters of this inflated**

vaccine "scheme" expected a 5 billion dollar profit in the first year of
operation.** (Direct quotations supporting this statement are given under the topic
heading, THE MONEY MOTIVE, at the end of the chapter.)" - p 98

"**A PARTIAL LIST OF DEATHS FROM SALK VACCINE**
Susan Pierce (age 7), Pocatello, Idaho, died April 27, 1955
Ronald Fitzgerald (age 4), Oakland, Calif., died April 27, 1955
Allen Davis Jr. (age 2), New Orleans, La, died May 4, 1955
Janet Kincaid (age 7), Moscow, Idaho, died May 1, 1955
Danny Eggers (age 6), Idaho Falls, Idaho, died May 10, 1955" - p 100

These days, there're the deaths from the COVID-19 vaccines. As of today, May
26, 2022, there're 12,397 deaths listed on "The Vaccine Adverse Event
Reporting System (VAERS) Results - Data current as of 05/13/2022" (https://
wonder.cdc.gov/controller/datarequest/
D8;jsessionid=8BA7E4C24EA2DF0C54069C6BD6A0) I've looked online for
such data before & found considerably higher results but I'll settle for 12,397 for
now. Imagine being pressured into getting vaccinated, told that it's necessary to
protect you from death AND the socially responsible thing to do. THEN you die.
Bit of a drag, eh? When I told a vaccination supporter about these CDC-
acknowledged deaths she shrugged them off as a sortof collateral damage. Tell
that to the loved ones of the deceased.

"Shortly after the Salk vaccination program was swung into action, the *American
Public Health Service* (June 23, 1955) Announced that there had been "**168
confirmed cases of poliomyelitis among the vaccinated,** with six deaths . . .
How many vaccinated children will eventually be reported as developing the
disease is as yet unknown . . ."

""**The interval between inoculation and the first sign of paralysis ranged
from 5 to 20 days and in a large proportion of cases it started in the limb on
which the injection had been given. In fact in the state of Idaho, according
to a statement by Dr. Eklund, one of the Government's chief virus
authorities, polio struck only vaccinated children in areas where there had
been no cases of polio since the preceding autumn; in 9 out of 10 cases the
paralysis occurred in the arms in which the vaccine had been
injected.** (News Chronicle, May 6, 1955)**

"According to the *Daily Telegraph* (June 18, 1955) **Mr. Peterson,** State Health
Director of Idaho, stopped further inoculations and stated "**We have lost
confidence in the Salk Vaccine.**" He also stated that **he "holds the vaccine,
together with the instructions for its manufacture, directly responsible for
the outbreak of polio and the deaths that had occurred.**"" - p 102

"In **Australia** when a few children died as a result of smallpox vaccinations the
government abolished compulsory vaccination in that country and smallpox
suddenly declined to the vanishing point. Australia had only three cases of
smallpox in 15 years as compared with Japan's record of 166,774 cases and

28,979 deaths from this cause in only 7 years (1886-1892) under compulsory vaccination and re-vaccination." - p 103

I was curious about finding verification for the above claim online. The question was this: "did australia ever abolish compulsory smallpox vaccinations?" My search for it was superficial so I only looked at the 1st 2 results - this is as close as I got:

"The Commonwealth quarantine of Sydney was controversial both within Sydney and the country more broadly. On 18 September 1913, the *Sydney Morning Herald* reported the New South Wales Board of Health resolved:

"*That ... the board is satisfied from [its experience with 700 cases of smallpox] that the disease now present in Sydney is an exceedingly modified form of smallpox, mild in its nature, and with no tendency to change its type and become more virulent ...*

"*That the danger of it becoming epidemic in the other States of the Commonwealth is negligible, seeing that compulsory vaccination laws have been in force there for very many years past.*

"*That the board has by its strong advocacy of the effectiveness of vaccination as a protective measure succeeded in inducing more than one-half of the population of Sydney and suburbs to protect themselves in this manner. ...*

"*That, in conclusion the board is of opinion [sic] that the proclamation issued by the Federal Government quarantining Sydney should be withdrawn, as it is unnecessary, and while in existence it is injurious in many respects to the State of New South Wales in particular, and to the Commonwealth in general.*"

- https://www.aph.gov.au/About_Parliament/Parliamentary_Departments/ Parliamentary_Library/FlagPost/2021/July/1913_Smallpox_Epidemic

"Vaccination wasn't always taken up readily and it wasn't always safe, either.

"In 1928, 12 children in the south-east Queensland town of Bundaberg died after receiving contaminated diptheria vaccines.

""Among those who died, three came from one family, while two more families each lost two children," Dr Hobbins writes in a 2011 academic paper.

""A Bundaberg correspondent opined that 'immunisation is as popular as a death adder'.

""Within days the events in Bundaberg had compromised diphtheria control programmes around the globe, including the complete termination of immunisation in Cape Town (South Africa) and across New Zealand.""

- https://www.abc.net.au/news/health/2020-09-22/vaccine-history-coronavirus-smallpox-spanish-flu/12673832

Obviously, neither of those articles verifies McBean's claim. That doesn't mean that I think that they refute them either. The 1st article is about lifting quarantine & *not about* repealing mandatory vaccination. Furthermore, their reasons for lifting it are that the authorities no longer considered smallpox to be a threat & that they considered the vaccinations a success. The 2nd article is about *diphtheria* vaccinations & *not smallpox* ones. That article states that vaccination "wasn't always safe" - the implication, as usual, is that science has gotten its shit together now & that it's safe - but what will be sd about vaccination now when it's reported about 100 yrs from now? I suspect that the supposed 'safety' of vaccines now will be soundly refuted - esp in the light of the thousands dead from COVID-19 vaccines.

"In 1926 two prominent English Professors of Pathology, **Doctors Turnbull** and **McIntosh** reported a number of cases of encephalitis lethargica (sleeping sickness—a form of polio) following vaccination. This led to the appointment of two commissions of the *British Ministry of Health*, to investigate the situation. Their reports published in 1928 revealed that there had been 231 cases and 91 deaths from this cause (post vaccinal encephalitis) in England and Wales." - p 104

One of my favorite writers is Mervyn Peake, author of The Gormenghast Trilogy. According to John Watney's biography of Peake:

"In 1917 there was a world-wide outbreak of *encephalitis lethargica*, or sleeping sickness. According to Dr Oliver Sacks, an expert on the subject, it started in Vienna in the winter of 1916-1917, and soon spread. As many as five million people may have died directly from its effects, but this figure, huge as it was, went unnoticed because of the greater losses of the war and the Spanish influenza that followed it. Worse, perhaps, than the immediate effects of the illness were its long-range possibilities. It was an illness that could be contracted at any age and lie dormant for decades, only to appear many decades later, in moments of stress. Somewhere, somehow, during those last few years of the war, it must have sought out and found, as it did many thousands of other children at the time, the active and busy young boy in Tientsin. There was nothing at the time to show that anything had happened—perhaps a day or two of lassitude—but, like many others of his age, Mervyn Peake was to carry dormant within him this destructive germ." - pp 28-29, John Watney's Mervyn Peake, St. Martin's Press, 1976

I have a copy of Oliver Sacks's bk on the subject entitled Awakenings (1973, 1976, 1982, 1983). In the index there's no entry for "vaccination". Therefore, alas, I once again struck out for finding independent verification for McBean's claim. I did an online search for "encephalitis lethargica following vaccination" & got quite a few interesting, but not exactly directly relevant, results:

"Herein, we describe a 48 years old man presenting with rapidly progressive cognitive decline and hyponatremia diagnosed with anti LGl1 AE, occurring shortly after the second dose of mRNA COVID -19 vaccine and possibly representing a severe adverse event related to the vaccination." - https://www.frontiersin.org/articles/10.3389/fimmu.2021.813487/full

That's relevant to encephalitis but not encephalitis lethargica specifically.

"Background

"Since introducing the SARS-CoV-2 vaccination, different adverse effects and complications have been linked to the vaccine. Variable neurological complications have been reported after receiving the COVID-19 vaccine, such as acute encephalopathy.

"Case presentation

"In this report, we describe a 32-year-old previously healthy man who developed acute confusion, memory disturbances, and auditory hallucination within 24 hours from getting his first dose of the COVID-19 Moderna vaccine.EEG showed features of encephalopathy, CSF investigations were nonspecific, and MRI head did not depict any abnormality. He received five days of ceftriaxone and acyclovir without any benefit.

"Discussion

"Extensive workup for different causes of acute encephalopathy, including autoimmune encephalitis, was negative. Also, Our patient improved dramatically after receiving methylprednisolone, supporting an immune-mediated mechanism behind his acute presentation. Accordingly, we think the COVID-19 vaccine is the only possible cause of our patient presentation, giving the temporal relationship and the absence of other risk factors for encephalopathy."

- https://www.ncbi.nlm.nih.gov/pmc/articles/PMC8420261/

"Can vaccines cause encephalitis?

"Vaccines are essential for preventing many diseases and they save millions of lives every year. It's rare for a vaccination to cause encephalitis, but it can happen."

"The National Vaccine Injury Compensation Program includes encephalitis as a covered vaccine injury. People who have been injured by vaccines can be compensated under this program without paying any legal fees or costs."

- https://www.vaccineinjurylegalteam.com/vaccine-reactions/encephalitis/

Naturally, I find this last item particularly interesting since it's referring to the current time of COVID-19 vaccinations.

Finally, the 4th thing I'll quote from here is the most relevant to the passage under scrutiny in The Poisoned Needle:

"POSTVACCINATION ENCEPHALITIS
With Special Reference to Prevention
By Charles Armstrong, *Surgeon, United States Public Health Service*

"Postvaccination encephalitis, a disease of unknown etiology, was first brought to the attention of the medical profession in 1924. Approximately 700 cases, with a case fatality rate of 40 per cent, have now been recognized. With the exception of 71 cases recorded for the United States during the past 10 years, reports of this complication have been largely confined to European countries, Holland, England, Germany, Sweden, and Norway having been most severely affected. Within certain of the affected countries a peculiar "spotty" distribution of cases has been noted. Within the United States the heaviest incidence so far recorded occurred during the autumn of 1930 in a city of 450,000 inhabitants, where 5 cases of postvaccination encephalitis developed among school children within a period of 13 days. These cases had been vaccinated by five different physicians who employed various types of single insertions. During the following fall two additional cases developed in this same city, again among children entering school." - PUBLIC HEALTH REPORTS, Vol. 47, July 22, 1932, NO. 39 - https://www.jstor.org/stable/4580503

Hence we have some verification of postvaccination encephalitis from a source that most wd find reputable.

"In England (1923-1925) an intensive vaccination campaign was carried out which caused a steady increase in the cases of *Encephalitis-Lethargica* and similar diseases until in 1924 there were 6,296 reported cases with unestimated thousands of unreported cases. Liverpool was at least 50% better vaccinated than the rest of England and had about 100% more encephalitis. The physicians tried to direct blame away from vaccination by claiming that the diseases were due to "latent infection," "inherited weakness," or "special susceptibility."" - p 107

"It has long been known that inoculations of all kinds frequently cause nervous diseases inlcuding paralysis. Smallpox vaccination often results in paralysis of one side of the body. That it also causes encephalitis lethargica (sleeping sickness) is now well established." - Herbert M. Shelton, N.D. - p 163

"A report in the *Lancet* (April 8, 1950) stated that **Dr. McClosky** of Melbourne had inquired into 340 cases of poliomyelitis which occurred in Victoria (from January to August, 1949) and found that 31 of them had occurred in children who had received diphtheria or whooping cough (pertussis) vaccine, or both, within three months of the onset of the disease. Of the 31 cases, 30 of them suffered severe paralysis which was most pronounced in the limb in which the injection was given.

"The *Medical Officer* (April 8, 1950) reported the findings of **Dr. D. H. Geffen** who stated that in the first four months of that year, in S. Pancras, London, 30 children under five developed infantile paralysis within four weeks of being immunized against diphtheria or whooping cough (or both) "the paralysis affecting, in particular, the limb of injection. In 7 other recently vaccinated cases, paralysis occurred but not in the limb that had received the injection.

"**Dr. Arthur Gale** of the Health Ministry reported (in the *Daily Express* for April 10, 1950) that he had observed 65 cases mainly from the Midlands' in which paralysis developed about two weeks after an injection; and in 49 of these, paralysis was confined to the limb in which the injection had been given." - p 106

"In Los Angeles (1946) after the largest smallpox vaccination campaign in the history of the city, an epidemic of polio broke out within two weeks after the "shots" and by the end of the summer there were 26 deaths from this cause and 1,900 reported cases. The County Epidemiologist estimated that there were about 15 times as many *unreported* cases which would bring the count up to 28,000 cases. The report for the entire United States for 1951 was only 28,395. An average vaccination campaign nets the doctors about $2,000,000 so this one must have been tremendous in profits. The amount of the "take" has never been revealed. This *Polio epidemic* triggered off a "bigger than ever" *March of Dimes* campaign and the people were "bled" again for money to support the medical racket that creates more disease in order to scare the people into more donations and so the vicious circle continues." - pp 106-107

As such, it seems that the fear-mongering of pandemic prediction is a self-fulfilling prophecy.

For me, one of the main questionings of genius that McBean presents is one in wch the whole notion of viruses is considered anew.

"The word poliomyelitis was coined from the Greek words "polios" meaning *gray*, and "myelos" meaning *marrow*, plus "itis" meaning *inflammation*. Thus the word inflammation of the gray matter of the spinal cord or brain or wherever gray matter is to be found. Every cell of the body contains a small amount of gray matter but it is more concentrated in the brain and spinal cord. Inflammation results from poisoning, irritation and obstructions that interfere with normal functioning. This damage, decay and disintegration of cell structure may terminate in paralysis or death without any invasion of viruses or germs. Isolated germs have never been known to *attack* and cause decay and disease of any part of the body. It is the corrosive *poison products of decay* that form the body of the vaccine that does the damage and causes the disease."

[..]

"But the physicians and serum makers prefer to imagine a virus that can fly around in the air and attack people. Any other kind of a virus could not be made to frighten people into making large donations and in being vaccinated in the hope of protecting against it. Without an "air-borne" virus to play with, the whole vaccine racket would collapse." - p 108

To a PANDEMIC PANIC critic such as myself that makes perfect sense. I'm in a 'high risk' group, being over 65 & having been diagnosed as diabetic type II but I've always maintained that blocking one's breathing & speaking apparatus is unhealthy. As such, I've worn a mask as little as possible during the QUARANTYRANNY. 'Strangely', the only respiratory illness I've experienced was something I'd call a mild head cold at the end of 2021 - after almost 2 & a half yrs exposure to this supposedly 'deadly plague'.

McBean credits vacination w/ being a major *cause* of polio but sensibly postulates other causes.

"Although vaccination is one of the quickest and surest ways of getting polio because poisons are injected directly into the unprotected blood stream, there are a number of other causes that should not be overlooked. Some of the most common of these are listed below and discussed briefly in this section:

"1—*Carbohydrates: Sugar* products such as ice cream, soft drinks, desserts, etc. Denatured products such as *flour, alcolohics*, [sic] etc.
2—*Cola* beverages that contain drugs and sugar.
3—*Poison sprays* on foods, poison preservatives in foods, etc.
4—*Operations* of all kinds, especially removal of tonsils.
5—*Fatigue* and all other practices that deplete the vital energy.
6—*Negative emotions* that generate internal poisons.
7—*Interference* of any kind that impedes circulation or normal functioning." - p 114

I assume that "*alcolohics*" is a typo but I reproduce it as is anyway since I prefer to quote accurately. It seems to me that 1 includes 2. As stated in the introductory paragraph, these causes are discussed in greater detail after the presentation of the list.

""Persons with bulbar type of polio (an especially fatal type) give a history of removal of tonsils and adenoids more frequently than do persons with other forms of poliomyelitis. If a person who has had tonsillectomy develops polio the likelihood of bulbar involvement is 4 times as great as one where the tonsils are intact." ("The Salk Story" Bayly. p. 24)

"Tonsils and adenoids are part of the lymphatic glandular system of purification and when they are removed the impurities that they normally handle are thrown back into circulation and cause all manner of diseases. The poisoning effects of the anesthetic and the shock of the operation also help to reduce the vitality and over tax the already poisoned body. The fact that the tonsils are enlarged and saturated with poison does not indicate that they should be removed; it merely indicates that the intake of poisons should be discontinued and the blood stream cleansed by fasting and proper diet. Inflamed tonsils and practically all other abnormal conditions heal themselves when the obstructions are removed." - p 116

Once again, that seems like common sense & clear-thinking to me. But I have to add that I had my tonsils removed 62 yrs ago & my appendix removed 20 yrs ago & I'm still alive & kicking. Nonetheless, I wonder what the quality of my health wd be if I hadn't gotten those operations (& had taken better care of myself).

"SUGAR is at last being recognized as a destroyer of health rather than an article of food as was formerly believed. Technically white sugar (C12H22O11) is classified as a drug and not as a food. Natural sugar in fruit is in a balanced combination with other constituants that enable it to be assimilated without damage but when it is separated into the white crystalline carbon substance called *sugar* it is converted into alcohol almost immediately after it is taken into the body and does the same damage that alcohol does. It dehydrates the cells and leeches the calcium from the nerves, muscles, bones, teeth and all tissues that are supplied with calcium and other alkaline elements. A serious calcium deficiency is a forerunner of polio." - p 116

I agree. The overingestion of sugar & alcohol have been 2 of the most self-destructive habits I've had. On the other hand, I've also had some exceptionally good times w/ them. Ha ha! That's a part of what it boils down to, isn't it? The things that humans tend to overindulge in usually bring pleasure - & this pleasure is desperately needed to counterbalance the suffering. I got into the sugar habit largely b/c I was living in such poverty that sugar was one of the only affordable reliefs. A 3 liter bottle of cheap soda goes a long way when you only have a few dollars to spend on food.

"(1) "Cola is loaded with habit-forming *caffeine* so that once the victim gets accustomed to the stimulant, he cannot very well get along without it." says **Dr. Royal Lee,** of the Lee Foundation for Nutritional Research."

[..]

"(2) Cola drinks contain as much as 10% **sugar** which is 9 times more than the body can metabolize.

"(3) Colas contain *artificial coloring and flavoring* matter. These coal tar products are harmful *drugs* and have no place in the diet.

"(4) The **most damaging ingredient in cola drinks is the PHOSPHORIC ACID.** This acid is a destroyer of the vital calcium supply in the body and even dissolves the enamel on the teeth. Phosphoric acid is made by treating phosphate rock with *sulphuric acid.*" - p 117

Now a naysayer might think that the description of the making of phosphoric acid sounds like a health nut's urban myth. So let's look it up, shall we?

"Phosphoric acid, H3PO4, is produced from phosphate rock by wet process or thermal process. 80% of the world's phosphoric acid is obtained by the wet process. The wet process consists of reaction, filtration and concentration steps. The phosphate rock is ground and acidified with sulfuric acid in the reactor vessel." - https://www.vaisala.com/en/industries-applications/chemicals-allied-products/wet-process-phosphoric-acid-production

"**Ralf R. Scobey, M.D.,** president of the Poliomyelitis Research Institute, Inc. Syracuse, New York (in the Archives of Pediatrics, Sept. 1950) lists 170 diseases of *polio-like* symptoms and effects but with different names such as: *epidemic cholera, cholera morbus, spinal meningitis, spinal apoplexy, inhibitory palsy, intermittent fever, famine fever, worm fever, bilious remittent fever, ergotism,* etc. There are also such common **nutritional deficiency diseases** as **beriberi, scurvy, Asiatic plague, pellagra, prison edema, acidosis** etc. No drugs, medicines or medical treatments have ever been able to cure any of these diseases and no germs have been isolated as the cause. But they all respond to fasting, cleansing, proper diet and improved circulation. The similarity of these diseases to polio is too obvious to go unnoticed. They are, in reality, all one disease with varying stages of intensity and different names." - p 120

To me, it seems that McBean keeps hitting the nail on the head. From having been a cholera research volunteer I remember vividly that people die from cholera b/c of fluid losses from diarrhea. Hence the treatment for it is influx of properly nourishing fluids. Otherwise, the person dies from shock. That's simple enuf, isn't it? The problem is that in the environment that the cholera occurs in it's the water that's polluted w/ shit, such as in the Ganges River, so it was drinking the polluted water that caused the problem in the 1st place. Therefore, drinking more of the same water to replace the fluid lost by diarrhea won't work as a solution.

"The *Lancet* (Jan. 9, 1915) mentions an epidemic of polio that was said to have been caused by *raw milk*. It has never been proved that raw milk has ever caused polio or undulant fever or any other disease but the *added poison preservatives* could cause many diseases. Pasteurization companies capitalized on this incident to stampede some laws into operation which would prohibit the sale of **raw** (unprocessed) milk. Pasteurization makes matters even worse because the heating process melts the butterfat in the milk and this coats the calcium globules and renders the calcium unassimilable to a large extent.

"From the *Journal of Biochemistry,* (Vol. 75, pp. 251-62, Oct., 1927) we read: "In the Nutrition Laboratory of Home Economics of the University of Chicago, in 1927, experimentation revealed that only about 27% of the calcium in *pasteurized* milk and 30% of *evaporated* milk was retained by the body."

"Rats fed on pasteurized milk grew at only half the normal rate and developed various ailments and died early. Calves fed on pasteurized milk died before maturity in 9 out of 10 cases." - p 122

Yet-another battle of the beliefs. When I lived in Canada in 1995 I went to someone's home in rural Ontaria where a farmer presented non-pasteurized milk to be tasted & gave a spiel on why it was considered healthier. I almost never drank milk so I wasn't really concerned w/ the issue. I tasted it, it seemed thicker?, somehow richer? The thing about the demonization of raw milk is that it's the same-old, same-old demonization of nature: *Nature is the Enemy.* Big business takes a resource, converts it into something only they can have control over & then sells it at prices that they determine. What's next? Probably the demonization of breast-feeding - making people afraid of the milk that their own body generates, making them dependent on some sort of formula *that they have to buy*, of course.

"Although the **real** cause and cure of polio is obvious, the promoters of polio find it more profitable to "look" for a cure than to "find" it. The *March of Dimes* millions would cease to flow into their pockets if they were ever to find the cure. Therefore, they keep up the pretense of *looking*—but always in a direction where they are sure *not* to find the answer.

"(1) At first they tried the *polio serum* and when they found it killed the people but did not stop polio they tried something else just as far away from the solution as they could get.

"(2) They *sprayed towns, crops and orchards with deadly poison DDT* insecticide, suggesting that polio *might* be caused by *flies*. The chemical companies reaped the benefits but the orchards were seriously damaged, crops were ruined and people were killed by this wild and unwarranted experiment. An article on polio in "*Your Health*" magazine (1954, Fall Quarter, p. 69) comments on the fly situation by saying that in **Iceland where there are no flies . . . there have been polio epidemics.** Experiments on eliminating flies by spraying whole communities with DDT have not had any effect on combating epidemics."" - p 124

"**CURARE,** which is arrow poison used by the Indians of the Amazon is used on polio patients. A doctor in the Los Angeles Health Department told me that the drug does not decrease the pain of the polio patients but it makes them unable to move or scream and they experiment on them then. Many of the patients die under the intense pain of this torture treatment." - p 125

I decided to look online for verification of this one & there's plenty of it.

"During the last decade considerable attention has been devoted to the treatment of muscle spasm, tightness or shortening of affected muscles in patients with acute anterior poliomyelitis. This symptom generally has been treated with some form of heat, either moist or dry. Hot fomentations (packs) or hot water (tub baths, Hubbard tank or pool therapy to the muscle. He noted relief of therapy) have been the usual methods of applying moist heat; radiant heat lamps, light cradles, infra-red or luminous bakers have been used to supply dry heat. Artificial fever has also been employed as a form of thermotherapy. It has been generally observed that application of heat provides considerable relief to the patient during the acute stage and many hold the opinion that repeated applications of hot packs cause muscular relaxation, reduction in muscle tenderness and lengthening of the shortened muscle.

"Evidence has accumulated that at least some of the muscle tightness is caused by heightened irritability of myotatic reflexes resulting from lesions in the brain stem although it also seems quite possible that the pathologic site responsible for a portion of the muscle tightness may be located within the muscles or at the neuro-muscular junction. However, the latter hypothesis requires substantiation.

[..]

"The use of curare has been advocated by Ransohof who reported rather spectacular results in the treatment of poliomyelitis patients given this drug. Reasons for advocating its use may be summarized as follows: (1) to relieve spasm in the acute stage; (2) to aid in preventing shortening in weakened muscles; (3) spasm is deleterious to the muscle. He noted relief of muscle tightness of the neck, back, knee flexor and plantar muscles as well as of pronator and intercostal muscles. He also noted restoration of the ability to swallow within a short time after administration of the drug. His system of treatment consisted of intramuscular injections two to three times daily of a standardized extract of curare known as intocostrin, followed by vigorous active and passive movement of involved muscles. Beneficial results reported following this treatment were so impressive that we decided to try it in some selected patients."

- "Use of Curare in the Treatment of Anterior Poliomyelitis" (From the Communicable Disease Section, Los Angeles County Hospital, and the Department of Medicine, University of Southern California, Los Angeles, Calif.) - Albert G. Bower, M.D., O. Leaonard Huddleston, M.D. *and* Deron Hovsepian, M.D. - p 160, *American Journal of Medicine*, Volume 8, Issue 2, February, 1950 - https://www.sciencedirect.com/science/article/abs/pii/0002934350903573

Not surprisngly, the language that McBean uses to describe curare's effects is quite different from that of the people quoted above. One wdn't expect the physicians to get into how the Amazonian natives use curare to poison arrow tips or darts so they can paralyze the muscles of creatures that they're hunting.

"" '4—Two Nobel Prize winners, **Dr, John Enders** and **Dr. Wendell M. Stanley** both have publicly indicated their uncertainties about the Salk vaccine . . ." - p 126

Everyone wants Nobel Prize winners on their side. Ever since Kissinger won one I've been a bit suspicious of them.

"In 1946, during one of our most serious polio epidemics in Los Angeles, I wrote a book called POLIO CONTROL in which I denounced the "virus theory" and named vaccination as one of the causes of polio. The medico-drug trusts had their stooges working for two years trying to get something legal on which to put the book out of circulation. I had had a lawyer go over the manuscript before I had it printed and I had the Post Office examine the printed book to be sure it was in order and nothing was said that was against the law. The lawyer said, "The book is documented and entirely within the law." When the "medical opposition" couldn't get anything on me lawfully they used unlawful measures and had their stooges in the Post Office *bar the book from the mails.* No jury trial was permitted and no testimony of mine was accepted. Their charge was that the **"book was dangerous because it disagreed with accepted medical belief."** So the book went out of circulation, but not before two editions had been sold to all parts of the world—mostly to drugless doctors." - p 128

Can you imagine something happening like that today? Maybe not w/ the mails, partially because in addition to the USPS there're alternatives like Fed Ex & UPS & Amazon, but it's already happening on Fecesbook & YouTube & Twitter, etc. Even I, inoffensive sweet old man that I am, get censored on YouTube.

"After the announcement from Philadelphia (1934) that **Dr. John A. Kolmer** had discovered a "protective (?) vaccine" against infantile paralysis, **Dr. W. Lloyd Aycock**, director of the Harvard Infantile Paralysis Commission and called "one of the most distinguished experts in this disease", gave a statement to the press that **Nature does a better job of immunizing against infantile paralysis than the artificial methods" which he branded as "Hazardous."**

"If we remove the *obstructions* and supply the body with the necessary building material, nature will heal practically any disease so long as too much of the vital organs or glands have not already been destroyed by disease or operations.

"What are the *obstructions* to be removed in the healing process? Obviously, drastic poisons such as vaccine serums, drugs, tobacco, alcohol, coffee, tea, cola drinks, etc. poison the system immediately and set up disease states. But other blockages form slowly in the tissues as a result of incompatible foods, wrong food combinations, and over eating." - p 130

I reckon that the average American looking at the above list will scoff at the possibility of not consuming everything there. I don't get vaccinated, I don't take any drugs, I don't consume tobacco, I only drink about 1 beer every mnth or 2, lately I've only been having coffee every 2 or 3 or 4 days, I don't usually drink tea, I only drink cola drinks when I go out to eat & not necessarily every time. That makes me an unusually abstaining person - &, yet, I'm still poisoning myself by McBean's standards - I reckon she's right but that's as far as I'm likely to go & only b/c that's what I choose to do.

By page 133 the Bibliography is reached. Don't trust any edition that doesn't provide this. There're 23 references given. A serious researcher wd be well-advised to read them all.

Then, starting on page 134, is the Addendum. It consists of entire articles written by people whose opinions McBean respects. As such, the style of the bk changes slightly. The 1st of these is by Rex U. Lloyd.

"The report of the Surgeon General of the Army, 1919, Vol. 1, p.38, gives the number of Admissions to Hospitals during the year 1918 on account of vaccinia, the disease caused by vaccination, as 10,830. Another report of the Surgeon General of the Army, 1918 gives the number of Admissions to Hospitals during the year of 1917 on account of vaccinia and typhoid vaccination combined as 19,608." - p 136

Impressive, eh? But, hey!, they're only soldiers, right? So why shd the general public give a fuck?! The thing is, the general public is also being subjected to this madness.

"The Supreme Court of Massachussetts handed down a decision, stating that 'If a person should not be willing in his case and the authorities should think otherwise, it is not in their power to vaccinate him by force."

"The verdict rendered by the prominent U.S. Supreme Court Judge and similar decisions reached by various State and Federal Courts should be sufficient to outlaw all compulsory medication and vaccination laws and make it a criminal offense for anyone to use force in an attempt to do so." - pp 136-137

I'm 100% in agreement that there shd **never be any forced medical attack on anyone for any reason whatsoever**. I'm 100% in favor of informed consent - & how many people are truly "informed" about the medical (mal)practices that drs promote to them?!

"We should recall the splendid examples set by prominent persons of great intellectual calibre who have been staunch anti-vaccinationists and non-conformists in regard to orthodox therapeutics. Bismarck. Salisbury. V. Zedwiz, Earl Dysart, Wm Lloyd Garrison, Premier Asquith, Gladstone, Voltaire, Victor Hugo, John Bright, G. K. Chesterton, George Bernard Shaw, Mark Twain, Elbert Hubbart, Robert Ingersoll, Alexander von Humbolt, Herbert Spencer, Alfred R. Wallace, W. Webb, Luther Burbank, Thomas Edison, Henry Ford." [tENTATIVELY, a cONVENIENCE] - p 137

"If the death rate of smallpox and fevers was so enormous, it was largely due to the medical treatment of that time. A number of eruptive diseases such as measles, chicken pox, scarlet fever etc. were regarded as smallpox before Dr. Seydenham differentiated between the various symptom-complexes. How great the number of deaths was from scarlet fever, measles, chicken pox etc. that were included in the smallpox epidemics will never be known. Dr. Russell T. Trall, the eminent Natural Hygienist, considered smallpox "as essentially . . . not a dangerous disease." He cared for large numbers of patients afflicted with small-pox and never lost a case. Under conventional medical treatment, patients were drugged heroically, bled profusely, were smothered in blankets, wallowed in dirty linen, were allowed no water, fresh air and stuffed with milk, brandy or wine. Antimony and Mercury were medicated in large doses. Physicians kept their patients bundled up warm in bed, with the room heated and doors and windows carefully closed, so that not a breath of fresh air could get in, and given freely large doses of drugs to induce sweating (Sudorifics), plus wine and aromatized liquors. Fever patients were put into vaporbath chambers in order to sweat the impurities out of the system. Given no water when they cried for it and when gasping for air were carried to a dry-hot room and after a while were returned to the steam torture. Many must have died of Heat Stroke!" - p 138

Keep in mind that the people inflicting those cures were the 'experts', the drs occupying the same positions of prestige as the drs of today. To 'cure' people they were subjected to extreme abuse of the body - it seems obvious to me that such abuse only weakens the body further instead of making it better. How long will it be before the parallel extremes of today's so-called 'medicine' will be looked back on as similarly deranged?!

"On June 9, 1955, Robert S. Allen said in his syndicated column: "Doctors and others on the staff of the National Institute of Health are not inoculating their own children with Salk Vaccine." Explanation of this has been carefully avoided up to this time, but here are the facts.

"To the bewilderment of NIH's big brass, their scientists were both capable and honest. Many have children of their own. None could be bulldozed into giving the "right answer" because that didn't jibe with facts. After experimenting on 1,200 monkeys (which cost the taxpayers $45 apiece) these scientists reported that Salk Vaccine was worthless as a preventative and dangerous to take." - p 142

"Dr. Sandler's research showed specifically that sugars and starches lower the normal level of blood sugar in the human body, producing a condition known as hypoglycemia. Also that the phosphoric acid in soda pop has a tendency to absorb the phosphorous and sulphate in the food before the natural process of metabolism can get it to the nerves which these substances feed and nourish.

"When the blood sugar becomes deficient, and the nerves are deprived of phosphorous and sulphate, certain nerve trunks fail to function properly and the victim loses the use of one or more limbs. Serum, usually pus from diseased animals, only aggravates this condition by increasing the toxin load for nature to throw off.

"So he told the world thru his hometown newspaper at Asheville. All North Carolina got it; when the 1949 season approached these Tarheels cut down on their physiologic contraband 90%. It was therefore no co[i]ncidence that polio decreased in that state in 1949, 90%. We have the N. C. State Health Department's figures. They say there were 2,498 cases of polio in 1948 and 229 in 1949. Need anything more be said?" - p 146

Some of the people whose writing is presented in this Addendum are religious. I'm adamantly anti-religious.. &, yet, I find myself agreeing w/ these folks w/o sharing their belief in God, I even like an expression that appears a few times: "choosing gold over god" b/c I agree w/ the basic sentiment of criticizing choosing money over valuing ethics.

"Second, our Creator, when He fashioned the human body, closed the bloodstream from outside contamination. All objects entering the circulation must pass through intricate, but efficient, natural filters which bar out impurities. Yet man, with his hypodermic gadget, circumvents these filters, regardless of the Almighty's forethought." - p 158

I reached a similar conclusion about the dangerousness of bypassing the body's filters before ever reading about such things & I *don't need to think about it in terms like "the Almighty's forethought" in order to perceive this bypassing as a very bad idea.*

"Let us look at the Salk concoction, which admittedly is made from the diseased dead entrails of paralyzed monkeys. Last Summer, some 440,000 children of a group of 1,500,000 had their clean blood streams contaminated by this monkey-juice which had not even been laboratory tested—let alone clinically tried out, Frankly, those vaccinations of last Summer did not constitute a real test. It was a complete hoax. One child in four received the Salk concoction—the others, colored water. The names of all youngsters—those poisoned and those who only got water—were kept secret from everybody by a select group of hirelings inside the March of Dimes. So it was easy to "shuffle the cards" anytime during the so-called "six months trial period" when scientists supposedly were checking the test tubes. If any child—who had received the Salk monkey-juice—died, it was easy to slip that child's card into the cards of those youngsters who had received "only colored water" and to claim the victim was a "poor unfortunate" who had not been vaccinated at all! If such a "set-up" is a sample of American medical fair play—it is time the laymen made some changes in the rules." - p 158

The problem I have w/ the above is that the alleged fixing of the results is *possible* but *unsubstantiated*. There's no proof offered whatsoever that such a 'shuffling of the cards' took place. I have no problem perceiving the Salk polio vaccine (or any other vaccine) as dangerous, I have no problem imagining that greedy people will fix results in order to guarantee that their product makes them a fortune, I *do have a problem w/ acting as if that's what actually happened w/o proof.*

"Let us see what a frank, honest medical man learned of small-pox vaccination. His name: Charles Creighton, M.A., M.D., of England. This licensed medical man was instructed by the Encyclopedia Britannica to look into the subject of small-pox vaccination and to write objectively about it as the facts warranted.

"Dr. Creighton delved into the work and for several years checked through the records in all civilized countries. His written report was eventually printed in the

Ninth Edition of that publication."

[..]

"As an example, Dr. Creighton shows that compulsory vaccination became a law in England in 1854. Infant deaths, from small-pox, rose from 380 in 1853 to 591 in 1854. The records of the Eastern Metropolitan Hospital (Homerton, England) tell a complete story: from its opening early in 1871 to the end of 1878—about eight years—of the 6,533 persons received as small-pox patients, 4,283 of them bore vaccination marks, 793 admitted they had also been vaccinated but bore no marks while only 1,477 of the 6,533 HAD NOT BEEN VACCINATED. Also, in Bavaria, in the year 1871—of the 30,742 small-pox cases reported, 29,429 had been vaccinated. Is there any wonder that England repealed its compulsory vaccination law? In addition, Dr. Creighton's figures indicate that in practically every small-pox epidemic since the invention of the small-pox vaccine in 1799—a full one hundred and fifty-six years—the start was always with vaccinated victims." - p 159

I trust this report but I wanted to quote the full Dr. Creighton small-pox vaccination entry from the 9th edition of the Encyclopedia Britannica so I looked for it online not expecting to find it (at least for free). However, the entire, rather long & apparently very thorough, entry is here: http://www.whale.to/a/ creighton4.html . B/c of the length of the entry, I'll only quote from it briefly below.

"Risks of vaccination
The risks of vaccination may be divided into the risks inherent in the cowpox infection and the risks contingent to the puncture of the skin. Of the latter nothing special requires to be said; the former will be discussed under the five heads of (1) erysipelas, (2) jaundice, (3) skin eruptions, (4) vaccinal ulcers, and (5) so-called vaccinal syphilis."

[..]

"**Jaundice**
(2) It is only within the last few years that jaundice has been recognized as a post-vaccinal effect; and at present there is only one accepted instance of it on the large scale.   This was the epidemic among re-vaccinated adults in a large shipyard at Bremen from October 1883 to April 1884. Owing to an alarm of smallpox, 1289 workmen were re-vaccinated between the 18th August and 1st September with the same humanized lymph preserved in glycerin; of these 191 had jaundice at various intervals down to the month of April following. Circumstantial evidence (agreement and difference) clearly traced the epidemic to the vaccination [See Lurman, *Berl. klin. Wochenschrift*, 1885, p. 20.]. In future an outlook will be kept for this effect of vaccination; at present it has no intelligible theory.   It may be noted that the lymph which caused the Bremen epidemic was mixed with glycerin.

"Skin eruptions

(3) The eruptions that follow vaccination are proper to cowpox infection. Although little is said about them in the accidental infection of milkers, they were very common in the practice of Estlin, Ceely, and others with primary lymph. The eruption is a kind of exanthem, or "secondary" of the local infection, and does not ordinarily appear before the second week.   One of its commonest forms is a patchy rose-rash, or macular roseola, not easily distinguishable from the macular roseola of syphilis (Parrot, *La Syphilis Hereditaire*, 1886, p. 33).

"Another form is lichen or dry papules, apt to scale; it may also occur as a vesicular eruption, and in the form of pemphigoid bullae or blebs. In one of Ceely's cases the eruption extended to the whole mucous membrane of the mouth and throat.   A peculiarity of the exanthem is that it may come and go several times before it finally disappears; and, like other skin eruptions, specific or non-specific, it may become inveterate.   The widespread belief that much of the eczema of childhood dates from vaccination is not by any means to be dismissed as a mere fancy.   The skin-disorders that followed vaccination in the first years of the practice were declared by Birch and others to be new in type. At present the vaccinal eruption, especially on the scalp, is sometimes distinguished by the size and form of the crusts, and by scars remaining for a time.

"Vaccinal ulcers
(4) Ulceration of the vaccine vesicle, or of the site of it, is one of the commoner forms of "bad arm."   It is a return to the native or untamed characters of cowpox on the cow's teats, or on the milker's hands or face, or in the child's arm after experimental inoculation with primary lymph.   It crops out not infrequently in everyday practice, and is probably dependent for the most part on the lateness at which the lymph was taken for vaccination, or on retardation of the process in the vaccinifer, or on emptying the latter's vesicles too much; however, it may result from picking the scab or otherwise dislodging it.   The ulceration usually proceeds to some depth in the form of a crater under the crust, and is attended with induration and rounding of the edges and induration of the base. According to Bohn (*op. tit.,* p. 166), it may alarm practitioners by its resemblance to syphilis. In other cases the crust is wanting and the ulceration has the distinct type of phagedena. The destruction of tissue in either case may be very extensive, going "down to the bone" and having as much as an inch or more of superficial area. Healing is frequently an affair of weeks, and may be aided by mercurial treatment. There are no statistics of this sequel of vaccination ; but the frequency or infrequency of it may be learned in conversation with any intelligent chemist whose shop is resorted to by the poor, or with a medical practitioner of average experience.

"Vaccinal syphilis
(5) It has been proved by many experiments, undesigned or otherwise, in Paris (1831 and 1839), Vienna (1854), and elsewhere, that an infant with congenital syphilis develops correct vaccinal vesicles, provided its skin be clear of eruption and the lymph have been taken at the usual time; also that the lymph taken from the correct vesicles of a syphilitic child produces correct vesicles in its turn, but

does not produce syphilis in the vaccinated child. The congenital taint is, in fact, irrelevant to the course of cowpox infection. So far as experiment and casual experience can prove anything, that has been proved ; the recent attempt to disprove it by an officer of the Local, Government Board (*Report* for 1882, p. 46) is vitiated by fallacies, and has no value against the overwhelming testimony collected thirty or forty years ago. What, then, is the meaning of the numerous outbreaks of syphilis in groups of children or adults vaccinated or re-vaccinated with lymph from one source ?

"A careful examination of these cases shows that syphilis at the source of the vaccine matter was in all cases an after-thought, that in most of the cases there was no evidence for it, and that in the remaining cases the evidence was so far-fetched as to be unlikely (apart from the known *a priori* improbability), or that the traces of constitutional infection found in the vaccinifer were subsequent to vaccination, and therefore capable of being explained as an effect concurrent with the more obvious symptoms in those vaccinated therefrom. The effects, however, were very much the same as in the venereal pox. The vaccine vesicle either became an indurated or phagedenic sore, as described in the foregoing section on vaccinal ulcers, or the scar opened into an indurated sore after the usual sub-crustaceous healing was complete, or became indurated without opening. The axillary and cervical glands were often indurated. In most of the epidemics there were a certain number of cases in which the effects were purely local, or confined to one only of the seats of puncture ; if these had not occurred along with others in a group, they would have been counted as ordinary vaccinal ulcers. But there were often secondary symptoms as well including the roseolar, lichenous, or (rarely) pemphigoid eruption, and not unfrequently condylomata *circa anum et genitalia*. In some epidemics (but not in all) there were, in a small minority of the cases, mucous patches on the tonsils, tongue, or lips, tending to ulcerate; and in some of the Italian outbreaks the infection spread among the mothers and other members of the households in the form of specific sores of the nipples, with or without constitutional symptoms. Affections of the bones and viscera do not seem to have followed; fatalities were not very common.

"It will be hard to persuade medical authorities that these secondary effects are not the result exclusively of the venereal pox. The evidence, however, does not allow us to assume any other specific infection than that of cowpox, which, as we know, has its proper secondary exanthem in the form of macular roseola, lichen, or pemphigus; the eruption has even been known to affect the mouth and, throat. The evidence from epidemics of vaccinal sore arms teaches us that condylomata, mucous patches of the tonsils, tongue, and lips, and even iritis, are also possible, although far from invariable, among the "secondaries" of the primary vaccinal ulcer. The most general fact that comes out in these epidemics is that the lymph was taken late from the vaccinifer, or that the vesicles of the vaccinifer were drained dry to vaccinate a large number, or that the same vaccinifer was used for arm-to-arm inoculation on two successive days. It is not difficult to see how, in those circumstances, the abbreviated cycle of humanized cowpox may be departed from and the native or untamed characters of cowpox infection reverted to. Cowpox, indeed, is parallel with the venereal pox, both in the circumstances of its becoming an infective ulceration (indurated or suppurating) and in its secondary or constitutional manifestations as an infection in man. But the "bad" lymph has hardly ever been used beyond the second remove; and there the parallel fails.

"The following is a list of the so-called syphilitic epidemics after vaccination, including those that have been considered spurious, because they were either anomalous in type from the point of view of syphilitic infection or had no obvious causal connexion with that Disease.

"Udine, 1814 (see Viennois, in *Syph. Vaccinale*, p. 221). Cremona, 1821 (see Depaul, Projet de Rapport," in *Syph. Vacc*). Grumello, 1841 (ibid). Coblenz, 1849 (Wegeler, in *Pruess. Vereins-Ztg.*, 1850, No 14; abstracts in *Schmidt's Jahrb.*, vol. lxvii., 1852, p62). Upper Franconia (the Huber case), 1852 (*Intelligenzbl. Der Bayre. Aerzte*, 1854; Bohn, *loc. Cit*). Lupara (Italian prov. Molise), 1856 (see Depaul, *loc. Cit.*). Dispon near Pesth, 1855-57 (*Oester. Zeitschr. fur prakt. Heilk.*, 1862; Bohn, *loc. cit.*, p. 322). Rivalta (Piedmont), 1861 (Pacchiotti, *Sifilide Trasmessa per Mezzo della Vaccinazione in Rivalta presso Acqui*, Turin, 1862). Torre de' Busi near Bergamo, 1802 (see Depaul, *loc. cit.*). United States (troops on both sides in the Civil War), 1861-65 (Jones, *Circular 11., Louisiana Board of Health*, Baton Rouge, 1884). Argenta near Ferrara, l866 (Gamberini, in *Gaz. des Hopitaux*, 1870, p 505). Morbihan (neighbourhood of Vannes and Auray), 1866 (Depaul, *Bull. de l'Acad. de Med.*, xxxii, 1866-67, p. 201; Bodelio, *ibid.*, p. 1033). Cardaillac (Lot), 1866 (*Bull. de l'Acad. de. Med*, 28th February 1867). Schleinitz (Styria), 1870 (Kochevar, *Allgem. Wiener Med. Ztg.*, 1870, Nos. 21 and 24 ; abstract in *Arch, fur Dermatologie und Syph.*, 1870). London (two series), 1871 (Hutchinson, *Med. Chir. Trans.*, liv., 1871). Switzerland, 1878 (*Bull, de la Soc. de la. Suisse Romande*). Algiers, 1880-81 (Journ. d'Hygiene, 25th August 1881). Lyck (East Prussia), June 1878 (Pineus, *Vierteljhschr. f. gericht. Med.*, 1879, p. 193). Asprieres (Aveyron), March 1885 (P. Brouardel, *Rapport*, Paris, 1886)."

Miss Lily Loat has this to say on Creighton's Encylopedia Brittanica entry:

"Dr. Creighton was asked about the year 1884 to write the article on vaccination for the ninth edition of The Encyclopedia Brittanica. He agreed to do so, but instead of contenting himself with the usual stock statements he went right back to Jenner's own writings and to contemporary documents. He searched the pro- and anti-vaccination literature of many countries and came to the conclusion that vaccination is a "grotesque superstition." He wrote to the editor of the Encyclopedia Brittanica and said: "If you want an apologetic article, I am not the man to write it." The editor promised to publish whatever he wrote and so in the ninth edition of The Encyclopedia the article on vaccination is an anti-vaccination article. About the same time Creighton wrote a little book called "Cowpox and Vaccinal Syphilis" and a year or so later a larger book called "Jenner and Vaccination."" - pp 215-216

Wd such a willingness to publish non-mainstream opinions exist today? In my own personal experience, a piece of mine called "Sentimental Journey" (https://medium.com/@idioideo/sentimental-journey-69219d7d289c), something critical of the medical industry, was rejected by Mark Young, the editor of *Otoliths* b/c he disagreed w/ its content.

Here's the beginning of Wikipedia's article on anti-vaccination. I find it to be completely transparently biased & of no use whatsoever from a scientific standpoint. It's just propaganda, no more, no less:

"**Vaccine hesitancy** is a delay in acceptance, or refusal of vaccines despite the availability of vaccine services. The term covers outright refusals to vaccinate, delaying vaccines, accepting vaccines but remaining uncertain about their use, or using certain vaccines but not others. The scientific consensus that vaccines are generally safe and effective is overwhelming. Vaccine hesitancy often results in disease outbreaks and deaths from vaccine-preventable diseases. Therefore, the World Health Organization characterizes vaccine hesitancy as one of the top ten global health threats.

""**Anti-vaccinationism**" refers to total opposition to vaccination; in more recent years, anti-vaccinationists have been known as "**anti-vaxxers**" or "**anti-vax**". Vaccine hesitancy is complex and context-specific, varying across time, place and vaccines. It can be influenced by factors such as lack of proper scientifically based knowledge and understanding about how vaccines are made or work, as well as psychological factors including fear of needles and distrust of public authorities, a person's lack of confidence (mistrust of the vaccine and/or healthcare provider), complacency (the person does not see a need for the vaccine or does not see the value of the vaccine), and convenience (access to vaccines). It has existed since the invention of vaccination and pre-dates the coining of the terms "vaccine" and "vaccination" by nearly eighty years. The specific hypotheses raised by anti-vaccination advocates have been found to change over time.

"Although myths, conspiracy theories, and misinformation spread by the anti-vaccination movement and fringe doctors leads to vaccine hesitancy and public debates around the medical, ethical, and legal issues related to vaccines, there is no serious hesitancy or debate within mainstream medical and scientific circles."

- https://en.wikipedia.org/wiki/Vaccine_hesitancy

It seems to me that the open-mindedness to a truly scholarly appraisal of vaccination by the editor of the 9th edition of the Encyclopedia Britannica has been eradicated from Wikipedia & its fellow travelers.

"The investigation had been carried out by Prof. A. Bradford Hill and Dr. Knowelden. These men reported in the *British Medical Journal* (July 1, 1950) that in the 1949 epidemic of poliomyelitis in England and Wales "cases of paralysis were occurring which were associated with inoculation procedures carried out within the month preceding the recorded date of onset of the illness.""

- p 164

"The news account, as published in the New York *Times*, June 12, 1951, says:
"The next step was taken on the basis of a preliminary report on a research study
started here last March, when findings published in medical journals by English
and Australian scientists came to the attention of health authorities in this
country.

""The foreign studies, which now appear to be substantiated by the health
department survey, indicated that of children who contracted paralytic polio a
considerable portion had been immunized within the previous month with
diphtheria toxoid, whooping cough vaccine, and that the site of the paralysis was
more apt to be in the limb injected."" - p 165

Shelton makes meaningful puns, a practice I enjoy.

"On the following day, June 13, the State Department of Hell(th) of New York,
issued a statement to the press in which it said that this proneness to
development of paralysis follows all injections of diphtheria toxoid and anti-
whooping cough vaccine. This means that all injections of any substance
whatsoever, are harmful and, if they are not the direct cause of the disease that
follows, as Mahoney insists, they are determining and complicating factors that
must be reckoned with in any study of the cause of these diseases." - p 166

The worst epidemic fear-mongering campaign I've witnessed in my 68 yr life has
been the one of the so-called COVID-19 pandemic starting at the beginning of
2020. This bk's good for informing me that such scare tactics didn't begin then
but, instead, have a long history.

"Before parents can be induced to submit their children to the dangers of
inoculation in this manner, and especially before they can be induced to permit
their children to be used for purely experimental purposes, they must first be
frightened out of their wits. There is nothing like an epidemic to frighten ignorant
people." - pp 168-169

"Hammond stated that the gamma globulin used in the polio experiment is the
same as that used in measles and that it is obtained from blood collected by the
American Red Cross from donors all over the country. Thus Clara Barton's
organization (one she repented ever having formed) is revealed as being in all
ways a stooge of organized voodooism." - p 170

Some names mentioned in this bk are ones I've run across many times before in
connection w/ medical history, such as Florence Nightingale & Clara Barton,
but're people I only have a vague knowledge of. Reading the above got me
searching for more about Barton online. Barton was the founder of the American
Red Cross, the International one already existed, but I didn't find anything about
her regretting ever having done so - but it's possible that the following quote may
shed some light on that:

"As criticism arose of her mixing professional and personal resources, Barton was forced to resign as president of the American Red Cross in 1904 at the age of 83 because her egocentric leadership style fit poorly into the formal structure of an organizational charity. She had been forced out of office by a new generation of all-male scientific experts who reflected the realistic efficiency of the Progressive Era rather than her idealistic humanitarianism. In memory of the courageous women of the civil war, the Red Cross Headquarters was founded. During the dedication, not one person said a word. This was done in order to honor the women and their services. After resigning, Barton founded the National First Aid Society." - https://en.wikipedia.org/wiki/Clara_Barton

"Hygienists do not discredit the presence of the three guilty viruses. We say they are merely a symptom of polio rather than the cause. And we openly admit our skepticism of any approach to the problem of polio that systematically or unwittingly ignores the *basic enervating and toxemic forces* that nurture the soil of its growth in the human body." - p 179

"It may stun you to learn that in the year 1939 the per capita consumption of white sugar in China, where only sporadic cases of polio were seen, was 3.2 lbs., whereas in the United States 103.2 lbs. per person were consumed. I understand from good authority that this pattern ran true in the epidemics of 1949 and 1952, which were the most severe in our history.

"This is not a rare and isolated Statistical item. The United States Army documented well the figures of our soldiers contracting the disease in the Philipines, in China and in Japan—with no outbreaks among the native children. The Surgeon General's Office has reports filed away concerning the 1945 outbreak amongst our troops in the Philippines, when 250 were afflicted with the disease, with 50 deaths resulting—*but with no outbreaks amongst the natives.* The only plausible answer here, and one which has been reported by a few honest investigators, exists in the fact that the American soldier was allowed to take his eating habits along with him. He was permitted to stuff himself with ice cream, soda pop, candies, pastries, etc. With all due respect for his great fighting ability he was coddled into ill health by the misguided generosity of the Army authorities, perhaps in recompense for some of the ghastly aspects of that systematic decimation of human beings called war. In one account I read the following: "Ice cream manufacturing equipment generally soon followed combat equipment."" - p 180

I can attest from my own experience that eating sugar has probably been my worst habit. As for the "misguided generosity of the Army authorities" that provided ice cream for the soldiers I doubt that they were concerned about the soldiers' long-term health & the sugar probably gave them a low-level rush that kept them energetic.

"Salk and all he represents in method and philosophy see health only in terms of suppression of symptoms. We *Hygienists* look upon health as a high standard of physiological efficiency." - pp 181-182

This issue of "suppression of symptoms" is one that's preoccupied me for a long time. It's always astounded me that it's common medical practice to prescribe people pain-killers so that they can cover up the cause of their problem rather than be aware of it so that they can treat it correctly & feel the pain subside naturally. A person who's hurt their leg who then takes pain-killers so that they don't notice so much that it's hurt are more likely to walk on that leg when they shd be resting it to allow it to heal.

"We say you just cannot beat disease by preventing its symptoms. True prevention must start with the elimination of cause." - p 182

That's common-sensical - but to be fair I don't think conventional medicine is only about treating symptoms, I think there's a fair amt of orientation around elimination of cause too.

"The world's greatest brains and scientists all agree **THERE IS NO SUCH THING AS A VIRUS** . . . merely a mythical word to confuse the public. Read Dr. A. J. Shadman's (MD) new book "Who Is Your Doctor And Why"." - p 184

If I weren't already so swamped w/ bks to read I might just add that one to the list. I'm intrigued by the idea that viruses might be a fiction. As I've already stated here, viruses cd be conceived of as just a replacement of the idea of disease being caused by evil spirits. It's not like the average person who believes in viruses has any good reason to do so, they've just been told that they exist & that they're the problem & they're so bombarded by this that they accept it as fact. Over the time of the QUARANTYRANNY I've become interested in seeing how many movies there are that promote the virus theory & there're far more than I'd ever realized before. The trope of viruses created in a lab for military purposes that get loose into the world causing massive deaths is extremely common in sci-fi/horror movies.

As for the claim that "The world's greatest brains and scientists all agree **THERE IS NO SUCH THING AS A VIRUS**"?! That's so obviously an exaggeration that it's not even worthy of much comment except to say that it's hard to say who the "world's greatest brains and scientists" are but even if we cd do that it strikes me as extremely unlikely that they'd all agree about much of anything.

Rabies is the next subject under review. This interests me b/c I, like probably the vast majority of people, have a fear of being bitten by a rabid animal at the same time that I've never witnessed one & have no good reason to think I ever will. In fact, it was partially b/c of this that I made a movie recently called "Rabies Shot PSA" meant to poke fun at exaggerated & unreasonable fears - on my onesownthoughts YouTube channel here: https://youtu.be/AK6xOoPALGQ - on the Internet Archive here: https://archive.org/details/rabies-shot-psa . At the time that I made it I wasn't aware that there're people who contest its very existence.

"Rabies was an old superstition — a relic of the times when devils ran to and fro between animal and man carrying disease.

"Pasteur, who had previously had a hemorrhage of the brain, changed this old superstition into a money-making disease.

"Rabies is now a pet child of the Vivisection Trust which works internationally.

"If vivisection has proven anything it has proven the impossibility of man contracting any real disease from a dog.

"How long will filthy lucre keep the facts from the fooled public?

"In early times, as recorded in articles available in old libraries, the kiss of a king would cure rabies. It was later discovered that a piece of the king's garment would be as efficacious.

"Still later the "mad stone" when applied over the area of the bite would "draw out the madness". Later some of the "hair of the dog that bit you" could either be chewed and swallowed or bound on the wound.

"A still later discovery was that which employed an extract of "wild cockroach".

"In 1806 a Mr. Kraus was awarded $1000, by the then rulers of New York territory for his scientific discovery which had kept rabies out of New York for over twenty years. His formula is a matter of record and consisted of the ground-up jaw bone of an ass or dog, a piece of colt's tongue and the green rust off a penny of George the First reign." - pp 185-186

Whether or not any of that invalidates the belief in rabies is beyond me. It does tend to reinforce what I've already stated that some of yesterday's science is today's foolishness - wch leads me to: some of today's science might be tomorrow's foolishness.

"Dr, Charles W. Dulles, lecturer on the History of Medicine at the University of Pennsylvania, who was pointed by the Medical Societies of the state to investigate rabies stated that he "inclined to the view that there is no such specific malady" because after sixteen years of investigation he had "failed to find a single case on record that can be conclusively proved to have resulted from the bite of a dog or any other cause." The report and Dr. Woods' letter were endorsed by Theophilus Parvin of Jefferson Medical College and President of the National Academy of Medicin[e]; Dr. Thomas G. Morten, Coroners Physician; Dr. Charles K. Mills of the University of Pennsylvania and Dr. Thomas J. Mays of the Polyclinic Hospital." - p 186

It's interesting, isn't it? I remember looking up rabies when I made my rabies movie & I don't recall reading the slightest doubting statement about its existence

- &, yet, reading the above makes it seem as if mainstream medical science had more or less reached the conclusion that it didn't exist.

"In a book entitled, "Bechamp or Pasteur," by E. D. Hume, there may be found much proof pertinent to our discussion. A notable failure of the Pasteur treatment was that of a young postman, named Pierre Roscol, who, with another man, was attacked by a dog supposed to be mad, but was not bitten, for the dog's teeth did not penetrate his clothing; but his companion received severe bites. The latter refused to go to the Pasteur Institute and remained in perfect health; but the unfortunate Roscal was forced by the postal authorities to undergo the treatment, beginning March 9th. On the following April 12th severe symptoms set in with pain at the point of inoculation, not at the place of the bite, for he had never been bitten. On April 14th he died of paralytic "hydrophobia" the new disease brought into the world by Pasteur." - p 188

Another interesting story. The implication is that the drs w/ the most convincing (& enforced) worldview are the ones who call the shots. That doesn't mean that there's any reality to what they're putting forth. It's another classic situation of *History belongs to the victor* but, as I like to add, *but only the loser believes it*. But back to the Salk polio vaccine.

""This dosing with 'Salk slop' is in the same category of so-called medical research which permits a surgeon to yank the tonsils from a helpless child or 'scare' an appendix operation from a grown-up. Not many years ago the 'big bugs' in the medico-scientific field had the effrontery to calmly state, in their public meetings, that the 'appendix' served no useful purpose in the body and the tonsils were the remains of our 'gills'—left over in our bodies when man emerged from the sea as a land animal. Now, students realize that the appendix helps to keep the cecum on the safe side by holding the putrefactive bacteria, in that blind sewer, at a safe level and that the tonsils play important parts in killing disease organisms that would otherwise get into the lymphatic circulatory system." - p 190

I do find the idea of 'vestigial organs' to be ridiculous. That's one of the reasons why, after seeing the effects of brainwashing on such a huge scale, one of my current sarcastic slogans is *Your Mind is a Vestigial Organ, Leave the Driving to Us* as a way of saying that I think that people aren't using their brains anymore, as if they don't need them - w/ the "Leave the Driving to Us" being a repurposed ad slogan meant to imply that the people who've stopped thinking are just letting other people do it for them b/c they're too lazy & conformist to do otherwise.

""Park, Davis (Chemical manufacturers) had a special reason for pushing its vaccine development program. The company had suffered a serious financial setback. It had gambled heavily on Choromycetin, an antibiotic, only to have the bottom drop out of the market after disturbing indications that the new wonder drug might, under some conditions, cause fatal blood disorders. The stockholders were asking blunt questions, and management was understandably on the lookout for a new product sufficiently spectacular to wipe out the memory of Chloromycetin. POLIO VACCINE SEEMED A GOOD BET, and in the early spring of 1953, after spending a little over $1 million on research, Park, Davis believed its own experimental vaccines were READY FOR TESTING ON CHILDREN."

"SALK VACCINE: THE BUSINESS GAMBLE By Spencer Klaw, p. 125-126, Sept, 1956 issue of **FORTUNE MAGAZINE.**" - p 193

Are things like the above so hard to understand? Big businesses make big investments w/ the expectation of making big profits. When something backfires what are they to do? They can go out of business, they can be bought out by another company, they can gamble on something else. Even when they're still cautious, even when it's in their best interests to produce the safest product possible they might still be fighting for their life & cutting corners in acts of desperation.

Then there's an index from 195 to 201A. I can't stress enuf that the bad editions don't have this. For scholars, for people who actually want to be able to refer back to what they've read, to be able to use a bk as a reference, indexes are crucial. After the index there's additional material, again by authors other than McBean that, if I understand correctly, were added specifically for the 1959 edition that this, apparently, is a reprinting of.

""In 1955 we were told by the National Foundation, the United States Public Health Service and 'lesser' Public Health Departments, Medical Societies through their political officers and appointees that ONE injection of the currently used polio vaccine whould give prolonged, possibly life-time protection. In 1956 we were told, after those receiving the initial shot developed polio, that a 'booster' or second shot was necessary to raise the protective antibody level to a height whereby 90 per cent protection against paralytic infection would be assured. In 1957 we were told, after those receiving the second injection also developed polio, that a third 'shot' was imperative to effect absolute protection. For some this was their fourth and even fifth boosters. Now in 1958 we are told that a fourth shot is "Expedient."" - p 202

Sound familiar? When the lockdown was originally proposed In March 2020 we were told it was probably for 2 wks, just to keep the hospitals from becoming overcrowded. That was somewhat reasonable. It was also a way to get people to accept it, people figured that 2 wks of oppression wdn't really matter that much, a quick inconvenience. It's now been 2 yrs, 2 mnths w/ no end in sight. That's NOT reasonable. Then there's the vaccine: only one shot seemed like something people cd bear so they acquiesced thinking it wd then be over. But, of course, it wasn't over b/c then there 'had' to be a booster, then another, then another. & how many people do you know who still got sick after being vaccinated? To say that *so many more people wd've died if we hadn't quarantined & vaccinated* is an unproveable assertion, forget the computer projections. To say that the people who died b/c they were vaccinated wdn't've died if they hadn't been vaccinated is more believeable to me.

"I think J. H. Tilden, M.D. puts it about right, when he said—"It is exceedingly difficult to secure an honest hearing for any criticism of authority. Established beliefs are well nigh invulnerable because they are accorded infallibility by the masses who are educated to believe that they will be damned for thinking, and because of this, few will tolerate opposition of any nature to anything they have been educated to believe. People who have their thinking done for them are always intolerant."

"It is a truism that we are living in a time the like of which has never been seen before. We are front and center in a battle for men's minds. Now as never before it is vital that we have the truth, and yet there is an active conspiracy not just to conceal the truth but even to change the facts. The communications media are concentrated in the hands of a few men who are thereby enabled to manipulate the emotions of the American public at will." - pp 203-204

That was written in 1959. I've maintained for a long time that after so-called World War II the battle was on for control of the masses thru propaganda w/ the main protagonists being the U.S.S.R. & the U.S.A.. It's always seemed to me that the U.S.A. was the more successful of the 2, maybe b/c tv became such a major factor, maybe b/c more talent was given more free reign. I was 5 yrs old thru most of 1959, I certainly wdn't've been aware of the socio-political atmosphere, my awareness, such as it was, wd've been extremely local. As such, for me, now, the yrs of 2020 on have been even worse than the 1950s. But let's remember that the 1950s were followed by the 1960s & the 1970s, 2 decades of very widespread resistance to the insanity of the Vietnam War & the Military-Industrial Complex & more. Let's hope that a similar resistance to the Medical-Industrial Complex starts soon. To me, if humans are going to continue to be thinking beings, this resistance is imperative.

"Address by Miss Lily Loat
A quotation from "PHILOSOPHY OF HEALTH" January, 1927.
Edited by J. H. Tilden, M.D.

"Our own fight against vaccination has been a long and arduous battle. While individuals and small groups were fighting for freedom in this matter as far back as the time of the passing of the compulsory vaccination act of 1853, the definitely organized struggle started with the passing of the harsh vaccination act of 1867, which aimed at compelling every parent of a child to have that child vaccinated within three months of birth. Those who refused could be ordered by the magistrates over and over again until the child attained the age of fourteen to have it vaccinated and could be fined for each refusal to comply with such magistrates' orders." - p 214

An interesting sidenote for me is that I recently watched the movie Mary Shelley's Frankenstein, "a 1994 science fiction horror film directed by Kenneth Branagh, and starring Robert De Niro, Kenneth Branagh, Tom Hulce, Helena Bonham Carter, Ian Holm, John Cleese, Richard Briers and Aidan Quinn. The film was produced on a budget of $45 million and is considered the most faithful film adaptation of Mary Shelley's 1818 novel *Frankenstein; or, The Modern Prometheus*, despite several differences and additions in plot from the novel". [..] "While performing vaccinations, Waldman is murdered by a patient, who is later hanged in the village square. Using the killer's body, a leg from a fellow student who died of cholera, and Waldman's brain, Victor builds a creature based on the professor's notes. He is so obsessed with his work that he drives Elizabeth away when she comes to take him away from Ingolstadt, which is being quarantined amid a cholera epidemic." (https://en.wikipedia.org/wiki/ Mary_Shelley%27s_Frankenstein_(film))

The way I remember the movie is that the "patient", as he's referred to above, was a man missing a limb who was being compelled by law into being vaccinated. It's possible, but not stated in the movie, that the limb was missing from the effects of a previous vaccination. Since he was adamantly against being vaccinated he killed the dr when the dr was forcing the vaccination on him. I consulted a webpage titled "The anti-vaccination movement that gripped Victorian England" By Greig Watson (BBC News Published 28 December 2019) (https://www.bbc.com/news/uk-england-leicestershire-50713991) for its timeline & it confirmed for me that there had been no mandatory vaccinations in England before or during the time of the writing of Frankenstein. *However*. it also informed that "1805 - First attempt at compulsory vaccination, in Italy, fails" & it's noteworthy that Frankenstein was written in Geneva where Mary Shelley was w/ Lord Byron's physician, John Polidori who might've been aware of the Italian attempt at mandatory vaccination.

Whatever the case, Mary Shelley's Frankenstein had its plot tweaked to essentially show the creation of Frankenstein's monster as a partial consequence of compulsory vaccinations that didn't actually happen in England until 35 yrs after the writing of the bk. I just looked thru my annotated edition of Frankenstein for any mention of that in the original bk & found nothing. I find this detail to be of substantial significance even though Victor Frankenstein starts his career in Ingolstadt, Bavaria, Germany, wch is where the forced vaccinations take place in the movie. This, however, is also historically inaccurate - there were no forced vaccinations in Germany at that time. Was there a subtext of anti-vaccination sentiment to the movie?

"That fight for freedom from medical tyranny in this particular matter has been waged in England for nearly sixty years and it is going on still.

"For many years it was confined mainly to the poorer classes. Only a few men of intellect and distinction championed our cause. It was natural that most of the disasters due to vaccination should fall on the poorer classes and that those classes should publish them while the upper and middle classes would be more likely to keep such things to themselves. But by degrees what might be called the artisan class, the smaller shopkeepers and the lower middle classes, became the backbone of the movement. They paid large sums in fines, they had their goods seized and sold when they could not or would not pay fines. Those who had no goods or would not let them be seized went to prison, some were ruined, and some emigrated to avoid ruin. There are men living in America today whose parents left England on account of the harsh vaccination acts." - pp 214-215

Is it so fantastic to think that the oppression of the so-called COVID-19 'pandemic' might be a parallel to the oppression described above?

"Four very important things happened between the passing of the vaccination act of 1867 and the passing of the act of 1898, which contained the first conscience clause.

"The first was the smallpox epidemic of 1870-72, which carried off 44,000 persons in England and Wales and proved to hundreds of thousands of people that vaccination is not a protection against smallpox, for that epidemic occurred when 97&1/2 per cent of the people over two and under fifty had either had smallpox or been vaccinated, as was stated by Sir John Simon, chief medical officer to the Privy Council, in his evidence before the select committee which in 1871 inquired into the vaccination act of 1867." - p 215

"In 1887 Dr. Edgar M. Crookshank, who at that time was professor of pathology and bacteriology at King's College, was asked by the government to investigate an outbreak of cowpox in Wiltshire. Sir James Paget drew his attention to Creighton's work, evidently hoping that Crookshank would refute it, but the results of his laborious investigations are contained in two large volumes entitled "The History and Pathology of Vaccination", in which he says that the credit given to vaccination belongs to sanitation and isolation and that nothing would more redound to the credit of the medical profession than to give up their faith in vaccination." - p 216

I just ordered copies of both volumes of Crookshank's bk. Given that the principles of vaccination haven't really changed that much over the centuries I suspect that I'll still find its critique valid.

"Among scientists our most notable supporters were Alfred Russel Wallace and Herbert Spencer."

[..]

"Of literary men, George Bernard Shaw is our most noted supporter." - p 217

Here's an excerpt from an open letter to the governor of California in 1956 from Eleanor McBean:

"In spite of all the high pressure sales talk and advertising technique, plus the false reports and garbled statistics designed to sell the people and their leaders this dangerous vaccine, enough of the truth has leaked out so that **17 states have already rejected their supplies of Salk vaccine.** Why should our great state be the slowest to wake up to this monstrous fraud?" - p 218

"Health Director Peterson of Idaho didn't mince words when he stated, to the press, that polio **struck only vaccinated childrem in areas where there had been no cases of polio** since the preceding autumn. "In 9 out of 10 cases the **paralysis occurred in the arms in which the vaccine had been injected.**"" - p 219

I didn't find online info about wch 17 states those might've been but I trust McBean's integrity enuf to believe her. The closest I got to such info was this:

"Massachusetts was the first to impose compulsory vaccination in 1809. Washington, DC, and eight other states later joined in requiring infant vaccinations.

"Other state officials opposed such mandates, and by 1930, Arizona, Utah, North Dakota, and Minnesota had passed laws against vaccination requirements for their residents.

"A total of 35 states did not have legislation for or against mandates, and instead allowed local authorities to regulate such actions." - https://www.webmd.com/vaccines/covid-19-vaccine/news/20211014/vaccine-opposition-not-new

"Upon demand of the National Foundation for Infantile Paralysis, whose racket he was seriously interfering with, he was railroaded by a Fede[ra]l Court in Florida after the Post Office Dept. was importuned to go far afield and split hairs to convict him of sending out "derogatory statements" about the Salk racketeers on a postal card.

"There actually is a PO regulation which says that such statements are all right in sealed mail, but "unmailable" on postal cards. The usual procedure is for postal inspectors to notify the "culprit" and have him sign a stipulation not to do it again. We know this from personal experience. In the Miller case this was not done, because of pressure from NFIP and those who own the serum trust—the House of Rockefeller.

"Miller was given a 2-year prison sentence and pleaded on probation for two years in July of 1955. It was admitted by the "prosecution" that its main object was to stop him from telling the truth about Salk vaccine. So, terms of probation were that he was not to send anything through the mails having the slightest reference to vaccine or medicine of any sort." - p 220

Wow. If it was illegal to send derogatory statements about the Salk vaccine on post-cards it shd've also been illegal to send out post-cards in favor of the vaccine. That wd be only fair, to just ban any publicly visible opinion on the matter whatsoever. As for "It was admitted by the "prosecution" that its main object was to stop him from telling the truth about Salk vaccine.": I suspect that the prosecution didn't express it that way.

"VACCINE ECONOMICS—A 2-year prediction in CAPSULE NEWS came true last week when a verdict for $147,000 was rendered against the Cutter Laboratories of California for the crippling of two children with Salk vaccine." - p 221

"In April 1955 more than 200,000 children in five Western and mid-Western USA states received a polio vaccine in which the process of inactivating the live virus proved to be defective. Within days there were reports of paralysis and within a month the first mass vaccination programme against polio had to be abandoned. Subsequent investigations revealed that the vaccine, manufactured by the California-based family firm of Cutter Laboratories, had caused 40,000 cases of polio, leaving 200 children with varying degrees of paralysis and killing 10." - https://www.ncbi.nlm.nih.gov/pmc/articles/PMC1383764/

"Salk vaccine made a barrel of money in 1956 for the killers.

"Pitman-Moore is one of three subsidiaries of Allied Laboratories. So great was the profit from Salk vaccine that Allied's 1955 profit ($1,292,136) was upped to $2,598,060 in 1956. An increase in profit (not just sales) of around 100% for the parent company.

"Eli Lilly made a profit after taxes of $16,328,081 in 1955. With the Salk vaccine sales in 1956 this profit was upped to $30,052,815; an increase of 90%.

"Sharpe & Dohme is one of 25 subsidiaries of Merck & Company. Salk vaccine sales so boosted Merck's business in 1956 that its net profit, after the tax bite, was $20,224,427 compared with $15,714,342 in 1955.

"Wyeth is one of 41 subsidiaries of American Home Products. Salk vaccine sales boosted AHP's profits from $20,536,619 (1955) to $31,250,355 in 1956. A 50% profit increase for the parent.

"Parke-Davis' management must have been below that of the others for this dealer in death only increased its 1956 profits from $14,322,015 (in 1955) to $17,645,728.

"All of these figures are taken from Moody's Manual of Industrials, the investors' bible. All are "after taxes" profits." - p 222

To be fair, all of those increased profits probably weren't exclusively from the polio vaccine. That sd, if anyone out there is naive enuf to think that the possibility of such enormous profits wdn't influence a company's "looking the other way" about safety issues is too naive for me to even bother to discuss this w/.

"September 13, 1958 — Five more polio cases were reported in Honolulu today. The five cases included a three-year-old girl who had received all three Salk vaccine shots. She is the 9th fully vaccinated Islander to come down with polio.

"Two others reported today had received two shots. One is suffering with paralysis of the neck muscles and the other has paralytic symptoms in her left foot. The two other victims reported both with mild forms of paralysis had not received Salk shots,

"September 18, 1958 — The Territory of Hawaii's 62nd polio victim of this year is a two-year-old Marine dependent who had received all three Salk vaccine shots. The victim is suffering from paralysis of the left leg. The boy received his third shot two weeks ago and is the 10th Island resident inoculated with three Salk shots to come down with paralytic polio." - p 228

"MOSCOW, June 17 — The Communist party organ Izvestia said today nearly two million Russian children have been given spoonsful of Soviet-produced live polio vaccine and not a single case of polio developed.

"The questions were deeply challenging and vitally important: Is a live-virus polio vaccine safe and effective?" - p 229

All in all, despite whatever quibbles I might have w/ it, I think Eleanor McBean & this bk are brilliant. She's a true paradigm challenger & I think she makes her case well. The problem is that I don't know a single pro-vaccine person who wd ever read it - they already 'know' that they disagree w/ it w/o even bothering to check out what its points are - & that's what makes pro-vaxxers different from myself, at least - I'll actually carefully read things w/ opposing opinions & consider their points. Until pro-vaxxers can do likewise I can only assign them to the ranks of the brainwashed puppet people.

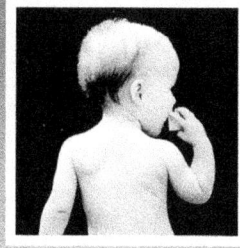

review of
Paul A. Offit, M.D.'s **Autism's False Prophets - Bad Science,
Risky Medicine, and the Search for a Cure**
by tENTATIVELY, a cONVENIENCE - July 1-20, 2022

I'd already read 2 bks on autism before reading this one. The 1st was Elizabeth Fein's Living on the Spectrum - Autism and Youth in Community. That one's by a friend of mine & I found it reasonable. The 2nd was Andrew J. Wakefield's Callous Disregard—Autism and Vaccines—The Truth Behind A Tragedy. Wakefield has been widely lambasted so I decided to give him a chance & to read what he has to say in defense of both himself & his medical findings & opinions.

I also decided to read Autism's False Prophets to give consideration to the opinions of one of Wakefield's critics & to learn more about what I take to be common medical opinions re autism. This, understand, was done in the spirit of at least making an attempt to be fair to a medical industry & philosophy that I generally have a highly critical opinion of. The inside front cover blurb gives the reader an idea of what to expect:

"Paul A. Offit, a national expert on vaccines, challenges the modern-day false prophets who have so egregiously misled the public and exposes the opportunism of the lawyers, journalists, celebrities, and politicians who support them."

The mood is immediately set that Offit represents 'objective truth' & that he's going to put those who disagree w/ him in their place as scoundrels. One thing that I appreciated about Wakefield's Callous Disregard is that he was more nuanced in his positioning, he gave substantial scientific backing for his opinions & conclusions, & he didn't depict everything in terms of good guys vs bad guys. Offit doesn't either, at least not all the time, but his approach immediately strikes me as arrogant.

To be direct, I've spent my whole life as what one might call a creative outlaw. The so-called 'straight' world (& I'm not referring to sexuality here) has been my adversary for as long as I can remember, since I was a small child. Offit strikes me as typifying, even exemplifying, the self-righteousness that characterizes people who've always been mainstays of the status quo. Offit's a BELIEVER: he knows he's 'right' &, from 'on high', he's going to tell the SHEEPLE that he is their leader & that to disagree w/ him shows a person to be a fool. Obviously, my attitude is showing that I strongly disagree. For me, Offit's narrative embodies an unquestioned paradigm - he's essentially a priest of the new religion, unquestioned science, science that's implied to be monolithic.. but isn't.

At the end of the back inside cover it's stated that "Paul Offit will donate all royalties from sales of this book to autism research." Even if I thought Offit is a poor man I wdn't be impressed by that b/c I'd have to think that autism research is the way to go, wch I don't. Non-institutionalized support groups, such as friends, are much more promising - & also a sort of 'vanishing species' in today's enforced social distancing environment.

However, I'm getting ahead of myself. Recently, I've been reading science bks, esp medical science bks, for the layperson. Some are written by people w/

standard accreditation, such as this one; some are written by autodidacts. I tend to prefer & identify w/ the latter. However, I've found ideas I value in all the bks. The challenge, for me, is to not *throw out the baby with the bathwater*. In other words, I'm not going to 'blanket-accept or reject' anything by anyone based on the degree to wch I can relate to their societal position. Nonetheless, I feel that it's important to try to analyze what subtexts I find underaddressed b/c I think they're more profoundly influential on the conclusions presented than will be otherwise apparent.

In his "PROLOGUE" Offit states that:

"I get a lot of hate mail.

"Every week people send letters and e-mails calling me "stupid," "callous," an "SOB," or "a prostitute." People ask, "How in the world can you put money before the health of someone's baby?" or "How can you sleep at night?" or "Why did you sell your soul to the devil?" They say "I don't have a conscience," am "directly responsible for the death and damage of hundreds of children," and "have blood on [my] hands." They "pray that the love of Christ will one day flood [my] darkened heart." They warn that my "day of reckoning is coming."" - p xi

Giving Offit the benefit of the doubt that he's telling the truth, I find this outpouring of contempt for him to be very unfortunate & uncalled for - despite my disagreeing w/ him over many things. A theme throughout the bk is that the people who disagree w/ him are ignorant religious people, people too stuck in the dark ages (although he doesn't exactly express it that way) to understand the enlightenment of his scientific position. NOW, I'll quote some of the reviews that're on Goodreads of Wakefield's <u>Callous Disregard</u>. Since Wakefield is presented by Offit as backed by the religious people, these attacking reviews are presumably written by pro-science people:

"poorly written book by a disgraced doctor with an introduction by a playboy centerfold. the sequel to this, Charlatan by pope brock is much better written and shows how medicine like wakefield's can succeed."

"I would give this book zero starts if I could. This book is incredibly difficult to follow and poorly written. It screams of a child caught doing something bad who is saying, "Yea, I was caught and did something wrong , but ...." Not worth your time."

"My contempt for this man and his book knows no bounds."

"Wakefield has caused an incredible amount of harm to the world with his theories, none of which have ever held up to scrutiny. He used a sample of 12 children and from that declared MMR dangerous and recommended cessation. That is about as irresponsible as it gets. His study has been reviewed and analyzed, and millions of children in several countries have been studied, and not once has he been validated.

The best thing that could happen is if his works are gathered up, put in a pile and burned, except for a single copy to be placed in a museum under lock and key. It can be a warning to us all what irresponsibility can lead to."

"Andrew Wakefield is a fraud and a hack. His falsified data and bogus science has caused thousands of deaths, yet people continue to believe him because it fits their Big Pharma and government "mind control" narrative. There's a reason why his medical license was revoked - he's a total disgrace to the profession and should be locked up for life."

- https://www.goodreads.com/book/show/10700469-callous-disregard

Now, obviously, I chose the negative reviews. Fortunately, there're positive ones as well. What I hope to show here is that the people who agree w/ Offit are just as likely to be as hateful & ignorant as the people who disagree w/ him. 2 of them call the bk "poorly written" - but I've read it & found it reasonably well-written. The challenge for the lay reader is that it uses a fair amt of specific medical jargon to explain Wakefield's findings. The average lay reader is likely to find that too difficult a hurdle. The reviewers feel free to call him "disgraced" & a "fraud and a hack". One makes the unsubstantiated claim that "His falsified data and bogus science has caused thousands of deaths", another calls for the burning of his bks & says he shd be "locked up for life." It seems likely to me that Wakefield has been subjected to far more abuse than Offit has & that Offit is one of the reasons for this abuse. In other words, Offit is far from innocent. NEITHER OF THEM SHD BE SUBJECTED TO SUCH HATEFULNESS.

Offit is from where I spent the 1st 40 yrs of my life. As such, that gives him a slight prejudice on my part in his favor b/c I can relate to his origins: "I grew up in a suburb outside of Baltimore." (p xi)

Offit describes childhood influences on his wanting to become a doctor. The narrative is of a typical sort: sensitive-caring-person-perceives-suffering-in-the-world-that-they-set-out-to-alleviate. I assume that for many people this narrative is genuine. For me, however, I'm somewhat distrustful of what *goes unsaid*.

"I also remember how hard it was for the other children in the ward, horribly crippled and disfigured by polio. As I got older, the image of those children remained. I wanted to protect them, to make them feel better, to champion their causes. So I became a pediatrician and later a pediatric infectious disease specialist." - p xiii

To some people, I may seem to be being horribly cynical - but, to me, I'm just being realistic: this image of the doctor as the champion of good is a bit too pat for me, it's too much a standard public relations poster boy thing. I wonder whether Andrew J. Wakefield has a similar tale to tell of his early inspiration to be a doctor. If he does, he didn't tell it in <u>Callous Disregard</u> - not that I remember, at least.

My questions about those kids in the hospital ward suffering from polio are: 1. How many of them were vaccinated? In other words, how many of them either got polio *despite the vaccination* OR how many got polio *b/c of it?* It's unclear to me whether Offit's story is from polio vaccine times or before. If it was 1956, as the narrative seems to imply, the Salk vaccine wd've been available for at least a yr. 2. How many of the children were *made sick or sicker in the hospital?* I don't see hospitals as necessarily a place where people are only helped - I see them as places where people are often harmed. Either way, *they're always charged* - & *that's* pretty damned important. Maybe Wakefield is greedy.. but how many doctors aren't?! Not many, I'd wager - & I doubt that I'd be able to count Offit among them. If he's lived at the level of poverty that I have for most of my life I'd be impressed - but there's no way that that's likely. According to https://networthpost.org/net-worth/paul-offit-net-worth/ his net worth is 1.6 million.

"I went to college in Boston and medical school in Baltimore before training in pediatrics at the Children's Hospital of Pittsburgh." - p xiii

Ok, there's the Baltimore connection again AND there's a Children's Hospital in Pittsburgh w/in sight of my current house. It's not, however, the same children's hospital. The one I can see was created anew in 2005, long after Offit was there. It helps abate my cynicism slightly when I can identify w/ physical particulars.

"Because modern medicine is incapable of preventing diseases, it's enormously frustrating. But the measles vaccine has been around for more than thirty years. It works and it's safe. Still, these parents had chosen not to protect their children." - pp xv-xvi

I find that language rather manipulative. 1st, I think that "Because modern medicine is incapable of preventing diseases, it's enormously frustrating" might be better expressed as 'Because modern medicine is incapable of preventing diseases, there's no good reason to trust it.' 2nd, the claim that the measles vaccine "works and it's safe" goes strangely against the 1st sentence, does it not? If "modern medicine is incapable of preventing diseases" then how exactly is it that the measles vaccine works? 3rd, saying that "these parents had chosen not to protect their children" b/c they reject the vaccine is out-&-out lying propaganda of the worst sort: parents who reject getting their children vaccinated do so b/c they think that the vaccine will do more harm than good, they haven't "chosen not to protect their children", they've *chosen to protect them* - they may be wrong - but they're not necessarily ignorant religious fanatics.

Now, keep in mind that Offit's PR hype presents him as "a national expert on vaccines". In fact, his specialty is working on rotaviruses: "We found if we inoculated rotaviruses into the mouths of baby mice, they would get diarrhea. For the next ten years we worked to determine the parts of rotavirus that made mice sick and the parts that evoked a protective immune response. Then we constructed a series of combination rotaviruses—made from cow and human strains—that protected mice without causing disease. We were confident we had made a vaccine." (p xiv) No doubt he was pd very well for this.

I'm reminded of a former friend of mine. She was born rich, she went to a university that cost something like 3 times as much a yr as what I made working. She got a well-paying job as a union organizer. Her heart's in the right place. After 10 yrs or so, as far as I know, she'd been pd over a half-million to organize a union *but, gee, hasn't done it yet*. It appears that only her heart's in the right place & the rest of her is out-to-lunch. Do you get my drift?! These people who fancy themselves to be so smart expect to be able to have an indefinite amt of time to do their job & to be pd very well for doing so.. *even if they fail*.

Now, imagine someone who isn't using the 'I'm-so-smart-&-what-I'm-doing-is-so-hard-you-really-wdn't-understand' card: a plumber comes to yr house b/c you need yr toilet fixed, you've been having terrible diarrhea b/c you're addicted to cheap Mexican fast food. You're paying the plumber $150 an hr to get that diarrhea-filled toilet up & running again. 10 yrs later he's still at it. He's now a millionaire. He explains to you w/ a very straight serious face why it's so damned hard for him to do. He's tried every plunger known to man but he's had to have some specially made in Wuhan, China, & he has to pay a research team to design them. Let's face it, a real plumber wd have the job done in half a day - & even then that wd be extreme. If you're lucky, he might even give you some worldy advice gratis about changing yr eating habits.

"During the next ten years, I saw several children come into our hospital with pneumonia caused by whooping cough, or severe skin infections caused by chickenpox, or meningitis caused by the bacterium *Haemophilus influenzae* type b (Hib), because their parents chose not to vaccinate them." - p xvi

You see? This is the way it works: the parents choose against vaccination & the children imediately break out in a disease, *it's the **decision** that causes it!* Eureka! All the drs have to do is get everyone to make the decision that *they* dogmatically present as the ultimate wisdom & *no-one will get sick again!* Too bad that totally flies in the face of all reality given that people who get vaccinated still get sick all the time - & it's not just "several" of them either!!

Even though I find Offit to be a transparently dishonest person b/c of the manipulative language he uses, I'll give him the benefit of the doubt again in regards to the following story:

"After I appeared on MSNBC, an extreme anti-vaccine activist called our home;

later, our eleven-year-old daughter asked whether I thought anyone would ever hurt me. While I was on a federal advisory committee to the CDC—one that had made recommendations about the use of the mercury-containing preservative thimerosal in vaccines—I got a death threat. A man from Seattle wrote, "I will hang you by your neck until you are dead!" I called the CDC, which sent the e-mail to the Department of Justice, whch sent it to the FBI. The threat was deemed credible, and for the next few years an armed guard was placed at the back of advisory committee meetings; for the first few months, he followed me to and from lunch, a gun hanging at his side." - p xvii

Now, it's funny: poor people often live in neighborhoods where one's life is more threatened than that on an almost daily basis - but, for us, the police are as much a part of the threat as the less legal violent people are. & we don't get our very own guard. I'm not in favor of Offit having his life threatened. Strangely, perhaps, I'm a 'live & let live' type - but the armed guard thing just highlights the *haves & the have-nots*.

"Some people who believe vaccines cause autism hate me because they think I'm in the pocket of the pharmaceutical industry, that I say vaccines are safe because I am paid to do it. To them, it is logical that I would spend twenty-five years working on a rotavirus vaccine—a vaccine that has a chance of saving hundreds of thousands of lives every year—so that I could lie about vaccine safety and hurt children." - p xvii

Again, I find the language manipulative. He essentially 'puts words in the mouths' of people who distrust him & disagree w/ him. Are they saying 'it is logical' that Offit wd 'spend twenty-five years working on a rotavirus vaccine' 'so that [he] could lie about vaccine safety and hurt children'? OR are they saying something more like this: 'Offit is being paid well by companies that have extreme profit motives to do something that he believes in but that may have the opposite effect than what it's purported to.'

What's *illogical*, to me, is that a person wd get wealthy spending "twenty-five years working on a rotavirus vaccine" "that has a **chance** of saving hundreds of thousands of lives every year" [**emboldening mine**]. 25 yrs & millions of dollars spent on something that produces a "chance". These claims by pro-vaccine people that "hundreds of thousands of lives" are saved every year have always struck me as hard to prove. I can say: 'I prayed to the Medical Industry God that 400,000 children wd not die this year &, *Lo & Behold!*, 400,000 children *did not die!* but there are always people who might point out that a lesser number of child deaths might be due to other causes such as improved nutrition, etc.. OR they might point out that the statistics used to 'prove' the claims are suspect.

"In 1916, polio became an American disease. In New York City alone, in one summer the virus paralyzed 10,000 people and killed 2,000. No one knew what was causing it." - xix

Statements like this are very dramatic & impressive but there's always the question of: *According to whose statistics?!* Furthermore, did any of those people paralyzed or killed have this happen to them *while under doctor's care?!* To me, there's far more of a likelihood of people *being harmed by doctors & hospitals* than ever seems to be generally admitted to - esp not by doctors & hospital staff.

"Doctors offered treatments that were equally absurd. They injected adrenaline or fresh human saliva into the spines of infected children, or they took spinal fluid from infected people and injected it back under the skin." - p xix

To me, the above "absurd" attempts at cures are essentially of the same ilk as vaccination. The initial smallpox vaccination, e.g., consisted of taking pus from cow pox, cutting the patient's arm, & smearing the pus in the wound. The basic theory of that practice hasn't changed much over the centuries - or, at least, not until the recent GMO (Genetically Modified Organism) variations. Offit is a BELIEVER, he believes in the type of vaccination he works on - & he ridicules a large variety of other possibilities. I'm not convinced that they aren't *all* ridiculous, including what Offit believes in.

"Today we see George Retan's polio cure as ill conceived and barbaric. But that's only because we know what causes polio and how to prevent it." - p xx

Really? The older I get & the more I observe the Medical Industry at work the more I see one 'cure' generating one or more side-effects that then have to be 'cured' by things that generate even more side-effects ad infinitum. The basic premise of Western medicine, at least, is that outside forces, usually from nature, attack & debilitate the body. It seems more accurate to me to say that most ailments are caused by self-inflicted toxins - or, at least, human-manufactured toxins - that cause the body to react in certain defensive ways that're best ameliorated by reduction of toxin-intake. In other words, nature isn't humanity's enemy as much as humanity is - & the Medical Industry's whole philosophical basis is a large part of what aggravates the situation.

"In 1998, a researcher in London scared parents by claiming that autism was caused by the combination measles-mumps-rubella vaccine known as MMR." - p xx

The researcher referred to here is Andrew J. Wakefield. I won't 100% back Wakefield b/c there was a tiny bit too much 'legalese' in the bk of his that I read, Callous Disregard, & he did, apparently, write that bk to present himself in a good light, but, still, Offit's presentation of what Wakefield did is the standard version. The following is an excerpt from my review of Wakefield's bk in wch I try to give a more accurate representation of what he, at least, claims was his actual position. I find it believable partially b/c it's too complex for most people to take in. Hence, it gets reduced to Offit's version.

"Wakefield & his fellow doctors got young patients referred to them who were experiencing gastrointestinal distress AND developmental regression that may've been autism. Some of the parents claimed that these problems started shortly after they rc'vd their MMR (Measles Mumps Rubella) vaccinations. Clinical study of the patients suggested that the vaccination in a minority of recipients might possibly be causing gastrointestinal problems & that these problems could be so severe that the children were unable to develop any further in a normal way but were, instead, regressing under the influence of extreme pain & inadequate digestion of nutrients. It was never claimed by Wakefield & co that the MMR vaccine *directly caused autism*. It was, instead, suggested that the connection between the MMR vaccine & the gastrointestinal problems in *some children*, not all, merited further investigation - as did the relationship between the gastrointestinal distress & arrested & regressed development. The patients were mostly very young. Wakefield took the position that single-disease vaccinations were safer b/c they were less of an assault on the body." - https://www.goodreads.com/review/show/4486376528

That's a different kettle of fish, isn't it?! It's interesting to note that Offit sd: "We found if we inoculated rotaviruses into the mouths of baby mice, they would get diarrhea. For the next ten years we worked to determine the parts of rotavirus that made mice sick and the parts that evoked a protective immune response." (p xiv) In other words, Offit noted a connection between a virus & diarrhea. Wakefield & fellow colleagues noted a connection between a virus (specifically a certain type of Mumps vaccine) & gastroinestinal distress that included diarrhea. Is there really that much of a difference?

As for Wakefield scaring parents? Since when do doctors do anything but scare parents when they're making dire predictions of death-dealing pandemics? In my opinion, what I call the recent "PANDEMIC PANIC" has shown fear-mongering pushed to an incredible extreme. Here're some relevant quotes from Wakefield's bk:

"an experimental vaccine combination – measles and rubella (MR) – that was administered to approximately 8 million UK school children over a 1-month period in November 1994. The justification for the campaign was a mathematically-predicted measles epidemic. The principal architects of the campaign were Salisbury and his boss, Dr. Kenneth Calman, the UK's Chief Medical Officer. Through an intense and frightening advertising campaign, 3 parents were motivated to get their children revaccinated by the threat of up to 50 deaths from measles." - p 77

"3 Minutes of JCVI meeting May 5, 1995; 6.5 para 2. *HEA measles/Rubella campaign report: The HEA* [Health Education Authority] *did acknowledge the view that the TV advert used had been a little frightening, and also that not enough information on the possible side-effects of the vaccine had been provided for some people.*" - endnote 3, p 82

Back to Offit:

"Today, autism also has many well-intentioned doctors who offer parents hope for an immediate cure. But, like Retan's, their therapies, which can cost tens of thousands of dollars, don't help and occasionally hurt those who are most vulnerable." - xxi

Again, that just sound like typical medicine to me - cures that don't help that cost outrageous amts of money. It's easy to at least give the impression of being "well-intentioned" when there's so much profit to be made. It's also easy to assuage one's conscience by believing that one is doing good - even if reality nibbles away at that delusion from time-to-time.

"The first to offer a cure for autism was Bruno Bettelheim, a Viennese-born psychoanalyst, Bettelheim believed he had found the problem: bad mothers. He reasoned that such mothers, whom he called "refrigerator mothers," caused autism by treating their children coldly, freezing them out." - p 3

[..]

"But a closer look at Bettelheim's school showed his claims of success were fraudulent. Worse: his accusations caused mothers to feel guilty and ashamed." - p 3

Interesting. I'm not really familiar w/ Bettelheim's work but I did quote an article of his that was published in *Scientific American* (Joey: A "Mechanical Boy" (Scientific American Offprint: Scientific American, March 1959)) in my movie Robopaths (on my onesownthoughts YouTube channel here: https://youtu.be/-PR7C8nFKtA - on the Internet Archive here: https://archive.org/details/Robopaths). No doubt there are all sorts of people whose emotional impact on others is negative, even devastating, & "refrigerator mothers" wd be among them - but I have to side w/ Offit here & find that causing "mothers to feel guilty and ashamed" is likely to be counterproductive. I also strongly doubt that one such psychological cause cd have such a profound effect as to cause autism or that there are so many "refrigerator mothers" that autism wd be so widespread as a result. Whether Offit's claim that "a closer look at Bettelheim's school showed his claims of success were fraudulent" is true or not is a different story.

"Since the mid-1990s, the number of children with autism has increased dramatically. Now as many as 1 in every 150 children in the United States is diagnosed with the disorder. Two phenomena likely account for the increase. First, the definition of autism had broadened to include children with milder, more subtle symptoms." - p 3

[..]

"Second, in the past children with severe symptoms of autism were often considered mentally retarded. Today, as the number of children diagnosed with severe autism has increased, the number with mental retardation has decreased." - p 4

Offit does provide endnotes w/o having numbers w/in the text refer to specific ones. The endnotes are just organized according to what chapter & page they're relevant to. As such, a scholar wishing to read substantiation of the above-quoted claim wd have to look at the 4 citations for chapter 1, page 4 & deduce whether they'd shed any further light on the subject. In this case, as far as I can tell, this is not the case. Nonetheless, I'm inclined to believe that increased diagnosis is an important factor in the apparent increase in autism. However, I'm also inclined to think that it's probably not the only factor. Instead, I tend to think that there're probably other socio-environmental factors at play - such as an increased mechanization of how humans are expected to behave - what I might call a "refrigerator society". Moreso, I think it's likely that there're physical factors of the most importance - esp ones, as Wakefield postulates, that cause health problems that inhibit development & result in regression.

In elucidating one of the treatments for autism that Offit later renounces he mentions a process of "facilitated communication":

"Using facilitators, who held children's hands while guiding their fingers to letters on a keyboard, Biklen believed that autistic children could communicate." - p 6

"The results were amazing. With the help of facilitators, children with autism typed out messages that filled their parents with hope" - p 7

My reviewer's note to myself was: "How do we know it's not the facilitators?" &, yes, that's the criticism that's later directed at this method. *SO*, here's an instance where I won't 'throw out the baby with the bathwater', meaning that even tho I'm highly distrustful of much of the basis for Offit's position this is an instance where I can agree w/ him.

"and she said, 'Do you know Stacy can write?' And I just cried. I couldn't believe it. I said, 'No, no, you're wrong. This is my kid. She's learned maybe six signs in her whole life. This can't be true.' So one day [the facilitator] came over to [my] house and she said, 'Stacy, I know you're excited. After all these years, you must have something you want to tell Mom.' And Stacy typed out 'I love you, Mom.' "" - p 8

Groan. On the one hand I'm glad that the parent(s) cd get some happiness, on the other hand it's so sad that it just turned out to be wishful thinking.

""It was devastating to watch," said Phil Worden, Betsy's legal guardian. "Because what you saw was that the words being typed out were the words the facilitator had seen. It was just so clear and unmistakeable. I was sitting there watching this and saying, 'My God, it's really true. This stuff is bogus.' "" - p 11

"Parker Beck was born in January 1993. He seemed normal until a few months after his second birthday. "Then he completely zoned out and removed [himself] from the rest of the world," said Victoria. "He wouldn't respond to his name anymore and he had no interest in playing." Parker had facial tics, banged his head against the wall, and only had two words in his vocabulary: "chuss," for juice, and "k-k-k," for cracker. Worse, he suffered from chronic diarrhea and abdominal pain." - p 13

Once again, there's the gastrointestinal problems. It seems believable & reasonable to me, as Wakefield stated, that gastrointestinal problems cd lead to poor digestion of needed nutrients & hence to developmental regression. What's not mentioned in the above is whether these problems started after vaccinations. According to a Children's Hospital of Philadelphia website, these were the recommended vaccines at the time:

"1994 - 1995 | Recommended Vaccines
Diphtheria*
Tetanus*
Pertussis*
Measles**
Mumps**
Rubella**
Polio (OPV)
Hib
Hepatitis B
* Given in combination as DTP
** Given in combination as MMR" - https://www.chop.edu/centers-programs/vaccine-education-center/vaccine-history/developments-by-year

According to a Nemours Children's Health website, children at ages 2 & 3:

"should have had these recommended vaccines:
- four doses of diphtheria, tetanus, and pertussis (DTaP) vaccine
- three doses of inactivated poliovirus vaccine (IPV)
- three or four doses of *Haemophilus influenzae* type B (Hib) vaccine
- one dose of measles, mumps, and rubella (MMR) vaccine
- three doses of hepatitis B (HepB) vaccine
- one dose of chickenpox (varicella) vaccine
- two or three doses of rotavirus vaccine (RV)
- four doses of pneumococcal conjugate vaccines (PCV13)
- one or two doses of hepatitis A (HepA) vaccine"

- https://kidshealth.org/en/parents/medical-care-2-3.html

Whether the 1994-1995 vaccinations listed above were for children this young I don't know. I'm assuming that that was the case. The point is that since Offit is pro-vaccination he neglects to mention whether or not Parker Beck had had vaccinations &, if so, what they were. If, e.g., Beck had had the MMR vaccine shortly prior to his 'zoning out' shortly after his 2nd birthday that wd add circumstantial evidence to the claim that vaccination might cause, in some children, gastrointestinal problems that might inhibit & cause regression of development.

Onward to other treatments that Offit considers bogus:

"One Dallas doctor paid $800 for four secretin infusions and charged $8,000 to administer them." - p 15

Ok, the guy's clearly a crook - but does that make him so different from his or her fellows?! Not as far as I can tell. I went to a hospital where an intern gave me an injection of a local anesthetic for abscess pain. He told me he'd never done it before. I asked him how long the effect wd last, he told me several days. It lasted for several hrs. I explained that I didn't have health insurance & that I had to be careful about how much things cost. He told me he wasn't worried about the money. *Well* he, personally, might not've cared about the money but the hospital management certainly did b/c they charged me something like $1,300 for this little service. That was an outrageous overcharge but completely normal in the Medical Industry world. I'm sure there're doctors who're satisfied w/o getting floor-length mink coats every wk but you wdn't know that from the usual bills.

"Between 1999 and 2002, investigators from three continents performed fifteen more studies of secretin. They injected autistic children with multiple doses of secretin, watched them for longer periods of time after receiving the drug, and included those with or without bowel problems. The results were the same. Not one study showed that secretin was effective in treating autism." - p 16

This time the endnote consulted was the relevant one: "Other secretin studies: P. Sturmey, "Secretin Is an Ineffective Treatment for Pervasive Developmental Disabilities: A Review of 15 Double-Blind Randomized Controlled Trials," *Research in Developmental Disabilities* 26 (2005): 87-89" (p 252) Even tho I'm not likely to ever consult a non-online source, I admit that the citing of one helps convince me that I'm not being conned (even tho I cd be). The problem I have w/ the above is that I've come to think that *injecting anything* is a bad idea, one that increases the chances of negative effects that result simply from an assault on the body that bypasses its natural defense barriers.

The chapter ends w/ a lead-in to Wakefield-bashing:

"Although the false promises of facilitated communication and secretin squandered research funding and drained parents' resources, these therapies weren't harmful. Unfortunately, events soon to unfold in England would dramatically raise the stakes. Parents of autistic children were about to lose much more than time and money." - p 17

"On February 28, 1998, Andrew Wakefield, a gastroenterologist working at London's Royal Free Hospital, held a press conference. He believed he had found the cause of autism." - p 18

That's not quite the way that Wakefield tells it. To quote again from my review of Wakefield's bk: "It was never claimed by Wakefield & co that the MMR vaccine *directly caused autism*. It was, instead, suggested that the connection between the MMR vaccine & the gastrointestinal problems in *some children*, not all, merited further investigation". It seems that such subtle distinctions are anathema to the oversimplification of Wakefield's enemies. Or, maybe Wakefield's misrepresenting himself as a defensive mechanism. Since I think that the way Wakefield tells it sounds reasonable I tend to believe it.

"In 1988, in a paper titled "Measles Virus DNA Is Not Detected in Inflammatory Bowel Disease," Wakefield admitted he had been wrong. Although his findings had unnecessarily scared the public about measles vaccine, Wakefield staked a higher ground. "Hypothesis testing and presentation of the outcome—either positive or negative—is a fundamental part of the scientific process," he said. "Accordingly, we have published studies that both do and do not support a role for measles in chronic intesinal inflammation: this is called integrity."" - p 19

For any mainstream medical person to talk about "unnecessarily" [scaring] "the public" is pure hypocrisy. The whole history of vaccination, certainly during the time of the recent quarantine justified by an utter terror campaign based around COVID-19, has been little but constant fear-mongering to justify profteering - & Wakefield hasn't been the 1st person to notice this, there's been resistance over & over again: from the time of Dr. Creighton's 1884 article on vaccination for the ninth edition of The Encyclopedia Brittanica to Eleanor McBean's The Poisoned Needle. I quote from my review of that regarding another person who objected to vaccination long before Wakefield appeared on the scene:

""In 1887 Dr. Edgar M. Crookshank, who at that time was professor of pathology and bacteriology at King's College, was asked by the government to investigate an outbreak of cowpox in Wiltshire. Sir James Paget drew his attention to Creighton's work, evidently hoping that Crookshank would refute it, but the results of his laborious investigations are contained in two large volumes entitled "The History and Pathology of Vaccination", in which he says that the credit given to vaccination belongs to sanitation and isolation and that nothing would more redound to the credit of the medical profession than to give up their faith in vaccination." - p 216"

Wd Offit have the integrity to ever publicly announce that his well-pd 25+ yrs of rotavirus vaccine research has been a waste of time that led nowhere productive? Of course, that's a question predicated on the idea that perhaps it really has been a waste of time & resources wch I seriously doubt that Offit considers it to be.

At Wakefield's press conference about the possible connection between the MMR vaccine & intestinal inflammation, he's quoted as saying:

""Their children had developed normally for the first fifteen to eighteen months of life when they received the MMR vaccination. But after a variable period the children regressed, losing speech, language, social skills, and imaginative play, descending into autism."" - p 20

& on p 253 there's an endnote that presumably matches up w/ this:

"Wakefield and listening to parents: A. J. Wakefield, "The Case Against MMR: Wary Parents Have Proved the Experts Wrong Before. They Will Do So Again," *The Independent* (London), January 22, 2001."

Now, what's a little odd to me about that is that Wakefield's press conference was in February, 1998, but the newspaper quoted above is from 2001. Why not quote from the articles written at the time of the conference? Now here's what Wakefield has to say about this press conference as quoted in my review of <u>Callous Disregard</u>:

""I approached the idea of a press briefing with some anxieties; by this stage, I had examined the issue of MMR vaccine in great detail. I had reviewed all of the published scientific literature about measles and MMR vaccine safety studies on the basis that, as part of investigating parental concerns and before calling into question MMR vaccine safety, it was essential to have done so. On a personal level, I was dismayed that I hadn't done this research before vaccinating my two older children. On a global level, it was clear that the safety studies had been wholly inadequate. "George" the whistleblower had major concerns about the attitude of many of his public health colleagues toward concerns over vaccine safety. The forced withdrawal of MMR vaccines that had been "spun" as being *completely safe* was testament to their failings." - p 83"

Now, at this point I probably seem like an apologist for Wakefield. I'm not. I can believe that he's greedy & that financial well-being might help grease the wheels of his research - but I can also believe that about most doctors, including Offit. What makes me more inclined toward Wakefield's position is that I just find it more believable & reasonable than I do Offit's. After all, the purpose of Offit's bk is to ridicule that he calls "Autism's False Prophets", it's *not* to defend his own rotavirus research wch he basically presents as objectively unassailable. *His own research isn't under scrutiny.* I imagine that he writes about it in scholarly medical journals & that if I were to read such things, assuming that as a layperson I cd get much understanding from them, I might find a basis to agree or disagree w/ them - but this isn't a debate, where opposing positions are presented for the judgment of the reader, it's a *roast* where Offit's science is taken for granted as correct & other people's science is presented as fraudulent. I'm not convinced.

"Wakefield also had a solution. If parents wanted to avoid autism, they should separate MMR into three vaccines. "There is sufficient concern in my own mind for a case to be made for vaccines to be given individually at not less than one-year intervals," he said." - p 21

Now, Offit's reaction to that is one that I don't necessarily think is true:

"Unfortunately, because the pharmaceutical companies that made MMR vaccine for British children didn't offer the vaccines separately, Wakefield was effectively recommending that children not be vaccinated." - p 22

If Wakefield's assertion that having a Measles, Mumps, & Rubella vaccine altogether is too much of an assault on a child's body is to be taken seriously than it must also be taken seriously by the companies that manufacture the MMR vaccine. In other words, the concern is *not* whether it's inconvenient for them to offer the vaccinations separately *but whether it's the right thing to do*. Wakefield's prompting the companies to do what he thinks is right, not for parents to not get children vaccinated. Personally, I part ways w/ him here: I distrust *the basic principle of vaccination* - but, then, that's just the opinion of one person who's not in a position to influence public policy (& who doesn't want to be).

"So great was the fascination with Wakefield that on December 15, 2003, a ninety-minute docudrama about his brilliance and courage, *Hear the Silence*, appeared on British television." - p 22

"In the movie, Wakefield is forced to withstand tremendous pressures: his files are stolen, his phone is tapped, and he hears heavy breathing on the line. Viewers learn later who is behind the intimidation: drug companies." - p 23

Is there a possibility that Offit & other detractors of Wakefield are envious? After all, it's compellingly romantic for Wakefield to be depicted as such a hero. Let's not forget that Offit begins his bk w/ depictions of himself receiving death threats - he, too, is being depicted (by himself) as a hero fighting the good fight for better health for children. It's just that in Offit's case the bad guys are religious nuts rather than the pharmaceutical companies. I'm anti-religious & find Fundamentalist Christinanity [pun intended] to be particularly disagreeable - *but when it comes to issues of health, I think Big Pharma is the #1 villain*, even the religious right can't compete.

"In the months following Wakefield's warning, the proportion of children receiving MMR vaccine dropped from nearly 90 percent to 70 percent and, in some areas of London, to 50 percent. As a consequence, small outbreaks of measles first appeared in upper-middle-class elementary schools in London. Other outbreaks followed, first in underimmunized areas of London then in Scotland and Ireland." - p 24

The above statement is totally that of a BELIEVER, there's no questioning involved, Offit presents himself as an authority not to be questioned. There're no statistics provided as to whether any people vaccinated got sick. The basic message is: *Do what the Doctor Gods tell you to do or **your children will get sick**.* Wakefield has been thrown down from Olympus. Now, I had German Measles when I was a kid. I remember having a temperature of 105°F, that's serious. That wd've been in the early 1960s. I read online that:

"In 1963, the measles vaccine was developed, and by the late 1960s, vaccines were also available to protect against mumps (1967) and rubella (1969). These three vaccines were combined into the MMR vaccine by Dr. Maurice Hilleman in 1971." - https://www.chop.edu/centers-programs/vaccine-education-center/vaccine-history/developments-by-year#:~:text=In%201963%2C%20the%20measles%20vaccine,Maurice%20Hilleman%20in%201971.

1963 was probably the yr I got sick. As such, I don't know if I wd've been vaccinated or not. It's possible that the vaccine wd've been too new for me to've gotten it. At any rate, I survived w/o any aftereffects known to me. It was an experience that lasted a few days, I was pretty out-of-it, my mom was very worried.

"In 2006, a thriteen-year-old boy became the first person in England to die from measles in more than a decade." - p 24

That means that someone died of measles in England *before* Wakefield's controversial press briefing - so Offit can't blame that death on him. We also don't know whether the person who died was vaccinated or not. The idea that medical science is going to prevent all deaths strikes me as preposterously arrogant & ego-inflated anyway.

"Wakefield wasn't alone in his belief that MMR might cause autism. Following his press conference in 1998, several scientists and physicians stepped forward to support him.

"John O'Leary, a professor of pathology at Coombe Women's Hospital in Dublin, examined intestinal biopsies" - p 25

"Hisashi Kawashima, a virologist at Tokyo Medical University in Japan, found measles virus in white blood cells taken from autistic children but not from other children." - p 25

"Vijendra Singh, a biologist at Utah State University in Logan" - p 25

"Kenneth Aitken, a clinical psychologist at the Royal Hospital for Sick Children in Edinburgh, Scotland" - p 25

"Walter Spitzer, an emeritus professor of epidemiology at McGill University in Montreal" - p 25

"John March, a veterinarian at the Moredun Research Institute in Scotland, also came on board." - p 26

"Physicians from the United States also weighed in. Marcel Kinsbourne and John Menkes, pediatric neurologists from California" - p 26

"Arthur Krigsman, a gastroenterologist from the New York School of Medicine" - p 26

That's an impressive amt of people coming out in support of Wakefield's hypothesis. Wakefield, however, is the one chosen as the scapegoat to try to put everyone else in a defensive position.

"Next, he took his case to the United States, immediately finding a powerful ally: Dan Burton, a Republican congressman from Indiana." - p 26

"Dan Burton was a fixture in Congress, first elected in 1982 and reelected to his thirteenth term in 2006. Staunchly conservative and a born-again Christian" - p 26

"Burton also joined the debate about the AIDS epidemic. After the AIDS virus entered the United States in the late 1970s, Burton became obsessed with the disease. He refused to eat soup at restaurants, brought his own scissors to the barbershop, and tried to introduce legislation requiring AIDS testing for everyone in the country—legislation that, not surprisingly, failed." - pp 27-28

While the depiction of Burton certainly makes me feel leary of him it also seems to me that by associating Burton & Wakefield Offit is making an ad hominem connection that is implied to go deeper than it actually does - in other words, Wakefield is NOT Burton, it's NOT Wakefield who's refusing to eat soup in restaurants out of fear of AIDS, etc. If Wakefield & Burton had an alliance over a shared belief in a connection between vaccines & developmental regression then that alliance might be one of uneasy bedfellows motivated on both sides by beliefs arrived at by different routes.

"Scott Bono, from Durham, North Carolina, was also convinced that MMR had caused his son's autism. "On August 9, 1990, Jackson [began] his journey into silence," said Bono. "That was the day he received his MMR immunization. He would not sleep that night. In the days to follow he would develop unexplained rashes and horrible constipation and diarrhea. His normally very healthy body was ravaged by an invader. Over the next weeks he would slip away, unable to listen or speak. What was the reason for the change? It is my sincerest belief that it was that shot."" - p 30

I tend to take these parental testimonials seriously. There's a reason why so many people flock to supporting Wakefield's research & I think that reason is that their own experience bears him out. I *don't think that all Wakefield's supporters are just trying to get rich off of lawsuits.* Instead, I think it's much more likely to think that the pro-vaccine 'experts', such as Offit, are trying to hold onto their exalted positions by denying the validity of people who talk from direct experience unmediated by the 'expert' propaganda.

"Unfortunately, Wakefield presented his evidence as if he were speaking to scientists, not congressmen and parents. He showed pictures of intestines of autistic children viewed from the end of a fiber-optic scope, intestinal cells containing small fragments of measles vaccine virus, and plastic gels containing measles virus proteins. He talked about follicular dendritic cells, ileocolonic lymphoid nodular hyperplasia, crypt abscesses, and hepatic encephalopathy, concluding, "We have a biologically plausible hypothesis, that ungraded chemicals from the gut may be getting through and impacting the rapidly developing brain during the first few years of life [causing] autism." Then he surprised the committee. He presented findings not only from the eight autistic children in his *Lancet* paper, but also from 150 children in whom he claimed MMR had caused autism in "all but four."

"John O'Leary, the Irish pathologist who had identified measles virus in the intestines of Wakefield's autistic patients, was next. He too presented as if he were speaking to a group of molecular biologists. O'Leary talked about TaqMan real-time quantitative PCRs, RNA inhibition assays, low-copy viral gene detections, fusion proteins, neuraminidases, hemagglutinins, subacute sclerosing panencephalitis, nucleocapsid genes, and black signals, concluding, "I am here to say that Wakefield's hypothesis is correct."" - pp 30-31

Note that in Offit's quoting of Wakefield he adds a bracketed insertion of the word "causing". That means that that *isn't* what Wakefield sd, but Offit is presenting it as if it's what one can reduce what he sd to. This seems to me to be in keeping w/ Offit's typical procedure of misrepresentation for scapegoating. Note that he also refers to "his *Lancet* paper" as if Wakefield were the only author, wch he wasn't. That also serves the purpose of scapegoating.

Note, also, that to Offit it's unfortunate that Wakefield & O'Leary presented their evidence in scientific language. As I understand it, their doing so resulted in the Congresspeople & parents not understanding what they were saying - but can't Offit at least give them credit for trying to accurately describe in the appropriate jargon what their findings were? After all, their testimony was presumably going into the Congressional Record, meaning that scientists cd consult it if need be. Their job wasn't necessarily to translate things into layperson's terms but to seriously describe the results of their research.

"On February 18, 2004—six years after Andrew Wakefield had published his paper in the *Lancet*—Brian Deer, an investigative reporter for the *Sunday Times* of London, called Richard Horton, the *Lancet*'s editor-in-chief. Deer said he had some shocking news about Andrew Wakefield." - p 37

Wakefield's Callous Disregard has a whole chapter dedicated to countering Deer's claims. Offit seems to accept Deer's position at face value. Once again, I excerpt from my review of Wakefield's bk. The 1st *italicized* portion is attributed to Deer:

""*The dynamite in The Lancet was the claim that their conditions could be linked to the MMR vaccine, which had been given to all 12 children.*

""False: *The Lancet* paper did not "claim that their conditions could be linked to the MMR vaccine." No such claim was ever made in the paper; on the contrary, it was explicitly stated in that paper that *no association* – let alone a causal association – had been proved between MMR and the syndrome described. It reported only that the parents said onset of symptoms started after MMR vaccination in 8 of the 12 cases." - p 186"

[..]

"Another person working against Wakefield was Professor Tom MacDonald, a scientist studying bowel disease.

""The GMC vetted MacDonald as a potential witness against me and his erstwhile colleagues. The attendance note of his meeting with GMC lawyers in 2005 reads:

""*He* [MacDonald] *believes Wakefield is a charlatan, who has been pursuing his own agenda since 1995, this being to win the Nobel Prize. He believes Wakefield's alleged link between measles vaccine and Crohn's was entirely fabricated in order to obtain publicity for this reason.*" - p 210

""He may also have failed to disclose conflicting agendas, one scientific (as above) and one personal; as related to me by John Walker-Smith, when MacDonald declined the invitation to transfer to the Royal Free with Walker-Smith, he had reportedly vowed to his boss to destroy my career. Deer has been useful to him in that respect." - p 211

"It seems to me that something is missing from this depiction of Deer & MacDonald - viz: Why do they hate Wakefield so much? What this hatred reminds me of the most is moments from my own life when people have attacked me, ostensibly for one reason, not b/c of the ostensible reason but b/c they were jealous that I'd had sex w/ someone they wanted to have sex w/ but cdn't. Does that seem petty? Sure it is.. but it's all too human. Maybe Wakefield was popular & Deer & MacDonald weren't. It just seems to me that there's more going on here than the bk addresses."

- tENTATIVELY, a cONVENIENCE review of Wakefield's Callous Disregard

The point is that Deer may or may not be what Offit presents him as: simply an investigative journalist. He may be someone w/ an ulterior motive. Offit finds fault w/ all of his purported "False Prophets" but he seems to accept the people he uses to support his case at face value. Why not question them all?

"Richard Barr had considerable experience suing pharmaceutical companies, first coming to public attention during his crusade against the anti-arthritis drug Opren. Opren had already been withdrawn from the United States, where it had been linked to more than seventy deaths and 1,000 cases of kidney and liver failure, The drug's manufacturer, Eli Lilly, eventually settled cases in Britain for $6 million. Unfortunately, because he had filed his claims after an agreed-upon deadline, Barr's clients received nothing. "My worst day as a lawyer," recalled

Barr "was when the first judgments were handed down in the Opren cases.["]" - p 39

I find that this passage from Offit, whether intended this way or not by him, validates grievances against pharmaceutical companies.

"Although the British press saw Brian Deer's allegations as tainting the notion that MMR caused autism, scientific studies had already gone a long way toward refuting Wakefield's claims. In his original *Lancet* paper Wakefield admitted that he had not proven an association between MMR and autism. To determine whether MMR caused autism, researchers would have to perform a series of epidemiological studies, comparing hundreds of thousands of children who did or did not receive the vaccine." - p 42

What's wrong w/ this picture?! Offit keeps referring to Wakefield as claiming that "MMR caused autism" but in the same paragraph states that "Wakefield admitted that he had not proven an association between MMR and autism" as if Wakefield has been somehow forced into this 'admission' when, in fact, it was part of what he stated all along! Offit wants it both ways: he wants Wakefield to've sd that MMR causes autism but then wants to act as if Wakefield somehow admitted to being a fool for ever having stated that that was so - the problem is that Wakefield *never did make that claim* so he's being criticized over & over for a claim he never made.

"Fourth, Wakefield believed that the combination of measles, mumps, and rubella vaccines has simply overwhelmed a young child's developing immune system. But the challenge of those three viruses is miniscule compared with that of vaccines that had been developed and administered in the past." - pp 44-45

I don't really see that claim as challenging much of anything. People got sick & died from vaccines in the past too.   The question is: Do triple vaccines, triple induced disease, challenge the immune system more than just one vaccine? It seems obvious that that wd be the case. If you're shot w/ a gun, run over by a lawn mower, & hit by a car will it be worse for you than just being subjected to one of those assaults?

"Indeed, children confronted more immunologic challenges by receiving only smallpox vaccine 100 years ago than they do while receiving 14 different vaccines today (200 versus 153). If immunological overload were the cause of autism, with fewer immunilogic challenges in modern vaccines, rates of autism should be decreasing, not increasing." - p 45

It seems to me that Offit's avoiding the issue: he compares a single smallpox vaccination to a triple MMR vaccination. The point seems to be that 3 vaccines at once is too many, not that Offit's comparative method makes the earlier vaccine more "challeng"ing. The mere fact that he admits that the smallpox vaccine was a challenge at all is simply confirming what anti-vax people are already saying.

I've already stated that it's my observation that the Medical Industry & its personnel are greedy - but Offit makes it out as if only Wakefield & other 'False Prophets' are the greedy ones.

"On December 22, 2006, Deer received a report titled "Freedom of Information Act Request: MMR Multi Party Action."" - p 46

"Barr's Science and Medical Investigation Team gave $800,000 to Andrew Wakefield to support his research. When Wakefield had been confronted by journalists two years earlier, following Deer's claim that he had received $100,000, he argued that the sum was closer to $50,000. Now, with a document from the Legal Services Commission in front of him, Wakefield didn't deny that the amount was far greater." - p 47

Offit's bk was written in 2008, Wakefield's was written in 2011. The contested funding is given a labyrinthian history in Wakefield's bk in wch he never refers to a sum even remotely close to $800,000. Here're some relevant excerpts from Callous Disregard:

"During the first half of 1996, I was asked for help by Richard Barr of Dawburn's law firm and lead attorney on the UK MMR cases. Specifically, I was asked to review the safety of measles-containing vaccines (MCV) and, separately, to design a study that would help determine whether there was or was not a likely case in law against the manufacturers of MCV. Barr's initial interest was in Crohn's disease as a possible adverse outcome, but autism in children with intestinal symptoms rapidly took center stage. I prepared a research proposal for Barr's submission to the Legal Aid Board (LAB), a means-tested, government-funded legal assistance program to which Barr was contracted for the vaccine work." - p 49, Callous Disregard

"I attached the letter of award that the LAB had written to the law firm Dawburns and confirmed that the first [pounds] 25,000.00 should be paid into a designated research account as was standard practice for research grants." - p 50, Callous Disregard

In other words, according to Wakefield, the money wasn't exclusively for him at all. In order to avoid conflict of interest problems, Wakefield claims that he always clearly expressed his relationship w/ Barr & the LAB funding to people higher in the hospital hierarchy.

"Confused by his lack of clarity over this supposed conflict, I wrote again on March 24, 1997. I enclosed all the documents relevant to the grant, including the proposal, the letters from the LAB, and the relevant protocols — once again there was no secrecy in terms of frankly telling the dean what was proposed." - p 52, Callous Disregard

"By May 1997, I had grown tired of the whole fiasco over the LAB funding and decided to seek funding elsewhere. Exasperated, I asked the finance department to return the funds in full to the lawyers." - p 60, <u>Callous Disregard</u>

That didn't happen, according to Wakefiled, the decision to continue to accept the funding was made by people other than himself.

Now, maybe I'm simply misunderstanding this whole fiasco. Maybe the money that Offit & Deer refer to is something other than the LAB funds referred to above. Whatever the case may be, Wakefield certainly never acknowledges receiving $800,000, as Deer supposedly alleged, nor do I recall his even referring to it in <u>Callous Disregard</u>. Given that Wakefield goes to great lengths to counter Deer's charges against him in his chapter 12 & I just glimpsed thru it again just now to look for allegations of an $800,000 payment, I find it strange to find no mention of it & no refutation of it. Is Offit lying? Was Deer lying? Did Wakefield omit it b/c he cdn't convincingly defend himself? I haven't the foggiest. It seems to me that Offit's lying is the least likely of the 3. & this is where we get to most damning evidence.

"Wakefield wasn't the only investigator to benefit from Barr's largesse. Barr's team had given John O'Leary more than $1 million to perform his studies." - p 48

"In the book, Aitken had postulated the cause of autism: "It now seems certain that the brains of persons who become autistic in their early childhood already had microscopic faults in their development in early intra-uterine life, probably first expressed among cells of the early embryo, in the first month." But after receiving money from Barr, Aitken appeared to change his opinion, now believing that autism might be caused by MMR." - p 48

"Arthur Krigsman, the gastroenterologist from the New York University School of Medicine, had declared during Burton's hearings that his findings were "independent" from Wakefield's; he had received $30,000.

"Marcel Kinsbourne and John Menkes, the California neurologists who had been first to support Wakefield's hypothesis in the American press, had received $800,000 and $90,000, respectively. Toward the end of their careers, both men had become expert witnesses for lawyers suing vaccine makers. (Marcel Kinsbourne would be heard from again)." - p 49

Looking at the endnotes, it appears that Deer is the only source for all those allegations above. *SO*, I decided to search for lawsuits against him to see if anyone other than Wakefield sued him. Apparently, none of the other people mentioned above did & Wakefield's multiple suits were thrown out of court w/ him having to pay costs. Then I looked at Deer's Wikipedia entry.

"In 1986, one of Deer's early investigations exposed research by British scientist Michael Briggs at Deakin University, Australia into the safety of the contraceptive pill. Deer's reports revealed that several of Briggs's studies were fabricated so as to give a positive profile for the products' cardiovascular safety. The research was largely financed by the German drug company Schering AG.

"In 1994, his investigation of the Wellcome Trust led to the withdrawal in the UK of the antibiotic, Septrin (also sold under the name Bactrim) and the sale by the Wellcome Trust of its drug company subsidiary.

"In 2005, the withdrawal of the painkiller Vioxx was followed by an investigation by Deer into the people responsible for the drug's introduction.

"In 2006, Deer's *Dispatches* documentary, "The drug trial that went wrong", investigated the experimental monoclonal antibody TGN1412. It was nominated for a Royal Television Society journalism award.

"In 2008, the media psychiatrist Raj Persaud was suspended from practising medicine and resigned his academic position after being found guilty of plagiarism following an investigation by Deer."

- https://en.wikipedia.org/wiki/Brian_Deer

It seems from the above-quoted Wikipedia entry that Deer's investigative journalism vis à vis the Medical Industry goes more against sd industry than for it. Therefore, it's harder to make a case that Deer's persecution of Wakefield is Big Pharma funded. That, in itself, makes the best case against Wakefield that I've encountered yet. I still find Wakefield's science to be convincing but his apparent financial motives, at least as alleged by Deer, are more than a bit hard to ignore.

"John March, the veterinarian who had claimed that animal vaccines were tested more extensively than MMR, had received $180,000." - p 49

March is quoted as saying:

"["]The ironic thing is [Barr's team] was always going on about how, you know, how we've hardly got any money compared to the other side who are funded by large pharmaceutical companies. And I'm thinking, judging by the amounts of money you're paying out, the other side must be living like millionaires."" - p 50

Now the source for that telling revelation is attributed to "personal communication" in an endnote on p 258. Is it to be trusted? I find it all-too-easy to believe that Big Pharma really does have far more money to use for influencing outcomes than Barr's legal backers. I can't help but refer to Dr. Marcia Angell's wonderful bk, The Truth About the Drug Companies - How They Deceive Us and What to do About It. Here're excerpts from my review of that:

""Prescription drug costs are indeed high—and rising fast. Americans now spend a staggering $200 billion a year on prescription drugs, and that figure is growing at about 12 percent per year (down from a high of 18 percent in 1999).1" - p xii"

[..]

""I witnessed firsthand the influence of the industry on medical research during my two decades at *The New England Journal of Medicine*. The staple of the journal is research about causes of and treatments for disease. Increasingly, this work is sponsored by drug companies. I saw companies begin to exercise a level of control over the way research is done that was unheard of when I first came to the journal, and the aim was clearly to load the dice to make sure their drugs looked good. As an example, companies would require researchers to compare a new drug with a placebo (sugar pill) instead of with an older drug. That way the new drug would look good even though it might actually be worse than the older one." - p xviii"

[..]

""By 1990, the industry had assumed its present contours as a business with unprecedented control over its own fortunes. For example, if it didn't like something about the FDA, the federal agency that's supposed to regulate the industry, it could change it through direct pressure or through its friends in Congress." - p 10"

You get the idea. I recommend Angell's bk far more than I'd recommend Offit's or Wakefield's.

"Following Andrew Wakefield's publication in the *Lancet*, rates of immunization with MMR declined and measles outbreaks swept across the United Kingdom and Ireland; hundreds of children were hospitalized and four children died from a disease that could easily have been prevented." - p 55

Ah, I was really starting to be swayed by the evidence in favor of Wakefield's greed & then the pendulum swings back to this kind of crap again. 1st, there's the reference to the *Lancet* article as Wakefield's alone, something completely false. 2nd, there's the implication that the publishing of this article in a scholarly medical journal in England wd, in itself, have an influence on the behaviors of the British masses in general. That's certainly not the case. Rates of immunization might've decreased b/c of more mass media attn pd to Wakefield's claims - but certainly not b/c of an article in a medical journal, something likely to influence doctors more than anyone else. As for 4 children dying from "a disease that could easily have been prevented"?! I find that an unlikely claim, one predicated on a total 100% belief wch I find completely unsubstantiated. All in all, to act like Wakefield is to blame for any of this just strikes me as ridiculous - he presented reasonable findings from his research, findings that're never accurately portrayed anywhere in Offit's bk, people responded to these reasonable findings w/ understandable doubts about a vaccine. Offit's portrayal of the situation is transparently one-sided: an anti-vax person cd say that while some children died from the measles, many more were saved from gastrointestinal distress that might've led to developmental regression. Maybe what both sides wd claim wd be accurate, maybe neither. Offit continues to mislead:

"Wakefield published his *Lancet* paper based on the findings of eight children with autism. He knew that the only way to prove his contention was to show that autism was more common in children who had received MMR. And he knew that he hadn't done that study. Correctly, Wakefield explained the limits of his study in the discussion section of his paper: "We did not prove an association between measles, mumps, and rubella vaccine and [autism]."" - p 55

Three reoccuring things are strange about Offit's presentation: 1. His hammering the idea in that the paper was Wakefield's alone, it wasn't; 2. His repeated statement that Wakefield made claims that he didn't followed by quotes from Wakefield that prove that Wakefield didn't make the alleged claims; 3. His insertion of key words in [brackets] that skew the meaning of the statements made by people quoted in a way that, apparently, *is not what they actually sd!* All 3 of those things are insupportable &, for me, at least, ruin Offit's argument.

Wakefield is depicted as having been presented by the mass media as a hero & the mass media is criticized for this "great theater".

""The media coverage told parents not only what to think, but also how to think about the MMR vaccine," wrote Tammy Boyce, author of *Health, Risk and News: The MMR Vaccine and the Media*. Boyce argued that the media's ritualistic mantra of balance—equally weighing one man's speculations with studies that clearly exonerated MMR—created a "charade of objectivity."" - p 56

Oh, well, anyone who's naive enuf to think that the mass media's normal functioning is much more than sensationalist fear-mongering that serves hidden interests is too naive to have much grip on what I, at least, consider to be 'reality'. I don't mean to say that Boyce is wrong, I mean to say *What's new?!* If the media were to promote vaccination use they'd be acting in the same irresponsible way, it hardly matters what side they take. As for the "studies that clearly exonerated MMR"? I'm still convinced that there're valid opinions on both sides of the vaccination-in-general fence & I seriously doubt that there's any science that I, personally, wd find to 'clearly exonerate MMR' or any other vaccine or even medicine in general - but, then, I'm NOT A BELIEVER. Setting broken bones, ok, injecting diseases &/or GMOs (or anything else).. nah.

"Learning little from his encounter with Andrew Wakefield, Richard Horton has published papers in the *Lancet* claiming that genetically modified foods damaged rat intestines, silicone breast implants induced harmful antibodies, and casualties sustained during the U.S.-led invasion of Iraq totaled 655,000 (about ten times the actual number). Like Wakefield's paper, all of these assertions garnered enormous media attention for his journal, and all have been clearly refuted." - pp 58-59

There's that "clearly" again. We've read "clearly exonerated" & "clearly refuted". I have my doubts about both. It seems to me that it might be more accurate to say something like: 'If you're a believer in certain things presented as axioms such as: medical science does the best job it can & is philosophically sound OR the

American military is doing its best to be a good policeman for the whole world then chances are you'll find studies exonerating the insertion of a foreign substance into breasts to make them bigger believable OR you'll find statistics minimizing the death toll in a country invaded on false grounds believable.' If not, well maybe you'll approach any tale told about them w/ a grain of salt. That sd, I don't mean to say that the *Lancet* is a reliable source. After all, there's also been the recent hydroxychloroquine mess:

"It's interesting to note that in *The Lancet*'s entry sidebar there's nothing discrediting the journal - this despite their having published at least 3 papers that've been retracted, the most recent of wch was something far more clearly fraudulent than what Wakefield's been accused of:

""In May 2020, *The Lancet* published a metastudy by Dr. Mandeep R. Mehra of the Harvard Medical School and Dr. Sapan S. Desai of Surgisphere Corporation, which concluded that the malaria drugs hydroxychloroquine and chloroquine did not improve the condition of COVID-19 patients, and may have harmed some of them.

""In response to concerns raised by members of the scientific community and the media about the veracity of the data and analyses, *The Lancet* decided to launch an independent third party investigation of Surgisphere and the metastudy. Specifically, *The Lancet* editors wanted to "evaluate the origination of the database elements, to confirm the completeness of the database, and to replicate the analyses presented in the paper." The independent peer reviewers in charge of the investigation notified *The Lancet* that Surgisphere wouldn't provide the requested data and documentation. The authors of the metastudy then asked *The Lancet* to retract the article, which was done on June 3rd 2020."

"- https://en.wikipedia.org/wiki/
The_Lancet#Metastudy_on_the_use_of_hydroxychloroquine_and_chloroquine_(
2020)

"Now what's particularly noteworthy about this latter scandal is that it seems to've largely slipped past the notice of the general public. Hydroxychloroquine still seems to be 'discredited' even though it's a cheap drug that's been long-since readily available all over the world."

- https://www.goodreads.com/review/show/4486376528

"Frank Pallone is a congressman who represents New Jersey's Sixth District. First elected in 1988, he has been a passionate supporter of Native Americans, working to protect the sovereignty of tribal governments. He's also an environmentalist. Because Pallone lives in a district that includes a string of towns on the Jersey shore, he's particularly worried about contamination of fish with mercury. In 1997, Pallone attached a simple amendment to an FDA reauthorization bill. Only 130 words long, the amendment would soon lead to chaos. Pallone gave the FDA two years to "compile a list of drugs and foods that

contain intentionally introduced mercury compounds and [to] provide a
quantitative and qualitative analysis of the mercury compounds in the list." The
bill—the FDA Modernization Act of 1997—was signed into law on November 21,
1997. Few in the press or public took notice." - pp 60-61

"Kyushu's chemical factory continued to pour mercury into industrial wastewater
until 1968. By 2001, mercury had poisoned 3,000 Japanese citizens and killed
600.

"*****

"FDA officials knew that mercury could damage the nervous system. When they
first responded to Frank Pallone's directive, they also knew that mercury-based
preservatives had been used in vaccines for decades. They were the ones who
had put it there."

[..]

"And in Queensland, Australia, in 1928, twelve children injected with a
contaminated diptheria vaccine died from bacterial abscesses and bloodstream
infections."

- p 62

Ultimately, Offit makes the claim that the version of mercury used as a
preservative in vaccines, thimerosal, is safe & not the same as the mercury that
poisoned & killed people. This conclusion is to refute the actions of concerned
citizens who wanted thimerosal removed from vaccines - claiming, in part, that
thimerosal is a contributing agent to autism. Hence, in Offit's mind, they're more
of "Autism's False Prophets".

In my review of Eleanor McBean's The Poisoned Needle, I address the
Queensland story:

""Vaccination wasn't always taken up readily and it wasn't always safe, either.

""In 1928, 12 children in the south-east Queensland town of Bundaberg died after
receiving contaminated diptheria vaccines.

"""Among those who died, three came from one family, while two more families
each lost two children," Dr Hobbins writes in a 2011 academic paper.

"""A Bundaberg correspondent opined that 'immunisation is as popular as a
death adder'.

"""Within days the events in Bundaberg had compromised diphtheria control
programmes around the globe, including the complete termination of
immunisation in Cape Town (South Africa) and across New Zealand.""

"- https://www.abc.net.au/news/health/2020-09-22/vaccine-history-coronavirus-smallpox-spanish-flu/12673832

"Obviously, neither of those articles verifies McBean's claim. That doesn't mean that I think that they refute them either. The 1st article is about lifting quarantine & *not about* repealing mandatory vaccination. Furthermore, their reasons for lifting it are that the authorities no longer considered smallpox to be a threat & that they considered the vaccinations a success. The 2nd article is about *diphtheria* vaccinations & *not smallpox* ones. That article states that vaccination "wasn't always safe" - the implication, as usual, is that science has gotten its shit together now & that it's safe - but what will be sd about vaccination now when it's reported about 100 yrs from now? I suspect that the supposed 'safety' of vaccines now will be soundly refuted - esp in the light of the thousands dead from COVID-19 vaccines."

- https://www.goodreads.com/review/show/4753078086

Now, do I think that thimerosal (mercury) is safe as something to be ingested? Offit has this to say:

"Before they put thimerosal in vaccines, Lilly scientists first studied it."

[..]

"Although thimerosal didn't treat meningitis, doctors found that it was safe." - p 63

Let's not forget that Offit was quoted above as having written: "Opren had already been withdrawn from the United States, where it had been linked to more than seventy deaths and 1,000 cases of kidney and liver failure, The drug's manufacturer, Eli Lilly, eventually settled cases in Britain for $6 million." In other words, Offit is honest enuf to admit that pharmaceutical companies, including Eli Lilly, can make serious mistakes that damage people's health. In other other words, it's reasonable for people such as myself to take the safety conclusions of pharmaceutical companies w/ some suspicion.

"Mercury is a naturally-occurring chemical element found in rock in the earth's crust, including in deposits of coal."

[..]

"Elemental or metallic mercury is a shiny, silver-white metal, historically referred to as quicksilver, and is liquid at room temperature. It is used in older thermometers, fluorescent light bulbs and some electrical switches. When dropped, elemental mercury breaks into smaller droplets which can go through small cracks or become strongly attached to certain materials. At room temperature, exposed elemental mercury can evaporate to become an invisible, odorless toxic vapor. If heated, it is a colorless, odorless gas. Learn about how people are most often exposed to elemental mercury and about the adverse health effects that exposures to elemental mercury can produce."

- https://www.epa.gov/mercury/basic-information-about-mercury

In short, mercury is generally considered toxic but there're variations. Even tho Offit makes the case that thimerosal has been proven safe & has justified its use as a preservative in vaccines that might deteriorate & become poisonous on their own otherwise, I have to wonder *why take the risk?*

"So Halsey decided to evoke the precautionary principle, exercising caution in the absence of evidence. Unlike Andrew Wakefield, who had made precipitous recommendations against the MMR vaccine, Halsey would not recommend the removal of a component of vaccines necessary to protect children from potentially fatal infectious diseases. He would merely propose that vaccine makers switch to single-dose vials free of preservatives." - p 66

Oh, c'mon! Really?! Wakefield proposed switching to single doses, something the pharmaceutical companies weren't set up to produce but cd hypothetically change to, & Halsey proposes switching to vaccines that don't require preservatives - also something that the pharmaceutical companies weren't producing but cd switch to. But Offit has to Wakefield-bash & praise Halsey.

"Few vaccine experts are more respected, more knowledgeable, or more dedicated than Neal Halsey. If he was concerned about something, people listened. "There was no question that it was Neal's concerns that drove a lot of this early on," recalled John Modlin, head of the vaccine advisory committee to the CDC. "He expressed a higher level of anxiety than the rest of us did. He wasn't convinced that there wasn't harm [caused by thimerosal]. And he was the driving force behind the AAP's decision." Halsey's initial proposal surprised his colleagues: he wanted to stop giving any vaccine that contained thimerosal." - p 67

Nonetheless, he finds Halsey's concern unjustified & into the "Autism's False Prophets" go the people who wanted thimerosal removed from vaccines.

"Kessler's removal of silicone breast implants from the market precipitated a flood of litigation. In one year, the number of lawsuits against Dow Corning increased from 200 to 30,000. Many of the lawsuits came from patient advocacy groups set up by lawyers to recruit clients. By 1994, breast implant manufacturers had been brought to their knees, forced to settle a class-action lawsuit for more than $4 billion, at the time the largest medical product settlement in American history. One billion dollars went to the lawyers. In May 1995, Dow Corning filed for bankruptcy." - p 75

Ok, it's more than a little hard for me to feel sorry for Dow Corning. Breast implants, to me, are a prime example of one of the stupidest things made possible by the medical industry. I've had 2 women friends who got breast implants to make their breasts larger. They were both strippers. Obviously, the idea was that they'd get bigger tips w/ bigger tits. Personally, I like breasts, I'm heterosexual so the attraction to breasts is all part of the lust package - & I've been lovers w/ women w/ small breasts & women w/ large breasts. Ultimately, I'm not going to pick who I'm lovers w/ on the basis of the size of their breasts, there're other more important factors such as how intelligent & imaginative they are. Of course there're physical factors too, there're physical characteristics that I'm attracted to & some that I'm repulsed by - but I wd never want a woman that I'm lovers w/ to get silicone breast implants, I'd prefer them to be who they naturally are. The obsession w/ having large breasts is, at least partially, a culturally induced thing that I think we'd be better off w/o - & making a big business out of exploiting women's insecurities over breast size is just despicable.

"Gabriel's conclusion was simple: "We found no association between breast implants and the connective tissue diseases and other disorders that were studied."

"Gabriel's study was the first of many. In 1995, Jorge Sánchez-Guerrero, a researcher in the department of rheumatology and immunology at Harvard Medical School, examined the records of 90,000 women and published their findings, also in the *New England Journal of Medicine*. Again, women with breast implants were not more likely to have connective tissue diseases. Six more studies followed. Researchers at Wayne State University, the University of Calgary, the University of Kansas School of Medicine, the Johns Hopkins School of Medicine, the University of Pennsylvania School of Medicine, and the Harvard School of Public Health all agreed with Gabriel and Sánchez-Guerrero: breast implants didn't cause connective tissue diseases." - p 76

WTF?! Didn't these places have anything better to do?! It's obvious that some serious money was put into all these studies. Why put so much money in if it weren't to protect the breast implant companies? Maybe Brian Deer shd investigate where *their funding came from*! Truthfully, just from the perspective of 'common sense', I find it hard to believe that inserting silicone (or anything else) into breasts wdn't have harmful side-effects. Once again, Offit loses his believability w/ me when he defends such things.

"Later, the Institute of Medicine (IOM) reviewed hundreds of epidemiological and biological studies and concluded that breast implants didn't cause connective tissue diseases. But plaintiffs' lawyers and breast implant recipients claimed that IOM was part of a conspiracy to mislead the American public—a conspiracy financed by breast implant manufacturers." - p 78

Of, those wacky conspiracy theorists, believing that big business will do unpleasant things to make more money! I wonder, *was it ever transparent as to where the funding came from?* Now Marcia Angell, the aforementioned author of The Truth About the Drug Companies is someone who I have the highest respect for & she, apparently, was one of the people who reinforced the breast implant manufacturers's claims to safety. If she says it, I'm inclined to believe her. She strikes me as a person of very high integrity. Then again, even she cd be wrong - or looking at the issue from a myopic perspective.

"Marcia Angell, executive editor of the *New England Journal of Medicine* and author of *Science on Trial: The Clash of Medical Evidence and the Law in the Breast Implant Case*, also came under attack. "I was subpoenaed twice," said Angell. "They wanted a large number of documents that don't even exist, alleging contact between me and the manufacturers. They thought the manufacturers paid me."" - p 78

Offit's bad logic continues to march onward:

"When public health agencies delayed the birth dose of hepatitis B vaccine, children suffered. Ten percent of hospitals, frightened by the notion of giving a thimerosal-containing vaccine to newborns, ignored recommendations and simply suspended the hepatitis B vaccine for *all* newborns. One three-month-old child born to a mother infected with hepatitis B virus in Michigan died of overwhelming infection, having failed to receive the vaccine in the nursery." - p 79

This statement is based on an unproveable hypothesis. Since the child wasn't vaccinated we don't know what wd've happened if the child *had been vaccinated* - it's as simple as that. To say that the child wd've lived is predicated in an unquestioned total faith in vaccines - something that, as far as I know, has no basis in reality whatsoever. People who're vaccinated sometimes get sick & die - sometimes from what they're vaccinated against, sometimes from things that the vaccination helped enable. When doctors, such as Offit, talk about an imaginary event as if they have absolutely certain predictive powers then it's time to recognize that they're Medical Industry Priests claiming prophetic power & to dismiss them as, yes, "False Prophets".

"If Mercury caused autism, they reasoned, perhaps removing it from autistic children could help. "With one in one hundred and fifty children now diagnosed with autistic spectrum disorder," they wrote, "development of mercury-related treatments, such as chelation, could prove beneficial for this large and seemingly growing population." Chelation therapy for austistic children—the administration of chemicals designed to bind to mercury and to eliminate it from the body—was born. (The word *chelation* is derived from the Latin *chelos*, claw.)" - pp 82-83

Of course, *that's* predicated on the "if": "If Mercury caused autism". IF that's the case then chelation might be the ticket. What *IF* autism is a product of an overall human-created environment? What *IF* there're many, MANY elements involved, all parts of a whole that's more & more dominant w/ every passing yr? What *IF* humans are our own ecological disaster & autism is a symptom? *AND* cancer? *AND* immune disorders? What if it's not as simple as just one thing causing another? I'm not saying this is the case, I'm just imagining a possibility. What *IF* the more technical control humans exert, the more new injuries squeeze thru the cracks? & what if the whole philosophical basis of the Medical Industry's technical control is one of the prime factors in this process? My intuition tells me this might very well the case, that the Medical Industry might be doing more harm than good, that autism (& so many other things) are a byproduct not only of vaccines but of every other technical invasion of the natural mind & body. Of course, that's 'just' my intuition - so why shd anyone trust it just b/c I do? No good reasons whatsoever.. so what does *your* intuition tell YOU?

"In the spring of 2003, less than one year after they had published their VAERS study, the Geiers published another in the *Journal of American Physicians and Surgeons.* Again, using the VAERS database, the Geiers found the more mercury children received in vaccines, the more likely they were to develop autism and speech disorders; worse, they were also more likely to have heart attacks and epilepsy." - p 85

Now, you can see where this is going: the Geiers are going to be exposed as poor sources, etc. What I wonder is: *what wd exposing every medical researcher to such scrutiny reveal about the weaknesses of their procedures?* Is it possible that there're *systemic* weaknesses in the whole basic framework? Just sayin'.

"In 2002, Handley's wife, Lisa, gave birth to a son, Jamie. For his first eighteen months Jamie was happy, playful, and engaging. Then he began a frightening descent into autism. He stopped making eye contact, stopped responding to his name, and spent days spinning around in circles." - p 86

At this point, in every one of these cases, I want to know: 1. was the child vaccinated?, 2. what vaccinations did the child receive?, 3. when were these vaccinations received? Offit doesn't give that info b/c he wants the reader to think such things are irrelevant. To those of us who think that vaccinations, esp vaccinations of developing children, are unnecessary & dangerous assaults on the body, every report that provides such information can be considered a statistic. If Jamie had just been vaccinated in the preceeding 2 wks maybe the vaccination had something to do w/ his regression. If such occurences repeat often enuf it's more than a little ridiculous to write them off as coincidence.

""If you look at [autistic] children," he said, "they have high testosterone, they masturbate at age six, they have mustaches, they're aggressive, and you can treat them by lowering their testosterone and removing the mercury."" - pp 88-89

I was immediately suspicious of this one. To me, if you're male & you have high levels of testosterone then that's only a 'problem' insofar as one's urges will be strong. Sure, that's a problem - but it's a problem that the individual shd learn to control, not something to be removed from them. I'm reminded of a friend w/ breast cancer. She was told her cancer was caused by high levels of estrogen & that a hysterectomy wd improve her situation. That sounded like a really bad idea to me. Nonetheless, that seems to be accepted theory these days. Being the hedonist that I am, just having more sex seems like a better cure. I wonder how many doctors recommend that?

"Verstraeten mined the VSD database in an effort to determine whether mercury in vaccines had caused harm. After his first pass through the data, he concluded that it had." - p 91

But you know by now, doncha?, that the reader's being set up for a fall here. Verstraeten is wrong & Doc Offit's going to tell you why.

"Paul Stehr-Green, an associate professor of epidemiology at the University of Washington School of Medicine and Public Health, was the first to see a flaw in Verstraeten's study. Stehr-Green reasoned that children who weren't getting vaccinated (and were therefore exposed to less mercury) might also be less likely to visit their doctor. "I think [this] impacts on [Verstraeten's] conclusions tremendously," he said. Stehr-Green was concerned that children who got more vaccines were more likely to be diagnosed with neurological problems not because they were actually at greater risk, but because they were more likely to come to the doctor." - p 92

Offit & Stehr-Green apparently reason that going to the doctor results in a greater quality of diagnosis, a greater likelihood of problems being detected. Another possibility, however, occurs to me - one that I think is equally conforming to Occam's Razor: the doctors & what they do to their patients are actually the source of the problems. Hence, the more the children go to the doctor, the more likely they are to get sick.

Robert F. Kennedy, Jr. gets thrown into the mix. Kennedy is an environmental lawyer who's gotten involved w/ questioning the safety of some vaccines. he wrote an article for *Rolling Stone* called "Deadly Immunity". I generally like what Kennedy has to say.

"Kennedy's article contained other inaccuracies. Kennedy wrote: (1) "[The CDC] withheld Verstraeten's findings, even though they had been slated for immediate publication, and told other scientists that his original data had been 'lost' and could not be replicated." Verstraeten published his study only after the problems with the preliminary data had been addressed." - p 94

"(8) "Four of the eight CDC advisors who approved guidelines for a rotavirus vaccine laced with thimerosal had financial ties to the pharmaceutical companies developing different versions of the vaccine." No rotavirus vaccine has ever contained thimerosal." - p 95

"Flooded with letters and e-mails correcting Kennedy's many mistakes, *Rolling Stone* issued a series of retractions on June 17, 22, and 24" - p 96

Ok, those critcisms of Kennedy seem potentially valid. Let's take the quote from p 95: Let's say that "Four of the eight CDC advisors" [..] "had financial ties to the pharmaceutical companies developing different versions of the vaccine." Is that reasonable? Given that I accept Offit's expertise on rotavirus vaccines I'll give him the benefit of the doubt that no "rotavirus vaccine has ever contained thimerosal." It's a matter of not throwing out the baby w/ the bathwater again.

"On August 26, 2004, Arnold Schwarzenegger, governor of California, trumped his fellow politicians by banning mercury-containing influenza vaccines from the state. (In 2004, only multidose influenza vaccines contained thimerosal as a preservative.) By April 2006, six states had followed his lead; in 2007, another seventeen states were considering similar bans. Because only limited supplies of thimerosal-free influenza vaccines are available, public health officials worried that banning thimerosal was equivalent to banning influenza vaccines for some children, putting them at risk of severe and occasionally fatal disease." - p 102

Oh, Lawdy! How much of this claptrap do we have to be exposed to? Have you ever had the flu?! It's no big deal. Really. I'm sure I've had the flu many times, it's been a minor inconvenience, something that's resulted in me spending more time in bed resting & drinking more water. A runny nose, a fever. Get over it people, all these things that the Medical Industry act as if they're guillotines just waiting to cut your head off are more like pieces of paper that you might get a cut from. The danger is ridiculously exaggerated & that's how the Medical Industry & the mass media keep you in a state of constant fear. How many people actually die from the flu? Now how many of them really died b/c they were in bad shape for some other reason? All of them? It seems likely to me. If there really is a chance of developmental disorder as a side-effect from thimerosal then I'd rather take my chances w/ the flu. SHEESH.

"By the late 1990s, when health officials had completely eliminated thimerosal, the number of children with autism was higher than it had ever been, exactly the opposite of what would have been expected if thimerosal caused autism. Stehr-Green concluded, "The body of existing data is not consistent with the hypothesis that increased exposure to thimerosal-containing vaccines is responsible for the apparent increase in the rates of autism." - pp 106-107

What I learn from all this back-&-forth between different studies that disprove each other is what I already knew: viz: that science is NOT monolithic, there is NO ultimate objective position. Instead, there're different ways of arriving at different conclusions, many of them in competition w/ each other & w/ differing philosophies underpinning them.

"One year later, in September 2004, Jon Heron, an epidemiologist from the University of Bristol in the United Kingdom, published a study in *Pediatrics*. Heron examined the records of 14,000 children who had received different amounts of thimerosal in vaccines between 1991 and 1992. He wanted to see if he could find a relationship between the amount of thimerosal babies had received and the risk of neurological problems. He did. The more thimerosal children received, the *less* likely they were to be hyperactive or to have difficulties with hearing, movement, or speech." - p 107

This brings up another pet peeve for me: the whole notion of hyperactive: Just b/
c adults can't deal w/ the level of energy that children have doesn't mean that
there's anything wrong w/ it. Have respect for energy - just b/c you don't have it,
just b/c you're a BORE doesn't mean that you have the right to reduce everyone
to your level of dullness.

"Like Heron, Nick Andrews found the more thimerosal children received, the less
likely they were to develop neurological problems like attention deficit disorder.
Again, the amount of mercury in vaccines didn't presage the development of
autism. Andrews concluded, "There was no evidence that thimerosal exposure
via vaccines caused neurodevelopmental disorders."" - p 108

What if so-called 'hyperactivity' & ADD *weren't defined as neurological disorders?*
That turns everything on its head. What if they were defined as something more
like *abundance of life energy*? Well, then the thimerosal wd be seen as causing
the neurodevelopmental disorder of *life energy suppression*, something easily
attributable to a poison.

"Fombonne had found that the number of children diagnosed with autism in
Canada had increased throughout the mid- to late 1990s. This increase occurred
at the same time that thimerosal had been removed from vaccines. Obviously,
removing thimerosal hadn't caused the increase. But what had? Fombonne had
an explanation: "Factors accounting for the increase include a broadening of
diagnostic concepts and criteria, increased awareness and, therefore, better
identification of children with [autism] and improved access to services." In other
words, Fombonne reasoned that there wasn't an epidemic of autism; rather,
broadening the definition of the disability to include mildly affected children, as
well as heightened awareness among parents and doctors, had accounted for
the increase." - p 109

Let's face it, the closer we look at it, the sicker everyone obviously is. Everyone
MUST have their own doctor, everyone MUST be on medications that fuck them
up even more & then require a new medication to counterbalance how fucked up
they are - like a see-saw - aren't see-saws fun? The gist of it is: you MUST
believe in the Medical Industry & you MUST let it tell you what to do - & if there
seems to be some in-fighting, don't let it bother you, the person who dumbs you
down the most *will be the winner* - so let that be a life lesson to you. But, hey!,
don't think I'm only being sarcastic here. Offit does have his redeeming moments
& the following one is one of them:

"Although vaccines have probably saved more lives than any other medical
intervention, they have come with a price—occasionally causing severe, even
fatal, side effects. Epidemiological studies have been the single most powerful
tool to show that vaccines, like all medicines, are imperfect.

"In 1998, the FDA licensed a rotavirus vaccine, and the CDC recommended it for
all infants. The vaccine, designed to prevent a common cause of fever, vomiting,
and diarrhea, had been tested in 10,000 babies before licensure. But after it had

been given to 1 million babies, the vaccine was found to be a rare cause of intestinal blockage called intussusception." - p 110

Thank you, Dr. Paul A. Offit, for admitting this. But as for vaccines "probably" having saved more lives than any other medical intervention? I have my doubts. At any rate, I'm more enthusiastic about the way cigarette smoking finally got a bad name. But what I want to know is: *What were those 10,000 parents thinking who let their kids be guinea pigs?!* Really, people, really. Anyway, notice how we come back to, yet again, a vaccine causing gastrointestinal problems. Isn't that what Wakefield sd? Yes, "vaccines, like all medicines, are imperfect." Not to mention doctors.

"In 1976, public health officials in the United States, fearing an unusual outbreak of influenza pandemic, immunized millions of Americans with what was called the swine flu vaccine. Unfortunately, some people immunized with the vaccine developed a rare form of paralysis called Guillain-Barré Syndrome. Epidemiological studies showed that the vaccine was the cause. One of every 100,000 people who got swine flu vaccine—400 people among 40 million—had been afflicted." - p 111

It's exactly that sort of thing that makes people like myself not want to get vaccinated. Pro-vax people tend to downplay collateral damage. Downplaying paralysis to a person who got vaccinated b/c they were trying to avoid getting sick based on the advice of an (imperfect) expert is sick in & of itself. Nature isn't likely to inject you w/ something that'll make you paralyzed unless it's from a poison from a creature attacking you. I seriously doubt that 400 people were paralyzed by an attacking creature during the time that this vaccination campaign took place. The Medical Industry makes *prophecy*, b/c that's all these pandemic predictions are, & then puts people at risk w/ something that's pretty iffy in the 1st place: vaccines, imperfect vaccines. I'd rather take my chances w/ nature.

"The vaccine, pioneered by Jonas Salk, was made by inactivating polio virus with formaldehyde. After it was licensed, five companies stepped forward to make it. One company, Cutter Laboratories of Berkeley, California, made it badly. Because Cutter hadn't completely inactivated its vaccine, more than 100,000 children were inadvertently injected with live, dangerous polio virus. Seventy thousand got mild polio, 200 were permanently paralyzed, and ten were killed. It was one of the worst biological disasters in American history." - pp 112-113

To Offit's credit, he brings up these stories; to his discredit, he always takes the mainstream position that while such a disaster is unfortunate there's an excuse for it but what wd've happened if the vaccinations hadn't taken place might've been worse. To my mind, this wdn't've happened w/o vaccination - it's as simple as that - therefore, the vaccination, regardless of what the excuse for the disaster is, was something that shd've never been inflicted on people in the 1st place. Polio vaccine critics have stated that the Cutter lab was used as a scapegoat b/c they were independent from bigger pharma conglomerates.

"Further epidemiological studies consistently showed that although lung cancer caused by cigarette smoking was rare, affecting less than 1 percent of those who smoked, it was real. The results of these epidemiological studies no longer allowed an industry that wished to believe smoking didn't cause cancer to hide behind laboratory studies that had proved worthless." - p 114

Again, there're conflicted findings: cigarette manufacturers supported research that sd cigarette smoking was ok; then anti-smokers produced research that smoking causes cancer. This type of conflict is typical in the science world - esp when big money's at stake. During the so-called COVID-19 pandemic research in France claimed that cigarette smokers were less likely to get COVID. I've seen 'news' of research in France claiming that red wine intake, in moderation, is healthy for you. I vaguely recall the same thing about research exonerating coffee. &, yet, I doubt that mnay people other than cigarette addicts, alcoholics, & coffee addicts wd take such research seriously. Nonetheless, this is science at work. From my POV, pro-vax people are like embattled cigarette manufacturers fighting to continue promoting a big business. SO, 400 people out of 40 million people getting paralyzed from a vaccination is as real as 1% of people getting cancer from smoking - & an equally valid reason for stopping vaccinations.

"Because everyone drinks water, everyone has small amounts of methylmercury in their blood, urine, and hair. A typical breast-fed child will ingest almost 400 micrograms of methylmercury during the first six months of life. That's more than twice the amount of mercury than was ever contained in all vaccines combined. And because the type of mercury in breast milk (methylmercury) is excreted much more slowly than that contained in vaccines (ethylmercury), breast milk mercury is much more likely to accumulate. This doesn't mean that breast milk is dangerous, or that water is dangerous. Not at all. It means only that anyone who lives on the planet will consume small amounts of mercury all the time." pp 114-115

But this isn't about breast milk mercury, is it? It's about mercury contained in something that's injected directly into the bloodstream, bypassing the body's filtration.

An assistant professor of pediatrics at the University of Colorado named Sarah Parker started getting what she considered threatening emails & phone-calls b/c she's pro-vax.

"Parker called the police to obtain a restraining order on one particular caller. "My impression after talking to the police [was that the callers] seemed to be very well trained in not using words that would get them in trouble with the law. So, they wouldn't make a direct threat, saying, 'I'm going to hurt you' or 'I'm going to hurt your children.' They wouldn't tell me what the 'or else' was. [But] I was worried that they were going to do something to my family. I quit answering my phone for about a year." The threats worked. Sarah never published another paper on vaccine safety." - p 117

As I've stated earlier in this review, I'm against harrassing or threatening anyone b/c of their opinions. I'm also against any mandatory medical intervention. What people who advocate vaccination have to understand is that as long as they're saying: 'This is what I think is right' then, IMO, it's good to have their input - but when things morph to: 'This is what you have to do regardless of whether you agree with me or not' then that's when things turn particularly ugly. Parents who think vaccines might paralyze or kill or developmentally disable their child are understandably not going to take well to being told that their children *have to be vaccinated* in order to go to school or whatever. They're going to take the people that they consider to be responsible for this to be serious enemies, murderers, sadists. I use this review as a way to show 1. that I'm informing myself about a variety of opinions, 2. to express my own opinions formed under these conditions. Even tho I disagree w/ much of what Offit says I wd never harass him over it. This review is my way of addressing the subjects. I'd expect the same courtesy from him & others like him. Perhaps that's what I consider to be 'civilized' behavior & I'm in favor of it. Mandatory medical interventions are NOT 'civilized', they're fascist. Threats, veiled or otherwise, are also fascist.

As for Parker not answering the phone for a yr? Before the Do Not Call list was created, I was getting 7 scammer phone calls for every call I got from a friend or work. I got on the list & that stopped it for awhile & it's never reached 7 a day again but now it can be 1 or 2 scammer calls a day & at least 1 txt msg scammer a day, etc. The effect is the same: I stopped answering my phone. Then I got caller ID, that helped. I only answer the phone if the caller's in my contacts list. The point is that while the malevolence of the veiled threats to Parker are nasty so are the constant invasions by the capitalists & their motive is, in a sense, even more ruthless & unmotivated by anything but a greedy drive to parasitize off of others *no matter what*. As an older person, I now get to experience that special breed of scammer scum that tries to find old people who're defenseless to rob, hoping for weak-mindedness or senility.

"The CDC also received a series of threatening e-mails. One stated, "Forgiveness is between you and God. It is my job to arrange a meeting.""

[..]

"The CDC contacted the FBI, instructed its staff on safety issues, hired more security guards, and showed employees how to respond if pies were thrown in their faces." - p 119

But what actually happened? Were any pies thrown in anyone's faces? There's plenty of mayhem in the US, that's for sure, but there's far more fear-mongering of things that don't actually happen - & that originates with the people that Offit seems to consider to be the 'good guys' as much, if not more so, than w/ the 'bad guys'. I'm more concerned w/ the mass murderers who have private arsenals & who commit their shootings at vulnerable places like schools & churches & shopping malls than I am w/ threats to the CDC.

"Stephen B. Edelson, director of the Edelson Center for Environmental and Preventative Medicine, claimed he could treat autistic children with high-intensity sound waves, calling his miraculous new therapy sonar depuration. (Sonar depuration therapy has never been formally tested.) Edelson said sonar depuration helped damaged brain cells to regenerate: "A classic example is you can take a six-year-old, remove half their brain, and within two years the child will be perfectly normal." (Children who lose half their brains don't grow them back.)" - p 120

Whew! Ok, it's not hard to be on the same page as Offit w/ this one. I didn't find "sonar depuration" online but I did find "depuration":

"**Depuration** of seafood is the process by which marine or freshwater animals are placed into a clean water environment for a period of time to allow purging of biological contaminants (such as *Escherichia coli*) and physical impurities (such as sand and silt). The most common subjects of depuration are bivalves such as oysters, clams, and mussels." - https://en.wikipedia.org/wiki/Depuration

As for "take a six-year-old, remove half their brain": *Don't try this at home!*. &, yeah, I find the claim that half a brain cd grow back in 2 yrs to be as preposterous as it gets. One can only hope that no-one was brain-dead enuf to fall for it. The endnote that I take to be providing the source for the quote is this:

"Stephen Edelson: Cited in U.S. Congress, House of Representatives, Committee on Government Reform, *The Future Challenges of Autism: A Survey of the Ongoing Initiative of the Federal Government to Address the Epidemic*, 108th Congress, First session, November 20, 2003 (Washington, DC: U.S. Government Printing Office, 2004), 137." - p 269

That sd, tho, strangely enuf there has been some research about brain regeneration, something that is generally considered impossible:

"Deconstructing birdsong may seem an unlikely way to shake up biology. But Nottebohm's research has shattered the belief that a brain gets its quota of nerve cells shortly after birth and stands by helplessly as one by one they die—a "fact" drummed into every schoolkid's skull. On the contrary, the often-rumpled Argentina-born biologist demonstrated two decades ago that the brain of a male songbird grows fresh nerve cells in the fall to replace those that die off in summer.

"The findings were shocking, and scientists voiced skepticism that the adult human brain had the same knack for regeneration. "Read my lips: no new neurons," quipped Pasko Rakic, a Yale University neuroscientist doubtful that a person, like a bird, could grow new neurons just to learn a song.

"Yet, inspired by Nottebohm's work, researchers went on to find that other adult animals—including human beings—are indeed capable of producing new brain cells. And in February, scientists reported for the first time that brand-new nerves in adult mouse brains appeared to conduct impulses—a finding that addressed lingering concerns that newly formed adult neurons might not function. Though such evidence is preliminary, scientists believe that this growing body of research will yield insights into how people learn and remember. Also, studying neurogenesis, or nerve growth, may lead them to better understand, and perhaps treat, devastating diseases such as Parkinson's and Alzheimer's, caused by wasted nerves in the brain."

- https://www.smithsonianmag.com/science-nature/birdbrain-breakthrough-64765165/

Don't get me wrong, I don't think this supports the claim of sonar depuration - it does, however, point to an interesting future. I, personally, wdn't mind have some new brain cells of the right kind (no cancerous cells need apply).

"But by far the most extensive network of physicians offering alternative therapies for autism belongs to Defeat Autism Now (DAN), a group based in San Diego, and part of the Autism Research Institute. Like Andrew Wakefield, DAN practioners believe autism is caused by toxic substances that enter the body through a leaky gut. (Studies have failed to prove that autistic children have leaky guts, and brain-damaging toxins have never been identified.)" - p 121

Since I find Elizabeth Fein's bk Living on the Spectrum to be reasonable I looked in its index for "Defeat Autism Now!" to see what, if anything, she wrote about them & found only this:

"Defeat Autism Now! changed its name to the Autism Research Institute, in response to requests from autistic people for whom the name felt like a "personal affront" (Edelson, n.d.)." - pp 146-147, Living on the Spectrum

"In June, 2007, Participant Productions, facing overwhelming evidence that vaccines didn't cause autism, stopped production on the movie version of Evidence of Harm." - p 127

That's interesting, the movie that wasn't. For me, tho, the issue has never been 'do vaccines cause autism' but can vaccines cause harm? Such as paralysis? Encephalitis? A general deterioration of health that can lead to developmental regression in the very young? Death? YES is the answer to all those questions IMO.

"Seidel was raised in Anaheim, California. Her father was a chemical engineer, her mother a music and special education teacher who worked with severely impaired children. In 1973, after graduating from high school, Seidel attended the University of California at Santa Cruz, majoring in English and Russian literature and Book Arts and later venturing to New York City, where she earned a master's degree in Library Science from Columbia University." - p 130

"Kathleen Seidel has almost single-handedly exposed the unsavory allegiances of those who proffer cures for autism." - p 228

Offit has no problem w/ the lack of academic-scientific credentials of Seidel b/c she's against the same people he is. On the other hand, he's more particular about the lack of academic-scientific credentials of some of the 'expert witnesses' used by the "False Prophets".

"Seidel decided to start a Web site and blog, calling it neurodiversity.com, the domain name coming from a family brainstorming session in 2001." - p 131

Again, I quote from Fein's bk:

"At the same time, a growing movement of individuals on the autism spectrum, often referred to as (part of) the *neurodiversity movement*, argue that autism is instead a natural and valuable aspect of human diversity, a cultural identity calling for accommodation rather than prevention. A cure for autism would thus be not a mercy but a genocidal suppression of difference, equivalent to "curing" someone's race, gender, or sexual orientation." - p 2, Living on the Spectrum

"First, the Geiers referred to VAERS as a mandatory reporting system for problems following vaccines. But VAERS is a passive system; people who believe that a vaccine might have caused a problem are encouraged, not mandated, to fill out a form. Because reporting is at best haphazard, most researchers (other than the Geiers) don't use VAERS to prove a vaccine has caused harm." - p 137

Ok, I'm not a researcher in the sense that Offit's using the word. I LIKE VAERS (Vaccine Adverse Event Report System). Here's the link to the government VAERS website: https://vaers.hhs.gov . 'Experts', like Offit, are contemptuous of it b/c it allows & encourages 'non-experts' to report things from their own direct experience instead of having the experts 'interpret' it for them. Someone might get a COVID-19 vaccine & then shortly thereafter develop Myasthenia Gravis. SO they go to a doctor & the doctor acts like associating the 2 things is ridiculous. The doctor doesn't want to acknowledge that a 'perfectly safe' vaccine that he's been promoting to, & giving to, his patients might have a possible side-effect of such a serious condition. The doctor is in denial, the doctor can't be trusted. The patient, however, *can be trusted* b/c they're the unfortunate person undergoing the experience. I LIKE VAERS.

"TAP Pharmaceuticals—which stands for Takeda Abbott Phrmaceuticals, a joint venture between Takeda Chemical Company in Osaka, Japan, and Abbot Laboratories in Abbott Park, Illinois—wasn't a stranger to controversy. Several years before, to compete with the rival drug Zoladex, a product of Astra Zeneca, TAP had given free samples of Lupron—a drug that cost about $20,000 per treatment course—and offered grants of $25,000 to any physician who agreed to stop using Zoladex. The Department of Justice saw this marketing practice for what it was: fraud. It charged fifteen employees and five physicians for fraudulently promoting Lupron. On October 3, 2001, TAP agreed to pay almost

$900 million to settle the government's criminal complaints. It was the largest settlement for health care fraud in U.S. history." - p 144

It seems time to quote my review of Marcia Angell's bk again:

""And TAP didn't stop there. In 1996, the company also tried to persuade a large Massachussetts HMO, Tufts Health Plan, to stay with Lupron by offering its medical director of pharmacy programs a $25,000 "educational" grant that he could use for anything he wanted. The company couldn't have chosen a worse target: The medical director of pharmacy programs, Joseph Gerstein, is someone I know to be among the least likely people to take a bribe. When Gerstein refused, the company upped the offer to $65,000. But this time Gerstein, who with the support of Tufts had alerted federal authorities, taped the conversation, and that led to the unraveling of the company's illegal activities." - p 131

"Ha ha ha. Oh, how I love seeing white collar criminals taken down. Of course, no-one goes to jail - while a friend of mine can get sentenced to 2 yrs in prison when he tries to legally purchase a case of beer not having an ID on him & gets into an altercation. HE goes to prison - but these corrupt shits sail clear as usual, a little bruised but still straight on course for more of the same. Nonetheless, thank you Joseph Gerstein for not beng bought.

""Many doctors become indignant when it is suggested that they might be swayed by all this industry largesse. But why else would drug companies put so much money into them? As Stephen Goldfinger, chairman of the APA's Committee on Commerical Support, said, "The pharmaceutical companies are an amoral bunch. They're not a benevolent association. So they are highly unlikely to donate large amounts of money without strings attached. Once one is dancing with the devil, you don't always get to call the steps of the dance."" - p 147"

The point is, as Angell so eloquently elucidates, that it's not just a case of a few corrupt criminal pharmaceutical companies, it's systemic, pharmaceutical companies in general are likely to use unscrupulous methods to increase their profits w/o making health their primary purpose. Hence, is it really any wonder that people find people like Wakefield believable? Offit may be 100% sincere in his pursuit of rotavirus vaccines but that doesn't mean that the more administrative/marketing/business arm of the whole industry shares his primary concerns. Hence, I find Offit's loyalty to the system he represents naive.

"And many parents swear by the wonders of chelating children, claiming dramatic improvement within days. But because a cell damaged by heavy metal doesn't recover—much less within a few days—this is simply not possible." - p 146

&, yet, at the same time that I can see that what Offit's saying seems reasonable, I've been told that things were "impossible" before by at least 2 doctors & they were both wrong. 1st, I was told by a doctor that steroids cdn't possibly be the cause of crippling pain that I had - but then I stopped taking the steroids & the pain went away. IMMEDIATELY. 2nd, I was told by a doctor that it was "impossible" for me to cure my hypothetical Diabetes Type II w/ diet & exercise & w/in 5 days of my proposed changes my symptoms were gone & w/in 2 wks my blood sugar was normalized. I'm sure that the same people who object to VAERS wd simply dismiss my testimonial. W/ these experiences in mind, I'm more inclined to believe the parents than I am the doctors. After all, doctors aren't perfect - in fact they can be downright arrogant assholes.

"When Tariq's calcium level dropped precipitously, his heart stopped beating. Tariq Nadama was the third person to die from EDTA therapy since 2002. But J. B. Handley's Rescue Angels wasn't about to quit. "We're not stopping," said one of them, Marla Green." - p 147

Now this child patient's death is tragic. The therapy being used was a bad idea. But Offit has a double standard: if kids die from vaccination it's tragic but vaccination in general is a-ok b/c Offit is a BELIEVER. However, if a kid dies from an alternate therapy that's a sure sign that the people administering that therapy are deranged & money-hungry. I fail to see the difference between what Offit supports & what the people he declares "False Prophets" support. They're both responsible for deaths, they both plow ahead anyway.

"In 1983, following a conviction for possession of illegal drugs, Kennedy was sentenced to two years' probation, periodic drug testing, mandatory supervision by Narcotics Anonymous, and 800 hours of community service. He satisfied his community service by working for the Hudson River Foundation, now called the Hudson Riverkeepers. Later, Kennedy became its chief prosecuting attorney." - p 149

The endnote on p 274 reveals that the drug in question was heroin. I think using heroin is a bad idea. Still, let's not forget that heroin was 1st manufactured & distributed as a *non-addictive alternative to morphine* by the prominent drug company, Bayer. Bayer sold it as such for 25 yrs. A very interesting bk to read on the subject of heroin is Alfred W. McCoy's The Politics of Heroin. It's one of my favorite bks. Kennedy's sentencing for possession of heroin seems a bit light for 1983 but he's from a rich family so he had a Get-Out-Of-Jail Card. But let's get real here: if Offit's trying to establish Kennedy as a sleazy character then we'd better throw in the pharmaceutical industry & the US government as the BIG PUSHERS b/c Kennedy's a speck of dust on the culpability spectrum in contrast to them when it comes to heroin - &, perhaps hypocritically, Offit doesn't seem to have much of a problem w/ the pharmaceutical industry or the US Government in general. There's that double standard again.

Chapter 8 is called "Science in Court" - but is that what it's really about? Is it science that was on trial? Or is it, instead, a particular manifestation of science? It seems to me that the latter is more accurate but that Offit wants to act as if challenging any science he supports is challenging *all* science. Let's not confuse the 2.

"In the summer of 2007, parents of children with autism took their case to court.

Called the Omnibus Autism Proceeding, it was an unusual lawsuit." - p 156

[..]

"They were suing the federal government in a federal court. That wasn't their preference. They would much rather have argued their cases in state courts in front of juries. In federal court they would have to convince a panel of three judges. But they had no choice; no one can sue a vaccine maker without first going through this unusual court." - p 156

[..]

"In 1986, following a series of lawsuits that threatened to end vaccine manufacture for American children, Congress passed the National Childhood Vaccine Injury Act. Included in the act was the Vaccine Injury Compensation Program. If parents felt their children had been harmed by vaccines, they sued the federal government for compensation, making their arguments in front of federally appointed judges. As a consequence, the number of lawsuits brought against vaccine makers declined dramatically." - pp 156-157

If that doesn't reek of lobbying I don't know what does. If lawsuits put an end to vaccination for children wdn't that just be due process? Instead, the government has the vaccine manufacturer's back - regardless of how much legal evidence might accrue that the vaccinations are harmful. 1986 was during the Reagan presidency. For those of you too young to remember what that was like, Reagan was a 'conservative' president (it's questionable what he was 'conserving'). To me, he was a glaring puppet president. Once again, I quote from Angell & my review of her bk about those yrs in relation to Big Pharma:

""You could choose to do well or you could choose to do good, but most people who had any choice in the matter thought it difficult to do both. That belief was particularly strong among scientists and other intellectuals. They could choose to live a comfortable but not luxurious life in academia, hoping to do exciting cutting-edge research, or they could "sell out" to industry and do less important but more remunerative work. Starting in the Reagan years and continuing through the 1990s, Americans changed their tune. It became not only reputable to be wealthy, but something close to virtuous. There were "winners" and there were "losers," and the winners were rich and deserved to be." - p 6

"Of course, the author is referring to her own professional class here; simultaneously there were punks & anarchists & other 'lunatic fringe' types whose priorities were definitely not w/ getting rich but were instead w/ Truth, Justice, & the Unamerican Way. I was solidly in that camp. How many of us were following legal developments such as what Angell details next I don't know, I certainly wasn't. But the Reagan administration in general was definitely high on the shit list.

""The most important of these laws is known as the Bayh-Dole Act, after its chief sponsors, Senator Birch Bayh (D-Ind.) and Senator Robert Dole (R-Kans). Bayh-Dole enabled universities and small businesses to patent discoveries emanating from research sponsored by the National Institutes of Health (NIH), the major distributor of tax dollars for medical research, and then to grant exclusive licenses to drug companies. Until then, taxpayer-financed discoveries were in the public domain, available to any company that wanted to use them." - p 7

"Hhmm.. Taxpayer money pays for research, results enter Public Domain. That seems reasonable to me. But it also seems reasonable for researchers to benefit from their hard work above & beyond just salaries. Surely, a compromise solution cd be reached in wch the research stays in the public domain but the researchers are still rewarded for their exceptional accomplishment. At any rate, the Reagan admin was about benefitting big business, not the public. & the following is still from his January 20, 1981 – January 20, 1989 reign.

""Starting in 1984, with legislation known as the Hatch-Waxman Act, Congress passed another series of laws that were just as big a bonanza for the pharmaceutical industry. These laws extended monopoly rights for brand-name drugs. Exclusivity is the lifeblood of the industry because it means that no other company may sell the same drug for a set period. After exclusive marketing rights expire, copies (called generic drugs) enter the market, and the price usually falls to as little as 20 percent of what it was." - p 9

"A justification for the original drug's high price is basically that the drug company had to spend a fortune on R&D (Research & Development). A significant part of this bk is spent debunking that as a PR myth."

- https://www.goodreads.com/review/show/3719233146

Note that there was much more going on in those yrs that protected Big Pharma than just the "National Childhood Vaccine Injury Act" that Offit refers to.

"Vaccine court was originally developed to handle one case at a time. But between 1999 and 2007, more than 5,000 parents filed claims that vaccines had caused their children's autism. This was twice the number of claims filed for all other vaccine-related injuries in the twenty years since the program had begun. Because of the number of claims and because the federal judges knew that it would be impossible to hear each case individually, they recommended that autism claims be tried together, like a class-action lawsuit. "There's never been another case like this," said Kevin Conway. one of the lawyers for the petitioners. With average individual awards of close to $1 million and thousands of petitioners, it was possible that a ruling in favor of the petitioners could exhaust the $2 billion available to compensate claimants. Much was at stake." - p 158

That's a fascinating practical problem. It's also unfortunately one not likely to result in justice for even the most valid of the claims. My general attitude toward lawyers is even more negative than mine toward doctors. But there are always exceptions, I've had friends in both professions.

"Thomas Powers, from the Portland, Oregon, law firm of Williams, Love, O'Leary and Powers, was the first to speak on Michelle Cedillo's benefit. Powers's firm had successfully filed lawsuits against the makers of silicone breast implants, as well as against Fen-Phen, a weight loss product associated with heart problems, and the Dalkon Shield, an intrauterine birth control device found to cause severe infections and infertility. Class-action awards for the Dalkon Shield had totaled more than $2 billion; for breast implants, nearly $5 billion; and for Fen-Phen, $21 billion (and counting). These awards had made Williams, Love, O'Leary and Powers one of the richest, most powerful law firms in the United States.

"Sylvia Chin-Caplan, from the Boston law firm of Conway, Chin-Caplan and Homer, also represented Michelle Cedillo. Chin-Caplan's firm specialized in claims before vaccine court, which allows lawyers to receive only 4 percent of awards. Trying cases before vaccine court isn't a very good way for personal-injury lawyers to make a lot of money. So unlike Thomas Powers and his partners, Chin-Caplan's firm was neither rich nor powerful, operating out of a modest three-story walk-up downtown." - p 160

I didn't find an endnote providing sources for the income info but I'm assuming that it's accurate enuf. If this had actually been a class action suit the lawyers cd've been considerably more motivated by greed b/c they'd, presumably (am I wrong?) be getting 4 percent of a HUGE settlement. As it is, these lawyers might've only made twice what I made in a *good* yr for what was probably less than a yr's work, a nothing by most people's standards.

""Michelle was born on August 30, 1994. She weighed eight pounds, roughly, and her Apgars [a ten-point scale of a baby's health measured one and five minutes after birth] were nine and nine. In other words, she was perfectly healthy. [The day] after she was born, she received a hepatitis B immunization. It contained 12.5 micrograms of mercury." - pp 162-163

I think giving a baby an injection of *anything* a day after their birth *is utter insanity*. Take someone at their greatest extreme of vulnerability & inject a poison into them that will supposedly protect them against disease & say that's safe?! I think not. To my mind, any sane parent wd be opposed to this. Of course, the 'experts' will use their best bedside intimidation manner to talk them into it.

"Vera Byers, who said thimerosal had caused a "dysregulation" of Michelle's immune system, also had her credibility challenged. Byers had described herself as a member of the faculty of the University of Nottingham and later the University of California at San Francisco (UCSF) as well as a member of the clinical team "that got Embrel approved." (Embrel is a drug used to treat diseases like rheumatoid arthritis and psoriasis.) But Byers wasn't on the faculty at either Nottingham or UCSF, and her name never appeared on the Biologics License Application to the FDA for the approval of Embrel." - p 167

Ok, that's downright strange. One wd think that a person so easily found out to be a fraud wd've been easily detected by the lawyers for the claimants & have been weeded out before the defense attorneys wd investigate them. After all, Byers's lack of claimed credentials wd be enuf to severely discredit the entire case.

"The testimony of Eric Fombonne refuted the petitioners' claim that Michelle's autism occurred after she had received an MMR vaccine." - p 170

But did it really? The testimony consisted of Fombonne analyzing Michelle's behavior on video before she was vaccinated & pointing out things about it that he claimed were autistic. It seems to me that such testimony might put an interpretative spin on things that seems convincing but might be questionable. It seems to me that what the defense cd've done was show video of a non-autistic child & looked for similar mannerisms. Lawyers & 'expert' witnesses are masters of manipulative language. Independent of that, Fombonne's tesimony didn't refute that there may've been negative effects from vaccinations she rc'vd before the taking of the video.

I saw a demonstration of NLP (Neuro-Linguistic Programming) one time in wch the NLP speaker inserted into his speech emphasized words telling the audience to do something like rub their eyes. Then he looked for whoever rubbed their eyes & told them they did it b/c he'd subliminally manipulated them. Of course, in order for it to be more convincing, *everyone* shd've rubbed their eyes & rubbing the eyes shd be a behavior that's not frequent & normal otherwise.

"Nicholas Chadwick testifed that Andrew Wakefield had not only ignored data that disproved his contention, but he also knowingly falsified them, If true, this revelation showed Wakefield had crossed the line from ill-conceived, poorly performed science to fraud." - p 174

I don't know what to make of that. Offit *does* qualify this w/ "If true" & that seems fair enuf. I consulted my review of Callous Disregard to see if there's mention of Chadwick & didn't find any. I did find this:

""In this study, measles virus genetic material was present in CSF from 19 of 28 (68%) cases and in one of 37 (3%) non-autism controls. Further tests confirmed that where there was sufficient amount of sample available, the genetic material was consistent with having come from the vaccine virus.

""The draft paper concludes by saying,

""*The data indicate that virological analysis of CSF is indicated in children undergoing autistic regression following* **exposure** *to live vaccine viruses.*

""The Paper's conclusions stop well short of any claim that the MMR vaccine causes autism. The most one can say from the findings of measles viral genetic material in CSF is that there is a strong statistical association between the presence of this virus and the autism group." - p 158

"As the reader has probably deduced by now, "CSF" = Cerebrospinal Fluid. As the reader has probably also deduced, the emboldening of **exposure** is the author's."

SO, yes, "If true," [MAYBE] "this revelation showed Wakefield had crossed the line" [..] "to fraud." - I'm still not convinced about the "ill-conceived, poorly performed science" part of the statement.

"The three judges in charge of the Omnibus Autism Proceeding aren't expected to reach a final verdict on whether vaccines might cause autism until 2009." - p 175

This being 2022, we can now find the results:

"In 2002, the NVICP, in consultation with a Petitioners Steering Committee, set up the Omnibus Autism Proceeding to aggregate these cases. They decided to examine six test cases that made one or more of the following claims about the vaccines-autism link:

"• Claims that MMR vaccines and other thimerosal-containing vaccines can combine to cause autism.
• Claims that center on vaccines containing thimerosal causing autism.
• Claims that MMR vaccines alone (with no mention of thimerosal) can cause autism.

"Three Special Masters examined the evidence for each of those claims. In 2009, they handed down their decisions. For each claim, the three Special Masters concluded that there were no links between vaccines and autism."

- https://en.wikipedia.org/wiki/Omnibus_Autism_Proceeding

The Cedillos appealed this decision to the United States Court of Appeals for the Federal Circuit. A decision was made upholding the original decision on August 27, 2010.

"Dr. Krigsman testified as to an autism-gastrointestinal dysfunction link and opined that the MMR vaccine can cause chronic gastrointestinal dysfunction. He testifed in particular that Michelle's gastrointestinal symptoms and ultimately, her autism, were caused by persistent measles virus from the MMR vaccine. Petitioners' theory of causation depended on the Unigenetics finding that the measles virus was present in Michelle Cedillo's body."

[..]

"In particular, in order to establish the unreliability of the Unigenetics testing, the government offered expert testimony and reports from, among others, Dr.

Stephen Bustin, a molecular biologist who was an expert in the UK litigation. In connection with those proceedings, Dr. Bustin was hired by vaccine manufacturers to evaluate the testing methods used by Unigenetics and to assess the validity of Unigenetics works. After analyzing Unigenetics equipment and notebooks, he concluded that the procedures used by Unigenetics rendered the testing unreliable."

- https://www.uscfc.uscourts.gov/sites/default/files/autism/ cedillo%20fed%20circuit.pdf

Ok, it's belaboring the obvious to say that an expert witness hired by vaccine manufacturers to determine whether evidence used against them was validly obtained isn't likely to find contrary to the employer's interests.

"We agree with petitioners that the government's failure to produce or even to request the documentation underlying Dr. Bustin's reports is troubling, but we think that in the circumstances of this case, that failure does not justify reversal."

[..]

"Finally, the Special Master specifically found that even if he were to disregard Dr. Bustin's expert reports and hearing testimony—and if he were to disregard all of the testimony from all of the experts that participated in the British litigation— he would have still concluded that the Unigenetics testing was not reliable. In doing so, he noted that the main points in his rejection of the Unigenetics testing were "(1) the fact that the laboratory failed to publish any sequencing data to confirm the validity of its testing, (2) the failure of other laboratories to replicate the Unigenetics testing, and (3) the demonstration by the D'Souza group that the Uhlmann primers were 'nonspecific.'""

- https://www.uscfc.uscourts.gov/sites/default/files/autism/ cedillo%20fed%20circuit.pdf

I don't have an opinion about what the causes of Michelle Cedillo's autism are. Mainly, I just feel saddened that her parents went thru 12 yrs of litigation to get no satisfaction, only more frustration.

"When parents became concerned that vaccines had caused their children's autism, scientists responded by performing a series of epidemiological studies. All showed the same thing: vaccines weren't at fault. But despite the singular, consistent, reproducible, and clear results of these studies—and consequent reassurances from national and international health groups—many parents remain fearful. Why?" - p 176

Why? Probably b/c children who were vaccinated continued to turn autistic & parents were desperately seeking an explanation & a way to avoid this from happening. Probably b/c the parents, mostly non-scientists, were unable to determine wch opinions they were given were most believable - in Offit's terms:

wch were good & wch were bad science. Probably b/c there was always the well-grounded suspicion that pro-vaccine results were produced by researchers funded by vaccine makers. That's like having a department of the police investigate another department of the police for criminal behavior - the public is right to expect nothing more than a little dramatic theater that solves nothing. One of the best ways of testing whether vaccinations lead to health problems is to simply not vaccinate every child whose parents are opposed to or wary of vaccination & to compare the vaccinated & unvaccinated childen for relative health.

"In the case of vaccines and autism, it isn't hard to find scientists on both sides of the debate. But, in truth, it isn't hard to find scientists on both sides of any issue, independent of whether it's a debate." - p 179

Hence, my opinion that *there is no such thing as THE SCIENCE*, meaning a single scientific position that all scientists are in agreement on. I keep saying this over & over to refute the people who say: What's THE SCIENCE? as if in any issue science has a single answer wch defines absolute truth.

"Scientists, bound only by reason, are society's true anarchists. Indeed, some of the greatest advances in medicine have been made by scientists who initially stood alone. For example, Barry Marshall, working at the Royal Perth Hospital in Western Australia, argued that an unusual bacterium called *Helicobacter pylori* caused stomach ulcers. No one at the time believed that bacteria could survive the harsh acid produced by the stomach, much less reproduce and cause disease. But Marshall was so convinced by his findings that he swallowed a Petri dish full of the bacteria, later developing severe inflammation in his stomach. In 2005, Barry Marshall won the Nobel Prize in Medicine." - p 185

Marshall's explanations of how he digested the Petri dish alone were priceless. But seriously folks, I consider myself to be a 'true anarchist' so I must be a scientist too. After all, I have multiple diplomas from t he Nuclear Brain Physics Surgery's Cool. What more accreditation do you need? Is there a Nobel Prize for Stubborn Persistence?

"In short, not all rogue scientists are good scientists. "History is replete with tales of the lone scientist working in spite of his peers and flying in the face of doctrines," wrote Michael Shermer, author of *Why People Believe Weird Things*. "Most of them turned out to be wrong and we do not remember their names. For every Galileo shown the instruments of torture for advocating scientific truth, there are a thousand or ten thousand unknowns whose 'truths' never pass muster."" - p 186

Doesn't it also follow that not every mainstream scientist is a good scientist? After all, mediocrity thrives best in not rocking the boat. A scientist who goes to a university & excels in believing whatever they're taught is not likely to apply the scientific method under conditions where doing so might threaten their middle-of-the-road comfort. Does Shermer include, under "weird things", the belief that cutting a person's arm & rubbing cow pus in the wound will prevent that person from getting smallpox? That seems like a pretty weird belief to me. Does he include the belief that some people have that people whose public image is that they're 'good' are unquestionably good? That the TV 'News' never lies? That a person who writes a bk from the POV that they have their finger on the pulse of objectivity must, therefore, be right?

& what about Giordano Bruno? He seems to be largely forgotten yet he was held in a Catholic dungeon & tortured for 8 yrs until he was burned at the stake in public on February 17, 1600 - all b/c he had opinions & observations similar to those of Copernicus & Galileo. Galileo didn't die until January 8, 1642. Instead of being tortured & burnt at the stake he recanted & denounced his astronomical observations to save his ass - & it's Galileo who's remembered - not Bruno. For that matter, Copernicus was before both of them.

ALSO, I seriously doubt that there're even "a thousand" but even more certainly "ten thousand unknowns whose 'truths' never pass muster". It seems to me that Shermer's fictitious & dramatic figuring serves a purpose of discrediting free thinkers & doesn't really have any statistical backbone to it whatsoever. Then again, I haven't read his bk - maybe I shd.

"Another trap for journalists is the lure of the single study. After Andrew Wakefield published his paper in the *Lancet* claiming that MMR caused autism" - p 186

Ok, sure the press is going to grab onto the simplest most sensationalist thing. They're not as interested in the 'truth' as they are interested in getting advertising to support themselves by 'reporting' on whatever rivets the boobs to their tube. One of these simple sensationalist things is to reduce Wakefield to the-guy-who-says-MMR-causes-autism. I barely care about the subject & I'm already irritated by Offit's repetition of this oversimplification. To quote the same passage *again* from my review of <u>Callous Disregard</u>:

""*The data indicate that virological analysis of CSF is indicated in children undergoing autistic regression following **exposure** to live vaccine viruses.*

""The Paper's conclusions stop well short of any claim that the MMR vaccine causes autism. The most one can say from the findings of measles viral genetic material in CSF is that there is a strong statistical association between the presence of this virus and the autism group." - p 158"

- https://www.goodreads.com/review/show/4486376528

SO, you see?, Wakefield, a person constantly misquoted as saying that the MMR vaccine causes autism, denies that this is true. Offit doesn't actually quote Wakefield as saying that, he simply asserts it as if it's a known fact. Is that 'bad science'?

"That's because scientists, even excellent scientists working at pretigious institutions, often get it wrong" - p 186

It really seems like Offit doesn't even learn his own lessons. Scientists are NOT voices of the Gods passing on absolute truth to those under them in the hierarchy. Instead. they're people, usually specialists, who give their informed opinion - *but they don't know everything*, & there're belief systems underlying those opinions that're just as fallible as any other, so any intelligent person shd consider their opinion, if it seems relevant to a subject of importance to them, & then make their own decision about what fits their own perceptions the best.

"After Eric Fombonne's testimony, the only other strategy left to the petitioners' lawyers was to question how scientists know things. They never disputed the fact that at the time ten separate epidemiological studies had exonerated MMR or that five had exonerated thimerosal; rather, they disputed the reach of those studies. Scientists are only human, they reasoned, they can't know everything." - pp 190-191

"Certainly it is true that scientists can't know everything, that the scientific method has limits, and that epidemiologial studies cannot detect extremely rare events. But to use these truths as a basis to claim that MMR and thimerosal caused autism, to build an industry based on mercury-binding therapies or chemical castration, and to sue the federal government and pharmaceutical companies for the harm they have caused is an unjustified and dangeous leap." - p 191

Fair enuf - but to question any opinion, research-based or not, on the basis that it's not adequately matching one's own observations is also fair enuf. I'm not going to believe an opinion from any scientist, so-called 'good' or so-called 'bad', just b/c they present that opinion as if they're a mouthpiece for objective truth. I'd have to add that building an industry based on vaccinations & a host of other acccepted medical practices is just as objectionable to me as "mercury-binding therapies or chemical castration". Offit distinguishes between them b/c he considers himself to be on the side of 'good' science, I consider them to be entirely too similar - AND "unjustified and dangeous".

"During the vaccine-autism controversy, Joe Lieberman, then a Democratic senator from Massachusetts; Christopher Dodd, another Democratic senator from Connecticutt; and Robert F. Kennedy Jr., a member of the most famous Democratic political family in America, all warned of the danger of vaccines— warnings that appeared on Don Imus's national radio program and in full-page advertisements in the *New York Times* and *USA Today*. Why? Given the wealth of epidemiological studies clearly showing vaccines didn't cause autism, why did these politicians stand up and tell the press and the public they did? A cynical view would be that they were paid to do it. Many Democractic politicians receive healthy contributions from the Association of Trial Lawyers of America (now the American Association for Justice), one of the most powerful lobbies in Congress." - p 192

Oh.. C'MON! This bk was copyrighted in 2008. A mere 12 yrs later, in 2020, the

Democratic Party was all about every Draconian medical measure that they shoved down the public's throat - including vaccination.

"The pharmaceutical industry has by far the largest lobby in Washington—and that's saying something. In 2002 it employed 675 lobbyists (more than one for each member of Congress)—many drawn from 138 Washington lobbying firms— at a cost of over $91 million." - p 198, Marcia Angell, M.D.'s The Truth About the Drug Companies - How They Deceive Us and What to do About It

Offit conveniently leaves out the Big Pharma lobbying force. It's interesting, to me at least, that Robert F. Kennedy, Jr, is one of the few Democrats who's remained true to his position that vaccines shd be better tested & that they shdn't be pushed thru w/ Emergency Authorizations.

However, I shd mention that Offit doesn't leave his argument where my quote from him ended. He points out that the Republicans also supported the notion that vaccines cause autism. He explains this rare bipartisan solidarity thusly:

"The more likely explanation for politicians' involvement in the autism debate is that they have been responding to their constituents—us."

[..]

"Their scaremongering has only encouraged some parents to subject their autistic children to potentially harmful therapies or to withhold vaccines that might save their lives." - p 193

& the scaremongering goes on & on to this day, except that now it's over a 'pandemic' the existence of wch I find to be so ridiculously exaggerated that it wd be insane if it didn't so obviously serve business interests. & this scaremongering originates from every propaganda source there is: the politicians, certainly, the mass media, certainly, but also the Medical Industry that Offit considers to represent 'good' science. How many thousands of people have encouraged their reluctant relatives to get vaccinated to 'save their lives' only to have them die shortly thereafter from 'rare but possible' side-effects?

"But Wakefield and Geier failed to recognize that science isn't about faith; it's about data." - p 194

Yeah, faith in data.

"The Church banned Galileo's offending book, forbade publication of his future works, and ordered him imprisoned for the rest of his life. But Galileo knew he was right; as he was led away from his Roman inquisitors, he muttered, referring to the earth: "Eppur si muove" (And yet it moves)."

[..]

"Later, when Brent Taylor presented his data in front of Burton's congressional committee, Burton denounced him in much the same way as the Church had denounced Galileo." - p 197

Gee, that's not a heavy-handed dramatic hero-creating comparsion or anything is it? A scientist presents data in Congress that goes against the beliefs of the congressman heading the committee. Did Burton have the scientist hauled away in irons to a dungeon where he might be tortured for those opinions? Did the scientist face being burnt at the stake?! Nope. He might've even gotten a nice meal out of it or some sort of honorarium. So why all the drama?! To make the scientist seem like a daring hero-in-search-of-truth instead of just another schlep.

"In a culture dominated by cynicism and hungry for scandal, many people believe that doctors, scientists, and public health officials cater to a pharmaceutical industry willing to do anything including promote dangerous vaccines—for profit." - p 198

I'm not sure that I'm a cynic, more of an observant social critic, & I'm definitely not "hungry for scandal" &, yeah, I believe people who get wealthy get so by doing things that aren't very ethical. I also believe that Big Pharma uses its enormous wealth to make things go the way it wants to so that it can make even more enormous wealth. As far as vaccines go: to me it's not so much a matter of Big Pharma *deliberately* making a dangerous vaccine - it seems more reasonable to think that Big Pharma spends money on research & production & expects a 'healthy' return on its investment. If something seems to threaten that then a 'healthy' dose of self-delusion goes a long way to justifying pushing the product anyway - by hook or by crook. If at the bottom of this there's an 'axiom' that *vaccines are good* then that self-delusion is easier to maintain w/o any cracks in it. To me, Offit exemplifies this - he can admit to all sorts of errors, including fatal ones, in medical procedure - but as long as he has his UNSHAKEABLE FAITH in vaccines, everything's ok.

"And it's not just the unseemliness of promoting lifestyle products that hurts pharmaceutical companies; some marketing practices have clearly evolved from aggressive to unethical. As a consequence, we don't trust pharmaceutical companies. Nor do we trust the doctors or scientists who work with them. Kenneth Rothman, an epidemiologist from Boston University, calls this "the new McCarthyism."" - p 199

What exactly is this "new McCarthyism"? A distrust of pharmaceutical companies & the doctors & scientists who work for them?! That's such a ridiculous & idiotic statement that it's almost insufferable. Just as Offit's comparing the negative reaction that Brent Taylor got from Dan Burton when testifying to Congress to Galileo's experience was a completely over-the-top drama queen move meant to make Taylor out to be some sort of near-legendary hero, so is this "new McCarthyism" bullshit. During the time of the House Un-American Activities Committee (HUAC), people were being blackballed so they cdn't make a living & they were being sent to prison *for their ideas*. These doctors & scientists who work for pharmaceutical companies are making megabucks, they're not going to prison - once again, Offit is trying to make them out as persecuted heros. They're not, they're just people doing a job that pays very well b/c there're huge profits involved. If someone calls attn to unethical aspects of this job from time-to-time they're not losing their job & being put in prison - life just goes on as business as usual for them.

"For example, when Andrew Wakefield published his study of autistic children in the *Lancet*, he should have acknowledged that he had previously received money from Richard Barr and that Barr represented some of these children in a lawsuit against pharmaceutical companies. The irony in Andrew Wakefield's case was that not only did he fail to inform the *Lancet*'s readership of his funding source, but he failed to inform his coinvestigators, most of whom later withdrew their names from his paper." - p 200

Wakefield more or less refutes all of that. Here's more from my review of Wakefield's Callous Disregard:

"On to the 1st paragraph of chapter 11, "Disclosure":

""I have been accused and ultimately found guilty of professional misconduct for not disclosing in *The Lancet* paper that I was a medical expert involved in assessing the merits of litigation against the manufacturers of MMR on behalf of plaintiff children possibly damaged by the vaccine. Notwithstanding the fact that – long before publication – details of my involvement as an expert in the litigation had been provided to my senior coauthors, the dean of the medical school, and the editor of *The Lancet*, it is a matter of fact that it was not disclosed in the publisher paper." - p 169

"Some, or much, of this bk borders on legalese - in other words, technical fine points that are potentially valid but still have the stink of lawyer-manipulation-speech about them. This essentially applies to both the charges against Wakefield, wch are all-too-easy to write off as an attempt to discredit something that might lead to exposure of both Big Pharma & the UK's medical establishment as not living up to their responsibilities, AND to Wakefield's defense of himself. My inclination is to favor Wakefield even when I find some of the logic tricky. Wakefield argues that the disclosure rules of *The Lancet* did not require the particular type of disclosure that he's accused of not presenting at the time of publication. These rules than changed to be stricter in the yrs that followed.

""What have been the practical consequences for this move to stricter disclosure requirements from 1998 to 2007 for *The Lancet*?"

"[..]

""With the stricter rules in place, between 1998 and 2007 the rate of disclosures per *Lancet* article went from one in two hundred to more than one in two articles."

- p 170

"OK, that's obviously a huge & very significant difference."

- https://www.goodreads.com/review/show/4486376528

As for the coinvestigators withdrawing their names from the *Lancet* paper? It seems likely to me that they knew wch way the wind was blowing & were saving their careers - that doesn't mean that they didn't originally support the findings of the paper. After all, contrary to Offit's depiction of scientists as Galileo-like heros undergoing McCarthy Era style persecution, most scientists are just people out to make a living, not heros.

"In the end, it doesn't matter who funds a scientific study. It could be funded by pharmaceutical companies, the federal government, personal-injury lawyers, parent advocacy groups, or religious organizations. Good science will be reproduced by other investigators; bad science won't." - p 201

I find that statement to be myopically self-delusional. If there're 20 studies funded by one biased source (pick any of the above) there's going to be reproducibility - & that won't make the result 'good' science. If the 20 studies are funded by 20 different sources w/ conflicting agendas & *there's still reproducibility* then that makes the data much more believable. But if one side in a legal argument produces research data that reinforces their position & the opposing side does the same what does it prove? To me? Only that science can be skewed every wch way to support ulterior motives & unquestioned axioms (i.e. belief systems) such as the faith that Offit has in mainstream medicine.

"Other aspects of our culture also determine how people process scientific information. During the past few decades, doctors have started to treat patients differently. No longer do they always take a paternalistic, I-know-what's-best-for-you-so-don't-worry aprroach. Doctors are more apt to encourage patients to actively participate in their own medical care." - p 202

As long as they continue to buy the drugs.

"When Lyn Redwood and Sallie Bernard searched the medical literature for clues to the causes of autism, they were doing only what many doctors encourage parents to do: participate in the care of their children." - p 202

But, really, doctors have co-opted people's self-care. These days, many people talk about "their doctor", they go to "their doctor" fairly often. Business is good. If they were dedicating more time to their self-care, going to "their doctor" wd be something they'd do very rarely. Instead, they go to "their doctor" & the doctor prescribes them pills b/c they're overweight &/or depressed &/or whatever. They shd be taking care of these problems on their own but they abdicate their responsibility & pay for the doctor to be responsible for them. But, then, surprise, surprise (NOT) the pills have side-effects & they need to take new pills to counterbalance the side-effects of the old ones & a vicious circle just goes on & on.. & the doctor profits while the patient wallows in their lack of self-responsibility.

"If they want to research thimerosal, they should read the hundred or so studies on mercury toxicity, as well as the eight epidemiological studies that examined whether thimerosal caused harm. This would take a lot of time."

[..]

"Instead they read other people's opinions about them on the Internet. Parents can't be blamed for not reading the original studies; doctors don't read most of them either. And frankly, few doctors have the expertise necessary to fully understand them, so they rely on experts who collectively have the expertise." - p 203

Such a system leaves plenty of room for error. Rather than study the epidemiological reports most people wd probably trust 1st-hand testimonials more. I was told by a doctor that it was "impossible" for me to have reactive arthritis as response to steroids. Still, stopping the steroids solved my problem. When I tell that story to friends they recount similar experiences. I don't care what the epidemiological reports say, if I suffer from steroids & cease to suffer after I stop taking them then steroids aren't for me. If someone reads about my experience online they can take it into consideration, maybe they won't have a bad experience, maybe they will.

"During the past century, vaccines have helped to increase the life span of Americans by thirty years, and they have a remarkable record of safety." - p 203

Life spans *have* increased. Not everyone attributes that to vaccines. The people who die from them certainly don't have their life spans increased!

"Perhaps the greatest human accomplishment of the past century was the remarkable increase in life expectancy. In a century the world changed markedly from having almost no countries with life expectancy more than 50 years to having many countries with a life expectancy of 80 years as life expectancy almost doubled in the long-lived part of the world. As an example, in the United States life expectancy at birth over the 110 years from 1900 to 2010 went from 47.3 to 78.7 (Centers for Disease Control and Prevention/National Center for Health Statistics [CDC/NCHS], 2012, 2013). At first, this increasing length of life resulted from declines in infectious disease and deaths concentrated among the young. After most deaths from infectious conditions were eliminated, cardiovascular conditions and cancer dominated the causes of death. These then became the targets of science and medicine in the second half of the last century.

"Due in large part to declining mortality from heart disease, life expectancy continued to increase in the last decades of the 20th century. Because heart disease primarily causes death among older adults, recent increases in life expectancy have occurred at older ages. Life expectancy has increased all the way up the age range, certainly up to 100 years. For instance, life expectancy at ages 65 and 85 increased by about 50% over the century (Bell & Miller, 2005)."

- https://www.ncbi.nlm.nih.gov/pmc/articles/PMC4861644/

While "declines in infectious disease and deaths concentrated among the young" wd probably be attributed to vaccines by many, others attribute it to improved sanitation & nutrition. Greater longevity can also be attributed to *keeping people alive as long as possible* regardless of dramatically decreased quality of life. Do you want to be blind & deaf living in a nursing home pushed around in a wheelchair, kept alive by a regimen of pills? I don't. That's not really my idea of what being alive is. I prefer to have as high a quality of life as I can manage & then to die w/o medical intervention. Maybe I'll die in my 70s as a result, rather than dragging my life out to my 90s.

"Harkin had been influenced by fellow Iowan Berkeley Bedell, who was convinced that his Lyme disease had been cured by eating special whey from Lyme-infected cows." - p 205

Offit considers this to be aternative medicine. To me, it's no different from the basic principle of vaccination. The idea of using something connected to the cause of suffering to relieve that suffering is widespread. Take this example:

"Palauans have a remedy for the venomous sting of the rabbitfish, which could be of more general pharmacological use. They rub the raw internal organs (or sometimes just the gallbladder) of the fish on the wound and the pain subsides in just a few minutes. The fact that the sting never causes pain if a person is attacked only once suggests that the reaction is of an immunological nature." - p 74, Daniel Nettle & Suzanne Romaine's Vanishing Voices - The Extinction of the World's Languages

I find the above example to be a little different, though. Something that a creature gives off as poison isn't poisonous to that creature. Therefore, it has an internal antidotal chemistry that can be exploited as such. Ideally, one wd just manage to be mindful enuf of the rabbitfish to never rub it the wrong way.

"science is the only discipline that enables one to distinguish myth from fact" - p 207

That, in itself, seems like a myth to me. Who are the ultimate mediators of absolute truth? No-one.

"Although the scientific method has almost single-handedly brought us out of the Dark Ages and into the Age of Enlightenment, it can be difficult to explain how it works." - 208

More of the same mythifying. One friend of mine considers this to be the Technological Dark Ages. If the so-called 20th Century was a part of this so-called Age of Enlightenment then there sure was a plethora of enlightened scientific cleverness put into making life miserable & genocidal. I'm glad we're not living in an age when heretics such as myself were crucified on the outskirts of cities. We can thank the untempered dominance of religious fanaticism for those days. Alas, now science has become the new religion for many people, they don't have to believe in God (& I'm w/ them there) to support fanatical dictatorships, science can be the new justification for cruelty. Militarized police forces for suppressing popular uprisings can be very scientific. On the other hand, artists and musicians may've played as large or larger a part in creating the so-called Age of Enlightenment than scientists did - but since Offit's not one he myopically credits only scientists. I'll take Pietr Bruegel the elder over Edward Jenner anyday.

& where does Cotton Mather fit in? He's credited w/ being the 1st European-descended American to innoculate - at the same time he was a witch-persecutor. Was he Dark Ages or Age of Enlightenment? It seems to me that he personifies the type of arrogance & pompous sense of entitlement that scientists can exemplify, that scientists can turn into Dark Ages behavior.

"When Andrew Wakefield reported the stories of eight children with autism whose parents first noticed problems within one month of their children's receiving MMR, he was observing something that statistically had to happen. At the time, 90 percent of children in the United Kingdom were getting the vaccine, and one of every 2,000 was diagnosed with autism. Because MMR is given soon after a child's first birthday, when children first acquire language and communication skills, it was a statistical certainty that some children who got MMR would soon be diagnosed with autism." - pp 209-210

Fair enuf.

"Another aspect of our culture—and one reason the MMR and thimerosal controversies gained immediate attention—is that it's easy to scare people." - p 214

& in that case, it scared people away from vaccines. These days, it's scared people into going along w/ every oppressive Medical Industry nonsense shoved down their throats - particularly wearing masks for yrs on end & getting vaccination after vaccination. Most of the SHEEPLE who're terrorized into such conformity sincerely believe that if they don't go along w/ the program *they will die*. Somehow, seeing HERETICS such as myself not wearing masks (most of the time) & not getting vaccinated *not even getting sick* they can't even understand that the HERETICS are living proof of the inaccuracy of the scaremongering.

"As a consequence, people are more frightened by things that are less likely to hurt them. They are scared of pandemic flu but not epidemic flu (which kills more than 30,000 people a year in the United States)" - p 217

Funny, I'm not scared of the flu at all - & I don't think most people wd be if they weren't so susceptible to Medical Industry-induced panic.

"Researchers have shown that autism is genetic by studying twins. They found that when one identical twin had autism spectrum disorder, the risk to the second twin was greater than 90 percent; in contrast when one fraternal twin had autism, the risk to the second twin was less than 10 percent. Because identical twins share the same genes and fraternal twins don't, these studies proved that autism was in large part genetic." - p 218

I can see the logic of that. But what will that be used to justify? Last yr I read a bk by The President's Council on Bioethics called <u>Beyond Therapy - Biotechnology and the Pursuit of Happiness</u> & wrote a review of it. The bk is almost 2 decades old now but the authors did have considerable prescience about the possibilities, negative & positive, of genetic modification. Here's a sample of my review:

"Definitions of "biotechnology" are given in a footnote on p 1:

""These range from "engineering and biological study of relationships between human beings and machines" (*Webster's II New Riverside University Dictionary*, 1988), to "biological science when applied especially in genetic engineering and recombinant DNA technology" (*Mirriam-Webster OnLine Dictionary*, 2003), to "the use of biological processes to solve problems or make useful products" (Glossary provided by BIO, the Biotechnology Industry Organization, www.bio.org, 2003). In the broader sense of the term that we follow here, older technologies would include fermentation (used to bake bread and brew beer) and plant and animal hybridization. Newer biotechnologies would include, among others, processes to produce genetically engineered crops, to repair genetic defects using genomic knowledge, to develop new drugs based on knowledge of biochemistry or molecular biology, and to improve biological capacities using nanotechnology. They include also the products obtained by these processes: nucleic acids and proteins, drugs, genetically modified cells, tissues derived from stem cells, biomechanical devices, etc." - p 1

""In this sense, it appears as a most recent and vibrant expression of the technological spirit, a desire and disposition rationally to understand, order, predict, and (ultimately) control the events and workings of nature, all pursued for the sake of human benefit." - p 2

"This was the type of rhetoric that I was expecting, one untempered by acknowledgment of *things that can go wrong*. Note no mention of bioweaponry, e.g.. Note no mention of "control" as imprisonment. But it isn't long before the authors redeem themselves:

""Biotechnologies are already available as instruments of bioterrorism (for example, genetically engineered super-pathogens or drugs that can destroy the immune system or erase memory), as agents of social control (for example, tranquilizers for the unruly or fertility-blockers for the impoverished)" - p 6"

""We want better children—but not by turning procreation into manufacture or by altering their brains to gain them an edge over their peers. We want to perform better in the activities of life—but not by becoming mere creatures of our chemists or by turning ourselves into tools designed to win or achieve in inhuman ways. We want longer lives—but not at the cost of living carelessly or shallowly with diminished aspiration for living well, and not by becoming people so obsessed with our own longevity that we care little about the next generations. We want to be happy—but not because of a drug that gives us happy feelings without the real loves, attachment, and achievements that are essential for true human flourishing." - p xvii

"How many people have given much thought to the above considerations? Entirely too few IMO (In My Opinion) - including those who clamor for "the science". Yes, let's have "the science" - but let's not act as if all science inevitably generates change for the better."

- https://www.goodreads.com/review/show/4029224855

"Then researchers found it wasn't only drugs that could cause autism; viruses could do it too. If mothers were infected with rubella virus (German measles) early in pregnancy, their babies were at higher risk of autism. More than any other clues, thalidomide and rubella showed that environmental factors could influence the devlopment of autism." - p 219

Offit is calling drugs "environmental factors", I'd call them "human-introduced toxins". Vaccines fit into that category.

"Somewhere in 2000, I was on the Internet and found a description of hyperlexia, the very early ability to read." - p 222

That's me! I remember being in my mother's womb & writing a treatise on the cave paintings I found there.

"Roy Ranger Grinker got his doctorate from Harvard University and is now professor of anthropology and director of the Institute for Ethnographic Research at George Washington University. He's also the father of an autistic teenage daughter, Isabel."

[..]

"Grinker is excited by the explosion in autism research. Recently, he received a grant from the National Alliance for Autism Research to conduct the first-ever epidemiological study of autism in Korea. Like Clark, Seidel, and Hotez, Grinker doesn't believe vaccines cause autism" - p 231

& like Seidel, at least, since Grinker's on the 'right side' of the debate, Offit isn't bothered by Grinker's lack of relevant academic-scientific training.

Offit quotes actress Jenny McCarthy on the *Oprah* TV show:

"I do have a theory [based on] mommy instinct."

[..]

"What number does it take for people just to start listening to what the mothers of children with autism have been saying for years—which is that we vaccinated our babies and something happened. That's it." - pp 235-236

Many people might scoff at the idea of "mommy instinct". Offit is certainly negative about McCarthy & sees her subsequent appearances on other TV shows & a bk she wrote & toured w/ as big moneymakers. I, on the other hand, consider instinct & intuition to be very important. Instinct is a sense acquired from evolution, science is learning acquired from research. IMO they're both valid but when one is unsure about what to trust the most & one's instinct is very strong in a particular direction, I, at least, might go w/ the instinct.

The Holy Ceiling Light knows TV shows are going for ratings & not responsible advice so I can't really disagree w/ Offit's criticisms there - but Offit's not really 100% critical of the mass media, he just likes the people on the 'right side' of the debate sticking it to the media so that vaccines won't be criticized.

"Jenkins knew that seventy-four children had died of influenza in 2007 and more than 300 had died in the previous four years. She mentioned this statistic in her letter and continued, "ABC will bear responsibility for the needless suffering and potential deaths of children from parents' decision not to immunize based on the content of the [*Eli Stone*] episode."" - p 245

What does the CDC have to say about what Jenkins 'knew'?:

"How many people died from flu during the 2007-2008 season?

"Exact numbers of how many people died from flu this season cannot be determined. Flu-associated deaths (which have laboratory confirmed influenza), are only a nationally notifiable condition among children; however not all pediatric influenza deaths may be detected and reported and there is no requirement to report adult deaths from influenza. In addition, many people who die from flu complications are not tested, or they seek medical care later in their illness when flu can no longer be detected from respiratory samples. However, CDC tracks pneumonia and influenza (P&I) deaths through the 122 Cities Mortality Reporting System. This system collects information each week on the total number of death certificates filed in each of the 122 participating cities and the number of death certificates with pneumonia or influenza listed as a cause of death. The 122

Cities Mortality Reporting system helps gauge the severity of a flu season compared with other years. However, only a proportion of all P&I deaths are influenza-related and, as noted, most flu deaths are not lab confirmed. Thus, this system does not allow for an estimation of the number of deaths, only the relative severity among different influenza seasons. For the 2007-2008 season, the proportion of deaths due to pneumonia and influenza was higher than the previous two years, but was similar to the 2004-2005 season."

- https://www.cdc.gov/flu/pastseasons/0708season.htm

It's interesting that Jenkins 'knew' such exact figures while the CDC didn't. It's also interesting that the Stone episode aired on January 31, 2008, & that it's a *fictional* show. According to the CDC, as quoted above, the 2007-2008 season was of the same severity as the 2004-2005 season, 3 yrs before the Stone show. But who am I to argue w/ the statistics of one of the "False Prophets" that Offit supports? Or shd I call them "False Offits". My point is that such assertions as that people wdn't've died if a fictional TV show hadn't featured a certain opinion is really stretching the 'scientific method' beyond breaking, a bit like popping a bubble-gum balloon.

"["]You can expect that if any child were to become seriously ill or die from a lack of inoculation in the years following airing of this episode of *Eli Stone* . . . then lawyers like myself will hold ABC responsible for the damage the televsion show caused."" - p 246

I find that even more ludicrous than sonar depuration & facilliated communication for the severely autistic. People don't die from not being inoculated. People can die from a lack of air to breathe, it's called suffocation; people can die from a lack of food to eat, it's called starvation - but I haven't been inoculated in 54 yrs & I'm still alive: if a lack of inoculation were to cause death, I'd be dead & I'm not. The whole premise of the scaremongering behind pro-vaccination positions is that if people aren't vaccinated they'll die. This is a fantasy, it's an unproveable assertion based on statistical projections, *prophecies*, & the people Offit supports are just as much "False Prophets" as the ones he rejects.

"A growing body of evidence now points to the genes that are linked to autism; and despite the removal of thimerosal from vaccines in 2001, the number of children with autism continues to rise." - p 247

Given the way that people bandy about prophecy in relation to such matters, what's to stop someone from saying something like: '*If thimerosal hadn't been removed from vaccines the number of autistic people would have been four times what it was!*'? Is that sensational enuf for you?

For what it's worth (& I'm not claiming to be a world expert on such things, only a person w/ a somewhat informed opinion), I think the notion of the 'autism spectrum' does more harm than good b/c it lumps together people w/ low functionality & high functionality. SO, for the moment, when I refer to a person as "autistic" in this paragraph I mean people w/ low functionality. I think it's worth considering autism as a developmental retardation (to use old school language) that can be brought about by anything that lessens the mind's functionality. That can mean genes, that can mean toxins, it cd mean vaccines - what it doesn't mean is that every possible cause has to be present in every case or that every possible cause even has to have the same outcome.

To Autism's False Prophets's credit, there're endnotes starting on p 249 (although, as I noted earlier, I wd prefer that they were more specifically referential), there's a selected bibliography from 283-284, there're acknowledgements on p 285, & there's an index from 287-298. *SO*, scholars take note.

In the long run, I didn't find this bk completely useless - but I do wholeheartedly question what strikes me as its unacknowledged subtext, an arrogance that the worldview it represents is axiomatic for everyone & we just need to 'see the light of reason'. I think Offit & the people Offit commends are just as much "False Prophets" as the people he condemns. As long as the basis of his self-righteousness is unproveable assertions, predictions based on systems as open to weakness as any other, then the predictions uttered as if they're absolute fact will be prime targets for scepticism on my part, a scepticism too profound for me to feel like I'm on the same side as Offit.

review of
Solomon E. Asch's **A Study of Change in Mental Organization**
by tENTATIVELY, a cONVENIENCE - July 22, 2022

B/c I'm interested in the opinions of psychologists on conformity, I originally wanted to read Asch's <u>Social Psychology</u> or some other major work potentially relevant to the subject.

"While Solomon Asch left many lasting impacts on the field of psychology, his studies on conformity also known as Asch Paradigms are by far his most recognized achievement. The purpose of these experiments was to prove the significance of conformity in social settings. Many following researchers were heavily influenced by Asch's research and studies. Among these was Stanley Milgram who was supervised by Asch during his PhD at Harvard University.

"Also among his greatest achievements is Solomon Asch's textbook, Social Psychology (1952) which is an embodiment of his theories. More publications by Asch include, Effects of group pressure upon the modification and distortion of judgment (1951), Opinions and social pressure (1955), Studies of independence and conformity: A minority of one against a unanimous majority (1956) and Social psychology (1987)."

- https://www.famouspsychologists.org/solomon-asch/

Alas, I didn't find any of the bks I was interested in for an affordable price so, wanting to read *something* by him, I settled for this simple article that's been turned into a bk & bound. The bk, as an object, is fine - but it's a bit too vague & superficial otherwise to be of much use to me.

THE PROBLEM: "It was the purpose of the following investigation to study the influence of growth on the relationships between a number of mental functions. A considerable body of knowledge exists at the present time which suggests that mental development is characterized, not simply and solely by increases in amount of ability, but also, and equally importantly, by processes of integration and differentiation. Indeed the study of such changes is at the present time the main concern of students of foetal and newborn behavior." - p 5

GROWTH & DECLINE: "The individual spends by far the greater part of his life span either growing or declined; rarely does he stand still. Thus, the period of mental growth is generally regarded to extend from the beginning of the organism's existence to a period variously estimated from 14 to 25 years. Decline, on the other hand, is considered to set in almost immediately after the limit of mental development has been reached." - p 5

It's been my own observation that many people listen to the music that most excited them as teenagers throughout the rest of their lives. I went on a 'date' w/ a woman roughly my age (a scientist) & various musics played on the sound system in the bar we were in until Jimi Hendrix came on. My 'date' sd something like: 'Finally, some *real* music.' I was disappointed. I love Hendrix's music but there's been plenty of music since his that I love too - for me my growth as an open-minded person didn't stop when I got out of high school.

I doubt that many mentally active adults wd take kindly to such an early age for decline to be reputed to set in. 25?! I hope I didn't start to decline that early. I'm 68 & while I'm declining in many ways physically I'm not convinced that my so-called "limit of mental development has been reached". Given that I keep track of my accomplishments for every yr (or, at least, those of a certain type), it's easy for me to look at each yr & see whether I show signs of having developed mentally. It seems to me that in 1978, when I turned 25, I was getting started rather than finishing. In the interest of keeping this review 'short' (by my standards) I'll give ONE example:

My review of Paul A. Offit, M.D.'s <u>Autism's False Prophets - Bad Science, Risky Medicine, and the Search for a Cure</u> (http://idioideo.pleintekst.nl/ CriticOffits.html) is something that I cdn't've written in 1978. I had to accumulate decades of experience, knowledge, & 'wisdom' 1st. In fact, I don't think I cd've written it until NOW. Is that a sign of "mental development"? I think so.

QUESTIONING: "In spite of the enormous amount of evidence concerning the changes which continually occur in behavior, students of mental organization have assumed, almost without exception, that the *relationships* between mental performances remain constant during the life of the individual, and that changes accompanying growth and learning leave these relationships unchanged.

"It is this assumption of the constancy of mental organization that we are questioning and for the investigation of which the present experiment was conducted." - p 6

That seems promising. It's not really clear to me why there wd be an "assumption of the constancy of mental organization" so it seems like a good idea to question it.

AGES: "The age levels studied were nine and twelve." - p 6

That's fine. I imagine there's more of an emphasis on child developmental psychology than there is on elderly developmental psychology. That sd, I'd be interested in reading a study exploring the changes a person makes from ages 60 to 70. They're probably out there so the question is: *Will I look for them & read them? OOOOKKKKAAAAAYYYY*, I just bought <u>A Handbook of the Psychology of Aging</u> edited by K. Warner Schaie and James E. Birren. The release date was 1990. I shd've probably gotten something more recent. The introduction of things like laptops & 'smart'phones has changed everything, even the way an elderly person takes a shit. When I want to take a shit I just tell Siri & 'she' takes c/o it for me. It's very convenient.

UNCERTAINTY: "A group of 203 unselected boys was retested, with a set of mental measures, four times, at yearly intervals. The ages covered were 14 to 18. Tests of cancellation, substitution, sentence completion, and opposites were studied. A comparison of the intercorrelations between different tests from one year to another leads to inconclusive results. According to the author, "we cannot say whether or not tests as a whole become more closely related from one year to the next. So many irregularities seem to be present throughout the course of the four years, that we are at a loss in even attempting to generalize in the case of many individual pairs of test correlations."" - pp 10-11

As inconclusive & ambiguous as that is I find it somewhat reassuring that a psychologist cd acknowledge that life can get in the way, so to speak. Perhaps I take it for granted that the most common human tendency is to have an attitude like this: I WANT ANSWERS! If you're a psychologist whose work is going to be applied by a big business in order to increase sales thru behavior modification then those answers are probably forthcoming.. but what will the application of them reduce life to?

TEST SUBJECTS: "However, every child included in the present experiment had taken all the tests in 1932 and in 1935. The children were largely of the lower middle class; all were Jewish; and most of them were born in New York City and received their training there. The degree of homogeneity of the group is indicated by the following summary of its characteristics:

1. *Age.* The average age of the group at the time of the present investigation was 11 years and 11 months, with a standard deviation of 8 months. At the time of Dr. Schiller's investigation they were almost exactly three years younger." - p 13

It's an interesting aspect of averages that, to take the above example, it's possible that *no-one was actually of the average age.* I'm not saying that's the case, I'm just thinking about averages in general. E.G.: there cd be a group of 200 boys, 100 of them aged 10 & 100 of them aged 12. Their average age might be 11 but none of them *wd actually be 11.*

SAMPLE TEST: "2. The sentence completion test contained 75 items. Each item consisted of a sentence, of which one or more words were left blank. The subjects were instructed to write words in the blank spaces so as to make the sentences sensible. Time: 40 minutes." - p 14

I'd probably love taking a test like this. Just imagine:

Jimmy looked strange after taking a random selection of _____ from his mom's medicine cabinet, the next thing we knew he was puking bullets.

Aaaa BbBB ccc DdDpbq ... XX
YY_____?

Professor Jones admitted he'd never _____ any of the books that he assigned!

The reason why I'm here today is because I was juggling _____ when I slipped & fell.

This sentence can't be a palindrome

because_____.

DECREASED GENERAL ABILITY: "We therefore conclude that there has occurred, in our group, a restriction in the size of the factor of "general ability," between the ages of nine and twelve." - p 24

Unfortunately, I can't blame that on TV b/c this was written in 1936.

WHAT IF THE INCREASED DIFFICULTY WAS TOO MUCH?: "Concerning differences between the nine- and twelve-year tests, we may note that their form was kept unchanged, and that their content differed solely in respect to difficulty." - p 24

WHY?: "The results of the present investigation show that between the ages of 9 and 12 there has occurred in our group a significant reduction in the relationships between a number of intellectual functions. The factor analysis confirms this finding and further indicates that a considerable portion of the reduction has occurred in the factor of "general ability." Further than this, our results do not allow us to interpret. They furnish no clue to the conditions that might have caused the changes. Whether schooling is primarily responsible for the results, or whether the higher level of maturity may have contributed to the change, we cannot now say. The study of the conditions producing changes in the relations between performances should become the next logical step of investigation." - p 25

Indeed. I wonder, if maturity is a possible cause for reduced general ability does that mean that maturity = arrested development? I tried to hold off on becoming mature for a long time but now I'm so mature it's sickening.

CONCLUSIONS: "1. Relations between performances are lowest at the pre-linguistic level.

"2. There is an increase in organization with the development of symbolic speech.

"3. At a period of time preceeding the age of 12, there occurs a decrease in the intercorrelation between certain intellectual performances. Whether this decrease is universal; when it begins; whether reversals also occur in the opposite direction in different groups; and more importantly, what the causes of those changes are—all these are questions which cannot be answered at the present time." - p 25

Obviously, conclusions 1 & 2 were reached in different studies since the test subjects for this study were past the age range of pre-linguistics & had long since developed symbolic speech. At any rate, it looks like the age of decline starts before 12 rather than at 14.

LAST SUMMARY PARAGRAPH: "8. The conclusion is drawn that mental traits change and undergo reorganization as the individual develops. Current theories of mental organization have neglected the possibility of change. A dynamic and experimental approach to the study of mental organization is suggested." - p 29

If it's true that as of 1936, "Current theories of mental organization have neglected the possibility of change" then I'm glad Asch came along. However, I find that a bit hard to believe. Why, countless studies have shown that if you shoot a swimming person in the head, e.g., that their mental organization changes as a result & their general ability to swim decreases.

Since a decline in general ability has already been shown by age 12 then it seems reasonable to conclude that Asch & co were far more deteriorated than the children being tested. As such, it seems fair to let the children test the adults next time & have their conclusions be taken into consideration when making decisions.

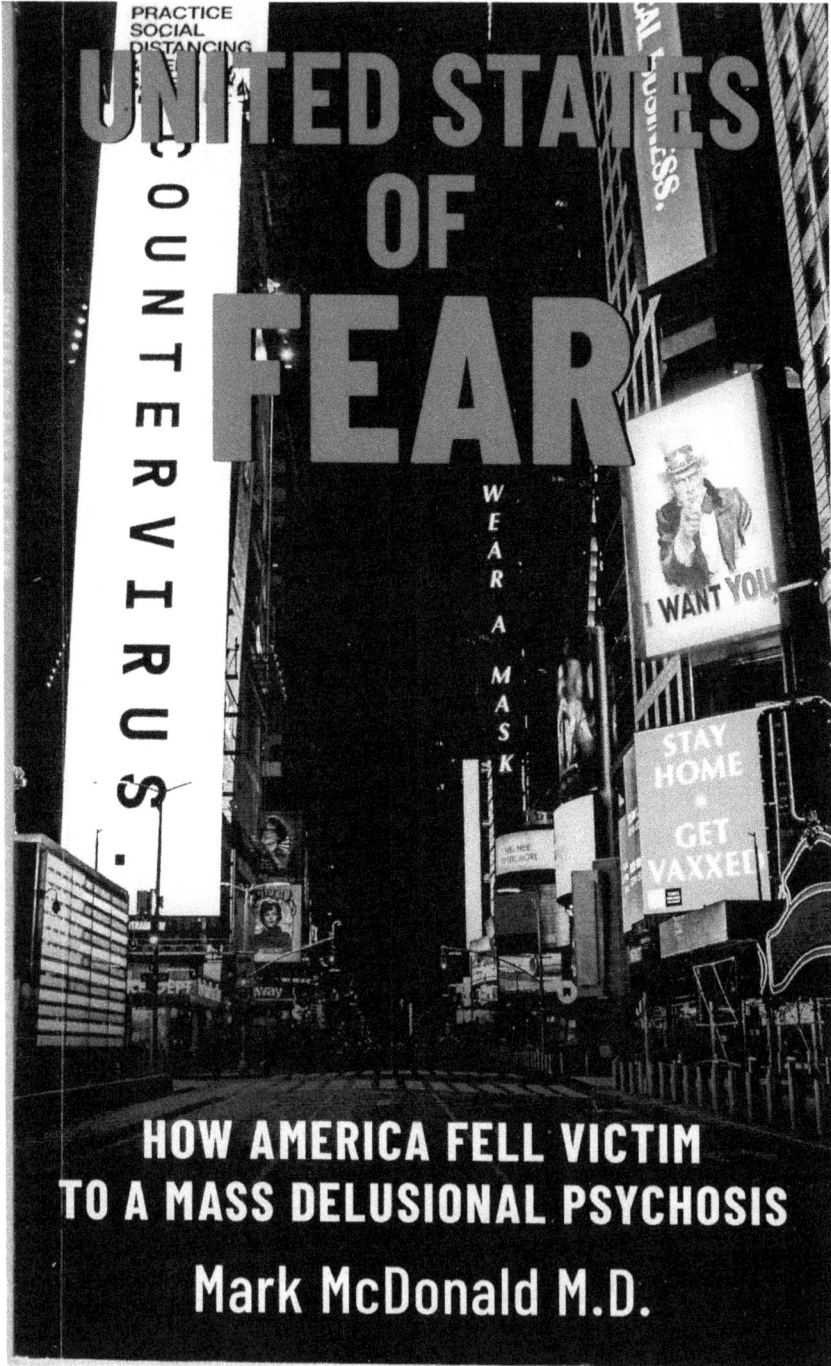

review of
Mark McDonald M.D.'s **United States of Fear**
by tENTATIVELY, a cONVENIENCE - July 24-27, 2022

I was raised in a conservative Republican Christian household. My parents hated each other's guts & separated by the time I was 9. I found them both to be completely delusional, self-serving, & in denial. I suspect that many of my peers had similar experiences & that that's partially why so many of us became radicalized. But everything changes all the time & what might've been more alert & aware in 1970 might've deteriorated into imbecility by 2020. W/ this in mind, I read this bk, as I have others recently, w/ an attitude of *not throwing out the baby with the bathwater*. In other words, while I have basic disagreements w/ it philosophically, I found other parts to be accurate & worth considering. In fact, I found it, at 1st, to keenly express my own feelings & observations.

"In the summer of 2020, the mother of a fifteen-year-old boy with ADHD declined to bring him to my office for his three-month medication follow-up visit. "We haven't been going out much recently,""

[..]

"After insisting that an in-person visit would be necessary for optimal care, I suggested she drop him off at the building instead. He was old enough to come upstairs by himself. This launched her into a hysterical rant about the dangers of her son walking through a public building and the health risk he would assume, not to mention the possibility of the family contracting a terrifying, highly contagious disease.

"Despite evidence to the contrary—that children are essentially immune to the Chinese Wuhan virus and even act as barriers to its spread—she insisted that I was uninformed, unreasonable, and cruel for demanding that her son come to my office in person to discuss his medications. A day later, I received another email, this one from her husband, apologizing on behalf of his wife, thanking me for helping care for their son all these years, but announcing that they would be seeking a new psychiatrist." - pp 1-2

This brief quote on its own opens multiple cans of worms for me. 1st, I question whether ADHD as a diagnosis isn't just another way of saying that a person has plenty of energy & very little focus. I don't think that's an illness & don't think it shd be medicated. I also think that medication adds a whole other set of problems. But that's really a different matter than what the story's about. *SO*:

2nd, I think that the mass media's fear-mongering has generated a fear of a pandemic that's just the usual thing, nothing to be afraid of. People grow & decline physically & mentally. Eventually, something comes along that pushes the physical decline over the edge into death. If you want to postpone that happening take better care of yourself: don't get fat, don't be addicted to anything. But you can't postpone it forever, eventually *you're going to die*. Get used to it, maybe it's just the way nature works & not really something to be afraid of.

3rd, That brings us to the mother who's apparently just another neurotic hypochondriac. Don't let people like this run yr life, they'll ruin it & make you sicker than anything they're exaggeratedly afraid of is ever likely to.

4th, I might as well mention that calling COVID-19 the "Chinese Wuhan virus" seems loaded on its own. I don't think viruses have nationalities. Whether COVID-19 originated in Wuhan, China, I don't know & I don't care. If it exists at all in the way that we're constantly being told it does doesn't really matter. *It's just another virus*, it'll do what viruses do (if they even exist) & that'll be that. It doesn't matter where it came from.

"When fear no longer protects from harm but simply inhibits one from living fully, it ceases to be helpful. When fear becomes the primary driver of decision-making, the quality of our decisions begins to suffer." - p 4

"In the extreme, when fear spreads throughout an entire society, the effect is paralyzing. Decision-making becomes irrational and reactive. The sensationalizing of outlier events and the pursuit of safety supplants sound public policy. Media begin covering the "fear story" and serving it to their audience on a regular basis." - p 4

& it's w/ statements like these that the good doctor & I are on the same page.

"Children and adolescents, for example, suffer more anxiety today than when I began my career. Although there are multiple reasons for this, the most important one, from my perspective as a clinician, is the rise of social media. Regardless of the country studied, the adolescent population shows a predisposition to internet and cell phone addiction, which strongly correlates with sleep deficit, anxiety, stress, and depression." - p 5

& that, too, seems reinforced by my own observations. The next bk that I'll be reviewing is Jaron Lanier's "**Ten Arguments for Deleting Your Social Media Accounts Right Now**" in wch I think he gives an excellent analysis of the way commercially-driven behavior modification affects social media users. But w/o even referring to Lanier's take on such things any further at the moment, I see social media as what I call the PEER PRESSURE COOKER. On the one hand, I think it's potentially wonderful that people can establish an international group of friends that they can share text & images & sound w/ so easily. On the other hand, these 'friendships' are superficial - if only b/c of the limits of technologically interfacing as opposed to actual in-person encounters - although I think the problem goes much deeper than that.

Humans, perhaps all living creatures, are herd or hive animals. It seems to me that conformist behavior is a sort of safety reflex - if one creaure runs from a possible danger, all of them do - but I doubt that any other creatures have developed a meta-awareness of this behavior & further developed a sense of how to manipulate it for reasons irrelevant to danger-avoidance. Some humans have: commercial interests know that if they change what's fashionable they can sell new clothes, commercial interests know that if they make people insecure about the quality of their appearance they can then sell them products like make-up & jewelry that're sold as making their user beautiful. The list goes on & on.. but what social media seems to do 'best' is offer people the 'security' of 'being on the right side': by conforming to mass opinions that seem to carry moral authority they're given the self-delusion that they, too, are on moral high ground. What the people thusly manipulated seem to miss is that being so thoughtlessly manipulable they can be pushed in any direction whatsoever as long as the delusion of moral high ground is maintained.

& here, I think, is where the anxiety & stress enters: the constant checking of social media accts on one's portable computer carries the danger that one might be 'on the wrong side', that one might be 'out of fashion', that one is somehow 'losing the competition'. A plunge off the precipice is always imminent & it's not always clear what 'has to be done' to prevent falling off into the abyss of being an unacceptable 'non-person'. Basically, the surest way of not falling into ignominy is to conform to whatever the majority opinions are, to not intentionally or accidentally express something that seems strange or unacceptable to one's immediate herd. This must be particularly hard on children & adolescents who're growing into socializing, just beginning to figure it all out. Adults have reached a somewhat rigid phase of socialization where their behaviors are more solidly in conformity w/ whatever social group they've developed into, the challenges are fewer b/c the fluidity is less.

"When the pandemic arrived in 2020, a new expression of fear emerged. Across the full patient spectrum, complaints of worry, insomnia, and drug cravings increased. Patients of mine who had been stable for months or years suddenly required medication dose changes. Former patients returned for therapy because they were struggling to cope. New patients nearly universally cited anxiety as their reason for seeking help." - p 6

& I've heard from an anxiety therapist friend of mine that GAD (Generalized Anxiety Disorder) has indeed been on the rise since the quarantine. What I object to in the above is even referring to the "pandemic" instead of the *lockdown* &/or the *quarantine* b/c, IMO, *it's the Draconian enforced restrictions on people's lives & the fear-mongering that's used to promote them that's the health problem.* The so-called "pandemic", if there even is one, is of considerably lesser importance. It's no surprise that weak & impressionable people wd get bogged down in anxiety when every day they're barraged w/ nonsense about their being beseiged by something *that's going to kill them if they aren't extremely vigilant.* But, don't misunderstand, the author, McDonald, & I are basically in agreement, I'm only being nitpicky about the use of the word "pandemic".

"In April 2020, the director of the Didi Hirsch Suicide Prevention Center announced the hotline had received 1,800 calls in March compared to only twenty in February." - p 6

& if suicide hotlines were really interested in preventing suicide they'd've immediately launched major lawsuits against the fear-mongering mass media - *b/*

c they're to blame - or, at least, made prominent public statements showing what bad affect all this pandemic publicity was causing. But, of course, these suicide prevention hotlines are part & parcel of the overall problem, not a solution to it.

"I noticed in my regular walks around the neighborhood that few people would greet or even acknowledge me. Many would avoid me entirely by abandoning the sidewalk in preference for using the street for their afternoon strolls. I never wore a mask outdoors, and I am certain this terrified them." - p 7

I remember walking thru my neighborhood early on (w/o a mask) when the stay-in-yr-homes mindset was strong & not seeing a single other person out on the streets. Even tho I was never paranoid enuf to think it was following me, for about 10 to 15 minutes of my walk a helicopter was flying overhead. That helicopter certainly added to the tension but I continued to walk to a park where I probably sat & read for awhile before returning home. There was a time when maybe 95% of the people I saw taking walks in the park wore masks even tho the most overkill health announcements didn't generally call for them when one was walking by oneself outside w/ no-one close around. The fear was preposterously exaggerated & completely unrealistic.

"Rather than explain the scientific basis for their decisions, politicians and unelected bureaucrats simply repeated empty slogans such as "Better safe than sorry" and "We're all in this together." In April 2020, New York Governor Andrew Cuomo famously said, in defending his statewide lockdown policy, "If it saves just one life, I'll be happy."

"This absurd fallacy became the basis for many destructive policies to come , most of which accomplished little or nothing and, in fact, actually cost many American lives. In 2020, nearly 40,000 Americans died in traffic accidents according to a National Highway Traffic Safety Administration report. Nearly all those lives could have been saved if we had reduced the speed limit to fifteen miles per hour. Of course, we would never do that because it would incapacitate the country." - pp 7-8

[..]

"In other words, there would be a cost. But the question of cost never came up when government officials chose to impose lockdowns." - p 8

& I agree w/ the author 100%. Imagine if we decided that the way to end crime was to strap everyone to a bed in a locked room & leave them there. They'd die but that sure wd lower the crime rate dramatically - except for the crime of confining them in the 1st place - &, to my mind, the imposition of the quarantine *is a crime*.

"Dr. Simone Gold, who went on to found America's Frontline Doctors, called me to ask if I would be interested in participating in a doctors' summit to bring attention to government and institutional mismanagement of the pandemic." - p

10

& these doctors were marginalized & ridiculed & censored but still managed to be widely listened-to anyway. I have a bk of Gold's that I intend to read & review eventually.

"One month later, I found myself standing alongside Dr. Gold and a dozen other physicians on the steps of the Supreme Court in Washington, D.C., speaking to a live audience of millions on the viral pandemic and the growing catastophe of the government response to it." - p 12

"Later that night, the original recording and all re-posts were taken down by Facebook, YouTube and Twitter. As this was occurring, President Trump and his son tweeted links to our talk. The tweets were taken down, and Don Jr.'s account was suspended." - p 13

& b/c Trump supported the Frontline Doctors they were more easily stigmatized as 'right-wing' - but since when is pointing out fear-mongering & manipulativeness by the government & mass media 'right-wing'?! If the so-called 'left-wing' were doing its job of resisting the police state it wd've been out there supporting them.

"Much of the nation had disconnected from basic reality. I began to call this condition a "mass delusional psychosis."" - p 14

& that's a description that I find wholly appropriate.

"People who feel a lack of agency over their own safety often resort to aggressive attempts to control those around them. Given that, the appearance of the "Karen" phenomenon in American society was not all that surprising. Countless videos were uploaded to social media showing angry, hysterical women screaming at others for not wearing a mask, often chasing them and even physically attacking them." - p 21

"There is a reason why the name "Karen" was chosen to describe people who accost strangers and shame them for "unsafe" behaviors such as standing too close or not wearing a mask outdoors: These people are overwhelmingly women." - p 27

"Wherever there was a mandate, there was a Karen. Two men were assaulted by a young woman outdoors in Manhatten Beach in August 2020 when she saw them without masks and threw coffee in one man's face. A month earlier, a woman in San Diego emptied a can of mace onto a couple eating hotdogs— maskless—at a dog park." - p 77

I can add my own story to those above:

"The other day I was at a big park & I saw an old guy, masked, walking his little dog. He threw a little ball for the dog to chase. A woman, also masked, started running toward the old guy, screaming at him something to the effect of "What do you think you're doing?!" & flailing her arms as if she were going to beat him. The old guy cringed and the woman stormed away. Then the old guy went back to throwing the ball for his dog to chase. That was part of my inspiration to coin the word "quarantiniac"." - pp 251-252, "**Unconscious Suffocation - A Personal Journey Through the PANDEMIC PANIC**" (http://idioideo.pleintekst.nl/ Book2020.09PANDEMIC.html)

Note that I don't use the term "Karen", wch I was previously unfamiliar w/ before reading this bk. I suspect that the term is in more common use in 'conservative' political circles that I don't have much to do w/. Even tho my own "Karen" story features a woman as the attacker, I hesitate to generalize the phenomenon to being mostly women. After all, this one incident is probably the only one I personally witnessed. Hence, I'll stick w/ "quarantiniac" to mean a person driven insane by the quarantine.

"Meanwhile, with everyone stuck at home in front of a screen, media had a captive audience." - p 22

Another, IMO, completely valid & important point. & the media hammered out more ridiculous overkill sensationalist fear-mongering than I've ever witnessed before - all to keep as many people as possible addicted to this fake 'news' in order to up their advertising revenues. After all, people just *had to tune into the next imaginary thing that was going to kill them!*

"Nor was there any mention of outpatient treatment options unless they were being attacked as ineffective and dangerous.

"These attacks began in early 2020 with hydroxychloroquine, a fifty-year-old medication that had been used safely and effectively throughout the world to prevent malaria in infants, pregnant and breast-feeding women, and the elderly. Any positive news of hope from this treatment was ridiculed" - p 23

I 1st heard of hydroxychloroquine from my liberal neighbor who ridiculed it something along the lines of 'Trump has been telling people they can use something like household bleach for COVID. It gives them heart attacks!'. I had no idea what he was talking about. His story had obviously come from the mass media. Later, I investigated this more & reproduced some of an article on the subject in my PANDEMIC PANIC bk:

"How the World's top medical journals were cynically exploited by Big Pharma

"Elizabeth Woodworth

"SUMMARY

"A publishing scandal recently erupted around the use of the anti-malarial drug hydroxychloroquine (HCQ) to treat Covid 19. It is also known as quinine and chloroquine, and is on the WHO list of essential medicines.

"The bark of the South American quina-quina tree has been used to treat malaria for 400 years. Quinine, a generic drug costing pennies a dose, is available for purchase online. In rare cases it can cause dizziness and irregular heartbeat. In late May, 2020, The Lancet published a four-author study claiming that HCQ used in hospitals to treat Covid-19 had been shown conclusively to be a hazard for heart death. The data allegedly covered 96,000 patients in 671 hospitals on six continents.

"After the article had spent 13 days in the headlines, dogged by scientific objections, three of the authors retracted it on June 5."

[..]

"- https://off-guardian.org/2020/06/23/the-deadly-hydroxychloroquine-publishingscandal/"

- pp 602-604, "**Unconscious Suffocation - A Personal Journey Through the PANDEMIC PANIC**" (http://idioideo.pleintekst.nl/Book2020.09PANDEMIC.html)

I have no idea whether hydroxychloroquine is of any use in treating respiratory problems but I do know that I wd've never recognized it as Quinine, something that I'd heard of many times, from my liberal friend's description.

"The Brown University Department of Pediatrics has found that babies born after January 1, 2020 show an IQ point loss of twenty points, presumably caused by the deprivation of home confinement and universal mask-wearing that impeded normal brain development." - p 24

I have mixed feelings about that one: on the one hand I think depriving a child of facial expressions is a form of child abuse & I think that people have generally become much more stupid since the widespread acceptance of cutting off facial expressions & proper breathing, on the other hand a 20 point IQ loss is so extreme that it's hard for me to not see that as yet-another fear-mongering exaggeration (wch is what I hope it is).

All of those things that I more or less completely agree w/ are in the introduction. It's when I got to "Chapter 1: **The Terrorization of Women: A Brief Cultural History**" that I found myself more at odds w/ the author.

"Men and women differ in other ways, too. Evolutionary psychology has shown there to be significant differences between the personalities of men and women. Numerous studies reveal that women report higher levels of neuroticism, extroversion, agreeableness, and conscientiousness compared to men. Moreover, these differences become apparent early in life and persist into old age. This large body of research shows that male/female differences are not necessarily culturally dependent, nor are they developmentally mediated. These differences reveal themselves in my private practice, where anxiety disorders are far more common in my female patients than in my male patients. Phobic disorders, in particular, present almost exclusively in women." - p 30

Before we go any further, I shd explain that I'm a feminist - but I also need to explain what I mean by that since I'm sure that "feminism" means many things to many people. In 1970, on the small suburban street that I lived on that had been a rural area a mere 15 yrs before, I was the only guy in the neighborhood w/ long hair. I had a friend who lived a few doors down who was the only girl who wore what wd've then been called "army boots" - in other words, boots not intended for women. As such, we both deviated from the normal gender roles we grew up w/. This wasn't rc'vd kindly by many people in my area, it was common for people to shout insults at me from their passing cars, such gems of wit as "Are yoooouuu a buoy or a guuurrrllll!" It always astounded me that these people were unable to tell the difference between the sexes. Anyway, my friend & I bonded over not wanting to conform to the established & highly oppressive gender role models. I had no desire to play sports, e.g.. We both wanted to look & act the way *we wanted to look* instead of the way other people told us to look & act. I had 3 queer friends, the girl wd've then been called a "Tom Boy", the guys were very neat & clean. I think for all of us it was naturally understood that we were all ok just the way we were. For the girls, this might've been feminism. The point is, we all supported each other, *we were allies*, not enemies. & that support for the girls from me was my feminism - it wasn't about claiming that we were exactly the same, it was about saying that we all had the equal right to be ourselves w/o other people lording over us.

Now since that time I've spent large portions of my life as an anarchist political activist. Sometimes this has involved a fair amt of meetings of political activists discussing proposed actions or just general concerns. I remember 2 meetings in particular that're relevant here: 1 was a meeting about sexual relations between women & men in our community. There were complaints about a particular guy who people found annoying in his ways of flirting w/ women; there was at least one guy who put on a display of contrition b/c of a specific sexual incident. During this discussion all sexual relations were described in socio-political terms. I brought up biology as an important factor. One angry woman from out-of-town insisted that biology was socio-politically determined too. I disagreed, thinking that physicality was key. Since people didn't seem to be inclined to discuss biology as a physical thing I told them I'd drop the subject & I left the room. One woman came out w/ me & expressed her solidarity w/ me & we both had a laugh about the melodramatic displays that we'd just left.

Two was the 1st meeting of an anarchist book club. One very academic woman dominated the entire meeting womansplaining to everyone that those of us who feel comfortable talking volubly shd be mostly quiet so that shy people wd be able to relax & express themselves. Basically, this was her telling the men to shut up so that insecure young women wd have more room for talking. Since I'm very comfortable w/ talking I was one of the ones who stayed mostly quiet to not interfere w/ this process. This establishment of the protocols went on for 3 very tedious hrs w/o a single word ever being sd about bks, our ostensible subject for discussion. I, personally, felt that the meeting wd've been more successful if we'd just broken up into smaller groups, partying style, where those of us who aren't shy wd've drawn out those who were. Instead, the meeting was dominated by one very inflexible woman seemingly blissfully unaware of how much she personified that type of person she was essentially accusing the men of being. I never went to another bk club meeting, I'm not sure there were any but if there were they didn't last long - the woman managed to kill any life they wd've naturally had otherwise.

I bring up that background to try to establish a little about why I agree w/ some of the author's claims & disagree w/ others.

"One of the most prominent environmental activists today is an eighteen-year-old Swedish woman named Greta Thunberg. This poster child for environmental awareness and activism has since childhood suffered from multiple mental illnesses, including panic disorder, obsessive-compulsive disorder, and anorexia. She also claims to be autistic. Her famously impassioned speech before the United Nations in 2019 excoriating world leaders for their "betrayal" of young people over climate change revealed her to be emotionally unstable and irrational.

"Alexandria Ocasio Cortez (AOC), at age twenty-nine the youngest woman ever to serve in Congress, apparently agrees, as she promised in 2019 that "The world is gonna end in twelve years if we don't address climate change."" - pp 33-34

I haven't pd much attn to either of these 2 public figures. I imagine that Thunberg was propelled to prominence b/c her youth & articulateness make good grist for the mass media theater mill. Both obviously touch a nerve, making some people feel that they're going to address a dire apocalyptic situation in a more straight-forward manner than the old established fogies. I don't think the world is going to end in 2031 but I do think that humanity is degrading the environment in ways that are having long-term negative effects that'll only get harder & harder to fix. I'm at a point in my life, however, where apocalyptic predictions seem more like dramatic displays meant to appeal to histrionic states than they are accurate.

"Racially motivated hate crimes against blacks are declining, while religiously motivated hate crimes, primarily against Jews, have dramatically increased. Moreover, the perpetrators of anti-Jewish hate crimes are predominantly black, as are those of the recently sensationalized anti-Asian hate crimes in urban areas scattered throughout the United States." - p 35

I seriously doubt that there're reliable statistics that support this or just about any other claims about who kills who for what reasons. It seems to me that the author is more than a bit caught up in the 'conservative' 'news' sources that share his philosophical opinions. As for "hate crimes against blacks are declining"? I hope that's true but to most political activists, myself included, it's hard to not see the murders of black people by police as hate crimes & they seem plentiful.

As for anti-Jewish hate crimes being primarily perpetrated by blacks? Well, that's news to me, maybe I'd find statistics to the effect convincing but the most glaring example in my neck of the woods was the Tree Of Life Synagogue mass shooting: "The congregation, along with New Light Congregation and Congregation Dor Hadash, which also worshipped in the building, was attacked during Shabbat morning services on October 27, 2018. The perpetrator killed eleven people and wounded six, including several Holocaust survivors. It was the deadliest attack ever on the Jewish community in the United States." (https://en.wikipedia.org/wiki/Pittsburgh_synagogue_shooting)

That shooting was by a white male: "A lone suspect, identified as 46-year-old Robert Gregory Bowers, was shot multiple times by police and arrested at the scene. Bowers had earlier posted antisemitic comments against HIAS (formerly, Hebrew Immigrant Aid Society) on the online alt-tech social network Gab. Dor Hadash had participated in HIAS's National Refugee Shabbat the previous week. Referring to Central American migrant caravans and immigrants, Bowers posted a message on Gab in which he wrote that "HIAS likes to bring invaders in that kill our people. I can't sit by and watch my people get slaughtered. Screw your optics, I'm going in."" (https://en.wikipedia.org/wiki/Pittsburgh_synagogue_shooting) I don't know whether Bowers' weapons were acquired legally or not.

"A related bogeyman pushed by media to frighten Americans is the phenomenon of mass shootings. Guns in general have been vilified by Democrats and liberal activists for many years." - pp 35-36

Personally, I wish guns had never been invented - but it's too late now. Contrary to McDonald, I find mass shootings to be a problem worth being concerned about. According to an article in the Washington Post:

"There have already been more than 300 mass shootings this year in the United States, according to the Gun Violence Archive. The shooting at a Fourth of July parade in Highland Park, Ill. that left six people dead and dozens injured was one of fourteen mass shootings over the long weekend. There have been just over 100 since a rampage at an elementary school in Uvalde, Tex., left 19 children and two teachers dead on May 24.

"Mass shootings, where four or more people — not including the shooter — are injured or killed, have averaged more than one per day so far this year. Not a single week in 2022 has passed without at least four mass shootings."

- https://www.washingtonpost.com/nation/2022/06/02/mass-shootings-in-2022/

According to the New York Times:

"From 1966 to 2019, 77 percent of mass shooters obtained the weapons they used in their crimes through legal purchases, according to a comprehensive survey of law enforcement data, academic papers and news accounts compiled by the National Institute of Justice, the research wing of the Justice Department."

- https://www.nytimes.com/2022/05/16/us/politics/legal-gun-purchase-mass-shooting.html

McDonald disputes this saying: "despite all evidence to the contrary, legal gun owners are frequently blamed for crimes committed using guns when, in fact, it is illegally owned guns that are used in the commission of crimes in nearly all cases." - p 36

It seems to me that McDonald has shifted the focus here: he went straight from talking about mass shootings as a bogyeman into most crimes being committed using illegal guns - bypassing the issue of whether most guns used in mass shootings are legally purchased or not. Now, I admit to doing what I think most people do: trusting statistics that support what I think is probably true & distrusting the ones that conflict w/ that. As such, I don't necessarily trust either the Post or the Times but I tend to believe the articles quoted above. Maybe I shdn't. But McDonald's shifting the focus is a bit too transparent: I can easily believe that most crimes are committed using illegal guns - but mass shootings are in a different category from most crimes - they're not committed for financial gain, they're usually just committed out of some exaggerated hate for a group that the shooter usually perceives as an enemy & this perception is often nurtured by hate media.

So what's to be done about it? As far as I'm concerned, we're not going to get rid of the guns, this is a country of gun nuts. As w/ any revolution that seems worthwhile to me, the gradual discrediting of hate groups & of gun mania is going to be a very slllooooooowwwwww one & it's not going to be accomplished by censorship or by abolishing the right to bear arms but by an increase in common sense - something that I find almost totally absent currently so I can't say I'm very optimistic.

"Between 1950 and 2010, not a single mass shooting occurred in an area where general civilians are allowed to carry a gun." - p 36

Again, McDonald seems to be shifting into irrelevant territory: after all, *it's also illegal to be committing mass murder* so regardless of whether the shooters's weapons are legal or not a crime's being committed. People who're in favor of more gun control just want the weapons to be harder to get. My proposal, & I think it's quite a good one, is to make the person who sells the gun to the shooter partially responsible for the crime. That wd make the vetting process a bit more serious.

"Now comes a highly contagious, lethal virus let loose on the world from a lab in Wuhan, China, that has saturated every inch of America." - p 37

This is where McDonald's beliefs become more iffy to me. I don't think it's proven that the hypothetical virus originated in a lab, I don't have any problem believing that that's possible, I just don't think it's proven. I also have my doubts about the "highly contagious, lethal" part of it. If it were so "lethal" I think more people wd've died from it who didn't have pre-existing health conditions. To me it's like this: if you're drunk & standing on the edge of a precipice & standing on one foot & waving yr arms about recklessly & a wind comes along & you fall over the cliff you don't blame it on the wind, the wind is just part of a whole, the being drunk & standing on one foot by a precipice is really the important part. Furthermore, the attribution of its origin to a chemical/biological warfare lab tends to support believing that it's more lethal than any other natural respiratory ailment that we've already encountered & lived or died thru. This belief in exceptional human-engineered deadliness increases the fear factor, something that McDonald is hypothetically opposed to.

It's when McDonald comes to gender roles that he & I part ways the most:

"according to the Deseret News, "When it comes to marriage, women are more satisfied if they are half of a religious couple embracing traditional gender roles." In my own experience, the more religious the couple, the truer this is." - p 40

The reason why I began this review w/: "I was raised in a conservative Republican Christian household. My parents hated each other's guts & separated by the time I was 9." is b/c I was foreshadowing reaching this point. My mom was very religious but my dad was more of a conman & probably didn't care about religion as anything other than a cover story to hide the commission of crimes. SO, in that sense, maybe McDonald was right when he says: "the more religious the couple, the truer this is" b/c I don't really think that I can credit both my parents w/ being religious. Nonetheless, my point is that I don't think I cd ever believe any generalizations about under what conditions women are "more satisfied". Some people are happy, others aren't. If I thought that there really were a believable system for guaranteeing this happiness we might see more people adhering to it but I think human emotions are generally in a state of chaos. My conservative family was certainly never happy & my liberal friends haven't necessarily been that much better. They all have their ups & downs & their philosophies don't assure much of anything.

"My suggesting that they surrender even a portion of their autonomy to their husband to achieve greater security would be viewed, at best, as an odd recommendation, and, at worst, as an insult to their capacities as women to survive on their own." - p 41

Obviously, I'm only quoting small bits in order to avoid much longer, but more thoroughly explanatory, quotes. This was preceded by the author's mention of his professional women patients priding themselves on their independence. This chapter concludes w/:

"Originally a liberal movement aimed at providing more freedom and opportunity for women, feminism evolved into a more radical movement that peddled a negative view of men as tyrants and toxic abusers.

"It is to this issue we now turn to understand why during the pandemic American men failed so dismally in their duty to provide a sense of safety and security for the women in their lives with consequences that have affected every one of us." - p 41

Ah.. This is where I part ways almost completely w/ McDonald. There're certainly women who see men as "tyrants and toxic abusers" who will nonetheless use us to do whatever they can't do, probably even turning on the charm for as long as it takes to try to get free labor before cutting off the charm again & putting up their "toxic abusers" defenses. That, however, is neither here nor there to the protective instincts that some of us have, both men & women, & how those protective instincts manifest themselves.

For me, when a couple unites, regardless of the genders, there's some trust involved & that trust enables a letting down of one's guard. It involves each partner wishing the other one well & being naturally inclined to do things that protect that wellness. How strong that relationship is, how strong the well-wishing is, can be measured by how naturally the protection of the wellness comes. I don't think this is only a function of the male. & there're limits, regardless of whether we want there to be, regardless of romantic hero narratives. No-one can protect anyone from everything. Still, at our best, we do our best. McDonald's conservative traditional gender roles are exactly what my neighborhood friend & I were rejecting when we were teenagers & I reject them still. Feeling some strength in autonomy isn't such a bad idea since even in the best of couples the other person isn't going to be there forever. Codependence can be a pleasure but it can also be destructive.

Chapter 2 is called "**Dereliction of Duty: How Feminized American Men Failed Their Women**"

"In late 2020, a woman was overhea[r]d commenting on how she felt leaving her house every morning and seeing men, both alone and with their wives, walking on the street with masks covering their faces: "This does not make me feel safe. On the contrary, it scares me. If I were married to one of these men, what would he do the moment a real threat appears? He would throw me under the bus and run for the hills. I'd be left to fend for myself."" - p 43

The author sees this as symptomatic of the loss of masculinity & the loss of women's faith in emasculated men. I see it more as neurotic. Personally, I'm irritated when I see anyone wearing a mask since I think it's just something imposed on us for no good reason & I see the people who go along w/ it as weak & gullible. I wore them when I went into stores that required them b/c I was already aggravated enuf by the state-of-affairs & didn't want to get into an argument w/ a store employee while trying to shop for food. I stopped going to the stores w/ the most outrageous 'safety precautions'. I defied the rules as often as I cd w/o turning it into a war. I stopped wearing them altogether as soon as the rules relaxed. But the woman's complaint about men she doesn't know & isn't married to is entrenched in so many "What if?"s that it's more of an irrational fantasy than it is anything else.

"Whenever I talk about the emasculation of American men, I see heads start to nod. Without fail, at the end of every talk, a number of women come up to thank me for calling out one of their greatest concerns. These women struggle to put into words what they have been feeling strongly for many years—a loss of masculinity in the American man." - p 44

Obviously, he's preaching to the converted. I look at the author's photo on p 103. He has short hair & a trimmed beard. He's wearing a suit jacket & a light colored button-down shirt. He projects a typical male image. But, to me, this image is a fraud. As a person who's been flagrantly unusual looking my whole life I know all too well that as soon as one deviates from the typical male image one comes under attack. I've been threatened & insulted from people in passing cars & groups of 2 or more men (more often at least 3) more times than I care to remember. It takes courage to maintain one's identity in the face of vicious conformist people. It takes no guts at all to look the way McDonald does. Men like McDonald aren't 'masculine', they're hiding behind a group identity, their 'courage' is rooted in knowing that there're millions more like them, that they're part of a gang. The average drag queen has more guts.

"A woman I know who recently began a new relationship learned that the man knows how to use a gun and keeps one at home. "I hate guns," she told me, "but when he told me that, it made me feel safe."" - p 45

Again, that just seems neurotic to me. If people want to have guns that's their business but I don't see it as increasing anyone's safety. It just increases the possibilities for mayhem. If you have a gun next to yr bed & someone breaks into yr house & shoots you before you get to it then the gun's useless. So where does the paranoia stop? When you sleep shd you sleep w/ the gun in yr hand 'just in case'? Shd you always carry at least one gun on you at all times 'just in case'? Shd you carry 2 or 3 or more guns on you at all times 'just in case'? What if one of them gets knocked out of yr hand & you need another as a back-up? The fear is endless. Having a gun makes people feel more powerful - but so does cocaine. Even if I were an expert marksman & carried a gun at ready at all times I cd still be shot in the back or shot by a sniper or blown to bits by a bomb or gassed or whatever.

"In 2017, over 7 million American men aged twenty-five through fifty-four were living in their parents' basement, not working and not looking for work." - p 47

Frankly, I find that statistic hard to believe. I don't know a single man of any age living in his parents' basement & not working. I assume they exist.. but 7 million?! At the back of this bk there're references. They aren't exactly endnotes, they're organized by chapter & subject. The source for this statistic is apparently this:

"https://www.scientificamerican.com/article/failure-to-launch-syndrome/

"*Scientific American* article notes the rise of home-dwelling men age 25-54 who live in their parents' basements, not working, with no desire to work. One reason: Safety." - p 115

I looked at the article. It's very superficial & there's no source provided for the statstics. Here's the relevant quote:

"No matter the culture or the label, failure to launch cases are mostly, but not all, young men. Numbers indicate the problem is increasing. Indeed, in 2014, over seven million American men ages 25-54 were neither working nor looking for work, up 25% from 10 years prior. And while the stereotype of a basement-dwelling man-child evokes labels of "loser," "dropout," or other unflattering descriptors, the phenomenon is more complicated than simplistic labels might indicate."

Given that I 1st left home when I was 17 to start hitchhiking around North America, I cd hardly get out of my mom's house fast enuf - but I didn't managed to completely get out until I was 22. I have girlfriends who left when they were 12 & 14. We were all in a hurry to get out into the world & away from the all-too-familiar oppression of home. So, yeah, I admit to being generally repulsed by guys who live w/ their parents after they've become (or delayed becoming) adults. Still, if I knew one I'd probably see his POV quickly enuf so I'm not going to completely pass judgement. I did have a woman friend who was living in her mom's basement when she got into her 50s. Alas, she's dead now - I don't think living in the basement helped. She had, however, "launched" & rebounded back to her mom's.

"The result of these cultural changes has been catastrophic for the mental health of both men and women. Although part-time work has long been found to have a beneficial effect on the psychological well-being of married women, a 2018 US Census report showed that two-thirds of working mothers have been working fulltime." - p 50

& why is that? In the 1950s, when I was born, it was most common for the man to be working & supporting the whole family & for the woman to be staying home & taking care of the children & the household. But if one or the other became dissatisfied w/ that arrangement what then? Divorce was certainly a commonly taken option. Then, if the father turned out to be a 'deadbeat dad' & didn't pay alimony or child support, as was the case in my family, the woman was forced by economic necessity to try to support the family on her own. Many of these women, having been in conservative traditional gender relationships by preference, suddenly found themselves ill-prepared for trying to make a living. As such, the family existed in dire financial crisis until another marriage cd replace the old one & the stepfather cd fill in the lacking income.

In the meantime, the children saw what was happening & the girls decided to not be caught in the same trap so they got jobs & learned how to make a living. Even then, if they were believers in the conservative lifestyle that McDonald espouses, they might've gotten married to a 'good conservative Christian' expecting it not to turn out as bad as their parents' marriage - only to have the pattern repeat again when the father jumped ship, bored & frustrated by the constraints. Having seen this happen 1st-hand, I'm not in the least convinced by McDonald's POV.

Even when both parents are working it turns out that they don't make enuf money. Sure, in the 1950s, dads cd support their families on their own but, as soon as the moms started working too, capitalism just made sure to exploit people even further so that working people still found themselves in desperate straits. There's nothing like artificially-induced inflation not accompanied by adequate wage increases to make working insufficient to make ends meet.

"Clearly, most women are not happy with the contemporary redefining of traditional gender roles in male-female relationships. Many are afraid to speak up—intimidated into silence by their fear of being seen as unsupportive of feminist doctrine." - pp 50-51

It's not so clear to me that a generalization about "most women" can be supportable. It seems more likely to me that McDonald travels in conservative circles in wch, once again, he's preaching to the converted. On the other hand, I'm sure there are women who're "afraid to speak up—intimidated into silence by their fear of being seen as unsupportive of feminist doctrine" - esp in dominantly liberal circles such as academia. It seems to me that it might be 'academic suicide' to be a woman professor who professes non-feminist opinions - except for in conservative Christian colleges. It seems to me that "traditional gender roles" aren't necessarily good for either men or women - while women who feel oppressed by them might think that they're only beneficial to men, my own observation is that the pressure on the male to BE the patriarch is potentially unbearable.

"Media frequently report that one in four college students has been sexually assaulted—an alarming statistic. If it were true, it would be hard to imagine any parent allowing their daughter to attend college. Yet it is simply another lie invented to malign and intimidate men and frighten young women." - p 53

On pp 119-120 there're multiple links under the heading of "RAPE CULTURE". One of them is: "**https://www.aau.edu/newsroom/press-releases/aau-releases-2019-survey-sexual-assault-and-misconduct** Association of American Universities 2019 survey of sexual assault." so I looked at that one.

" 1   The overall rate of non-consensual sexual contact by physical force or inability to consent since a respondent enrolled as a student at their school was

13 percent, with the rates for women and transgender, genderqueer, and non-binary (TGQN) students being significantly higher than for men.

"   1   For the 21 schools that participated in both the 2015 and 2019 surveys, the rate of nonconsensual sexual contact by physical force or inability to consent increased from 2015 to 2019 by 3.0 percentage points (to 26.4 percent) for undergraduate women, 2.4 percentage points for graduate and professional women (to 10.8 percent), and 1.4 percentage points for undergraduate men (to 6.9 percent). The changes for TGQN students were not statistically significant (which were 23.1 percent in 2019 and 14.6 percent in 2019 for undergraduate and graduate/professional students, respectively)"

I find the above statement to be rather confusing. Oh, well. I have plenty of friends who attended university, I don't recall any of them recounting being raped except for one gay male friend who described himself being taken advantage of by one of the professors. In the rare instances when I have heard of rapes of people I don't know occurring it's been more or less exclusively frat boys raping sorority girls - making it seem that the more conservative traditional gender role models involve guys getting girls drunk &/or incapacitated by rufies & then gang-raping them. Maybe that's a misrepresentation.

What I think of is an ex-girlfriend of mine who attended university & anonymously posted a list of every male on campus as "Potential Rapists". She was definitely a feminist of a confrontational nature but, really, I had to laugh at what she did (& she laughed too) b/c it was so over-the-top as to be ludicrous. It was a calculated provocation, meant to pull people's chains, & I'm sure it worked - but calling someone a "potential rapist" just means they have a penis, it doesn't mean they're ever likely to actually BE a rapist.

"Authors Suzanne Venker and Phyllis Schlafly describe in *The Flipside of Feminism* exactly how women have become less happy as they have gained more freedom, more education, and more power." - p 54

Show me a time when people have been absolutely happy & satisfied b/c of the system they're living in & I'll show it to you in a mirror as propaganda. In other words, nah, I don't believe it. Some people are happy, others aren't; the conditions under wch this occurs are a bit of a crapshoot. Personally, I think I'd be happiest having a woman partner who doesn't feel oppressed by me & isn't oppressive *to* me, who's very knowledgable & curious, & who feels strong & confident. That, to me, is a loving attitude.

"The surrender of real courage by men inevitably produces fearful women, and fearful women channel their fear into controlling others.

"I encountered a consequence of this dysfunctional dynamic on a local level in early 2020, when the home-owner's association (HOA) in my neighborhood in Los Angeles closed the nearby park after receiving a report that several small children were seen rolling on the grass. To make the park "safe," the entire ground was sprayed with a disinfectant. The disinfecting of the grass was simply a pretext to close the park, however, because once the spraying had ended, the park remained closed for nearly an entire year." - p 55

Yes, I find that completely insane - but why blame it on emasculated men not keeping 'their' women under control? I'm friends w/ many women whose jobs involve taking c/o young children. At least 3 of them think that the whole quarantine is as insane as I do. One of them is an anarchist, 1 of them has become more sympathetic to conservative POVs after being liberal her whole life. 2 of them have been forced into following rules that they think of as reprehensible b/c they'll lose their jobs, wch they're dependent on, if they don't. In some cases, their male partners aren't 'weak' as much as they are BELIEVERS in the PANDEMIC. Hierarchy, & a lack of individual power, a very conservative thing, is what keeps them 'in their place.' I wish they had more power so that they'd be able to hold their own against the authorities that make the decisions at their work places.

"Chapter 3 **Fanning the Flames: The Role of Media and Government**

"In March 2020, newspapers and television news programs launched tracking boards, updated daily, that announced the reported cumulative death toll from the Wuhan virus. Early on, evidence suggested these numbers were wildly inflated, as even the CDC noted that only five percent of all deaths attributed to the virus had no additional causes listed on the death certificate. Yet they were reported on page one of printed news and in the opening segment of broadcast news every day for months, providing the false impression that healthy Americans were dying throughout the country in alarming numbers and that the viral pandemic had become the greatest current threat to public health.

"Motorcycle accidents, suicides, and even drug overdoses were all categorized as deaths caused by the virus, so long as the victim had a positive nasal swab before, and—in some cases—after death." - p 57

As soon as McDonald gets back to the subject of the fear-mongering surrounding the so-called pandemic & gets away from blaming it on non-traditional sex roles I find myself agreeing w/ him. The constant pushing of death statistics into the public's mind & the ridiculous inaccuracies of these statistics served no good purpose whatsover - *but it DID serve a purpose*: viz: to produce a terrified & cowering populace easily manipulated into anything stupid.

"Fear-porn," as it came to be called , served both the economic and political interests of its purveyors." - p 58

"Even with corrupt data collection, it became harder and harder to find enough dead people to fuel the ongoing fear pandemic, so the media pivoted and replaced daily death and hospitalization trackers with a new statistical category: "Case Numbers."" - pp 59-60

& here's my own comment from July 3, 2020, from my own bk:

"Could I be any more sick of this? We're entering a shutdown again. It's basically being blamed on young people who've been celebrating summer & being released from their cages. I'm 100% on their side. Here's an excerpt from a relevant article with my own comments interpolated:

"After a week of record-shattering COVID-19 case numbers,

"By which it's meant people testing positive. It DOESN'T MEAN THAT ANY OF THOSE PEOPLE ARE ACTUALLY SICK. That's the statistics manipulation that I'm most sick of. Is that a form of COVID-19 sickness too?"

- p 652, **Unconscious Suffocation - A Personal Journey Through the PANDEMIC PANIC** (http://idioideo.pleintekst.nl/Book2020.09PANDEMIC.html)

"In May 2020, the President of Tanzania famously reported having tested a goat and a pawpaw fruit. Both the goat and the pawpaw were positive. Neither developed symptoms." - p 60

That's almost too bad - I wd've enjoyed seeing a pawpaw fruit coughing w/ a runny nose. Still, it's good to know that pawpaws are a deadly carrier b/c they grow in SouthEastern PA. Obviously, that part of the state shd be nuked - just to be on the safe side. Can it be arranged for Governor Wolf to be present at the ceremony at Ground Zero?

"The phenomenon of social contagion is illustrated clearly in Abigail Shrier's book *Irreversible Damage*, which investigates the explosion of transgenderism among adolescent girls. What had always been an extremely rare event—a girl announcing she was actually a boy—has recently become a common occurrence in junior high and high schools throughout the US. In a permissive environment that encourages "affirmative" receptivity to such transitions, girls download and share videos on social media that legitimize transgenderism as the best explanation for common adolescent angst." - p 63

Oi Veh! Here's a subject that's a lose-lose for me no matter what I say. The 1st person I ever knew who was getting a so-called 'sex-change' was a guy I knew in Baltimore around 1977. He explained to me that he had a girlfriend who wanted to be a man & that he wanted to be a woman. To my mind, I didn't see why it wasn't adequate just to role-play. They cd dress in drag. She cd wear a strap-on dildo & fuck him in the ass. At least that way they'd keep their sensitive genitalia that they were born w/. Instead, he was going to have his penis cut off, he'd get a fake vagina & fake breasts. He wdn't be able to give milk, he wdn't be able to be impregnated. A man w/ his penis cut off, who has fake breasts, who dresses in drag *is not a woman*, he'll never be a woman. But, then, it was his business, not mine, so he shd do what he thought best.

I've worn dresses before, I did it for the fun of it, I did it b/c I wanted to wear something different, I had no desire to 'be a woman' & didn't think wearing a dress wd make me one. IMO there's no 'law of nature' that says that men have to dress a certain way or that women have to dress a certain way. When I was a boy, men, for the most part, had short hair, women, for the most part, had long hair. These were societal norms, both sexes cd grow their hair long or cut it short. To my mind, there was nothing intrinsically masculine or feminine about either. Hence, I grew my hair long starting in early 1968 at age 14. I endured an enormous amt of abuse as a result but I had the strength of character to not let assholes dominate me. To be brief, I don't think men have to be a particular way, we're still men - even if we're so-called effeminate, we're still men. Same goes for women: all the societally dictated norms of behavior are irrelevant.

Alas, these days it seems that the 'politically correct' position is taking a stance I wd've previously associated w/ the most oppressive conservatism - w/ a twist. Viz: Now if a man wears a dress *he's a woman*. That's it, all biological differences are irrelevant. Ok, that's not quite right, men can wear dresses w/o being instantly declared a woman. The point is that gender role models seem even more oppressive to me despite some things having opened up enormously.

Furthermore, as a person who's highly critical of the Medical Industry I see the widespread emphasis on gender restructuring via surgery to be a wet dream for sadistic mad scientists - under the guise of being 'understanding' of people's gender needs we have surgeons willing to do just about anything, as long as there's lots of money involved, that gives them the opportunity to maim people, to show their 'skill'. It was bad enuf when people were encouraged to believe they were ugly & then convinced to get dangerous tummy-tucks & face-lifts, now it's even worse. There's even the idea that vaginas can be ugly & shd get plastic surgery to make them 'beautiful'. Insanity. It seems to me that alotof very rich people get off on the idea of making people infertile so that the population growth will drop off in an approved direction. So, is it 'understanding' or is it an ulterior motive of weeding out genes?

"Some of my speeches were even removed by YouTube for "spreading misinformation about the pandemic"—exactly as though my ideas and arguments were the equivalent of a contagious disease. All of the data in my speeches were taken from either government publications or peer-reviewed medical journals. But that didn't matter because the findings conflicted with the official narrative." - p 66

I certainly can't disagree w/ him about that one. The PANDEMIC PANIC has been used as the ultimate excuse for censorship & YouTube has been a prime offender. Even if we give the people calling the shots credit for desperately trying to be socially responsible they're ultimately just being short-sighted fools. Open the door to censorship by oppressing opinions contrary to the monolithic narrative regarding the so-called pandemic & it'll make censoring ideas & opinions about *anything else* much easier to get away w/. Goodbye freedom of speech.

& it's not even a matter of people using or not using their brains - there're the algorithms. One algorithm sd that I cdn't monetize a movie of mine b/c I used a certain song in it & the copyright holder of that song wd want money. *There was no music in the movie at all - not even in the background!* SO, I appealed it, pointing out that the song wasn't there. I won the appeal but whether I won or not was completely determined by the claimant - if they'd insisted that the song was there, they wd've won. Reality didn't matter, their litigious position was all that counted - & this type of idiocy, this AU (Artificial Unintelligence) is everywhere, all the time.

"I argued that there is no scientifically based medical reason to ever place a mask on a child in school. In fact, I declared it child abuse." - p 67

AGAIN, I'm w/ him 100% here. The extent of the psychological damage to children, esp young children, by enforced mask wearing & social distancing & the lot is untolled but bound to be substantial. I honestly wd like to see the maniacs resposible for this inflicted suffering be put on trial - & I'm an anarchist so my anger is far beyond its usual bounds.

"FDA Commissioner Scott Gottlieb joined Pfizer's boad of directors in June 2019. Mark McClellan, FDA head from 2002-2004, sits on the board of Johnson & Johnson. Pfizer and Johnson & Johnson both received emergency use authorization from the FDA to bypass safety and efficiency testing in bringing its experimental vaccine products to market in early 2020." - pp 68-69

A quick check to verify one of the claims above yielded:

"Former FDA Commissioner, Dr. Mark B. McClellan, to Join Johnson & Johnson Board of Directors

"New Brunswick, N.J. (October 14, 2013) –Johnson & Johnson (NYSE: JNJ) announced today that Mark B. McClellan, M.D., Ph.D., Senior Fellow in Economic Studies, and Director of the Initiative on Value and Innovation in Health Care, Brookings Institution, will join the Board of Directors on October 15, 2013. Dr. McClellan will serve on the Regulatory, Compliance & Government Affairs Committee and the Science, Technology & Sustainability Committee of the Board.

"As former commissioner of the U.S. Food and Drug Administration (FDA) from 2002 to 2004, and as the former administrator of the Centers for Medicare & Medicaid Services for the U.S. Department of Health and Human Services from 2004 to 2006, Dr. McClellan has more than two decades of public service and academic research experience. From 2001 to 2002, he served as a member of the President's Council of Economic Advisers and senior director for health care policy at the White House. During President William J. Clinton's administration, Dr. McClellan held the position of deputy assistant secretary of the Treasury for economic policy."

- https://www.jnj.com/media-center/press-releases/former-fda-commissioner-dr-mark-b-mcclellan-to-join-johnson-johnson-board-of-directors

&, yes, it does seem like there might be some conflict of interest involved.

"Government health authorities initiated what I consider to be the single greatest act of harm in 2020: closing schools and businesses." - p 69

Understand: I'm quoting the things I agree w/ & those I disagree w/ for a reason: viz: that I think it's important to listen to other people's opinions thoroughly instead of dismissing them or accepting them *en masse* b/c of one reason or another that might not really validate or invalidate everything. People have deeply entrenched philosophical opinions that might skew some things one way while not really effecting others.

"The RAND Corporation reported in September 2021 that not only did lockdowns not save lives, "To the contrary, we find a positive association between shelter-in-place (SIP) policies and excess deaths." - p 70

AGAIN, I agree - from personal experience I know how harsh this time has been for me & I extrapolate from there to how harsh it must be for people worse off than me - either b/c of mental instability or b/c of other health problems that've gone unaddressed b/c of the way emphasis on COVID has dominated the health care systems. When people ask me why I think masks are bad for people's health the question amazes me: How can these people be so oblivious to having something strapped across their mouths?!

"Masks have served no one but those in positions of authority and power. They have only made America sicker—physically, emotionally, and psychologically. As early as April 2020, the CDC admitted that mask wearing offered essentially no benefit in preventing the spread of the virus. Their function has been purely symbolic, an emblem of fear, anxiety, and compliance. Even the *New England Journal of Medicine* concurred, describing them as nothing more than "talismans."" - pp 70-71

& here's info about a movie of mine mocking mask-use:

625. "**Wearing Masks Saves Lives**"
- 1080p
- 1:18
- shot & edited July 9-11, 2020 (Vision)
- on my onesownthoughts YouTube channel here: https://youtu.be/oQBsMuVgnmU
- on the Internet Archive here: https://archive.org/details/wearing-masks-save-lives

There's an age-restriction on it on YouTube that's ostensibly b/c it includes pictures of dead people. These same pictures are available elsewhere on the internet w/o age-restriction. It seems probable to me that the age-restriction is really to limit the influence of my political satire.

Alas, McDonald continues to harp on women as a root cause of this whole mess. Personally, I'd fault the government figures & conniving billionaires above all.

"In fact, the experimental Wuhan virus vaccines—made available without FDA approval under an Emergency Use Authorization—are associated with the deaths of more people in the first half of 2021 than all vaccines combined over the previous ten years.

"Yet the forced march toward the universal vaccination continues. The demand for government control over every aspect of our lives has increased a hundredfold over the past year and a half, driven largely by anxiety and fear. Women, in particular, have been willing agents in this movement because their traditional source of security—men—has largely disappeared." - p 72

& yet it's largely men in positions of authority who're encouraging this anxiety & fear so it seems more accurate to me that the women McDonald's referring to are doing what he wants them to - following orders from domineering males.

"In New Jersey, Atilis Gym was fined $1.2 million for refusing to shut down under the state's business closure mandate." - p 78

More destructive bullshit that only helps the rich who're trying to monopolize business as much as possible. In Pittsburgh, there's a similarly defiant restaurant called the "Crack'd Egg". I've eaten there. Not only has it been fined but it's been defamed as a racist place b/c the typical propaganda lie is that anyone who's against the lockdown must be a hateful right-winger. Strangely, the manager of the restaurant is a black guy.

"Countless videos emerged in 2020 from Trader Joe's stores, where small groups of maskless protestors descended on different locations in a display of solidarity against the mask mandate. Without fail, staff refused to ring them up." - p 78

Trader Joe's had been the supermarket that I shopped at. Unfortunately, they pushed the quarantine insanity further than anywhere else I'd ordinarily go to. Customers were not only required to wear masks but were only allowed in in small numbers. B/c of this, people had to wait outside in a line waiting to get in. Apparently, waiting outside in the rain & snow & cold was healthier than risking being around other people, w/ a mask on, in the store. I remember one particularly aggravating time when I was waiting to be let in when 2 people left. That meant there was 'room' for 2 new people so when the person in front of me started to go in so did I. But, Oh No!, the employee called that to a halt. Apparently the math was too complex for him. I made a movie there on May 4, 2020 & soon thereafter stopped going there at all. I didn't return until their mask mandate ended.

619. "<u>You'll Never See A Billionaire Standing In A Food Line</u>"
- clandestine car iJones camera: ProjectileObjects; all else: tENTATIVELY, a cONVENIENCE
- 4:55
- 1080p
- shot May 4, 2020 (Vision), edited finished May 5, 2020 (Vision)
- on my onesownthoughts YouTube channel here: <u>https://youtu.be/I_r8F-oMtYE</u>
- on the Internet Archive here: <u>https://archive.org/details/youll-never-see-a-billionaire</u>

"Even the airline industry followed suit, jumping ahead of the federal government in April 2020, and ordered all passengers into mask compliance. They later lobbied the CDC and FAA to make masks a federal transportation requirement. That requirement went into effect on February 1, 2021, has been extended twice, and shows no sign of ever going away." - pp 78-79

April 18, 2022:

"Federal officials stopped enforcement of a federal mask mandate Monday in transportation settings after a federal judge struck down the requirement, raising public health concerns and prompting several airlines to announce that face coverings are optional on domestic flights." - https://www.washingtonpost.com/transportation/2022/04/18/mask-mandate-transportation-airplanes/

Thank goodness, one piece of the overall insanity defeated.

"Chapter 4 **The Way Forward: Working Our Way Back to Sanity**"

[..]

"The path forward will be different for each person, but several recurring themes will almost certainly emerge.

"For men, it will require a display of courage. They will need to stand up to hysteria and refuse to allow it to drive decision-making. Self-emasculation must end.

"For women, it will require emotional restraint. They will need to confront, to the extent they have succumbed to it, the hyper-emotionality that has led them to respond to the current crisis in unhealthy ways." - p 81

It seems to me that I, personally, have displayed courage throughout this entire madness & that the result has been an almost complete hatred & shunning of me by both men & women. Here's a sample of my psychological approach to this dilemma:

665. "<u>Check out of the Mass-Formation Roach Motel</u>"
- 37:40

- 1920 X 1440, 30fps, Stereo
- on my onesownthoughts YouTube channel here: https://youtu.be/vWlyZ5vKdAY
- on the Internet Archive here: https://archive.org/details/roach-motel

Was it something I sd?

"Just as with my physician colleagues, my therapist colleagues have been a profound disappointment to me. Unable to overcome their own fear and irrationality, they have abandoned their profession—and their patients—and refused to offer their professional help in the only setting that truly works for most patients: face-to-face." - pp 84-85

I feel a similar disappointment w/ political activists: succumbing to the mass psychosis, they've immediately lost all trace of any ability to analyze & critique the control systems of this society - turning into mouthpieces & slaves for the Medical Industry Police State.

"Men are frequently told to control their physical and sexual aggression, but what about women? Are their natures equally flawed, or are they born perfect, with no need to control their natures? When women are allowed or even encouraged to develop hysteria and express uncensored, unrestrained hyper-emotionality, they can wreak havoc on society. Smothering, intrusive, nanny-state behavior can predominate, with an emphasis on a fear-driven obsession for safety and a disregard of the need for intrepid, risk-taking behaviors that display courage." - p 87

Alas, I wish I disagreed w/ this more than I do.. but it's been my frequent observation that anarchists, both men & women, refer to patriarchy as an ultimate evil while never criticizing matriarchy one whit. &, yet, "anarchy" means "an" (without) & "archy" (rule by). That includes matriarchy (rule by women or by the mother-figure) as much as patriarchy (rule be men or the father-figure). I find that women always expect to be considered the equal of men in every way *except when it comes to taking responsibility for something harmful*: that's usually blamed on men. That's a double standard. But, in my own experience, women are often extremely harmful - so being in denial of that is something that just adds to the harm. As for courage? I've known plenty of women who are very courageous.

"Men need to start pursuing being men again, stop apologizing, and push back against the fear-driven women—in reality a vocal minority of mostly white liberal affluent women—who are wrecking our world." - p 88

It's almost a relief to so vehemently disagree w/ him here after agreeing w/ some of his last quote. I don't think that being a man has to be any particular thing so I can't agree there. It's also more than a little difficult to blame women for "wrecking our world" when wars, to cite an obvious example, are primarily a creation of men - & men & women alike are creating the PANDEMIC PANIC - but I think men probably have even more responsibility for it at the administrative &

business levels.

"If no local group exists, then you must create one. Telegram and Signal are
excellent phone apps that can be used to organize people into a group and share
meeting updates. These two apps also have the added benefit of being largely
secure from hacking and government spying, inaccessible to the NSA." - p 89

Alas, such independence from control by government organizations doesn't
usually last very long. Consider this Reclaim the Net report from March 21,
2022:

""On Friday, Brazil's Supreme Court Justice Alexandre de Moraes ordered the
banning of Telegram because of its failure to comply with Brazil's laws and for
allowing the spread of "fake news."

"On Sunday, he revoked the suspension, saying that the company had complied
with censorship requests to ban accounts.

"After the suspension, Telegram's founder Pavel Durov apologized for his
company's "negligence" in complying with court orders.

""Over the past 24 hours, we've integrated technical means to flag specific posts
in one-to-many channels as potentially containing inaccurate information,"
Telegram announced.

""These notices can now be added to the end of any message on Telegram and
will also remain visible when those messages are forwarded from the channel to
private or group chats.

""To better identify these posts, we are establishing working relationships with
important fact-checking organizations in Brazil, such as Agência Lupa, Aos
Fatos, Boatos.org and others."

"the Los Angeles Times reported that thousands of first responders have
announced they are refusing the vaccine, and a group of Los Angeles police
officers had filed a lawsuit against the city, demanding that any officer who has
recovered from infection be made exempt from the vaccine mandate." - p 90

The sadness goes on & on.. The mere fact that these folks even have to go thru
the trouble of taking legal steps to resist this unjustifiable medical intrusion into
their lives is what's truly sickening.

"With the advent of Zoom school in 2020, parents became aware first-hand of the
focus not on the teaching of critical skills but rather the indoctrination pervading
both public and private schools. Sexual politics, critical race theory, and
revisionist history are now the norm throughout the country. Many teachers are
simply professional activists in disguise." - p 91

That, of course, is the conservative perception of the state of things. As for teachers being "professional activists in disguise"? That doesn't really seem accurate to me. I've been friends for many decades w/ activists & I'm one myself. Are any of us "professionals"? When I think of "professionals" I think of people who're highly pd - very few, or NO, activists I know are pd at all, ever. Instead, we dedicate our lives & our, often limited, funds to trying to create a more just society. As soon as money enters into that there's always the chance of corruption. Therefore, the idea of a "professional activist" is a sort of oxymoron to me. It's the people who're activists b/c they're passionately committed that I respect; people who're pd to pursue a particular political agenda might not do so if there weren't something in it for them. I'm sure there're plenty of teachers who sincerely try to do their best for their students - wch includes protecting the vulnerable - but I doubt that there're (m)any who're getting pd extra to bias their politics, therefore being "professional activists". Instead, they might misguidedly just push their own agendas w/o realizing that this might be uncalled-for & unethical.

As for "[s]exual politics, critical race theory, and revisionist history"? That's certainly a conservative perception. To address these in reverse order: What, exactly, is meant by "revisionist history" here? Usually, it refers to the claims of holocaust deniers that millions of Jews weren't systematically & genocidally murdered by nazis. I doubt that that's very widely taught institutionally anywhere. I don't think McDonald is referring to that. Instead, he's probably referring to liberal history or, in particular, feminist history that he might think of as inaccurate.

I can provide an example that I doubt that McDonald wd be aware of: Delia Derbyshire. Derbyshire was born in 1937 & worked for the BBC starting in November, 1960. In 1963 she made an electronic arrangement of Ron Grainer's theme song for the TV program *Doctor Who*. According to Wikipedia, "Grainer attempted to credit her as co-composer, but was prevented by the BBC bureaucracy because they preferred that members of the workshop remain anonymous." (https://en.wikipedia.org/wiki/Delia_Derbyshire) A recent documentary about her made a claim to the effect that 'she invented electronic music as we now know it.' Given that she was born in 1937, that's feminist revisionist history at its most despicable. In 1886, Thaddeus Cahill developed the teleharmonium, wch he patented a yr later, arguably the 1st electronic instrument. Even if we don't include the teleharmonium, the theremin was invented in October 1919 & patented in 1928 & the ondes martenot was invented in 1928 - both of them electronic instruments, both of them immediately composed for & both of them before Derbyshire was born.

As such, 'electronic music as we know it' must mean 'electronic music as people who don't know it pretend to know it'. Even if we were to ignore the theremin & the ondes martenot & the music composed for them, electronic music has a long history that predates the rather simple electronic version of the Doctor Who theme - all of this music being considerably more original & complex. So, yes, there IS revisionist feminist history that makes false claims for women just b/c they're women.

On the other hand, certainly there're women who've been neglected & uncredited b/c patriarchical attitudes undeservedly consigned them to inferior roles. Rosalind Elsie Franklin, e.g., probably deserved more credit in connection w/ the discovery of DNA.

Still, back to McDonald, I suspect that much of what he considers to be "revisionist history" is probably something I'd consider to represent history made in defiance of the history of the victor.

Then there's "critical race theory":

"**Critical race theory (CRT)** is a cross-disciplinary intellectual and social movement of civil-rights scholars and activists who seek to examine the intersection of race, society, and law in the United States and to challenge mainstream American liberal approaches to racial justice. The word *critical* in its name is an academic term that refers to critical thinking, critical theory, and scholarly criticism, rather than criticizing or blaming people. CRT is also used in sociology to explain social, political, and legal structures and power distribution through the lens of race. For example, the CRT conceptual framework is one way to study racial bias in laws and institutions, such as the how and why of incarceration rates and how sentencing differs among racial groups in the United States. It first arose in the 1970s, like other critical schools of thought, such as critical legal studies, which examines how legal rules protect the *status quo*." - https://en.wikipedia.org/wiki/Critical_race_theory

It seems to me that whenever I've run across the term recently it's been implied to be some sort of deranged perception of social realities. Really, I find it astounding that there're people who still act like racism is some sort of self-serving fantasy, mostly on the part of black people. I remember going into a restaurant, I think it was on my 43rd birthday, run by a Korean man. The clientele were all apparently upper-middle-class white people. I was carded, seemingly in the hope that I wdn't have an ID on me so that the owner cd have an excuse for denying service to a scruffy-looking person obviously of a 'lower class' than the desired patrons. The only other person I saw carded was a black woman, who looked to me to be in her 50s. She was the only black person in there & she was obviously not under 21. It seemed clear to me that the owner didn't want to serve us. When the black woman was carded she looked very flustered & embarrassed - she knew exactly what the msg was: *no black people welcome here.* That was 1996, not so long ago. *SO,* Critical Race Theory? I'm all for it - as long as it doesn't involve oversimplistic racist generalizations of any kind. People need to understand that it wasn't that long ago that Jim Crow laws existed in the United States, that black & white people were separated as if we were practically different species - instead of just fellow humans.

As for "[s]exual politics"? That's trickier for me. I've been informed by at least one friend working at a school that gender agendas are being pushed to very young children. To me, children shd be left alone to discover & nurture their own sexual identity w/o being pushed & prodded by adults - that includes not imposing a heterosexual narrative *as well as not imposing any other narrative.* Protecting kids from persecution for belonging to a sexual minority is one thing, promotion of a 'politically correct' narrative is quite another. 'Political correctness' is just another oppressive norm.

"In 2020, after a full year and a half of mandatory masking of children, many districts are moving toward mandatory vaccination as a condition of re-enrollment." - p 91

The reader of this review will've long since realized that I find that completely egregious. I think there're valid reasons for finding vaccination dangerous & that it's 1st & foremost the decision of the parent or guardian whether children under their care shd be vaxxed. I don't believe for a second that any school board, health department, or political entity shd have the right to override the decisions of parents or guardians. Furthermore, I don't think that vaccination even fulfills the grand preventative claims made for it. Many of my friends have been vaccinated, many of them have also gotten sick. Vaccination has never been proven to my satisfaction to be anything but an idiotic risk taken in the name of shaky theory at great profit to the vaccine industry.

"No matter how emotionally healthy a child's parents are, it is not possible to protect any child from pathologic fear and anxiety in this sick and abusive environment. It's time for parents to remove children from the existing school system.

"Many are now doing exactly that. Homeschooling is taking off in the United States. Between just 2020 and 2021, the number of homeschooled children nearly doubled from three million to over five million." - p 92

It seems to me that Christian families have opted for homeschooling more than most b/c of an objection to the separation of church & state. Christians don't want their children exposed to atheistic opinions. Even tho I'm an atheist, I can be sympathetic to that b/c I can understand parents wanting to preserve their values in their children. I went to public schools from 1959 to 1971. I hated them. Nonetheless, I don't think they did me any lasting harm & I'm sure I learned many useful things. I don't have any children but if I did I'm sure I wdn't want them in schools that require masks & social distancing b/c I think that the psychological harm done to a person in formative yrs by those things is a very serious matter. Nonetheless, I think that homeschooling carries a different danger, the danger of making the child's environment too claustrophobic. One thing that I think was healthy for me about public schools was simply that I was in a social environment where I got to make friends & to learn about people other than myself & my immediate family & neighbors.

So what's the alternative to schools where the administrative policies are harmful to the students? Perhaps a communal homeschooling where parents share responsibility for teaching & child-care. There wd have to be agreements among the parents to not proselytize. As an anarchist I think I cd agree to not teach anarchism as long as other parents wd agree to not preach Christianity, e.g.. It wd be only fair to try to make the education provided be acceptably neutral to all the parents. Alas, creating such an alternative system wd be quite a burden on parents who might already be overburdened just trying to support their families.

"The most high-profile case in recent years occurred in 2017 at Evergreen College when biology professor Bret Weinstein was forced to resign after challenging a "no whites day" as racist." - p 93

I'm not familiar w/ this case. The idea of a "no whites day" is clearly racist to me regardless of what justification for it is provided so there's no way I'd comply w/ it either. Then again, I wd've never been hired in the 1st place. I hope that some non-whites objected to the "no whites day" as much as some whites did. If they didn't, then the state of anti-racism is pathetic.

"All of the major medical journals, including JAMA, the Lancet, and the NEJM, have disgraced themselves by publishing fraudulent articles to support the pharmaceutical industry and political interests. They can no longer be trusted as reliable sources of information any more than the *New York Times* or CNN." - p 96

I don't mind aspersions being cast on any of the above but I think that excluding Fox 'News' is an unacceptable omission. Furthermore, a bk that roundly disgraces the pharmaceutical industry is Marcia Angell, M.D.'s "**The Truth About the Drug Companies - How They Deceive Us and What to do About It**" & she was the chief editor of NEJM (*New England Journal of Medicine*) for 20 yrs.

"Freedom of speech and a free press are essential to a functioning democracy and a free people. There is no example in history of a dictatorship or totalitarian regime that allowed for either. Today, the United States is no longer a free country by these standards. The level of censorship of individual citizens and the press is unprecedented." - p 97

I agree that censorship is at an insane high. I'm not sure that the US was ever as much of a 'free country' as it's made out to be. It seems more likely to me that conservatives are unhappy about this censorship NOW b/c it's being applied to *them* but that they probably found censorship to be just dandy when it was applied to anarchists & communists & other political (&/or sexual, etc) orientations that they find objectionable. By the same token, it seems that 'leftists' are now fine w/ censorship against conservatives b/c, HEY!, 'everybody knows the conservatives are the bad guys' & it's not the 'leftists' who're being censored as much at the moment. I'm against censorship for anybody but, at the same time, I think it's up to the individual's conscience whether they'll support things they find egregious. Hence, if I were a printer I'd probably decline printing any txt calling for genocide of ANYBODY for any 'reason'.

"It really is up to the individual to start thinking for himself, or, as Henry David

Thorough wrote, "Others will think for you without thinking of you."" - p 101

1st, to get the obvious out of the way, I'll give the author the benefit of the doubt & assume that the misspelling of "Thoreau" as "Thorough" is one of those idiotic spellcheck mistakes that're one of the many reasons why I don't use such apps.

2nd, I'm surprised to see McDonald quote Thoreau at all since he's been a favored figure for non-traditional political thinking. That, however, is a sign of our topsy-turvy times: it seems that conservatives are more inclined to quote literary figures who were previously considered anathema to traditional values. George Orwell & his "**1984**" comes to mind. After all, Orwell fought w/ the Workers' Party of Marxist Unification (POUM – Partido Obrero de Unificación Marxista) in the Spanish Civil War. He was an anti-fascist & the side he was opposing was in favor of the king & the church. Traditionally, conservative values support hierarchies, such as royalty, & religion. As such, conservatives pointing our similarities to "**1984**"'s dystopia & current conditions *more than leftists seem to* is a bit shocking. The same goes for Thoreau.

The references go from pp 105-134. They're extensive. I've marked quite a few of them to be checked & quoted. This extends my reviewing task considerably but I'll take a go at it.

https://didihirsch.org/media/recent-coverage/l-a-suicide-hotline-sees-rise-in-coronavirsu-related-calls-counselors-feel-the-pain/

As is not too uncommon w/ URLs this link didn't work but I did find this similar article:

"A Los Angeles suicide crisis hotline has received more than 1,500 calls about coronavirus, and calls about COVID-19 have increased 75-fold over the past month, DailyMail.com can disclose.

"Didi Hirsch's Suicide Crisis Line, which runs one of the largest suicide line call centers in the country, warned the number would grow exponentially over the coming weeks, as people buckle under the pressure of lost loved ones, lost jobs and the nationwide shutdown.

"The charity said top concerns among callers were 'anxiety, stress, fear of eviction, inability to paying utilities and take care of family, unemployment, health concerns or losing loved ones to the virus.'"

- https://didihirsch.org/media/recent-coverage/daily-mail-los-angeles-suicide-hotline-has-received-upwards-of-1500-calls-in-march-75-times-the-previous-month-over-fears-of-getting-coronavirus-and-related-anxiety-about-eviction-inability-to/

*****

https://www.cdc.gov/coronavirus/2019-ncov/hcp/clinical-care/
underlyingconditions.html

That link is still good. Here's a relevant excerpt:

"   **1. Higher risk** for severe COVID-19 outcomes is defined as an underlying
medical condition or risk factor that has a published meta-analysis or systematic
review or complete the CDC systematic review process. The meta-analysis or
systematic review demonstrates good or strong evidence, (depending on the
quality of the studies in the review or meta-analysis) for an increase in risk for at
least one severe COVID-19 outcome.

        Asthma

        Cancer

        Cerebrovascular disease

        Chronic kidney disease*

        Chronic lung diseases limited to:

        Interstitial lung disease

        Pulmonary embolism

        Pulmonary hypertension

        Bronchiectasis

        COPD (chronic obstructive pulmonary disease)

        Chronic liver diseases limited to:

        Cirrhosis

        Non-alcoholic fatty liver disease

        Alcoholic liver disease

        Autoimmune hepatitis

        Cystic fibrosis

        Diabetes mellitus, type 1 and type 2*‡

        Disabilities‡

Attention-Deficit/Hyperactivity Disorder (ADHD)

Cerebral Palsy

Congenital Malformations (Birth Defects)

Limitations with self-care or activities of daily living

Intellectual and Developmental Disabilities

Learning Disabilities

Spinal Cord Injuries

(For the list of all conditions that were part of the review, see the module below)

Heart conditions (such as heart failure, coronary artery disease, or cardiomyopathies)

HIV (human immunodeficiency virus)

Mental health disorders limited to:

Mood disorders, including depression

Schizophrenia spectrum disorders

Neurologic conditions limited to dementia‡

Obesity (BMI ≥30 kg/m2 or ≥95th percentile in children)*‡

Primary Immunodeficiencies

Pregnancy and recent pregnancy

Physical inactivity

Smoking, current and former

Solid organ or hematopoietic cell transplantation

Tuberculosis

Use of corticosteroids or other immunosuppressive medications"

*****

https://www.theguardian.com/world/2021/aug/12/children-born-during-pandemic-have-lower-iqs-us-study-finds

This link worked. The problem I have w/ the article is that they say "pandemic" instead of "lockdown". The pandemic, such as it is, has nothing whatsoever to do w/ the conditions described. Those conditions were completely artificially created by the quarantine & the engineers of that deserve the blame.

"Children born during the coronavirus pandemic have significantly reduced verbal, motor and overall cognitive performance compared with children born before, a US study suggests.

"The first few years of a child's life are critical to their cognitive development. But with Covid-19 triggering the closure of businesses, nurseries, schools and playgrounds, life for infants changed considerably, with parents stressed and stretched as they tried to balance work and childcare.

"With limited stimulation at home and less interaction with the world outside, pandemic-era children appear to have scored shockingly low on tests designed to assess cognitive development, said lead study author Sean Deoni, associate professor of paediatrics (research) at Brown University."

*****

https://www.pewresearch.org/fact-tank/2021/03/16/many-americans-continue-to-experience-mental-health-difficulties-as-pandemic-enters-second-year/

This link works. I have the same complaint about it as I do about the quote above.

"One year into the societal convulsions caused by the coronavirus pandemic, about a fifth of U.S. adults (21%) are experiencing high levels of psychological distress, including nearly three-in-ten (28%) among those who say the outbreak has changed their lives in "a major way." The share of the public experiencing psychological distress has edged down slightly since March 2020 but remains elevated among some groups in the population. Concerns about both the personal health and the financial threats from the pandemic are associated with high levels of psychological distress."

*****

https://thehill.com/policy/energy-environment/426353-ocasio-cortez-the-world-will-end-in-12-years-if-we-dont-address

"Rep. Alexandria Ocasio-Cortez (D-N.Y.) on Monday said she thinks that there is an urgency needed in addressing man-made climate change, warning that it will "destroy the planet" in a dozen years if humans do not address the issue, no matter the cost.

"During an interview at the MLK Now event in New York City honoring Martin Luther King Jr., Ocasio-Cortez told interviewer Ta-Nehisi Coates that younger Americans are looking for bold solutions to climate change, and are not concerned about the cost.

"{mosads}"Millennials and people, you know, Gen Z and all these folks that will come after us are looking up and we're like: 'The world is gonna end in 12 years if we don't address climate change and your biggest issue is how are we gonna pay for it?' " Ocasio-Cortez asked Coates."

Promises, promises.. another politician typically giving an empty sensational speech that's unlikey to result in much of anything. She promises the end of the world by 2031, don't hold yr breath.

The problem w/ such an 'end-of-the-world' scenario is that it's really predicting an end to human-life-as-we-know-it. Given that humans are purported to be the cause of the problem it wd seem that the end of human-life-as-we-know-it might be the optimal solution for everything else on the planet.

*****

https://www.cdc.gov/nchs/nvss/vsrr/covid_weekly/index.htm?fbclid=IwAR3-wrg3tTKK5-9tOHPGAHWFVO3DfslkJ0KsDEPQpWmPbKtp6EsoVV2Qs1Q

The most important parts of this webpage are graphs wch I can't cut & paste here. As such, I'll make an admittedly biased & abridged selection of the 1st 3 lines of the "Place of Death (Table 2)" chart:

"Year in which death occurred: 2020-2022"; "State: United States"

"Place: Decedent's home; All Deaths involving COVID-19 [1]: 94,306; Deaths from All Causes: 2,878,140; Deaths involving Pneumonia [2]: 66,794; Deaths involving COVID-19 and Pneumonia [2]: 25,411; All Deaths involving Influenza [3]: 2,116; Deaths involving Pneumonia, Influenza, or COVID-19 [4]: 138,878

"Place: Healthcare setting, dead on arrival; All Deaths involving COVID-19 [1]: 1,000; Deaths from All Causes: 23,284; Deaths involving Pneumonia [2]: 584; Deaths involving COVID-19 and Pneumonia [2]: 252; All Deaths involving Influenza [3]: 20; Deaths involving Pneumonia, Influenza, or COVID-19 [4]: 1,352

"Place: Healthcare setting, inpatient; All Deaths involving COVID-19 [1]: 688,044; Deaths from All Causes: 2,714,555; Deaths involving Pneumonia [2]: 703,652; Deaths involving COVID-19 and Pneumonia [2]: 433,043; All Deaths involving Influenza [3]: 7,689; Deaths involving Pneumonia, Influenza, or COVID-19 [4]:

966,669"

When interpreted in the typical way, the "Healthcare setting, inpatient" wd be those patients only inpatients b/c of extreme duress &, therefore, the likeliest to die. Under that assumption, the statistic that all deaths involving COVID-19 at home total 94,306 while the same type of death involving inpatient care totals 688,044 wd seem to make sense. However, an admittedly perverse take on this might conclude that if one were in the hospital w/ COVID-19 one might be 7.3 times as likely to die. I wonder what the survival rate statistics are?

Otherwise, HEY!, it's worth noting that the inpatient "Deaths involving Pneumonia, Influenza, or COVID-19 [4]: 966,669" while the "Deaths involving Pneumonia [2]: 703,652" & the "Deaths involving Influenza [3]: 7,689". If one subtracts the influenza deaths from the combined deaths one gets 966,669 - 7,689 = 958,980. If one further subtracts the pneumonia deaths from that figure one gets 958,980 - 703,652 = 255,328 exclusively attributable to COVID-19 if one doesn't factor in other comorbidities. Now, if one takes the "Deaths involving COVID-19 [1]: 688,044" again & subtracts the 255,328 from that, one notices that the more definite figure for deaths attributable to COVID-19 is 432,716 *less*.

Now that's only in "Healthcare setting, inpatient" so let's take a large overview from a part of the chart not so-far quoted:

"Total - All Places of Death; Deaths involving Pneumonia, Influenza, or COVID-19 [4]: 1,434,881; All Deaths involving COVID-19 [1]: 1,021,212; Deaths involving Pneumonia [2]: 927,785; All Deaths involving Influenza [3]: 12,096"

Following the same procedure, we subtract 12,096 from 1,434,881 to yield 1,422,785 from wch we subtract 927,785 = 495,000. Yes, *Fun with Math!* Maybe I'm overlooking something important, maybe my math is wrong - both are possible. If I'm right, however, that makes 495,000 deaths in the US exclusively attributed to COVID-19 over a 2+ yr period. That's 526,212 deaths *less* than the deaths involving COVID-19 = 1,021,212 figure given.

Now, let's compare that to the death-from-COVID predictions from the same time. I found "**7,198,770 reported COVID-19 deaths** based on **Current projection** scenario by November 1, 2022" (https://covid19.healthdata.org/global?view=cumulative-deaths&tab=trend) Get to work, people!, 6,703,770 of you have to die in the next 3 mnths! OK, that's not fair, that's a GLOBAL prediction, not one just for the US. Now the CDC elsewhere has this to say about as-of-August-20, 2020: "The national ensemble predicts that a total of 1,035,000 to 1,048,000 COVID-19 deaths will be reported by this date." (https://www.cdc.gov/coronavirus/2019-ncov/science/forecasting/forecasting-us.html) Let's take the smallest of the 2 figures, no sense in being alarmist, right? That only yields 540,000 that have to die in the next 24 days, that's not so bad, now, is it?!

But, HEY!, why just rely on these old fuddy-duddies? I have a *really smart* former friend who emailed me on March 21, 2020: "you can't downplay this virus any more than you could downplay the Spanish Flu. 50 to 100 million died of the Spanish Flu (2.5% mortality rate)". Well, there you have it! A completely reliable projection from a *really smart guy*: 50 to 100 million people might die from COVID-19 b/c it's just like the so-called Spanish Flu. (Ok, that's not exactly what he sd but you get the idea, I'm being sarcastic b/c I think this guy is a pompous asshole.)

*****

https://www.usatoday.com/story/news/factcheck/2020/04/24/fact-check-medicare-hospitals-paid-more-covid-19-patients-coronavirus/3000638001/

I found this one particularly interesting b/c multiple 'fact checkers' mostly substantiated a claim by Senator Scott Jensen (also a physician) that hospitals get paid more if Medicare patients are listed as having COVID-19 and three times as much money if they need a ventilator. Jensen, who's gotten in trouble for his outspokenness, is quoted as stating:

"Hospital administrators might well want to see COVID-19 attached to a discharge summary or a death certificate. Why? Because if it's a straightforward, garden-variety pneumonia that a person is admitted to the hospital for – if they're Medicare – typically, the diagnosis-related group lump sum payment would be $5,000. But if it's COVID-19 pneumonia, then it's $13,000, and if that COVID-19 pneumonia patient ends up on a ventilator, it goes up to $39,000."

*****

https://www.cdc.gov/mmwr/volumes/70/wr/pdfs/mm7024e1-H.pdf?fbclid=IwAR153webZWliXPJyr0SUM7QGKV2J-5XoFpwVfvmNBZC2Lvi6NyIBya6uTBM

It took me awhile to get the URL right bc the lower-case 'el's & the 'one's look very similar in the bk (& the lower-case "el"s & the upper-case "eye"s look identical in the font I'm typing this in) but I got it right in the above so I saved you the trouble of trying to copy it from the bk (when this review was online that helped). Here's a brief excerpt from the beginning of that:

"Beginning in March 2020, the COVID-19 pandemic and response, which included physical distancing and stay-at-home orders, disrupted daily life in the United States. Compared with the rate in 2019, a 31% increase in the proportion of mental health–related emergency department (ED) visits occurred among adolescents aged 12–17 years in 2020 (*1*). In June 2020, 25% of surveyed adults aged 18–24 years reported experiencing suicidal ideation related to the pandemic in the past 30 days"

*****

https://www.pe.com/2021/03/30/southern-california-suicides-down-during-coronavirus-pandemic-but-not-among-young-eople??utm_email=E44D9493A4BD7421C57244AB74&g2i_eui=l1hHBZmxw9%2F3BUTOPT%2B8ItN%2BinSLPkLj&g2i_souce=newsletter&utm_souce=listrak&utm_medium=email&utm_term=Read%20more&utm_campaign=scng-ivdb-localist&utm_content=curated&fbclid=IwAR0ok8GLiXYapcV6wYmdz26EK_t-nF3O15_zrVgua8sYrxTCFlpPeEPq1wU

Copying that URL from the bk was a bit challenging but I was reasonably meticulous about it. It didn't yield a result but I was recognized as a probable subscriber wch means that McDonald is presumably a subscriber. I tried doing a search for "southern california youth suicides 2020" but that didn't work so I gave up.

*****

https://thefederalist.com/2020/10/12/cdc-study-finds-overwhelming-majority-of-people-getting-coronavirus-wore-masks/?fbclid=IwAR1j81e7lhGi15T1Q2ImbMmH2QvdZ6-dsc8xnAdJjnUa9_iqlyMw5lw_Ics

That URL didn't work for me either & might be, once again, b/c I typed it wrong but I did find this:

"The Centre for Evidence-Based Medicine at Oxford University summarized six international studies which "showed that masks alone have no significant effect in interrupting the spread of ILI or influenza in the general population, nor in healthcare workers." Oxford went on to say that "that despite two decades of pandemic preparedness, there is considerable uncertainty as to the value of wearing masks." They prophetically warned that this has "left the field wide open for the play of opinions, radical views and political influence."

"A study of health-care workers in more than 1,600 hospitals showed that cloth masks only filtered out 3 percent of particles. An article in the New England Journal of Medicine stated, "[W]earing a mask outside health care facilities offers little, if any, protection from infection" and that "[T]he desire for widespread masking is a reflexive reaction to anxiety over the pandemic."

"There are many other credible studies showing lack of mask efficacy, such as studies published in the National Center for Biotechnology Information, Cambridge University Press, Oxford Clinical Infectious Diseases, and Influenza Journal, just to name a few."

- https://thefederalist.com/2020/10/29/these-12-graphs-show-mask-mandates-do-nothing-to-stop-covid/

*****

https://www.reuters.com/video/watch/idOVCCIUZCJ

"Tanzanian President John Magufuli has poured scorn on coronavirus test kits imported to his country after saying that a goat and a pawpaw had returned positive results for COVID-19. Emer McCarthy reports."

*****

https://www.nejm.org/doi/full/10.1056/nejmp2006372

That URL didn't work either but I found the article easily enuf. Here're a few relevant excerpts:

"We know that wearing a mask outside health care facilities offers little, if any, protection from infection. Public health authorities define a significant exposure to Covid-19 as face-to-face contact within 6 feet with a patient with symptomatic Covid-19 that is sustained for at least a few minutes (and some say more than 10 minutes or even 30 minutes). The chance of catching Covid-19 from a passing interaction in a public space is therefore minimal. In many cases, the desire for widespread masking is a reflexive reaction to anxiety over the pandemic.

"The calculus may be different, however, in health care settings. First and foremost, a mask is a core component of the personal protective equipment (PPE) clinicians need when caring for symptomatic patients with respiratory viral infections, in conjunction with gown, gloves, and eye protection. Masking in this context is already part of routine operations for most hospitals. What is less clear is whether a mask offers any further protection in health care settings in which the wearer has no direct interactions with symptomatic patients."

[..]

"It is also clear that masks serve symbolic roles. Masks are not only tools, they are also talismans that may help increase health care workers' perceived sense of safety, well-being, and trust in their hospitals. Although such reactions may not be strictly logical, we are all subject to fear and anxiety, especially during times of crisis. One might argue that fear and anxiety are better countered with data and education than with a marginally beneficial mask, particularly in light of the worldwide mask shortage, but it is difficult to get clinicians to hear this message in the heat of the current crisis. Expanded masking protocols' greatest contribution may be to reduce the transmission of anxiety, over and above whatever role they may play in reducing transmission of Covid-19. The potential value of universal masking in giving health care workers the confidence to absorb and implement the more foundational infection-prevention practices described above may be its greatest contribution."

*****

https://www.businessinsider.com/scott-gottlieb-goes-from-fda-commissioner-to-pfizer-board-member-2019-6

June 28, 2019:

"Former FDA Commissioner Scott Gottlieb just joined the board of directors at drugmaker Pfizer, about two months after stepping down as the head of the US drug regulator.

"Gottlieb resigned from the Food and Drug Administration in April after two years on the job, saying he wanted to spend more time with his family.

"During his time at the FDA, Gottlieb won rare bipartisan praise, changing the image of the FDA. The FDA approved a steady stream of new drugs, hitting an all-time record last year, and pushed to crack down on youth e-cigarette use and tobacco products."

People have been known to notice before the completely coincidental connection between FDA employees, drug-approval, & new high-paying jobs at pharmaceutical companies. Of course, no-one wd be so foolish as to think that any conspiracy is involved.

*****

https://www.nature.com/articles/d41586-020-01003-6

"Ferguson is one of the highest-profile faces in the effort to use mathematical models that predict the spread of the virus — and that show how government actions could alter the course of the outbreak. "It's been an immensely intensive and exhausting few months," says Ferguson, who kept working throughout his relatively mild symptoms of COVID-19. "I haven't really had a day off since mid-January."

"Research does not get much more policy-relevant than this. When updated data in the Imperial team's model indicated that the United Kingdom's health service would soon be overwhelmed with severe cases of COVID-19, and might face more than 500,000 deaths if the government took no action, Prime Minister Boris Johnson almost immediately announced stringent new restrictions on people's movements. The same model suggested that, with no action, the United States might face 2.2 million deaths; it was shared with the White House and new guidance on social distancing quickly followed"

Ha ha. These alarmist predictions always make the news but are NEVER accurate. Funny how that works. Note that Fergusen's own case of COVID-19 was so mild it didn't even interfere w/ his working. Cd he secretly be one of the dead he predicted?

*****

https://www.nber.org/system/files/working_papers/w28930/w28930.pdf

That URL led to an error msg that read "Blocked Plug-in". SO, I tried https://www.nber.org/system/files/working_papers/w28930.pdf instead in case the repetition was a mistake. That led to the more usual "Page Not Found" msg. I tried a search on the National Bureau of Economic Research website for "Rand corporation lockdowns" & that didn't work so I tried "lockdowns ineffective". I didn't find what I was looking for then either so I moved on.

*****

https://nypost.com/2020/07/26/woman-maces-couple-for-not-wearing-masks-at-california-dog-park-video/

"A woman allegedly Maced a California couple for not wearing masks while eating at a dog park — a disturbing incident caught on videotape.

"The injured pair, Ash O'Brien and hubby Jarrett Kelley, said the lady, who was wearing a black mask, ambushed them with a can of the burning substance after ripping them for not donning face coverings and eating at the Dusty Rhodes Dog Park in Ocean Beach outside San Diego on Thursday afternoon.

"First she "automatically started saying stuff about us not wearing a mask when we were social distancing — there was no one near us," O'Brien told the local ABC-TV affiliate.

"O'Brien said the lady also took them to task for eating at a picnic table at the park while they were there with their 3-month-old puppy, claiming it was against the rules.

""If we knew there was a no-food policy, we wouldn't have brought it into the park," O'Brien told the TV station.

"The angry woman soon stalked off — only to come back a few minutes later and attack them, O'Brien said.

""She just came up without saying anything and just stuck the Mace can right in front of my face," O'Brien said. "My husband, being a good guy, walked in front of her and was like, 'Hey, calm down please don't do this' — and then she grabbed him and just starting macing him. She used the entire can on him.

""We drove to the hospital. He got treated and everything," O'Brien said — adding that her husband suffered injuries to his face while she was burned on her arms.

"A woman filming the incident shouted at the attacker, "What are you doing? You cannot be serious! You just Maced them!" — as a female weeps in the background, according to footage posted by the Daily Mail.

""What's wrong with you, lady?!" added the woman who caught some of the encounter on tape.

"The injured couple filed a police report afterward.

""I want her to go to jail — she assaulted my husband, and I'm angry about it," O'Brien told ABC."

That's the entire article. Ordinarily, I'd only excerpt a small portion but this one's short & such a good example of the madness that it's worth a full quote. Note that a woman was attacking a woman 1st & then the man who tried to protect the 1st victim & that then a 3rd woman shot footage of the incident & remonstrated w/ the attacker. That makes 2 sane women out of 3, a majority.

*****

https://www.latimes.com/california/story/2021-09-13/vaccine-exemption-requests

"City officials pushed back a deadline last week for employees to seek an exemption to the vaccination requirement, instead giving workers until the end of Monday to indicate that they plan to pursue an exemption. Across all city departments, more than 6,200 employees did so."

*****

https://www.wsj.com/articles/california-leftists-try-to-cancel-math-class-11621355858

"If California education officials have their way, generations of students may not know how to calculate an apartment's square footage or the area of a farm field, but the "mathematics" of political agitation will be second nature to them. Encouraging those gifted in math to shine will be a distant memory."

*****

https://thefederalist.com/2020/07/01/3-years-ago-bret-weinstein-endured-the-precursor-to-todays-riots

"Cancel culture has existed in this country for years on college campuses. Nowhere was that more apparent than at Evergreen State College, where campus-wide riots arose in 2017 when a biology professor named Bret Weinstein found himself on the wrong side of the student body over questioning an allegedly optional event.

"The Evergreen campus had, for years, engaged in a voluntary "day of absence," in which students and faculty of color would leave to highlight the important role they play on campus. However, in 2017 it was decided that, rather than non-whites leaving, it should instead be the white students. The idea was that the white and non-white students would engage in separate, specifically tailored events discussing race on campus.

"Weinstein was uncomfortable with this idea, and decided to send an email to the faculty and staff, stating: "There is a huge difference between a group or coalition deciding to voluntarily absent themselves from a shared space in order to highlight their vital and underappreciated roles....and a group encouraging another group to go away. The first is a forceful call to consciousness, which is, of course, crippling to the logic of oppression. The second is a show of force, and an act of oppression in and of itself."

"The reaction of the student body was aggressive, to say the least. Weinstein was labeled a white supremacist, and protests broke out across the school. Along with calls for sweeping reforms in regard to how race was handled on campus, the students were calling for Weinstein's firing. Physical alterations occurred between Weinstein and protestors, with no aid from campus security. Weinstein alleges campus security was told not to intervene, neither to stop violence nor calm protestors."

Wow.

*****

https://bariweiss.substack.com/p/med-schools-are-now-denying-biological

"During a recent endocrinology course at a top medical school in the University of California system, a professor stopped mid-lecture to apologize for something he'd said at the beginning of class.

""I don't want you to think that I am in any way trying to imply anything, and if you can summon some generosity to forgive me, I would really appreciate it," the physician says in a recording provided by a student in the class (whom I'll call Lauren). "Again, I'm very sorry for that. It was certainly not my intention to offend anyone. The worst thing that I can do as a human being is be offensive."

"His offense: using the term "pregnant women."

""I said 'when a woman is pregnant,' which implies that only women can get pregnant and I most sincerely apologize to all of you."

"It wasn't the first time Lauren had heard an instructor apologize for using language that, to most Americans, would seem utterly *inoffensive*. Words like "male" and "female."

"Why would medical school professors apologize for referring to a patient's biological sex? Because, Lauren explains, in the context of her medical school "acknowledging biological sex can be considered transphobic.""

To say that I find that ludicrous wd be an understatement. I like people to just be who they feel they are w/o bending to societal pressure - unless that means being a non-consensual aggressor such as a murderer or a rapist. But until a man who's undergone a 'sex change' operation actually gives birth I think I'll just stick to common sense & biology & continue to say that a person has to have a functional womb, etc, in order to become pregnant. (NOT) Sorry, but to me, test-tube babies don't count as 'pregnancies'. Call me old fashioned.

*****

https://thebl.com/us-news/gofundme-cancels-grieving-fathers-fundraiser-for-son-who-died-from-vaccine.html

I didn't find this story at the above URL but I did find it here:

https://fee.org/articles/why-gofundme-deleted-this-grieving-father-s-fundraiser-after-his-son-s-death/

"Ernesto Ramirez Jr. was one of hundreds of American children taken too early during the COVID-19 pandemic. Only, his life wasn't claimed by COVID-19.

"His father and media reports say Ernesto died five days after taking a vaccine that was supposed to protect him.

"**'We Clearly Have an Imbalance'**

"Two months after Ernesto's death, the Centers for Disease Control and Prevention (CDC) announced that atypical levels of heart inflammation had been observed in some patients following COVID-19 vaccination, particularly in young men receiving a second dose of a mRNA vaccine."

Oh, well, more collateral damage, right? I mean statistically who cares, right?! It's such a small amt of deaths. Of course, the dead person's family & friends care - even I CARE & I didn't know the boy. Do you think that if this happened to Fauci's kids maybe the reaction wd be a little different?!

*****

https://apnews.com/article/joe-biden-health-coronavirus-pandemic-us-supreme-court-8d397f378c01c369f0d8618a4d9b3a83

"WASHINGTON (AP) — President Joe Biden may have averted a flood of evictions and solved a growing political problem when his administration reinstated a temporary ban on evictions because of the COVID-19 crisis. But he left his lawyers with legal arguments that even he acknowledges might not stand up in court.

"The new eviction moratorium announced Tuesday by the Centers for Disease Control and Prevention could run into opposition at the Supreme Court, where one justice in late June warned the administration not to act further without explicit congressional approval."

What the hell (or shd it be WTF?), obviously Bidentity Crisis hasn't gone far enuf. I pronounce that I, as an exemplary citizen, no longer have to pay for anything, EVER AGAIN, b/c I might die someday & I *want to Party Hearty* between now & then.

*****

https://www.wsj.com/articles/bitcoin-comes-to-el-salvador-first-country-to-adopt-crypto-as-national-currency-11631005200

"SAN SALVADOR, El Salvador—Tiny and impoverished El Salvador's move to become the first country in the world to adopt bitcoin as legal tender got off to a bumpy start, as the government took its bitcoin e-wallet offline for several hours after tens of thousands of people tried to download the app, overloading servers.

"The administration of President Nayib Bukele, 40, plans to spend more than $225 million on the rollout, including a $30 credit in bitcoin to those who take up Chivo—local slang for "cool"—the government-run e-wallet that can be used for purchases in bitcoin or U.S. dollars."

*****

In conclusion, I'm not going to give this bk a star rating b/c even tho I agree w/ large portions of it & find the references at the back to be very useful, I don't want to give the impression that I share the conservative gender role opinions that McDonald has, I don't.

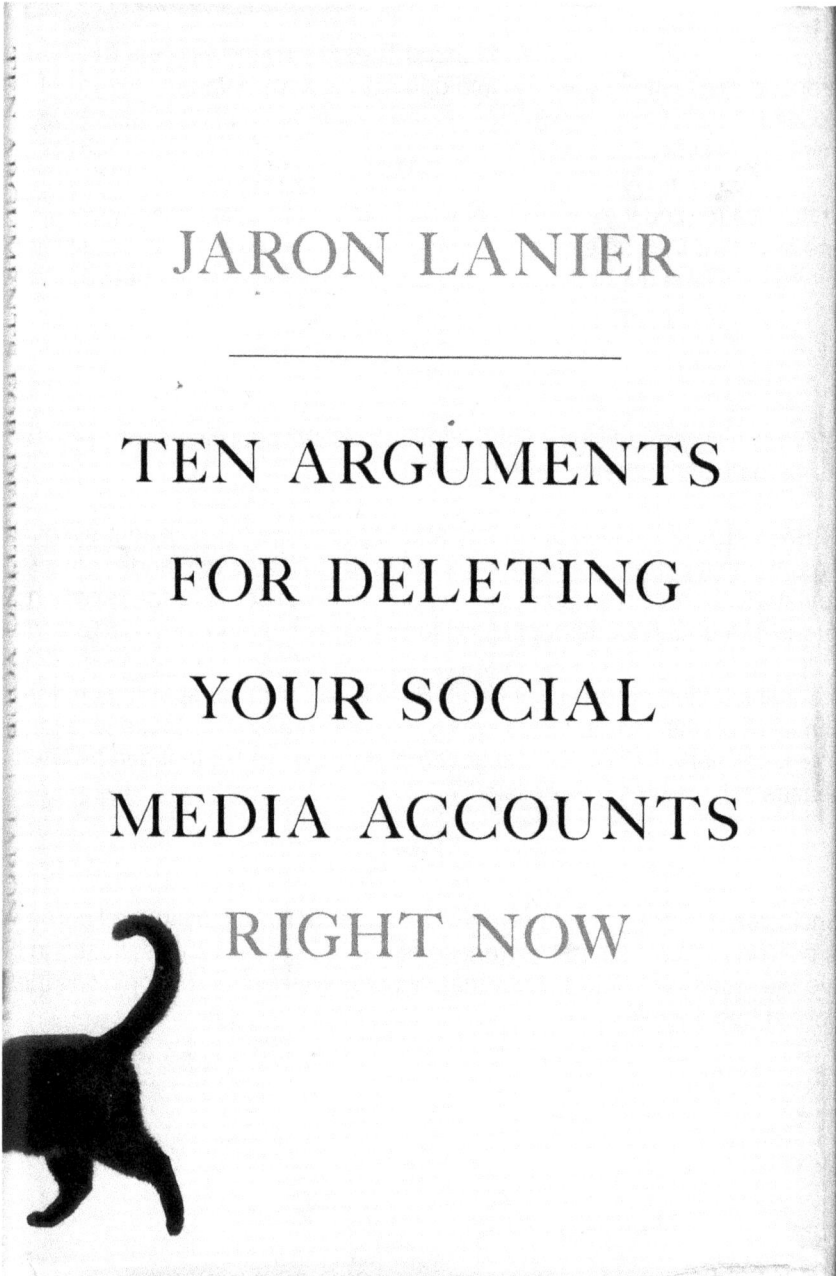

review of
Jaron Lanier's **Ten Arguments for Deleting Your Social Media
Accounts Right Now**
by tENTATIVELY, a cONVENIENCE - July 28-30, 2022

I use common online resources for presenting my writing (Goodreads, Medium), my (M)Usic (SoundCloud, Internet Archive, BandCamp, YouTube), my movies (YouTube, Internet Archive, Vimeo). I have had accts that I suppose wd be called Social Media ones w/ Facebook, Instagram, MeWe, ello, International Union of Mail Artists, LinkedIn, Twitter, Visual Music Village.. I've been a so-called "Mail Artist" since 1978 & communicating & exchanging w/ people internationally (or 'patanationally as I sometimes prefer to call it) is very important to me. *SO*, as postal rates got less & less affordable, having communication thru email, messaging, Facetime, WhatsApp, Skype, & Social Media, etc, became welcome & promising. While an exchange of actual physical objects thru snail mail was generally more exciting, one cd still see images & hear sounds & read txts thru these electronic mediums. I'm an electronic kindof a guy anyway, I make electro-acoustic music & make movies so it worked for me fairly well.

But, still, it wasn't quite right & the more widely used the Social Media, the more wrong it seemed to be. I've rarely been enthusiastic about popular things, I like the obscure, the esoteric, when things are popular, w/ a few exceptions, they're usually so LCD (Lowest Common Denominator) that there's very little there of substance to hold my interest - usually it's just aggravating, a dumbing-down that I resist for the sake of maintaining my precious intelligence (such as it may be). & Facebook was the worst. Just about the only online place where I made any quality friendships was Goodreads & that just seems to deteriorate more & more.

I don't remember when I joined Facebook but it was probably around 2007. At 1st I just posted the occasional picture of myself in chronological order - providing a simplified timeline of my growing-up. I rarely posted anything, I didn't make much of an effort to seek out friends. When people friend-requested me I asked them why they wanted to. If they replied I'd probably accept the request but if they didn't I just ignored it. Then there was the girlfriend from 2010-2012. As 1st she didn't care about FB either, then she started getting obsessive about it. I remember one particularly aggravating time when we sitting in bed & conversing & whenever I sd anything she'd type in into FB b/c she found what I was saying entertaining & wanted her FB 'friends' to see how witty *she was* since she wasn't contextualizing it as my half of our conversation.

Anyway, I started using FB to post links to anything new I put online, to advertise events I had upcoming, etc. It was useful for that but very-few-to-none of my local FB friends had any interest in the events I was promoting (including, appropriately, the UNDERAPPRECIATED MOVIEMAKERS FESTIVAL I organized in 2018: https://youtu.be/KwlfRxmRU3E) but it still seemed useful in a more & more central way. I'm prolific, so it was fun to be able to post links to new work on a fairly regular basis.

By 2017, my new girlfriend exposed me to Instagram Poetry as the worst poetry that either of us had ever seen. I hadn't used Instagram, I wasn't really interested, but I created an acct so that I cd post fake Instagram Poetry in the hope that people might actually like it. I even created an Instagram Poetry website: http://idioideo.pleintekst.nl/InstagramPoetry.html . Since thinking up

Instagram Poetry was so easy it became a kind of unhealthy obsession that took over my mind & I stopped doing it fairly quickly so that I'd stop thinking in such stupid oversimplistic phrases. To the limited extent that I used the acct at all, I created a fake personality, a person who was interested in the Kardashians, e.g., popular figures I knew next-to-nothing about & had no interest in whatsoever. Then I didn't post anything at all for about 4 & 1/2 yrs.

THEN, on March 16, 2020, I posted my 1st observation about the so-called pandemic on Fecesbook - calling the 'health crisis' into question as an exercise in mind control. I was immediately attacked & these attacks continued for the next 5 mnths. I was somewhat astonished, & certainly dismayed, by how narrow-minded people had become in the throes of fear. FEAR: it seemed so obvious to me that the fear was being generated by the mass media, that the so-called 'pandemic' was grossly exaggerated, & that it was being used opportunistically to create a situation where big business wd make insanely huge profits at the expense of the vast majority of people who were being beaten-down under the guise of 'for your own good'. Inititally, this insanity was presented as if it'd be over in a couple of wks. It's now been over 2 yrs, 4 mnths, & it's still going strong, & life has changed dramatically. People have become mind-bogglingly stupid, obedient, & gullible.

During those 5 mnths I continued to post observations about the current times that expressed accurately what I thought was happening. These were often in the form of text panels that were meant to be something that had a strong graphic presence that cd also be easily shared (although few people did so). I cowrote a bk w/ a few friends who had a similar take on the PANDEMIC PANIC (http://idioideo.pleintekst.nl/Book2020.09PANDEMIC.html) in wch I reproduced these graphics & I made a movie that strung them together: 665. "**Check out of the Mass-Formation Roach Motel**" - 37:40 - 1920 X 1440, 30fps, Stereo - on my onesownthoughts YouTube channel here: https://youtu.be/vWIyZ5vKdAY - on the Internet Archive here: https://archive.org/details/roach-motel . Despite frequent personal attacks & trolling-baiting & an almost total shunning of me at a local level, I didn't generally engage in interpersonal sparring but instead made more meta-observations & critiques. Friends who became abusive I simply unfriended. Eventually, the FB environment seemed so toxically conformist that I quit the many FB groups that I'd been a part of, announced my intention to stop posting & did so on August 16, 2020, 5 mnths after my 1st posting regarding the so-called 'pandemic'. Now, I post something only on March 16 & August 16 - still continuing my critical positioning - not allowing comments. Psychologically, it was a huge relief to essentially get off Facebook since every day brought out new hatefulness from the zombies there.

In early 2022, I revisited Instagram & posted yet-another critical statement about the quarantine. I was surprised to see that every Instagram Poem I'd uploaded there had been removed, the result was I had no posts other than the new one. When I went back a day or 2 later, that, too, was removed. I had no posts at all. Not having seen Instagram as a place where much talent manifested itself & utterly disgusted by the apparent censorship I closed out my acct.

In the meantime, despite the vast promise of increased social connectivity, I've been finding my online social life proscribed & limited more & more, often by subtle ways that reek of covert warfare: Goodreads changed my name from "tENTATIVELY, a cONVENIENCE" to "Tentatively, Convenience", they're getting rid of the Creative Writing section: reducing participants more & more to just free marketing promoters for Amazon - as if Amazon's not rich enuf already; YouTube makes problems for me of various sorts - such as a copyright claim that I'm using 'someone else's music' when there's no music in the movie-in-question at all (I can thank algorithms for that one) - some movies get blocked altogether, others get age-restricted; a poet's dream blog I liked posting to hasn't shown any sign of life for 4 yrs.. It's obvious where this is leading: *I was ripe for Lanier's critique.*

Lanier starts off promoting the mind-set of cats as one to take in relation to high-tech.

"Cats are smart, but not a great choice if you want an animal that takes to training reliably. Watch a cat circus online, and what's so touching is that the cats are clearly making their own minds up about whether to do a trick they learned, or to do nothing, or to wander into the audience.

"Cats have done the seemingly impossible: They've integrated themselves into the modern high-tech world without giving themselves up. They are still in charge. There is no worry that some stealthy meme crafted by algorithms and paid for by a creepy, hidden oligarch has taken over your cat. No one has taken over your cat; not you, not anyone."

[..]

"This book is about how to be a cat. How can you remain autonomous in a world where you are under constant surveillance and are constantly prodded by algorithms run by some of the richest corporations in history, which have no way of making money except by being paid to manipulate your behavior. How can you be a cat, despite that?" - p 2

"Something entirely new is happening in the world. Just in the last five or ten years, nearly everyone started to carry a little device called a smartphone on their person all the time that's suitable for algorithmic behavior modification. A lot of us are also using related devices called smart speakers on our kitchen counters or in our car dashboards. We're being tracked and measured constantly, and receiving engineered feedback all the time. We're being hypnotised little by little by technicians we can't see, for purposes we don't know. We're all lab animals now." - p 5

&, yes, I think he's spot on & that this is a **very important** observation wch it wd behoove more people to be aware of.

Now's probably a good time to interpolate a little personal history. When I was 20

I was pd (a very small amt) to be a research volunteer for 15 days at Johns Hopkins Hospital's Phipps Clinic in Baltimore. That wd've been 1973. The research was for NASA & was about preparing to live in space stations. Essentially, it was a test of the effectiveness of behavior modification strategies for keeping astronauts mentally & physically active while confined to the space stn environment & cut-off from ordinary physical & social life. The volunteers were college students (I was a music major at a community college at the time) selected in groups of 3 friends.

The simulated space stn was built inside the Phipps Clinic & raised above the floor & otherwise isolated so that no sound from the outside world wd seep in. It was rectangular. Imagine the rectangle as having its width more narrow than its length. There was a hallway running thru the middle of the bottom 2/3rds of this. Off the sides of the hallway were the 3 private quarters of the 'astronauts'. The hallway led to a communal room that constituted the top 1/3rd of the rectangle. Off of both this communal room & the hallway at its upper right extremity was a project room where the astronauts were kept busy working on making pot-holders, if I remember correctly.

Each individual living space was a cube roughly 8 ft X 8 ft X 8 ft very efficiently arranged to contain everything the 'astronaut' needed. There was a bed, a shower that had a fold-out toilet seat, a kitchen, a work table w/ a computer, & an exercise area in front of a wall that, when in use, had flashing button-lights that the exercising person was supposed to tap when they lit-up. The bed had a wooden cover that was to be locked over it when one was finished w/ sleeping. The work-table had a drawer that was also locked except when in use. These lockings & unlockings were controlled by a remotely viewing person & weren't directly controllable by the occupant of the room.

In one corner of the room was a camera aimed to cover everywhere in the space but the shower off to its left. In the upper rough middle of the room, to the left of the work table & the right of the bed, was a hook on wch the occupant was to hang cards w/ #s on them that indicated what activity was planned next. There was a 'tree' of possible activities that followed a plan that was restricted for behavior modification purposes. I forget the exact range & paths of possibilities but they went something like this: on getting out of bed, one had to close the lid of the bed, wch was then locked remotely by the unseen observer. The 'astronaut' cd then choose to go to the toilet, which was always available as an option followed by making a meal, wch was probably also always available.

After that, one cd hang up a # card that indicated that one wanted to do a math problem on the computer wch wd lead to a series of them appearing on the computer screen. A certain # of these had to be completed correctly, perhaps 100, before one cd move on to a different activity. A possible activity was reading or studying. I had brought William S. Burroughs's "**The Wild Boys**" w/ me b/c I saw it as an anti-control novel, a subject I took very seriously. I imagine it's obvious that that choice was calculated to be a counterpoint to the control that I knew was happening in the study. In order to get out of the room one had to both do the math problems & exercise. The exercise consisted of tapping the wall-lights as they flashed on w/in a certain amt of time, perhaps one had a 3 second window. Again, one had to tap them correctly, perhaps 100 times, before the exercise was over. There were 4 lights, arranged as the corner points of a rectangle defined by one's bodily reach - 2 upper, 2 lower. After going thru this routine one cd either stay in the room or ask to go into the hallway w/ a destination of either the main communal room or the project room. Only one 'astronaut' cd be in the hallway at the same time. Each room had a telephone that communicated w/ the other 2 private quarters. If I recall correctly (this was almost 49 yrs ago), we had to consult w/ each other & agree to meet in the communal rooms before being able to get into the hallway. Perhaps we had to agree on the order of going there since no more than one person cd be in the hallway at the same time.

A # card wd then be hung by the 'astronaut' visible to the surveillance camera that announced the intention to go to the communal room. & the room door wd be remotely unlocked so that we cd leave & locked behind us. Once we were gathered in the largest of the 2 communal rooms it was a relaxed environment. There were probably board games we cd play & a kitchen. One of the other volunteers, Ken Moore, had brought in a Moog synthesizer for us to play. I remember his being surprised that I wasn't more interested in it. I must've already had more of an interest in electro-acoustic music rather than pure electronic music. Ken & I are both still making electronic music now that we're in our late 60s. Going into the project room was probably something that we cd do collectively or not but, while there, we had to work on making pot-holders. Of course, that was designed as a non-challenging activity that wd keep us busy - actual astronauts wd have a different activity.

I deliberately toyed w/ the experimenters by deviating from their plan. When in my efficiency, I tore up tufts from the slightly shaggy carpet & used them in an artwork I was making, maybe something as banal as a portrait, I don't remember now. I figured that by slightly 'damaging' the physical environment I might be perceived as 'losing my mind'. The volunteers cd only leave the simulated space stn before the designated 15 day limit under threat of non-payment & the researchers weren't supposed to intervene in the environment at all (after all, in an actual space stn we wd've been physically inaccessible). Therefore, I was testing whether the observers wd intervene if I started manifesting any unusual behavior.

There were hidden hallways for the experimenters to come into so that they cd deliver requested supplies. I imagine this was how our food supplies were replenished. Perhaps we'd have to request food w/ a # card & then they'd place that food in the back of a drawer from their hidden hallways, perhaps that drawer was then unlocked for us so that we cd access the food, perhaps our refrigerator was already completely stocked, I don't remember.

The relevance of this story, aside from the obvious behavior modification aspects, is that I enjoyed this experience. I was young enuf to not be overwhelmingly bothered by the sexual deprivation, as I wd've probably been a mere few yrs later. I DID find the procedures & available activities stimulating. This was probably the 1st time I'd had access to a computer & I was good enuf at math to find solving the problems pleasurable instead of frustrating. The physical exercise was a slight challenge but not too much of one so that I was ever defeated by it. MOST IMPORTANTLY FOR THE PURPOSES OF THIS TELLING, *I enjoyed being under surveillance.* I saw widespread & constant surveillance as a major factor of the probable near-future but it didn't bother me - perhaps b/c I felt like I was performing for an unseen audience. Of course, enjoyment of this was predicated on no malevolent results of the surveillance existing.

NOW, 49 yrs later, the surveillance is far more widespread than what occurred in this study. At the same time, it's not so overtly claustrophobic unless one's in prison. *I want control over when & when not I'm under surveillance.* Obviously, that's not the way it works, the importance of surveillance to those making it happen is to have people be surveilled at the time when we don't want to be. Still, I tell this story to try to illustrate why some people might enjoy being under surveillance - it's like being the star of one's very own 'reality' tv show - definitely NOT something that I'd want to do now. I make movies of my life & I make most of them publicly available - *but it's more under my control than surveillance is.*

"Now everyone who is on social media is getting individualized, continuously adjusted stimuli, without a break, so long as they use their smartphones. What might once have been called advertising must now be understood as continuous behavior modification on a titanic scale." - p 6

"Most users of social media have experienced catfishing* (which cats hate), senseless rejection, being belittled or ignored, ouright sadism, or all of the above, and worse. Just as the carrot and stick work together, unpleasant feedback can play as much of a role in addiction and sneaky behavior modification as the pleasant kind." - p 12

* "**Catfishing** is a deceptive activity where a person creates a fictional persona or fake identity on a social networking service, usually targeting a specific victim. The practice may be used for financial gain, to compromise a victim in some way, as a way to intentionally upset a victim, or for wish fulfillment." - https://en.wikipedia.org/wiki/Catfishing

Aside from the poor comparison between "the carrot and stick" (one tool) & unpleasant & pleasant feedback (2 separate tools) I agree w/ the above.

"People are keenly sensitive to social status, judgment, and competition. Unlike most animals, people are not only born absolutely helpless, but also remain so for years. We only survive by getting along with family members and others. Social concerns are not optional features of the human brain. Thay are primal.

"The power of what other people think had proven to be intense enough to modify the behavior of subjects participating in famous studies like the Milgram Experiment and the Stanford Prison Experiment. Normal, noncriminal people were coerced into doing horrible things, such as torturing others, through no mechanism other than social pressure." - p 16

I agree here too. I often use the term "robopath" wch I find once again useful. Nothing has exemplified the above-described behaviors so much as the malevolent mass conformity of people on Fecesbook during the so-called 'pandemic' of 2020 into the misty future.

"Negative emotions such as fear and anger well up more easily and dwell in us longer than positive ones. It takes longer to build trust than to lose trust. Fight-or-flight responses occur in seconds, while it can take hours to relax.

"This is true in real life, but it is even more true in the flattened light of algorithms.

"There is no evil genius seated in a cubicle in a social media company performing calculations and deciding that making people feel bad is more "engaging" and therefore more profitable than making them feel good. Or, at least, I've never met or heard of such a person.

"The prime directive to be engaging reinforces itself, and no one even notices that negative emotions are being amplified more than positive ones. Engagement is not meant to serve any particular purpose other than its own enhancement, and yet the result is an unnatural global amplification of the "easy" emotions, which happen to be the negative ones." - p 18

That, too, seems fair enuf - but I wonder how the all-too-common mood elevator drugs are affecting that. In other words, are there millions of people whose ability to even have negative emotions, when appropriate, is diminished by drugs meant to make their life more bearable even if delusional?

"Seems like a good moment to coin an acronym so I don't have to repeat, over and over, the same account of the pieces that make up the problem. How about "Behaviors of Users Modified, and Made into an Empire for Rent"? BUMMER.

"BUMMER is a machine, a statistical machine that lives in the computing clouds. To review, phenomena that are statistical and fuzzy are neverthless real. Even at their best, BUMMER algorithms can only calculate the *chances* that a person will act in a particular way. But what might be only a chance for each person approaches a certainty *on the average* for large numbers of people." - p 28

"Spying is accomplished mostly through connected personal devices—especially, for now, smartphones—that people keep practically glued to their bodies. Data are gathered about each person's communications, interests, movements, contact with others, emtional reactions to circumstances, facial expressions, purchases, vital signs: an ever growing, boundless variety of data.

"If you're reading this on an electronic device, for instance, there's a good chance an algorithm is keeping a record of data such as how fast you read or when you take a break to check something else." - p 30

& perhaps all this data might seem inconsequential to some but it can all be used for manipulation - wch, of course, is Lanier's point. One might also consider the possibility of deliberately providing misleading data in order to increase the chances of skewing the behavior modification.

In 1981, I cofounded a telephone network whose # was AGENT-11. Callers were encouraged on an answering machine msg to say the words "heroin" & "assassination". That meant that those words were repeated every time the message played & there was a chance that callers wd repeat them too. I had recently learned that the NSA's surveillance of phone calls searched for those words to help find potential targets for increased attn. The idea was to try to overload the system or, at least, make it less functional. Cd people do something like that now w/ computer surveillance?

A game I like to play is something I call "Fauxed Rage": that's fake road rage, in wch I, as the driver, pretend to get very upset over something that I really find trivial or not upsetting at all. Then, in my exaggerated response, I wish all sorts of ludicrous horrors on the not-really-offending other driver. What about expressing fake & possibly randomized negative & positive emotions in response to things on social media?

"If owning everyone's attention by making the world terrifying happens to be what earns the most money, then that is what will happen, even if it means that bad actors are amplified. If we want something different to happen, then the way money is earned has to change." - p 34

Jaron Lanier's net worth is $5 million according to https://www.celebritynetworth.com/richest-celebrities/authors/jaron-lanier-net-worth/ . His proposal strikes me as solidly capitalist, something not surprising from someone that rich. But what wd happen if money were devalued? & I don't mean by inflation, we already have that & it seems to serve capitalism as much as anything else given that it enables people to have the delusion that their lives have improved w/ increased wages that're simultaneously accompanied by increased expenses. What I mean is: What if money were no longer the central enabler of survival? What if the billionaires cdn't buy ANYTHING w/ all their wealth? What if just being a living creature were what was valued the most? What if genuine kindness, an imagination, generosity were what were rewarded? & I don't mean rewarded w/ money, I mean rewarded w/ respect & kindness & imagination & generosity in return. That's more the direction I'm interested in going in.

"The most dispiriting side effect of BUMMER policy-tweaking is that each cycle in the arms race between platforms and bad actors motivates more and more well-meaning people to demand that BUMMER companies take over more and more of our lives. We ask remote, giant tech companies to govern hate speech, malicious falsified news, bullying, racism, harassment, identity deception, and other nasty things. Well-intentioned activists demand that corporations govern behavior more and more. "Please tell us what we can say, oh rich young programmers of Silicon Valley! Discipline us!" The bad actors who wish to discredit democracy using the BUMMER machine win even when losing ground to well-meaning activists." - p 36

Is that what's happening? I'm a "well-meaning activist" & I'm not wanting or asking for any of the above. I perceive the people that Lanier refers to as "well-meaning activists" as easily manipulated fools who've fallen completely for crowd manipulation psychology that's designed to turn them into the robopaths that they now are. As such, the question for me is: How much were they ever genuinely "well-meaning activists" in the 1st place rather than just poseurs &/or sub-culture conformists? Ultimately, all the censorship that people are clamoring for these days serves one primary purpose: to justify limiting critical free speech - by highlighting the assholes as a justification, other people who are actually trying to say something important can be swept under the same smothering rug. The real defeat of 'fake news' wd essentially require getting rid of EVERY mainstream 'news' outlet, there's not one of them that isn't lying & supporting a hidden agenda. While there's genuine news to be found thru them too one has be very perspicacious to sort thru all the garbage that's there to prop up the existing power structure.

"I'm a father, and I want the children my daughter interacts with to be immunized. Immunization is a common good, a gift we can give each other. It is one of the greatest inventions in human history." - p 59

Note that he says: "I want the children my daughter interacts with to be immunized": it's not just a matter of wanting his own children to be immunized. The alleged purpose of immunization is to make the immunized people no longer vulnerable to disease. If that were true then there wd no longer be any personal health worry about what's happening w/ other people as to whether they're immunized or not. But, strangely & illogically, the people who most believe in vaccination also react fearfully to the unvaccinated. The object of fear doesn't even have to be sick! One also has the completely idiotic idea of "asymptomatic" to eliminate even that shred of sense. Lanier, being a fat cat, takes it for granted that he has the right to dictate to other people.

"But I know other parents—educated, upper-middle-class American parents—who won't even consider vaccinating their kids. Some of them are "left" and some "right." It's not just that they think immunization is bad; they believe that it's evil, alien, and icky. They think it causes autism. They can't get conspiracy theories out of their heads. You might think I'm being elitist when I am more appalled that "educated" parents, who are more likely to be affluent, foment dangerous nonsense, but isn't the whole point of education supposed to be that it diminishes people's susceptibility to dangerous nonsense?

"I have tried to engage with these parents, and that's when they show me their BUMMER feeds." - pp 59-60

I find the above to be very loaded. What the parents he's criticizing have in their heads are "conspiracy theories", what he has in his head is 'solid scientific knowledge' - but his idea of "BUMMER" is easily dismissed by people as also a 'conspiracy theory'. When Lanier says he tries to "engage with these parents" what he means is he's proselyzing to them for his religion, science - & he does it for the same reasons that all the other proseltyzers do: he's sure he's right & the other parents are wrong. But why is he so arrogant as to believe that he has the right to try to browbeat parents to raise their kids the way *he wants them to*? B/c he has the same mindset as what I consider, to use his terminology, the other 'bad actors' to have: viz: a sense of superiority that, in his case, is grounded in that very education that he refers to when he says "but isn't the whole point of education supposed to be that it diminishes people's susceptibility to dangerous nonsense?" To wch I answer: NO, it isn't. The purpose of education is to educate but it's also to socialize the students into whatever class ideology it's grooming them for - & if that class ideology includes a conceited sense of superiority that makes the student feel the right to impose a medical system on EVERYONE then so be it - it's b/c Lanier is obvious to himself as a 'good guy' who deserves to lord it over others. He didn't get to be a millionaire w/o that built into his personality. The ultimate question here is: Whose kids are the healthiest? The proof is in the pudding.

"There wasn't anyone sitting in a tech company who decided to promote anti-vaccine rhetoric as a tactic. It could just as easily have been anti-hamster rhetoric. The only reason BUMMER reinforces this stuff is that paranoia turns out, as a matter of course, to be an efficient way of corralling attention." - pp 60-61

Is "anti-vaccine" speech really "rhetoric"? If so, why isn't he also referring to pro-vaccine speech as "rhetoric"? Perhaps it's b/c he wants to discredit anti-vax positions by making them seem like empty argumentation tactics. It seems obvious to me that anti-vax arguments are made w/ the same passion as pro-vax ones by people who are equally interested in doing what they think is healthiest. Since Lanier's on one side of the argument he's sure he's right & they're wrong. What if they both have valid points? It seems that Lanier can't even allow that as a possibility. That's his conceit.

Actually, I don't think it cd've been "anti-hamster rhetoric", it's not nearly that arbitrary. Vaccination has been a controversial practice for hundreds of yrs. Many of the people against it have been doctors who've had substantial experience seeing its effects 1st-hand. It becomes most controversial when people try to make it mandatory. &, no, I don't think anti-vax speech is being used to corral attn - *I think it's Lanier who's being paranoid here.* Are you a pot-head, Lanier? If you are, you might want to give it up.

"Speaking through social media isn't really speaking at all. Context is applied to what you say after you say it, for someone else's purposes and profit.

"This changes what can be expressed. When context is surrendered to the platform, communication and culture become petty, shallow, and predictable. You have to become crazy extreme, if you want to say something that will survive even briefly in an unpredictable context. Only asshole communication can achieve that." - p 65

"communication and culture" [also] "become petty, shallow, and predictable" when they're shaped by the LCD (Lowest Common Denominator) - as far as I can see the LCD *isn't* a desire for challenging, complex experiences it's for the simplest possible things that give one the delusion that one 'understands': it's for music that people w/ minimal imagination & talent can dance to w/o worrying about whether they look too weird to their peer group. I don't think "[o]nly asshole communication can" make an impression when "communication and culture" [has] "become petty, shallow, and predictable", I think it's the asshole communication that makes it that way.

"What if deeply reaching a small number of people matters more than reaching everybody with nothing?" - p 68

Exactly. & that's where non-LCD communication & culture comes in. I like to think that my reviews, e.g., are in that category.

"I note that news sites that are trying to woo advertisers directly often seem to show *spectacularly* greater numbers for articles about products that might be advertised—like choosing your next gaming machine—than for articles about other topics.

"This doesn't mean the site is fudging the numbers. Instead, a manager probably hired a consulting firm that used an algorithm to optimize the choice of metrics services to relate the kind of usage statstics the site could use to attract advertisers. In other words, the site's owners didn't consciously fudge, but they kinda-sorta know that their stats are part of a giant fudge cake." - p 68

Gosh, isn't that a CONSPIRACY THEORY? You must be one of those pro-Trump Anti-Vax subhumans!

"the most common form of online myopia is that most people can only make time to see what's placed in front of them by algorithmic feeds.

"I fear the subtle algorithmic tuning of feeds more than I fear blatant dark ads. It used to be impossible to send customized messages to millions of people instantly. It used to be impossible to test and design multitudes of customized messages, based on detailed observation and feedback from unknowing people who are kept under constant surveillance." - p 78

& here we really are at the crux of the matter, aren't we?! If anyone were to ask anyone else where their 'news' comes from & what things they've read recently & *why* they read those particular things wd they be able to answer? How many people read a particular thing b/c a promotion for it has appeared on their computer as a result of a long chain of data-analysis that suggested that this is 'the type of thing they want to read'? Now, let's look at a list of the 10 most recent bks I've read & reviewed before this one:

Eric Ambler's "**The Light of Day**"
Earl Derr Biggers's "**7 Keys to Baldpate**"
Harlan Coben's "**Six Years**"
Lawrence Block's "**Grifter's Game**"
Eleanor McBean's "**The Poisoned Needle - Sup[p]ressed Facts about Vaccination**"
John Brunner's "**A Maze of Stars**"
Members of the Detection Club's "**The Floating Admiral**"
Paul A. Offit, M.D.'s "**Autism's False Prophets - Bad Science, Risky Medicine, and the Search for a Cure**"
Solomon E. Asch's "**A Study of Change in Mental Organization**"
Mark McDonald M.D.'s "**United States of Fear**"

One cd probably deduce a psychological profile of me from this but wd it be accurate? The 1st 4 bks are crime fiction from my personal library chosen *roughly* in alphabetical order of the author's name w/ the authors all being people I hadn't previously read. But it gets more complicated than that. I read Coben's bk before I read Block's b/c Coben's was a smaller paperback & easier to carry around. I was only reading Coben at all b/c a friend of mine loves his writing so I decided to read it even tho I expected it to be what it was: somewhat formulaic pop thriller writing. Then I switched to an anti-vaccination bk written by an autodidact whose researching skills & dedication struck me as remarkable. THEN, science fiction b/c I love SF & was getting a bit bored w/ the crime fiction. THEN, back to crime fiction again. I had read work by some of the people in the Detection Club so I cd've rejected this bk on those grounds but I chose to treat the Detection Club as its own entity. Then I read a pro-vax bk by a mainstream doctor. THEN, a short child psychology study. THEN, an analysis of the COVID-19 quarantine from the POV of a conservative psychiatrist who believes the fearfulness of women is the problem b/c it's unchecked by the emasculation of the men in their lives. This is even a narrow range of bks for me but how many people do I know who wd read even this diverse a selection? None, I suspect. Did I choose these bks as a result of algorithmically created prompts? I don't think so - as far as I can tell, I'm still able to pick what I read w/o being trapped in a socio-political-economic stereotype. I seriously doubt that any of the people that I know who're so adamantly pro & against things that reek of subculture-required beliefs can make the same claim validly - including Lanier - who seems only slightly less intellectually buried alive than other 'hip' liberals.

"Here is one thing I discovered about myself: I don't mind being judged if the judges put in real effort, and a higher purpose is being honestly served, but I *really* don't like it when a crowd judges me casually, or when a stupid algorithm has power over me." - p 85

"More and more people rely on the gig economy, which makes it hard to plan one's life. Gig economy workers rarely achieve financial security, even after years of work. To put it another way, the level of risk in their financial lives seems to never decline, no matter how much they've achieved. In the United States, where the social safety net is meager, this means that even skilled, hardworking people may be made homeless by medical bills, even after years of dedicated service to their profession." - p 93

Doesn't that make it obvious that the *medical bills are a problem?!* The Medical Industry, IMO, is a juggernaut that sucks as much as it can out of its patients w/o necessarily providing much actual health care in return. Doctors & hospitals can literally kill the patients & be not only protected from consequences but even praised as heros AND make a fortune off of the procedures leading up to the death. & vaccinations are very much a part of the whole racket.

"The free services that you get are disguised versions of services someone like you would otherwise be paid to provide. Musicians use BUMMER to promote themselves for free, and yet a smaller percentage of musicians are doing well enough to plan families—which is a reasonable definition of "security"—than during the era when music was sold on physical disc. Recording musicians; language translators . . . who's next?" - p 101

Indeed. Musicians even became 'non-essential' during the quarantine - showing, to me, at least, the coworkings of BUMMER & the Medical Industry (wch includes vaccination). I put my work online, for no financial reward, partially just to help IT survive. I doubt that I'll be holding on much longer.

"*Car drivers instead of horsemen.* Indeed, the new roles that came into being because of such tech disruptions were often more creative and professional than the old ones. *Robotics programmers instead of ironworkers.* This meant that more and more people gained prestige and economic dignity.

"BUMMER reversed the trend. Now if you bring insight, creativity, or expertise into the world, you are on notice that sooner or later BUMMER will channel your value through a cloud service—probably a so-called AI service—and take away your financial security, even though your data will still be needed. Art might be created automatically from data stolen from multitudes of real artists, for instance. So-called AI art creation programs are already practically worshipped. Then, robotic nurses might run on data gabbed from multitudes of real nurses, but those real nurses will be working for less because they're competing with robotic nurses." - p 102

It's not clear to me why Lanier thinks that car drivers & robotics programmers were really improvements on horsemen & ironworkers. Does he even know what an ironworker is? I haven't seen any robots building I-beam skeletons for skyscrapers - have I missed that? & I don't see why BUMMER is any different from the use of the Spinning Jenny to put Cottage Industry craftspeople out of business so that all the wealth cd be even further accumulated by the controllers of the industrial devolution. In other words, the same process has been at work for a long time: there are those whose main concern is monopolizing the profits from production by syphoning it off from the producers & reselling it under their own financial system. Is it that Lanier doesn't want to diss capitalism so he sticks to the particulars of BUMMER? After all, capitalism is his feeding trough - what wd he do w/o it?

"Meanwhile my patriotic, hawkish conservative friends now find themselves aligned with a leader who would almost certainly not be in office were it not for cynical, illegal interventions by a hostile foreign power." - p 115

This bk's being copyrighted 2018 means that he's referring to Rump, the Idiot King. I reckon Lanier's referring to the alleged Russian backing of Trump, something that, as far as I know, is unproven &, therefore, another Conspiracy Theory! Yes, Lanier, who mocks the 'Conspiracy Theories' about vaccination has no problem w/ latching onto the Conspiracy Theories about Trump. Has there ever been a president who wasn't backed by "cynical, illegal interventions" of some kind or another? It seems intrinsic to the electoral process, particularly the electoral college. I mean the struggle to get 'your figurehead' into the position where YOU get what you want must be fiercely competetive - of course, some lobbyists just back both contestants to be on the safe side.

"A year after the election, the truth started to trickle out. It turns out that some prominent 'black' activist accounts were actually fake fronts for Russian information warfare. Component F. The Russian purpose was apparently to irritate black activists enough to lower enthusiasm for voting for Hillary. To suppress the vote, statistically." - p 120

More Conspiracy Theory. Wd we've been any better off under Hilary Clinton's presidency? Despite the widely Trumpeted mass media accts of Trump's opposition to masks & the like he was still the one who enabled such medical coups by signing into being the EUA (Emergency Use Authorization), a practice initiated by Obama. Trump managed to avoid taking responsibility by enabling state's rights, enabling governors to make the ultimate decisions about how extreme the quarantine wd be in their states, enabling the power-hungry Democrats to go completely dictatorial. Now we've got Bidentity Crisis in there & Roe vs Wade has been overturned & Biden, too, can claim innocence - after all he didn't appoint the Supreme Court judges that made the decision - &, once again, state's rights are enabled - both Trump & Biden have brought the post-Civil War South's wet dream into existence: the right to not be completely ruled by the federal government.

Don't misunderstand, I'm not dissing conspiracy theory, I'm dissing Lanier's very selective use of the term to criticize others while somehow appearing to consider himself to be free of them. &, once again, there's a post-2018 tie-in between political corruption & the vaccinators that Lanier loves so much. After all,

Sanders & Warren both disappeared pretty quickly as potential presidential candidates b/c of the so-called 'pandemic'. It seems to me that that's not exactly a coincidence.

"Activists might feel confident they are getting their message out, but it is indisputable that black activists have severely lost ground politically, materially, and in every way that matters outside of BUMMER."

[..]

"One example of Component F in the 2016 U.S. election was an account called Blactivist, which was run by the Russians. A year after the elections, the true power behind Blactivist was revealed and reporters asked genuine black activists what they thought about it."*

*"https://www.the guardian.com/world/2017/oct/21/russia-social-media-activism-blacktivist"

- p 122

"it is indisputable that black activists have severely lost ground": That's like saying in 2014 that there's no way Trump cd ever become president. It's "indisputable"? &, yet, Black Lives Matter started in 2013 after the murderer of Trayvon Martin got off. In 2020, after the police murder of George Floyd, Black Lives Matter became one of the most widely supported black activist mvmnts I've ever seen & 2020 saw an incredible number of protests in its support. As for "Blactivist"? Never heard of it - so who exactly was it important to? I'm certainly not aware of every manifestation of political activism that might be relevant to my concerns but I was very active in the Western Pennsylvania Committee to Free Mumia Abu-Jamal, I did organize a screening series on the subject of the "Suppression of Black Radicals in the US", etc, etc, so I'm not the most ignorant person on the subject either.. & I've never heard of Blactivist. So, maybe, just maybe, that was the type of online thing that's more important to people who think that some social media group is more important than the 'real world'.

At the top of the *Guardian* article that Lanier links to, there's a small banner that says: "This article is more than 4 years old". Of course, that wdn't've been the case when Lanier wrote this bk. It does, however, point to the possibility that something that was 'topical' in its day might've been proven wrong sometime after.

""Blacktivist" – a social media account coordinating and promoting the march online – was not run by black American activists, but instead, it is now believed, was operated by an agent of Russia attempting to interfere with US politics."

[..]

"Recent disclosures have revealed that Russian trolls and bots manipulated social media sites to spread false and inflammatory news in an apparent effort to stoke political divisions on a large scale. Facebook admitted last month that a Russian influence operation had purchased $100,000 worth of ads to spread divisive messages about racial injustice, LGBT rights, immigration and other hot-button subjects, and Congress is now investigating. It appears that Twitter, Google, YouTube, Instagram, Pinterest and other sites helped spread the content.

"This week, a Moscow-based news outlet called RBC uncovered the work of a troll factory that infiltrated US social networks, with Russians posing as Americans and making payments to legitimate activists in the US, directly funding protest movements. The revelations suggest that the Russian operation went beyond spamming online comment sections and spreading false news – and that a sophisticated interference campaign manipulated, controlled and created real-world events."

- https://www.the guardian.com/world/2017/oct/21/russia-social-media-activism-blacktivist

MAYBE, such a 'discrediting' serves purposes in & of itself. After all, the issues are certainly real so the issue might not be whether Russians were puppet-mastering people but whether the 'revelation' that this is the case might be an attempt to discredit completely valid mvmnts. The dirty tricks are everywhere, perpetrated by a variety of 'bad actors'. In the last few yrs I'd say that the Democratic Party, desperate to get rid of Rump BAMN (By Any Means Necessary) & to put one of their own puppets in there, has been willing to pull any & every dirty trick in the bk.

There is a redacted document available online called "Report On The Investigation Into Russian Interference In The 2016 Presidential Election Volume I of II" by Special Counsel Robert S. Mueller, III. This document contains the following:

"The first form of Russian election influence came principally from the Internet Research Agency, LLC (IRA), a Russian organization funded by Yevgeniy Viktorovich Prigozhin and companies he controlled, including Concord Management and Consulting LLC and Concord Catering (collectively "Concord"). The IRA conducted social media operations targeted at large U.S. audiences with the goal of sowing discord in the U.S. political system. These operations constituted "active measures" (активные мероприятия), a term that typically refers to operations conducted by Russian security services aimed at influencing the course of international affairs.

"The IRA and its employees began operations targeting the United States as early as 2014. Using fictitious U.S. personas, IRA employees operated social media accounts and group pages designed to attract U.S. audiences. These groups and accounts, which addressed divisive U.S. political and social issues, falsely claimed to be controlled by U.S. activists. Over time, these social media accounts became a means to reach large U.S. audiences. IRA employees travelled to the United States in mid-2014 on an intelligence-gathering mission to obtain information and photographs for use in their social media posts."

- p 22, https://www.justice.gov/archives/sco/file/1373816/download

As the plot develops it provides this:

"Many IRA operations used Facebook accounts created and operated by its specialists."

[redaction claiming: "Harm to Ongoing Matter"]

"groups (with names such as "Being Patriotic," "Stop All Immigrants," "Secured Borders," and "Tea Party News"), purported Black social justice groups ("Black Matters," "Blacktivist," and "Don't Shoot Us"), LGBTQ groups ("LGBT United"), and religious groups ("United Muslims of America").

"Throughout 2016, IRA accounts published an increasing number of materials supporting the Trump Campaign and opposing the Clinton Campaign. For example, on May 31, 2016, the operational account "Matt Skiber" began to privately message dozens of pro-Trump Facebook groups asking them to help plan a "pro-Trump rally near Trump Tower."

"To reach larger U.S. audiences, the IRA purchased advertisements from Facebook that promoted the IRA groups on the newsfeeds of U.S. audience members. According to Facebook, the IRA purchased over 3,500 advertisements, and the expenditures totaled approximately $100,000.56

"During the U.S. presidential campaign, many IRA-purchased advertisements explicitly supported or opposed a presidential candidate or promoted U.S. rallies organized by the IRA (discussed below). As early as March 2016, the IRA purchased advertisements that overtly opposed the Clinton Campaign. For example, on March 18, 2016, the IRA purchased an advertisement depicting candidate Clinton and a caption that read in part, "If one day God lets this liar enter the White House as a president – that day would be a real national tragedy." Similarly, on April 6, 2016, the IRA purchased advertisements for its account "Black Matters" calling for a "flashmob" of U.S. persons to "take a photo with #HillaryClintonForPrison2016 or #nohillary2016." IRA-purchased advertisements featuring Clinton were, with very few exceptions, negative."

- pp 22-23, https://www.justice.gov/archives/sco/file/1373816/download

Now documents like this that have the appearance of thoroughness & 'objectivity' are easy, perhaps all-too-easy, even for sceptics like myself, to find convincing. But 'justice' in the US has a long history of bias against what political activists consider to be justice. Is "justice.gov" exempt from this? Of course not. The political interests of the ultra-rich are always going to prevail. Expenditures of "approximately $100,000" are enormous to me but small-time in contrast to the political lobbyists in the US. $100,000 is the equivalent of what a pharmaceutical company might bribe a few well-placed doctor-administrators, small change for the company as long as the doctors then promote the use of their product to produce millions in profits.

"Believing something only because you learned it through a system is a way of giving your cognitive power over to that system. BUMMER addicts inevitably at least tolerate a few ridiculous ideas in order to partake at all. You have to believe sufficiently in the wisdom of BUMMER algorithms to read what they tell you to read, for instance, even though there's evidence that the algorithms are not so great." - p 129

Indeed. & it's precisely these algorithms that've helped create such fervent BELIEVERS in pro-vax & anti-vax positions. Lazy pseudo-scholars just go to the links that algorithms determine are appropriate to their demographic & get opinions that weren't really theirs to begin w/ reinforced. That's one of the reasons why I'm a big proponent of going to scholarly used bookstores & browsing for bks of possible interest. But, BEWARE!, even in those conditions the biases of the store managers & other personnel will prejudice things. E.G.: my 14th bk ("**Unconscious Suffocation - A Personal Journey through the PANDEMIC PANIC**": http://idioideo.pleintekst.nl/Book2020.09PANDEMIC.html) was shelved in the paperback fiction section of a used bookstore by the manager b/c she's a hypochondriac quarantiniac who will be hiding behind 2 masks, plexiglass barriers, & social distancing long after the rest of us will have gotten past the madness.

"I've been using both the term "spiritual" and the term "religious," and here's why: Religions generally are connected with specific truth claims, while Spirtuality might not be. Spirituality can usually coexist a little more with Enlightenment thinking." - p 130

& that's a common positioning taken by people who consider themselves 'enlightened' & who, therefore, also consider themselves to be "spiritual". To me, the difference in degree described above is not really enuf to stop 'spirituality' from being similarly egregious to "religion". I quote from an interview conducted w/ me by writer Alan Davies & published both online & as a hard-copy bk:

"When I hear people speak about their own 'spirituality' I have much the same reaction - I'm immediately suspicious. People often refer to 'spirit' instead of 'god' as a way of expressing their investment in a 'non-material 'higher' 'essence" 'free' of religious baggage. In other words, 'I believe in the spirit as the pure guiding force of all good impulses'.

"What I find that this usually means is something more along the lines of 'I can disguise my own sleazy ulterior motives by camouflaging them w/ references to an intangible higher authority that I am supposedly deeply in tune w/.' In other words, bullshit, dogshit, cowshit, humanshit, eatshit, seenoshit, hearnoshit, speaknoshit. In OTHER other words, I'm not sure I've ever met a 'spiritual' person who didn't strike me as a fraud.

"So what does that say about the concept of the 'spirit' & the 'spiritual'? At the risk of overquoting Wikiwhatever, "spirituality" is presented in the opening paragraph of its Wikipedia definition as:

""Spirituality can refer to an ultimate or immaterial reality; an inner path enabling a person to discover the essence of their being; or the "deepest values and meanings by which people live." Spiritual practices, including meditation, prayer and contemplation, are intended to develop an individual's inner life; such practices often lead to an experience of connectedness with a larger reality, yielding a more comprehensive self; with other individuals or the human community; with nature or the cosmos; or with the divine realm. Spirituality is often experienced as a source of inspiration or orientation in life. It can encompass belief in immaterial realities or experiences of the immanent or transcendent nature of the world."

"My questions are:

"If spirituality is an "immaterial reality" why is that "ultimate"? Why is there a hierarchy in wch immateriality is 'better' than materiality?"

- http://idioideo.pleintekst.nl/InterviewiNTERVIEW.html

"Usually Google has had a way of coming up with the creepier statements, but Facebook has pulled ahead: A recent revision in its statement of purpose includes directives like assuring that "every single person has a sense of purpose and community."" - p 133

Ha ha! What about those of us who have had a sense of purpose our whole lives that we've stuck to w/ the strength of visionaries? Shd we change to the purpose Fecesbook creates for us? & what about those of us who find so-called 'community' to be just as often a PEER PRESSURE COOKER clique or mob than it is its more positively touted 'sense of togetherness'? I'll express the lone wolf's mutual aid solidarity w/ other people over specific issues such as anti-racism & class war but I'm not going to merge w/ a subculture that can be jerked around by crowd psychology - & that's *ALL* subcultures, *ALL* communities. At least Lanier can recognize the 'creepiness' of such a mission statement as that of FB.

"Google's director of engineering, Ray Kurzweil, promotes the idea that Google will be able to upload your consciousness into the company's cloud, like the pictures you take with your smartphone. He famously ingests a whole carton of longevity pills every day in the hope that he won't die before the service comes online. Note what's going on here. The assertion isn't that consciousness doesn't exist, but that whatever it is, Google will own it, because otherwise, what could this service even be about?

"I have no idea how many people believe that Google is about to become the master of eternal life, but the rhetoric surely plays a role in making it seem somehow natural and proper that a BUMMER company should gain so much knowledge and power over the lives of multitudes." - pp 133-134

Do I detect some professional jealousy here? I knew of Kurzweil several decades before I ever heard of Lanier b/c I used his Kurzweil machine that enabled the reading of bks aloud to blind people way back in 1980, an invention I'm still quite fond of. At any rate, I think Lanier's warning is well-advised at the same time that it just seems to me that Kurzweil probably reads enuf SF to've come across that consciousness-upload idea in quite a few stories.

"**Mind uploading**, **whole brain emulation**, or **substrate-independent minds**, is a use of a computer or another substrate as an emulated human brain. The term "mind transfer" also refers to a hypothetical transfer of a mind from one biological brain to another. Uploaded minds and societies of minds, often in simulated realities, are recurring themes in science-fiction novels and films since the 1950s.

"**Early and particularly important examples**

"A story featuring an artificial brain that replicates the personality of a specific person is "The Infinite Brain" by John Scott Campbell, written under the name John C. Campbell, and published in the May 1930 issue of *Science Wonder Stories*. The artificial brain is created by an inventor named Anton Des Roubles, who tells the narrator that "I am attempting to construct a mechanism exactly duplicating the mechanical and electrical processes occurring in the human brain and constituting the phenomena known as *thought*." The narrator later learns that Des Roubles has died, and on visiting his laboratory, finds a machine that can communicate with him via typed messages, and which tells him "I, Anton Des Roubles, am dead—my body is dead—but I still live. I am this machine. These racks of apparatus are my brains, which is thinking even as yours is. Anton Des Roubles is dead but he has built me, his exact mental duplicate, to carry on his life and work." The machine also tells him "He made my brain precisely like his, built three hundred thousand cells for my memory, and filled two hundred thousand of them with his own knowledge. I have his personality; it is my own through a process I will tell you of later. ... I think just as you do. I have a consciousness as have other men." He then explains his discovery that the electrical impulses in the brain create magnetic fields that can be detected by a device he built called a "Telepather", and that "[t]hrough this instrument any one's mental condition can be exactly duplicated." Later, he enlists the narrator's help in constructing a new type of artificial brain that will retain his memories but possess an expanded intellect, though the experiment does not go as planned, as the new intelligence has a radically different personality and soon sets out to conquer the world."

- https://en.wikipedia.org/wiki/Mind_uploading_in_fiction

&, yeah, sure, there're going to be people who'd want to 'own' such a process & make themselves even more rich & powerful than they already are. & then there're going to be people who manage to avoid being captured in such a trap. Personally, I have no desire to be uploaded in such a way whatsover. While it seems to offer immortality, it strikes me as offering a very long-term prison that wd be a hell that competes w/ or surpasses the one(s) that religious people imagine. As far as I'm concerned, things are constantly changing & when I die that'll be just-another change. Maybe nothing that I currently conceive of as 'me' will be even remotely recognizable anywhere (esp to some semblance of what I call "me" now) but, WHATEVER!, somehow things that I consider to be important about 'me' will probably be in the mix.

"This is madness. We forget that AI is a story we computer scientists made up to help us get funding once upon a time, back when we depended on grants from government agencies. It was pragmatic theater. But now AI has become a fiction that has overtaken its authors." - p 135

I prefer the term (that, as far as I know, I coined) "AU" (Artificial Unintelligence). This gives me an opportunity in a bk review to promote one of my (currently 692) movies that hardly anyone ever watches:

648. "**Artificial Unintelligence**"
- made in June 2021 as a collaboration between tENTATIVELY, a cONVENIENCE + Dick Turner + Carrion Elited
- 1920 X 1440 - 30fps stereo
- 12:27
- on my onesownthoughts YouTube channel here: https://youtu.be/Iw-rTflW-EQ
- on the Internet Archive here: https://archive.org/details/artificial-unintelligence

My movies will put hair on yr chest so if that's not what you want use them wisely.

"Should machines be given "equal rights," as is so often proposed in tech culture? Indeed, Saudi Arabia has granted citizenship to a "female" robot, and with that citizenship, rights not available to Saudi human women."*

*"https://www.washingtonpost.com/news/innovations/wp/2017/10/29/saudi-arabia-which-denies-women-equal-rights-makes-a-robot-a-citizen/" - p 137

Given that corporations already have legal rights as if they're individual humans it's not that far-fetched that the same thing might happen w/ machines.

"Corporations express the collective investment goals of shareholders. The legal stricture known as fiduciary responsibility confines all but closely held corporations to this singular goal. By shutting off other values to focus solely on pursuit of profit in inherently amoral economic competition, corporations are by their nature amoral as well.
- "How corporations became 'persons'
- The amazing true story of a legal fiction that undermines American democracy."

- Tom Stites, May 1, 2003
- http://www.uuworld.org/ideas/articles/157829.shtml"

- http://idioideo.pleintekst.nl/tENTroboNotes.html

It's highly unlikely that any machine, no matter how it's programmed, will have the sense of ethics that humans hypothetically instinctually do. But are humans naturally ethical? Or is that something induced by education? I'm sure there're people who'd argue the latter - thusly adding to the case for giving machines 'rights'.

"Techies can become isolated through extreme wealth and might seem unreachable, but actually we miss you. When techies engage in fixing problems they helped create, they become connected again and that feels good." - p 143

I feel yr pain, Lanier - & b/c I'm such a compassionate (SPIRITUAL?) person I'll give you a chance to fix both our problems: give me one million dollars so that I don't have to suffer from my $8,000.00 a yr Social Security income & I'll play music w/ you. Deal?

& then we get to something I really wasn't expecting:

"Social media was playing a role in making the world newly dark and crazy, and I was asked about that. This book arose from things I thought to say when confronted. I must thank the journalists who forced this issue, including Tim Adams, Kamal Ahmed, Tom Ashbrook, Zoë Bernard, Kent Bye, Maureen Dowd, Moira Gunn, Mary Harris, Ezra Klein, Michael Krasny, Rana Mitter, Adi Robertson, Peter Rubin, Kai Ryssdal, Tavis Smiley, Steven Tweedie, and Todd Zwillich." - p 145

& why wd that surprise me? B/c Kent Bye is one of the journalists listed. Kent started making a documentary about me in 2001 but gave up after 7 mnths b/c he'd wanted to make a single-issue doc & I presented too complex of a subject. Still, he shared his footage w/ me & it's featured heavily in the following movies of mine:

599. "<u>Street Ratbag No5 Release Event</u>"
- shot by Kent Bye & tENTATIVELY, a cONVENIENCE in the fall of 2001E.V.
- centered around the manufacture & release of the "Street Ratbag No5" magazine edited by Rita Rodentia & RATical
- featuring performances by Guitarists Anonymous (tENTATIVELY, a cONVENIENCE + Daryl Fleming + Greg Pierce) + the Hip Criticals (HipHopcrates + Little Orphan Anarchy with guest appearance by tENTATIVELY, a cONVENIENCE)
- also featuring a movie by Mark Dixon of Think Tank VIII & an excerpt from a movie of tENTATIVELY, a cONVENIENCE's CircumSubstantial Playing & Blindfolded Tourism
- edit finished June 25, 2019E.V. by tENTATIVELY, a cONVENIENCE

- 3:43:50
- UNLISTED on my onesownthoughts YouTube channel here: https://youtu.be/ 3hEepzUGkss

598. "**Dixmont**"
- a tour of a closed-down mental hospital called Dixmont on February 17, 2002E.V. conducted by Zack Jones + scans & readings of a printed matter interview conducted by Rita Rodentia with Zack for the *Street Ratbag* magazine, issue 6
- featuring additional input from Erok & Angry Ron, urban explorers who joined us on the tour
- (almost) continuous camcorder: tENTATIVELY, a cONVENIENCE
- intermittent camcorder: Kent Bye
- all else: tENTATIVELY, a cONVENIENCE
- edit finished June 19, 2019E.V.
- 2:15:05
- on my onesownthoughts YouTube channel here: https://youtu.be/ yEWrSNKzg60

596. "**tENTATIVELY, a cONVENIENCE Interviewed by Kent Bye February 16, 2002E.V.**"
- shot by Kent Bye; edited & titled by tENTATIVELY, a cONVENIENCE who also enlarged upon the interview - especially by including footage shot by Kent Bye of his Frame of Reference puppet theater (1975)
- 45:34
- on my onesownthoughts YouTube channel here: https://youtu.be/6E8U-O0b21U
- on the Internet Archive here: https://archive.org/details/tENTInterviewKent

595. "**DUET February 17, 2002E.V.: Michael Pestel & tENTATIVELY, a cONVENIENCE**"
- shot originally by Kent Bye for tENTATIVELY, a cONVENIENCE's "Guitarists Anonymous Withdrawal Aids". It was also intended to be used in a documentary by Kent about tENT. The documentary was never finished. This version is the entire duet with most of the talking removed.
- 53:31
- on my onesownthoughts YouTube channel here: https://youtu.be/ hYMMgww0gLg

262. "**B.T.O.U.C.**"
- made under the name of Tim Ore
- cameras: David Yaffe, Lizard Media Systems, Craig Considine, Kent Bye
- conception, didActing, scanning, editing, etc: **tENTATIVELY, a cONVENIENCE**
- This is basically a drastically upgraded version of 036 that's different enough to get its own number.
- 1/2" VHS cassette, 35mm slides, mini-DV -> DVD
- 37:53
- '82-'84 / '01-'02 / march '06
- on my onesownthoughts YouTube channeel here: https://youtu.be/

_HxXGIMriKA

220. a. "**A Slide Show by tENTATIVELY, a cONVENIENCE**"
- slides originally assembled in this order in november of 1999 by **tENTATIVELY, a cONVENIENCE**
- transferred to mini-DV by Kent Bye in the fall of 2001
- narration from **tENTATIVELY, a cONVENIENCE**
- in-computer editing & additional photo insertions by Kent Bye
- 35mm slides & voice -> mini-DV -> computer -> 1/2" VHS cassette
- 43:53
- november, 99 / fall, 01
b. version with 2 titles added to end on June 16, 2019 for the internet
- 44:09
- on my onesownthoughts YouTube channel here: https://youtu.be/tWqjOGPyf98

So, thank you, Kent, I'm glad you're still out there being active. & then there's

"Thanks to Jerry Mander; this book's title is a tribute to his work." - p 145

&, yes, Jerry Mander's "Four Arguments for the Elimination of Television"? Yes, I agree - it's an excellent & very important bk. I used readings from it in 2 movies of mine. Here's a relevant quote:

""It is no accident that television has been dominated by a handful of corporate powers. Neither is it accidental that television has been used to re-create human beings into a new form that matches the artificial, commercial environment. A conspiracy of technological and economic factors made this inevitable and continue to."
- page 113, Four Arguments for the Elimination of Television - Jerry Mander, 1978"

- http://idioideo.pleintekst.nl/tENTroboNotes.html

As for "**Ten Arguments for Deleting Your Social Media Accounts Right Now**"? From my POV Lanier nails it when he critiques the way algorithms in social media are behavior modification tools used to narrow human behavior into economically motivated channels. Where I think Lanier fails is in not critiqueing capitalism in general & the Medical Industry in particular. As such, I can only give this bk a 3 star rating.

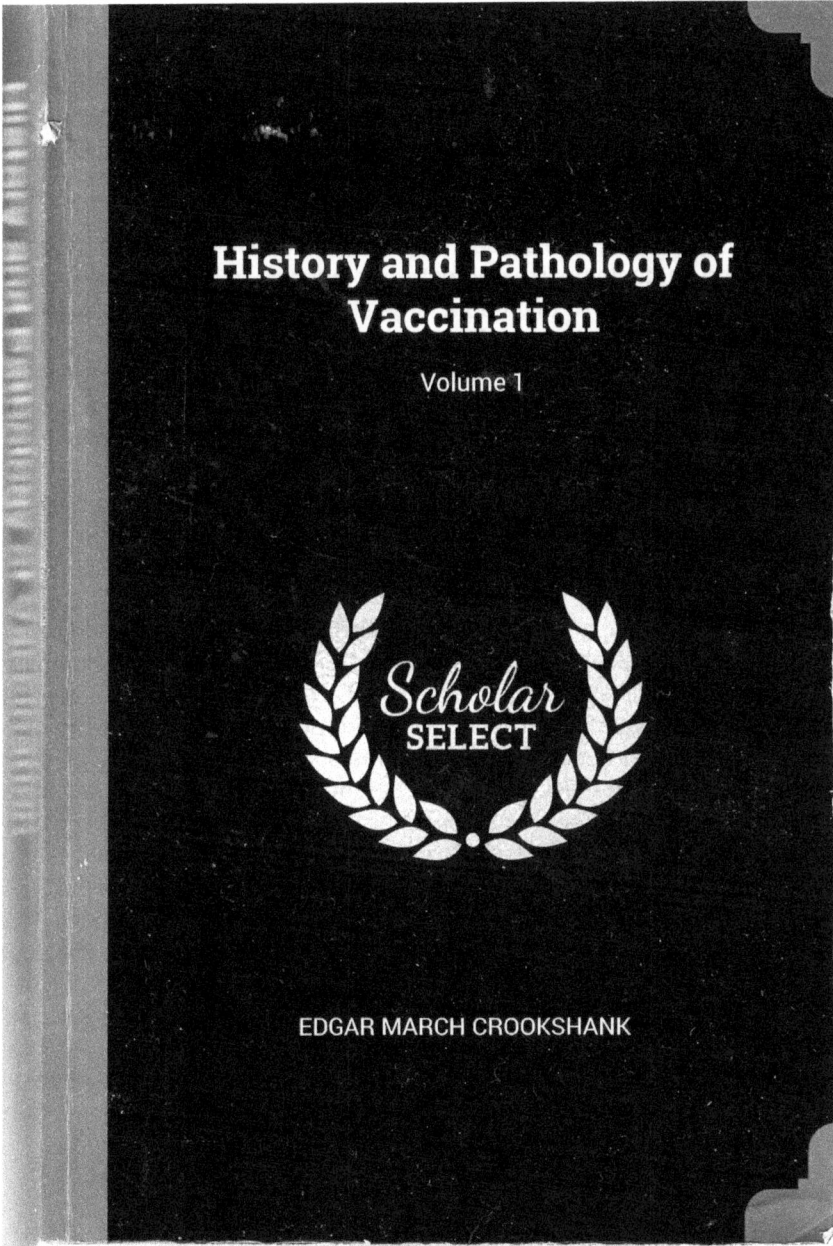

review of
Edgar March Crookshank's **History and Pathology of Vaccination - Volume 1**
by tENTATIVELY, a cONVENIENCE - October 1-10, 2022

As I've explained often enuf in other reviews, I've been what I might call "vaccination-neutral" for most of my life. As a child in the 1950s & '60s I probably rc'vd whatever the standard vaccinations were in my area. I'm sure that included one for polio & one for some sort of pox or another (I still have the scar for that one on my arm). I got them, like most children, b/c my parents had me get them. My parents, like children, did that b/c drs told them too.

The last vaccination I got was probably in 1968 when I was 14. My Junior High School lined the students up like domesticated animals going to the slaughter & injected us w/ a hypodermic that had a large silver cannister, they changed out the needle for each new injection. That one was for a disease that the birds that were temporarily roosting across the street from the school in some woods were supposedly carriers of, avian flu, I suppose. One of the students fainted when injected, the other students made fun of him. At the time, I thought the whole thing was ridiculous, I wasn't afraid of getting a disease from the birds or afraid of birds in general. I went for walks in the woods all the time, I felt at home there, the wildlife were creatures whose presence I enjoyed. I stopped getting vaccinated after that, 54 or 55 yrs ago.

I've gotten the flu, probably usually at times when I'd weakened my body w/ excessive drinking of alcohol. I don't remember even being particularly bothered by it, it's been almost like a vacation from my body's usual state: my nose runs, I cough, I get a mild fever, I lay around.. it's lasted perhaps 4 to 7 days, no big deal. I don't remember when flu shots became such a common thing as they are now, it seems to me that most people just got the flu & rode it out w/ only mild complaining. Now, of course, the flu shots are heavily promoted at drug stores & elsewhere every yr - & w/ this heavy promotion has come an increased fear of the flu - I'm still not afraid of it. If my body ever gets so weak that the flu threatens to kill me off then maybe it's time to go, I'm not expecting to live forever. At any rate, I'm 69 yrs old & the thought of getting a flu doesn't worry me at all.

The extremely Draconian pressure on people to get COVID-19 vaccinations & the apparently never-ending booster shots has never struck me as a positive or necessary public health measure. SO, having, like most people, esp most pro-vaccination people, very limited knowledge of the history of vaccination, I decided to learn more about it & came across mention of this bk & its author in the process.

"**Edgar March Crookshank** (2 October 1858 – 1 July 1928) was an English physician and microbiologist.

"Biography

"Crookshank studied at King's College London and qualified for medicine in 1881. He served briefly as an assistant to Joseph Lister, a physician noted for his work promoting antiseptics and sterile surgery. In 1882, Crookshank served as a doctor with the British armed forces sent to Egypt as a result of the Urabi Revolt; he was decorated for his service at the Battle of Tel el-Kebir.

"On return from Egypt, Crookshank toured Europe in 1884 for further medical training. In Berlin, he visited the laboratory of Robert Koch and learned methods of isolating bacterial strains to investigate infectious diseases.

"When he returned to London, Crookshank wrote a textbook, *An Introduction to Practical Bacteriology Based on the Methods of Koch*, which was published in 1886. Subsequent editions were published under differing titles in 1887, 1890 and 1896, and a French translation by H. Bergeaud was published in Paris as soon as 1886.

"In 1885, Crookshank founded one of the world's first bacteriological laboratories for human and veterinary pathology in London.

"Crookshank was also interested in the use of photography to study bacteria and published *Photography of Bacteria* in 1887, the first text in English devoted solely to the photography of bacteria. In the introduction to this book he wrote that the photographs were "intended to convince scoffers of the essential truth of the new Science, that specific, often morphologically distinct, microorganisms were the cause of particular infectious diseases".

"During this time he became interested in the study of infectious diseases in animals and in 1886 was awarded the chair of Comparative Pathology and Bacteriology at King's College London. In his new role he was asked to investigate an outbreak of cowpox in Lechlade, Gloucestershire.

"His investigations led him to reconsider the use of cowpox-derived vaccines to immunize against smallpox, a treatment developed by Edward Jenner nearly a hundred years earlier. His conclusion was that such vaccines were ineffective in preventing smallpox because the two diseases (cowpox and smallpox) were "totally distinct". Instead of a cowpox-derived vaccine, he advocated the use of a more dangerous vaccination using attenuated smallpox. In 1889, he published a two-volume treatise on the subject, *A History and Pathology of Vaccination*. Vaccination policies were a divisive topic at the time and in the ensuing controversy that resulted from his publication, Crookshank quit his chair at King's College London in 1891. He continued to speak out on health matters but never worked in a laboratory again."

- https://en.wikipedia.org/wiki/Edgar_Crookshank

I quote so much of the above Wikipedia entry in order to show that Crookshank certainly had what wd ordinarily be considered acceptable qualifications for writing a bk on vaccination. He was a physician & microbiologist whose studies were international. He studied infectious diseases in particular. He established one of the 1st bacteriological laboratories. He had multiple bks to his credit before writing the one under review. He was awarded the chair of Comparative Pathology and Bacteriology at King's College London. But there's something in particular about the Wikipedia entry that I want to call attn to - viz: "he advocated the use of a more dangerous vaccination using attenuated smallpox." There's no citation substantiating this claim that this was "a more dangerous vaccination" - this is, quite simply, propaganda. Note that the entry also claims that "Vaccination policies were a divisive topic at the time" - the implication seems to be that they're no longer "a divisive topic" when, in fact, vaccination is still a heavily contested procedure to this day.

This bk is 466pp long & is only the 1st of 2 volumes. As such, very few people are ever likely to read it in its entirety. Even this review will be so long that not many people will even read just this. I devote so much time & energy to writing the review b/c I think it's an important subject. My 'take-away' is that the drs under scrutiny here were an arrogant bunch infatuated w/ their own sense of superiority. I take the following definition of hubris from an investment article but I think it's as applicable to drs as it is to financiers:

"Hubris is the characteristic of excessive confidence or arrogance, which leads a person to believe that they may do no wrong. The overwhelming pride caused by hubris is often considered a flaw in character." - https://www.investopedia.com/terms/h/hubris.asp

HUBRIS, an important concept here. For hundreds of yrs vaccination has been heavily promoted by people who're sure they can do no wrong - & when vaccination backfires, as it does over & over again, the severe health problems & deaths caused by it are dismissed as flukes the result of improper vaccinating procedures or collateral damage - there's always an excuse to justify the casualties - always backed by the claim that if it weren't for the vaccinations the situation wd be even worse, the deaths even more numerous. I'm not convinced that that's true. I AM convinced that drs get away w/ murder.

The title page provides these qualifications & the date of publication:

"EDGAR M. CROOKSHANK, M.B.

"Professor of Comparative Pathology and Bacteriology in, and Fellow of, King's College, London

"Author of papers on the etiology of scarlet fever; anthrax in swine; tuberculosis and the public milk supply; and the history and pathology of actinmycosis; in reports of the Agricultural Department of the Privy Council, etc.

"Author of a Manual of Pathology, etc."

[..]

"1889"

It's worth noting that Crookshank wd've only been 30 when the bk was published. One cd admire the strength & dedication of one so young or write him off as too young or..?

I came to learning about this work thru reading 1st about Dr. Charles Creighton (22 November 1847 – 18 July 1927):

"Creighton was an anti-vaccinationist. He has been described by historian Roy Porter as the anti-vaccination movement's "most ardent and distinguished spokesmen." Creighton argued that vaccination was poisoning of the blood with contaminated material, which could provide no protection from disease.

Two articles he wrote for the *Encyclopædia Britannica* on pathology (1885) and vaccinations (1888) cast doubt on the existence of germs and the efficacy of vaccines. He was widely condemned for these views by leading medical journals. He continued to express his unorthodox and unpopular anti-vaccination views in *The Natural History of Cowpox and Vaccinal Syphilis* (1887) and *Jenner and Vaccination* (1889).

Creighton was an active member of the London Society for the Abolition of Compulsory Vaccination."

- https://en.wikipedia.org/wiki/Charles_Creighton_(physician)

One story has it that Crookshank was encouraged to write his History and Pathology of Vaccination b/c it was expected that he wd counter Creighton's influential publications on the subject. Instead, after careful & prolonged study, he came to agree w/ Creighton.

"My interest in this subject was further stimulated by Sir James Paget, who very kindly examined one of the milkers casually infected from the cows, and while so doing drew my attention to a copy of Dr, Creighton's work on Cow Pox and Vaccinal Syphilis, then just published. The question naturally arose, whether my observations supported or refuted the conclusions arrived at by Dr. Creighton as the result of his historical researches." - p vi

"I gradually became so deeply impressed with the small amount of knowledge possessed by practitioners, concerning Cow Pox and other sources of vaccine lymph, and with the conflicting teachings and opinions of leading authorities, in both the medical and veterinary professions, that I determined to investigate the subject for myself. From antiquarian booksellers in Paris, Berlin, and in this country, I succeeded in a very short time in obtaining a large number of works dealing with the early history of vaccination." - p viii

Note that by the time Crookshank wrote this bk, vaccination had been going on for roughly 100 yrs & yet the author writes about "the small amount of knowledge possessed by practitioners". Note also that Crookshank mentions "the conflicting teachings and opinions of leading authorities". This latter reinforces my observation that **there is no such as THE SCIENCE**, meaning that there's no such thing as a monolithic opinion about *anything*, certainly NOT vaccination, upon wch all drs &/or scientists are in agreement. This isn't just a matter of the 19th century, it continues to be the case in the 21st century. If it ever ceases to be the case it won't be b/c of scientifically valid consensus (there is no such thing, it's antithetical to the variety of ways that hypothetical conclusions are reached), it'll be b/c of an absolute enforcement of conformity, an inquisition, essentially.

By p xv we get to a list of plates. Alas, as the back cover of this bk warns:

"As a reproduction of a historical artifact, this work may contain missing or blurred pages, poor pictures, errant marks, etc. Scholars believe, and we concur, that this work is important enough to be preserved, reproduced, and made generally available to the public. We appreciate your support of the preservation process, and thank you for being an important part of keeping this knowledge alive and relevant."

Indeed, the reproduction is of a library copy of the bk, apparently last taken out in 1988 when the bk was 100 yrs old. Given that I have at least a few bks in my personal library that're older than that & that those bks are in much better shape this one apparently was in its original form I have to wonder whether the poor quality of the reproduction is b/c of inept copying & printing technique. For one thing, the original illustrations were in color &, in some cases, were fold-outs. Both of those aspects are eliminated here, probably to save money. Still, the resultant poor quality of the reproduction is *so poor* as to be almost incomprehensibly reduced to high contrast. I'm sure they didn't have to be that way & I blame it on the ineptness of the technician who rephotographed the pages. Even worse is the problem of MANY of the right-side pages being so poorly reproduced that they're often unreadable b/c so much of the image is missing. I did my best to read this as thoroughly as I cd but I'd get a better edition if one were available & affordable (wch might not be the case).

Chapter 1 begins:

"The practice of Small Pox inoculation is one of very great antiquity. It had been found by experience that a person was not, as a rule, seized with Small Pox a second time ; but when, and how, the method of artificially inducing the disease was discovered, or where this preventative treatment was first employed, is quite unknown. It has been suggested, not unnaturally, that as the Arabian physicians were acquainted with the nature and treatment of Small Pox, they were probably the first to whom it occurred to produce the disease by inoculation. Avicenna, who is said to have lived at Bokhara, on the east coast of the Caspian Sea, has been credited with this discovery" - p 1

It's my moderate contention that nature inoculates. In other words, by existing in whatever our natural environment is, in my case the temperate northeast of the us@, one becomes inoculated *w/o the 'need' for human intervention*. However, I think this inoculation is limited by how gentle one is in relation to the rest of the environment. I'd managed to never get poison ivy until I was in my 50s when I started attacking my lawn w/ ferocity. IMO, the lawn defended itself, putting me out of commission for awhile w/ poison.

The point is that, IMO, human inoculation is overkill, an example of hubris in action, something more likely to cause harm than good. The cow milkers who feature heavily in this history shd've just *stopped milking the cows when they showed pox on their udders*. After all, by continuing to milk them they not only caused the cows unnecessary suffering they risked receiving diseased milk. I've heard it sd that cows need to be milked daily in order to relieve them of their accumulated milk - but it seems likely to me that the milking of them for human consumption is what creates this unnatural problem. Cd exploiting animals for this purpose be done away w/ altogether? Might that then solve the pox transmission problem better than other solutions?

As for Avicenna? My interest always increases when he's mentioned - partially b/c he's credited as being one of the 1st historically noted thought experimenters. See my movie: "Avicenna's Floating Maori": https://youtu.be/gPOFMfbtRmc .

Crookshank tries to trace the history of inoculation thru dr's reports. Here's a description from Circassia in 1711:

""She took **three needles** fastened together, and prick'd first the pit of the stomach ; secondly, directly over the heart ; thirdly the navel ; fourthly, the right wrist ; and fifthly, the ankle of the left foot, till the blood came. At the same time, she took some matter from the pocks of the sick person, and applied it to the bleeding part, which she covered, first, with *angelica* leaves dri'd, and after with some of the youngest lamb-skins ; and having bound them well on, the mother wrapped her daughter up in one of the skin coverings, which, I have observed, compose the *Circassian* beds, and carried her thus packed up in her arms to her own home ; where (as they told me) she was to continue to be kept warm, eat only a sort of pap made of cummin flour, with two-thirds water and one-third sheep's milk, without either flesh or fish, and drink a sort of tisan, made with *angelica, bugloss* roots, and *licorish*, which are all very common throughout this country ; and they assured me that with this precaution and *regimin*, the Small Pox generally came out very favourably in five or six days."" - p 3

Crookshank provides positive reports:

"That although at first **the more prudent were very cautious in the use of this practice** ; yet the happy success it has been found to have in thousands of subjects for these eight years past has now put it out of all suspicion and doubt ; since the operations having been perform'd on persons of all ages, sexes, and different temperaments, and even in the worst constitution of the air, yet none have been found to die of the Small Pox ; when at the same time it was very mortal when it seized the patient the common way, of which half the affected dy'd. This he attests upon his own observation." - p 6

So far, so good. Judging from these early observations it seems possible enuf that the theory upon wch inoculation rests may've been proven by application. However, there are various complications that arise as the testimonials are accumulated, as distinctions are made, as witnesses are called into question.

One of the many things I found interesting was the expression "buying the Small Pox":

""This method of procuring the disease is termed buying the Small Pox on the following account. The child to be inoculated carries a few raisins, dates, sugarplums, or suchlike ; and showing them to the child from whom that matter is to be taken, asks how many pocks he will give in exchange.["]" - p 9

The child to be inoculated from the diseased pus from another child puts a happy face on the whole deal (under adult guidance) by offering sweets in exchange for the disease. I'm reminded of when I was a child & my mom bought a comic bk for me whenever I was 'well-behaved' in receiving an iron shot at the dr's. On the one hand, I can see how "buying the pox" is a neat psychological trick for turning a bad thing into a good one. On the other hand, there's a capitalist sickness to it all that I find disturbing - paying a sick person *to spread their sickness* - certainly there's something suspicious about that.

"In other parts of Africa a similar custom existed. Here also it was called *buying the Small Pox*, and there was a superstition that inoculation would be of no avail, unless the person from whom the variolous matter was taken, received a piece of money or some article in exchange." - p 10

Interesting. Is the idea that the sick person's 'evil spirit' will be appeased by a present & ameliorate its potency in appreciation?! Whatever the case, more positive accounts accrue:

"["]Otherwise this practice is so innocent and so sure that out of a hundred persons inoculated not two die ; whereas, on the contrary, out of a hundred persons that are infected with the Small Pox the natural way, there die commonly about thirty.["]" - p 11

However, ultimately, there wasn't any more unanimity than there is today:

"Dr. Hecquet published a thesis entitled *Raisons de doute contre l'inoculation* [*Reasons for doubt against inoculation*], and this, together with the reports of failures in Boston, U.S.A., and the great mortality of the natural Small Pox in London, which was attributed to the new practice, soon brought inoculation into disrepute in France, and the proposed experiments were abandoned." - p 16

"In Berlin, the practice was soon discountenanced, for Meckel inoculated his children, and on repeating the experiment on others had three deaths, two being in the family of a baron. Dr. Muzell inoculated six children, of whom three died, and the three who were recovered were much disfigured. In 1774, Dr. Baylies inoculated seventeen persons : one death occurred, which, in order to silence an unfavorable report, was attributed to a "putrid fever, of which the eruption was only symptomatic."" - p 18

This isn't the only bk I've read that references such cover-ups:

"According to **Thomas Morgan** in his **Medical Delusions** (p. 48-49) "Jenner soon discovered that vaccination did not give immunity from smallpox, including some who had been vaccinated by himself and had died from it. Not wishing to bring vaccination into disrepute, he endeavored to suppress reports, and in writing to a friend said, **'I wish my professional brethren to be slow to publish fatal reports after vaccination.'** and in 1810 he wrote: 'When I found Dr. Woodworth about to publish his pamphlet relative to the eruption (smallpox) cases at the Smallpox Hospital, I entreated him in the strongest terms, both by letter and conversation, not to do a thing that would so disturb the progress of vaccination.' (Barron's **Life of Jenner**).

""The foregoing plainly proves that Jenner himself was aware of the utter uselessness of vaccination; but, having received the bounty from the government . . . he preferred to resort to all kinds of schemes rather than acknowledge its failure.

""From its inception until the present day, the vaccination scheme has been an endless record of lies, deception, fraud, juggling statistics, and falsifying death certificates in order to preserve vaccination from reproach and to secure its continuation . . . and all this after more than a century of terrible experience, which has demonstrated that vaccination has killed more than smallpox, besides crippling and disfiguring millions more."" - p 35, Eleanor McBean's <u>The Poisoned Needle - Sup[p]ressed Facts about Vaccination</u> [ISBN 9780787305949 edition]

But back to Crookshank's <u>History and Pathology of Vaccination</u>:

"**In America.**—In 1721, Small Pox visited this country after *an absence of nineteen years.* The Rev. Cotton Mather copied the accounts of inoculation given by Timoni and Pylarini in the *Philosophical Transactions of the Royal Society,* and sent them to the practioners of Boston. Dr. Zabdiel Boylston was induced to inoculate his child, and two of his negro servants, and in six months he had inoculated 244 persons.

"It may be interesting to note that out of these 244 cases, in six there was no effect at all, and six died in consequence of the inoculation, but the deaths were attributed to other causes." - p 20

"Boylston's experiments, and particularly his fatal cases, excited a great deal of opposition to inoculation, and many pamphlets were published in defence of, and against, the practice. But the following manifesto gave a severe check to his operations :—

""*At a meeting by Public Authority in the Town-house of* Boston *before His*

Majesty's Justice of the Peace and the Select-Men ; the Practioners of Physick and Surgery being called before them concerning Inoculation, *agreed to the following conclusion :—*

" A Resolve upon a Debate held by the Physicians of *Boston*, concerning Inoculating the Small Pox, on the twenty first day of *July*, 1721. It appears by numerous Instances, That it has prov'd the Death of many Persons soon after the Operation, and brought Distempers upon many others, which have in the End prov'd fatal to 'em." - p 21

"Boylston expressed his disapproval of this report upon the cases of natural Small Pox.

""It is a thousand pities our Select-Men made so slight and trifling a Representation of the Small Pox, that had always prov'd so fatal in *New England*, as they seem to have done in this Advertisement."

"It would appear from this, that in order to make converts to inoculation, it was necessary to keep alarming accounts of the natural Small Pox, before the eyes of the public." - p 22

&, IMO, the same exaggerated fear-mongering exists to this day in relation to COVID-19 wch, regardless of how many fatalities are actually occurring, is constantly presented as a likely bringer of mass death.

"**In Ireland.**—Inoculation was first performed at Dublin in 1723. In this and in the three following years, twenty-five subjects were inoculated, and as three out of this number died, the results were not very encouraging. Two of these deaths occurred in one family, in which five children had been inoculated." - p 28

Imagine having 2 of yr 5 children die after having a dr perform a procedure purported to protect them from sickness - do you think you'd trust the dr after that? From my POV these drs are arrogant sadistic murderous maniacs who deserve the death penalty much more than, say, a poor person who cdn't pay the fine for having murdered someone (the practice of early Irish law).

"Maitland performed a second inoculation, in the following May, upon the son of Dr. Keith, with a favorable result. This was soon generally known in London, for the news spread rapidly, and excited the greatest interest among people of all ranks. Nevertheless, inoculation made very little progress, for it was regarded with so much fear and suspicion, that several months elapsed before a third trial took place in London. In fact, inoculation was regarded as of such a dangerous nature, that an attempt was not again made until there was an opportunity of inoculating some criminals in Newgate, who were promised a full pardon if they submitted to the experiment. They accepted the offer, and were accordingly inoculated by Maitland on the 9th of August, 1721. None of them had the disease severely ; in fact there were only sixty pustules on the one in whom the operation produced the most effect. A seventh criminal was experimented upon by the Chinese method." - p 36

The word "experiment" is one that recurs frequently throughout this bk, over & over the drs are *experimenting*, NOT treating. It seems to me that this is correct & accurate. However, it has to also be acknowledged that these experiments are not only carried out on prisoners & orphans but also on the aristocracy & on the children of the drs as well - in at least one case w/ a fatal result. As for "only sixty pustules"?! That seems extremely unpleasant to me.

"In the account given by Maitland of these cases, we have the first intimation of a danger arising from this practice, which a century later was the strongest argument for not only abandoning it, but also for suppressive legislation. The first of these patients was Mary Batt, two years old, the daughter of a Quaker, inoculated October 2nd, 1721. This child, having only twenty pustules, soon recovered."

[..]

"[']The case was in short this: Six of Mr. Batt's domestic servants, who all in turn were wont to hug and caress this child whilst under the operation, and whilst the pustules were out upon her, never suspecting them to be catching (nor indeed did I), were all seized at once with the right natural Small Pox of several and very different kinds."" - p 37

That's interesting, one source resulting in several kinds of Small Pox. One might think that instead of there being different KINDS that there might instead be different REACTIONS depending on the specific physiology of the afflicted. At any rate, it demonstrates that inoculated people are contagious.

"In 1772, inoculation was more generally adopted. The Princess of Wales ordered it to be practiced upon some charity children, and the successful result induced her Royal Highness to have the two young princesses inoculated." - p 38

Oh, well.. that's a clear instance of the poor being used as guinea pigs for the safety of the rich.

"In spite of the fatal cases, an advantage was claimed for inoculation, in that it had been calculated, that of all those affected with Small Pox in the ordinary way, about one in six died, whereas the deaths from inoculation contended for by the anti-inoculators amounted to not more than one in fifty. In 1731, a pamphlet was published exposing the fallacies in Dr. Jurin's and Dr. Scheuchzer's statistics, and claiming that the advantage of the inoculated Small Pox over the natural disease was fictitious. The writer maintained that by inoculation the variolous infection was spread far and wide, and a considerable increase of mortality by Small Pox occasioned ; thus the lives saved to the persons themselves inoculated fell short of the lives lost from the increased infection." - pp 42-43

&, thus, the debate continues to this day - but, as w/ all history, *history belongs to the victor:* i.e.: to the people who make the most money & control the most media reporting on the subject. These days, pharmaceutical companies seem to me to be the victors - but let's not forget those age-old tools of mind-control, the clergy:

"Mr. Sutton also employed a clergyman, who preached sermons, and wrote exaggerated accounts of his results. The Rev. Robert Houlton, the advocating clergyman, attributed the success to Mr. Sutton's treatment. He said "that not one person out of a thousand inoculated by Mr. Sutton had more variolous pustules than he could wish["]" - pp 47-48

How many pustules wd you wish for? Personally, I'd wish for none. I'm reminded of the 20th century's war-mongering preacher, Billy Graham - the point isn't whether the practice being preached is actually backed by the religion used to back the preaching but whether there's money to be made for the preacher.

"But the writers who first described the custom which prevailed in different parts of the world, of "buying" or "ingrafting" the Small Pox, were unacquainted with the details which were essential for the performance of the operation with comparative safety"

[..]

"**Practice of the Brahmins.**—In Hindostan, the operation was performed only at certain seasons of the year, and a preparatory regimen was enforced. Probably, the Brahmins selected the subjects for inoculation, as well as the subjects from whom they took the variolous matter." - p 52

There're so many nuances & cultural variants to the inoculation process - failures can be blamed by proponents of one system on the deviations of another - but what if they all fail at some point or another? That's when a basic flaw in the practice in general seems likely.

"**Practice of Burgess.**—In spite of the precautions which had been recommended by Jurin, inoculation still continued to be followed occassionally by bad results. It was by no means a safe operation, and in order to diminish the risks, Mr. James Burgess published, in 1766, an account of the necessary preparations and management, with additions and improvements." - p 56

"In fact, the course of preparation could be summed up in three words —"temperance, quiet, and cheerfulness." The patient, being in a proper state of body and mind, would then pass safely through the distemper" - p 57

Perhaps abstaining from things that have demonstrably caused harm to oneself ("temperance"), not inducing stress ("quiet"), & being positive in general ("cheerfulness") might be healthy on their own.

"The following is a short account of the progress of inoculation of the Empress :
—

"Previous to inoculation she abstained from "animal food at supper, and at dinner ate only as was easy of digestion." The day before inoculation "she took 5 grains of mercurial powder." Sunday, 12th, late in the evening, she was inoculated with *fluid matter* by one puncture in each arm, and "on the succeeding night was very restless, and complained of pains in different parts of her body." . . . On the 14th, "she passed a tolerable night, certain signs of infection appeared on the places of incision ; a little pain was felt under the arm." . . . On October 15th, "the giddiness and the pain under her arm ceased. The places of incision became more red." On the 16th, she "complained of heaviness in her head at intervals; . . . four grains of the mercurial powder were given" - p 74

"mercurial powder": another complex issue in wch there is not monolithic THE SCIENCE but, instead, a variety of opinions. I've written about the use of mercury for medicinal purposes in other reviews of medical bks. Here's a sample:

"Now, do I think that thimerosal (mercury) is safe as something to be ingested? Offit has this to say:

""Before they put thimerosal in vaccines, Lilly scientists first studied it."

"[..]

""Although thimerosal didn't treat meningitis, doctors found that it was safe." - p 63

"Let's not forget that Offit was quoted above as having written: "Opren had already been withdrawn from the United States, where it had been linked to more than seventy deaths and 1,000 cases of kidney and liver failure, The drug's manufacturer, Eli Lilly, eventually settled cases in Britain for $6 million." In other words, Offit is honest enuf to admit that pharmaceutical companies, including Eli Lilly, can make serious mistakes that damage people's health. In other other words, it's reasonable for people such as myself to take the safety conclusions of pharmaceutical companies w/ some suspicion.

""Mercury is a naturally-occurring chemical element found in rock in the earth's crust, including in deposits of coal."

"[..]

""Elemental or metallic mercury is a shiny, silver-white metal, historically referred to as quicksilver, and is liquid at room temperature. It is used in older thermometers, fluorescent light bulbs and some electrical switches. When dropped, elemental mercury breaks into smaller droplets which can go through small cracks or become strongly attached to certain materials. At room temperature, exposed elemental mercury can evaporate to become an invisible, odorless toxic vapor. If heated, it is a colorless, odorless gas. Learn about how

people are most often exposed to elemental mercury and about the adverse
health effects that exposures to elemental mercury can produce.""

- https://www.epa.gov/mercury/basic-information-about-mercury

"In short, mercury is generally considered toxic but there're variations. Even tho
Offit makes the case that thimerosal has been proven safe & has justified its use
as a preservative in vaccines that might deteriorate & become poisonous on their
own otherwise, I have to wonder *why take the risk?*"

- review of Paul A. Offit, M.D.'s Autism's False Prophets - Bad Science, Risky
Medicine, and the Search for a Cure by tENTATIVELY, a cONVENIENCE - July
1-20, 2022 - http://idioideo.pleintekst.nl/CriticOffits.html

Back to Crookshank's History and Pathology of Vaccination:

"However, instances sometimes occurred of accidents and even death ; but
these were attributed to other causes, in order to save the new method from
reproach, a fact which was many years afterwards commented upon by Moore(1)
in the following words :—

"An empiric never hesitates at making positive declarations, and is never at a
loss for pretexts to cover failures. Should an infant at the accession of the
variolous fever be carried off by convulsion, he denies with effrontery that the
Small Pox was the cause, and invents another upon the spot. Should the
confluent Small Pox and death ensue, he soon detects that his instructions were
not strictly complied with, but some important error was committed in regimen ;
or, that the patient was too much or too little exposed to the air. In fine, the fault
may be in the parents, in the nurse, or in the inoculated, but is never allowed
fairly to fall upon the inoculator."

"_____

"(1) Moore. *The History of the Small Pox,* p. 209. 1815." - p 80

Now this is where what my experience wd call "reality" enters the picture. The dr,
the inoculator, knowing that he's more articulate than his subject (& less affected
by any emotion other than whatever's condusive to self-preservation), can always
fall back on deniability. I'm w/ Moore, I don't buy it - the proof is in the pudding &
the pudding's dead. When any kind of con-man speaks be sure to have yr
Bullshit Detector turned on. Drs & lawyers are just as often con-men as used car
salesmen.

Chapter IV begins w/ this heading: "*HAYGARTH'S SYSTEM FOR PREVENTING
SMALL POX.*":

"The history of Small Pox inoculation has been given in the preceding pages,
from its first employment in England to the time of the general adoption of the

Suttonian method. This practice, though so long continued, had not only failed to exterminate Small Pox, but, on the contrary, there can be little doubt that it actually assisted in spreading the disease. Instances occurred in which Small Pox was introduced by inoculation into towns which were and had been for many years perfectly free from the natural disease, and an epidemic followed." - p 81

Now, is there any reason to suppose that the continued use of viral matter hasn't also continued the same problems? SO, now there's the relatively new territory of GMOs (Genetically Modified Organisms) being used in 'vaccines' to hypothetically stimulate an antibody response. Perhaps the GMOs are less likely to produce the disease hypothetically protected against, I can imagine a solid case being made for that. Nonetheless, what we have, once again, is *experimentation* w/ a large number of serious & fatal side-effects. Haygarth (1784):

"["]Let us reflect how widely and fatally this poison is dispersed among all ranks of people. It may be conveyed into any house unobserved from a great variety of families, adhering to clothes, food, furniture["]"

[..]

""The clothes of a patient generally contain the largest quantity of variolous poison. However, all the enumerated articles and many more that come out of an infectious house or from an infectious person find their way unsuspected into all families of a certain rank. The poison is quickly and universally dispersed among the lowest class of people whose poverty renders them dirty."" - p 85

"" ' About 1718, a ship from the East Indies arrived at the Cape of Good Hope. In the voyage, three children had been sick of Small Pox ; the *foul linen* about them was *put into a trunk and locked up*. At the ship's landing, this was taken out and given to some natives to be washed. Upon handling the linen, they were seized with the Small Pox, which spread into the country for many miles, and made such a desolation that it was almost depopulated.' "" - pp 85-86

"SO, onto A People's History..: This book, of course, is reknowned as a major work refuting the History of the Victor that constitutes most history as 'learned' by most people - if only because the Victor has more resources available for the dissemination of their worldview (thanks to the thieving of their 'victory'). Even before I'd read the book, stories from it had disseminated into my life somehow. Take, for example, the story of the smallpox blankets:

"When the war ended in 1763, the French, ignoring their old allies, ceded to the British lands west of the Appalachians. The Indians therefore united to make war on the British western forts; this is called 'Pontiac's Conspiracy' by the British, but 'a liberation war for independence' in the words used by Francis Jennings. Under orders from British General Jeffrey Amherst, the commander of

Fort Pitts gave the attacking Indian chiefs, with whom he was negotiating, blankets from a smallpox hospital. It was a pioneering effort at what is now called biological warfare. An epidemic soon spread among the Indians."''

- my review of Howard Zinn's A People's History of the United States as published in Street Ratbag #5

The point is that for any of you who might naysay the opinion that the British deliberately used Small Pox as a weapon against the tribes they were negotiating w/ in 1763 one can at least use Crookshank as a source that spreading Small Pox thru linen was something known as early as 1718. It's also referenced a bit more specifically in this bk in a quote from 1830:

"In 1830, Dr. Sonderland, of Barmen, claimed to have produced vaccine in cows by infection from human Small Pox. An account of these experiments was published in the *Medical Repository*' - p 293

"" 'The simplest and surest mode of producing Cow Pox in the cow, and this proving indisputably the identity between the contagion of Cow Pox and that of human Small Pox, is to follow the procedure here laid down. Take a woolen bedcover which has lain on the bed of a Small Pox patient who has died during the suppurating stage, or is sufering from the disease in a considerable degree"

[..]

"And afterwards hang in such manner in their stall that its exhalation may rise upward and be inhaled by them. In a few days the animal will fall sick and be seized with fever[' ']" - p 294

In case you haven't already noticed, an unacknowledged subtext running thru this bk is that the drs *are deliberately spreading diseases*. Even when they're not successful it's what they're trying to do.

"Although attempts to confirm Dr. Sonderland's experiments failed in the hands of Ceely in England, and Macpherson and Lamb in India, and at Alfort, Berlin, Weimar, Bergen, Dresden, Kasan, Utrecht, and Stockholm on the Continent ; nevertheless, his aphorisms were accepted in support of the theory, the popular doctrine of the present day, that Cow Pox is Small Pox, modified by transmission through the cow. Had Dr. Sonderland and his followers been acquainted with the characters of the *natural* Cow Pox, and had they appreciated the fact that a vesicle with the physical characters of the vaccine vesicle, could be produced on the human subject, by management of variolous lymph without the intervention of the cow, they could hardly have come to such a conclusion." - p 296

"Attempts to infect cows by enveloping them with the sheets and blankets of Small Pox patients were without result. Ceely nevertheless persevered" - p 297

Haygarth continued:

""Mankind are not necessarily subject to the Small Pox ; it is always caught by infection from a patient in the distemper, or the poisonous matter, or scabs that come from a patient, and may be avoided by observing these

""RULES OF PREVENTION.

""I. Suffer no person who has not had the Small Pox to come into the infectious house. No visitor who has had any communication with persons liable to the distemper, should touch or sit down on anything infectious.

""II. No patient, after the pocks have appeared, must be suffered to go into the street, or other frequented place." - p 91

Many people who are against vaccination stress sanitation & hygiene as the main thing necessary for the prevention of epidemics. Haygarth is making a case for this. This, obviously, is the logic involved w/ quarantine. According to the CDC:

"The practice of quarantine, as we know it, began during the 14th century in an effort to protect coastal cities from plague epidemics. Ships arriving in Venice from infected ports were required to sit at anchor for 40 days before landing. This practice, called quarantine, was derived from the Italian words *quaranta giorni* which mean 40 days." - https://www.cdc.gov/quarantine/historyquarantine.html

In the above case, quarantine seems sensible. I think, however, that what I prefer to call the QUARANTYRANNY that we've been subjected to for the last 2.5 yrs is complete overkill. Using a potentially valid precaution as an excuse for imposing a medical police state is insidious.

""IV. The patient must not be allowed to approach any person liable to the distemper till every scab is dropt off, till all the clothes, furniture, food, and all other things touched by the patient during the distemper, till the floor of the sick chamber, and till his hair, face, and hands have been carefully washed. After everything has been made perfectly clean, the doors, windows, drawers, boxes, and all other places that can retain infectious air should be kept open till it be cleared out of the house."

"As every restriction is attended with inconvenience, Haygarth proposed that a reward should be given for attention to the rules, and this was to be secured by annexing to them a

""PROMISSORY NOTE[']" - p 92

That seems like another variation on "Buying the Small Pox" - in another words an attempt to both lessen an epidemic & provide a reward in otherwise miserable circumstances.

"The practice of inoculation was to be altogether subsidiary to the plan of stamping out the disease by isolation, the latter system however was regarded by many as visionary, it was not generally adopted, and when the promise of perfect and everlasting security was made by the promoters of Cow Pox inoculation, Haygarth's system was ignored and lost sight of." - p 97

Here we reach another crux: offer a seemingly simple way out & people will almost inevitably choose it over something that takes discipline - the obvious problem being that the promise of the vaccinators of lifetime immunity had no basis in reality whatsoever. Hence, this promise set actual healthy practices back.

Chapter V is headed: "*THE TRADITIONS OF THE DAIRYMAIDS*" & provides a key history that I'd already read about elsewhere but that I was glad to re-encounter in Crookshank.

"Adams in his work on Morbid Poisons writes : "Shall we forget that to the barbers we owe the bold use of mercury, to the Jesuits, of the Peruvian bark, which they learned of the Indians" [meaning a source of quinine used as a remedy for malaria] ", that an African showed us the value of quassia, that a Greek slave taught a woman the art of inoculation, the blessings of which were for a time almost lost by our fancied improvements and ill-directed cautions? Lastly, shall we contrast all this with the manner in which a Jenner has availed himself of the neglected traditions of cowherds and dairymaids?"" - p 98

Point taken, it's wise to pay attn to the 'folk remedies' of the people most directly experienced w/ the diseases under study.

"In some parts of the country, a belief existed among those who had the care of cattle, that a disease of cows, which they called Cow Pox, when communicated to the milkers, afforded them protection from Small Pox.

"It is not without importance to consider when, and how, this belief arose. Pearson and Jenner were both of opinion that it originated simultaneously with the introduction of Small Pox inoculation." - pp 98-99

"But, although Cow Pox and natural Small Pox have been known from time immemorial, there is no evidence to show this belief originated simultaneously with the early experience of these diseases. How was it, we may ask, that the tradition arose as a result of Small Pox *inoculation?* It was evidently failure in attempting to inoculate Small Pox on the arms of those who had recently contracted Cow Pox, which gave rise to gossip among the dairymaids, and laid the foundation of the popular tradition." - p 99

Once again, various conflicting opinions refute the notion of a monolithic THE SCIENCE. For one thing, there're multiple types of Pox categorized by various sources that aren't in agreement w/ each other. Pox, or Pocks, are eruptions on the skin that occur in a large variety on humans & other animals. Observations of these pocks are made to distinguish between them. According to ScienceDirect:

"Ten poxviruses infect humans (Table 159.1).2 Except for the 'extinct' variola virus and the increasingly important molluscum contagiosum virus, the poxvirus diseases are zoonoses. With rare exception these zoonotic poxviruses fail to establish a human chain of transmission. Most human poxvirus infections occur through minor abrasions in the skin. Orf, molluscum contagiosum and monkeypox viruses cause the most frequent poxvirus infections worldwide; the incidence of molluscum contagiosum and monkeypox virus infections is on the rise, the former as an opportunistic infection in late-stage AIDS and the latter as a zoonotic infection in central Africa. Individuals with atopic dermatitis may be predisposed to poxvirus infections such as molluscum contagiosum, orf or cowpox."

- https://www.sciencedirect.com/topics/medicine-and-dentistry/poxviridae

However, according to Wikipedia:

"Four genera of poxviruses may infect humans: *Orthopoxvirus, Parapoxvirus, Yatapoxvirus, Molluscipoxvirus. Orthopoxvirus*: smallpox virus (variola), vaccinia virus, cowpox virus, monkeypox virus; *Parapoxvirus*: orf virus, pseudocowpox, bovine papular stomatitis virus; *Yatapoxvirus*: tanapox virus, yaba monkey tumor virus; *Molluscipoxvirus*: molluscum contagiosum virus (MCV). The most common are vaccinia (seen on the Indian subcontinent) and molluscum contagiosum, but monkeypox infections are rising (seen in west and central African rainforest countries). The similarly named disease chickenpox is not a true poxvirus and is caused by the herpesvirus varicella zoster." - https://en.wikipedia.org/wiki/Poxviridae

Is that a disagreement or just a different way of expressing the same thing? In other words, are there 10 poxviruses but only 4 genera? & do you find "chickenpox" *not being a pox* somewhat confusing? & how does syphilis fit in? Is it or is it not a poxvirus in the sense that the above listed are?

"The 'Great Pox', syphilis, is a systemic disease with many clinical manifestations caused by the spirochaete *Treponema pallidum*. It is usually transmitted sexually but congenital infections can occur and, in certain parts of the world, endemic nonvenereal disease due to *T. pallidum* exists. Controversy exists as to the historical origins of venereal syphilis.The most common theory, the 'Columbian Theory', is that Columbus brought it back from the New World in 1493. The second theory, or pre Columbian theory, is based on the fact that European medical literature in the 1200–1300s describes certain forms of 'leprosy' which were highly contagious, could be transmitted sexually and from mother to child in utero and were said to respond to mercury. This form of 'leprosy' may, in fact, have been syphilis. The least known theory, the Evolutionary Theory, postulates that the different *Treponema* spp. evolved from a single organism responding to changes in the environment. The first use of Salvarsan in 1909 was a breakthrough in the therapy of syphilis." - https:// sfamjournals.onlinelibrary.wiley.com/doi/full/10.1046/j.1365-2672.2001.01494.x

This issue of whether there's even such a thing as "Cow Pox" & whether it can be used to vaccinate against Small Pox is a central one to this bk.

Putting that issue aside, there's the question of whether the dairymaids & cowherds actually had a folk tradition stating that handling cow pox inoculated the handler against small pox & whether this tradition even held true in practice - OR did, as speculated above, did this 'tradition' come into existence as prompted by the justification for inoculation provided by drs?

If you're looking for evidence of Mad Scientism (in the popular negative sense) look no further than the history of inoculation & vaccination in wch experimenters became frustrated when they cdn't make healthy people sick:

""Although some people cannot, from the peculiar nature of their constitutions, take the Small Pox ; but that cannot be the reason of so many persons in one part of the country, and no other, being incapable of taking the Small Pox.["]" - p 101

Drs forbid (as opposed to "Heaven forbid") that people simply be *healthy*. It appears that not much attn has been pd to what makes "the peculiar nature" of constitutions "incapable of taking the Small Pox"! Instead, they're treated like suspect abberations. So-called "Asymptomatic" people are similarly demonized today.

As for the Cow Pox?:

"One cow having it will communicate it to a whole dairy." - p 102

It's not initially even noted that Cow Pox doesn't spread from cow-to-cow as much as it does from cow-to-milker-to-cow. The pox on the teats come into contact w/ the milker's hands who then moves onto another cow's teats. It hurts the cow to have its teats handled when they have sores on them but the milkers do it anyway. If they simply didn't milk cows w/ sores the likelihood of the disease spreading wd greatly decrease. Duh.

"When a number of Cows on a farm are at the same time affected, the infection seems generally to have originated in the constitution of some one Cow, and before the milker is aware of the existence of the disease, the infectious matter is probably conveyed by the hands to the teats and udders of other Cows. Hence they are infected. For if the disease in the Cow first affected are to be perceived in a certain state, and obvious precautions be taken, the infection does not spread, but is confined to a single beast." - p 103

Ah, I spoke too soon.

""Mr. Moore's candour begins to show itself about the ninth page, where he admits this Cow Pox to be erroneously attributed to that gentle Animal. ' No Cow that is allowed to suckle her own Calf, untouched by the Milker, ever had this complaint.' He concludes therefore, that the Vaccine Disease is some pollution, imposed upon the harmless Animal by contact of the Milker. This I can readily believe to be the case.["]" - p 210

That commentator takes things a step further: not only is the Cow Pox inappropriately named it's not a matter of the cow transmitting it to the human but vice-versa! The udders get abraded from excessive milking by hand & these abrasions receive infections from the milkers.

"Not only was the tradition well known to inoculators, but we are also informed that there were many who did not believe it ; for it was equally well known that many who had contracted Cow Pox had subsequently suffered from Small Pox. It was owing to this that when Jenner mentioned to his professional neighbors the subject of the prophylactic power of Cow Pox, their reply was not very encouraging.

""We have all heard" (they would observe) "Of what you mention, and we have even seen examples which certainly do give some sort of countenance to the notion to which you allude ; but we have also known cases of a perfectly different nature,—many who were reported to have had the Cow Pox having subsequently caught the Small Pox. The supposed prophylactic powers probably, therefore, depend upon some peculiarity in the constitution who has escaped the Small Pox, and not on any efficacy of that disorder which they may have received from the cow. In short, the evidence is altogether so inconclusive and unsatisfactory that we put no value on it, and cannot think that it will lead to anything but uncertainty and disappointment."" - pp 105-106

Note the reference to "some peculiarity in the constitution who has escaped the Small Pox" reiterating the earlier passage's "peculiar nature" of constitutions "incapable of taking the Small Pox". That seems like an important idea to me: what if every body is different enuf for these generalized prophylactic treatments to be specifically unviable? There're certainly varieties of what foods we eat, wdn't there then be varieties of constitution based on that & other factors?

"Dr. Pulteney had heard of an instance in which Cow Pox had been contracted intentionally by contact.

""A very respectable practitioner informed me that of seven children whom he had inoculated for the Small Pox, **five had been previously infected with the Cow Pox purposely, by being made to handle the teats and udders** of infected Cows ; in consequence of which they suffered the distemper. These five, after inoculation for the Small Pox, did not sicken ; the other two took the distemper."" - p 107

There are so many variations. It seems to me that this one is a bit more natural & a bit more gentle. Instead of pricking the experimental subject's skin & then injecting diseased matter into the wound the children simply handle the cow's affected parts. There's still also the recurring issue of whether the "Cow Pox" is simply the appearance of pocks on a cow that may not be ultimately that different from any other pox in terms of primary origin - this gets further explored later. After all, these pocks are simply expulsions of unwanted material by the body thru the skin - how these pocks appear on the cow are certainly going to be different than how they appear on a human child - but one can't necessarily conclude from that that it is *or isn't* the same disease.

A dr named Edward Jenner features prominently in the history of vaccination (as differentiated from inoculation). There're many who glorify him & many who excoriate him. The reasons for this are important to the narrative of this bk.

"Instances had occurred of persons having had the Cow Pox about 1750, and one woman, eighty years of age, asserted that as long as she could remember, the opinion prevailed that *people who had the Cow Pox cannot take the Small Pox;* and that people purposely exposed themselves to it to protect themselves from the Small Pox." - pp 109-110

SO, here we have testimony that supports the idea that 1. Cow Pox is a different disease than Small Pox, 2. That exposure to Cow Pox acts as a vaccination against Small Pox, 3. That this practice is a folk remedy that predated Jenner's advocacy of it. This, however, isn't THE SCIENCE any more than other conflicting opinions that're enumerated elsewhere. One thing that emerges is the position that Jenner got entirely too much credit for this particular vaccination procedure.

"I shall give in full all the evidence which I have been able to collect from different sources, with a view of establishing Jesty's experiment as an historical fact, for Jenner regarded the account of it as an invention to deprive him of the merit of discovering Cow Pox inoculation. Baron, the biographer of Jenner, turned a deaf ear to anything which he considered might detract from Jenner's credit, and only referred in his biography to Jesty's *alleged* vaccinations" - pp 110-111

Human nature. Why do people assume that any particular acct is uninflected by bias & ulterior motive? Usually, IMO, b/c it's easier to do so than it is to factor in a large variety of conflicting opinions & then relate those to one's own personal experience. People are generally too busy w/ other more personal things to care that much, ultimately, whether a subject they're only superficially paying attn to has resistence to oversimplication if studied honestly. Hence, we have the day & age where people unquestioningly accept & regurgitate the dogmatic opinions fed them thru their subculturally approved 'news'-feeds. I'm trying to provide a deeper more nuanced take on things in this review but, like anyone else, I have my limits.

I can't say I agree w/ the following logic but I do find it interesting:

"["]there appeared to him little risk in introducing into the human constitution matter from the cow, as we already without danger eat the flesh and blood, drink the milk, and cover ourselves with the skin of this innocuous animal."" - p 114

I noted earlier that I seriously doubt that the reproduction quality of this bk needed to be so bad. A case in point is an image of Elizabeth Jesty who appears in an oval frame & who's shown consisting mainly as a white blob w/ 2 tiny black pinpoints for eyes.

The Jenner section is long, from page 125 to 249. Since he's generally credited w/ establishing vaccination in England he's singled out for particular scrutiny.

"'["]I cannot take that disease, for I have had Cow Pox.' This incident riveted the attention of Jenner."

"That such an event occurred is extremly probable, for the famous tradition was part of the stock gossip of the dairymaids, and was well known to many practioners in dairy districts." - p 127

That's contradicted by something quoted earlier: "it was equally well known that many who had contracted Cow Pox had subsequently suffered from Small Pox." So what's the truth? Maybe it's true in some cases & not in others. That, apparently, is most accurate. People who want to believe one thing do so & explain away contradictory evidence.

Jenner's initial acclaim was based on observations of the cuckoo bird. Perhaps this appreciation of his talents led him to overzealously seek further fame.

""It proved the very singular fact that the infant cuckoo reared from the egg in the sparrow's nest expelled the young of that bird by placing them upon its shoulder, on a depression, which Nature gives for the purpose, on the back of the unfledged cuckoo, and throwing them out of the nest.["]" - p 129

Jenner goes on to make assertions about Cow Pox & the relation between humans & domesticated animals & the transmission of disease.

"["]He went over the natural history of Cow Pox ; stated his opinion as to the origin of this affliction from the heel of the horse["]" - p 132

"["]Domestication of animals has certainly proved a prolific source of disease among men.["]" - p 133

What about the reverse? What about humans as a source of disease among domesticated animals? Jenner's opinion that Cow Pox originated w/ 'horse grease' proved to be, once more, a topic of hotly contested disagreement. No THE SCIENCE here. Jenner's contention was that substituting Cow Pox for Small Pox in inoculation-turned-vaccination was safer than administering Small

Pox. Alas, his proofs of this were shakey & exaggerated. Nonetheless, he attracted adherents to the dogma.

""I think the substituting of Cow Pox poison for the Small Pox promises to be one of the greatest improvements that has ever been made in medicine : for it is not only so safe in itself, but also does not endanger others by contagion, in which way the Small Pox has done infinite mischief.["]" - p 140

Here the assertion is that a person contaminated w/ Cow Pox can't transmit it to another person. Does that seem logical or likely to you?

"Another friend endeavored to persuade him to seize this an an opportunity of acquiring fame and fortune. But Jenner declined, and in a letter in answer to his friend the reason is made apparent. Jenner preferred retirement in the country, because he knew that his theory would be rigidly tested in London, and he was not prepared to face failures." - p 142

In The Poisoned Needle Jenner is depicted as wreckless w/ other people's lives & greedy:

"**Edward Jenner** inoculated his 18 months old son with swine-pox, on November 1791, and again in April, 1798 with cow-pox. The boy was never well after that and died of tuberculosis at the age of 21.

"In **Baron's Life of Jenner,** (Vol. II, p. 304) we learn that, "On the 14th of May, 1796 . . . Jenner vaccinated James Phipps, a boy about eight years old, with the matter taken from the hand of a dairymaid infected with **casual cowpox."**

[..]

"The inoculation didn't "take" so on the strength of this one experiment and its questionable interpretation, Jenner based his claim that **one** vaccination would **"forever** secure a person from smallpox." No time had elapsed to prove whether it would last a lifetime or a month or not at all; but without any proof or any scientific basis or evidence for its practice, the doctors and the government adopted it and made it compulsory, no doubt, seeing the gold mine in profit it would yield.

"**James Phipps was declared immune to smallpox but he too, died of tuberculosis at the age of 20.**" - p 29, Eleanor McBean's The Poisoned Needle - Sup[p]ressed Facts about Vaccination [ISBN 9780787305949 edition]

McBean's acct of Jenner's behavior makes it seem as if his motive was to get rich & as if he wasn't likely to stand much opposition from the London-based government b/c they knew the financial gains that were at stake. Neither acct of Jenner are very flattering. Regardless, Jenner had defensive stances in place:

"Jenner had previously been confronted with the statement that there were undoubted instances of Small Pox occurring after Cow Pox, and he had met this argument by the assertion that there are two kinds of Cow Pox" - p 145

"Dr. Woodville, as well as Dr. Pearson, were very curious to try the new inoculation ; and after patiently waiting their wish was gratified, for the welcome news was received that Cow Pox existed in London dairies." - p 152

"["]On Wednesday, I called again at the cowhouse to make further inquiries, which I was very much pleased to find two or three of the milkers were infected with the disease, one of whom exhibited a more beautiful specimen of the disease than that which you have represented in the first plate.["]" - p 153

One thing that became impressed upon me by reading these various drs' accts is an 'objective' state-of-mind that seems more immersed in the experiments they conducted than the more health-minded values they might be expected to hypothetically represent. As such, they cd be happy that Cow Pox, a disease, cd be active & consider the manifestation of the disease on the hands of a milker to be "beautiful". Wdn't it have been preferable for the cows & the milkers to not be diseased?

"On March 12th, 1799, Pearson sent a letter enclosing an infected thread to two hundred practitioners, requesting them to try its effects and report the results.

"Pearson also sent the virus to Paris, Berlin, Vienna, Geneva, and to Hanover, Portugal, America, and supplied the army." - p 157

Now, while the ostensible reason for disseminating this disease so widely was to enable vaccination it seems to me that what was more clearly accomplished was simply the spread of the disease itself.

"But more trouble was in store for him. George Jenner happened to be in London, and became greatly alarmed at the part played by Pearson and Woodville, particularly the former. He wrote off post-haste to his uncle, warning him that Pearson would become the chief person known in the business, and that, if he did not go at once to London, his chance of obtaining fame and fortune would be lost for ever." - p 159

These statements of Crookshank's don't go unsubstantiated, they're all supported by primary source materials such as, in this case, the letter that G. Jenner wrote to Dr. Jenner. My not providing every citation isn't b/c they're not there it's b/c I'm already quoting the bk to an extreme. Interested parties can get a copy of it for themselves for confirmation.

I imagine that it's self-evident by now why I wd quote the above passage: the promotion of vaccination is a competitive "business" perhaps more than it is a selfless enterprise in the pursuit of public health. Jenner wanted to be *the* man dominating the vaccination industry.

"In the same month, Woodville published his *Reports*, in which he concluded that Cow Pox manifested sometimes as an eruptive disease of great severity, for three or four cases out of five hundred had been in considerable danger, and one died. Baron says that these results proved well-nigh fatal to the cause of vaccination." - p 161

Hence providing at least some contradiction to an earlier claim that Cow Pox was safe as a vaccination matter. Nonetheless, competition between Pearson & Jenner continued to be fierce as they each vied to be vaccination's head honcho.

"In London, in the meantime, Dr. Pearson had determined to organize an Institution for inoculation of Cow Pox. He appointed a vaccine board, of which the chief place was occupied by himself, and the Duke of York consented to become a patron." - p 164

"Ultimately, Jenner succeeded in inducing the Duke of York and Lord Egremont to withdraw from the Vaccine Institution, formed by Pearson, and thus, according to Baron, Jenner defeated the ambitious designs of those who sought for high patronage." - p 167

Was that in the best interests of public health or in Jenner's best interests alone?

""We, whose names are undersigned, are fully satisfied upon the conviction of our own observation, that the Cow Pox is not only an infinitely milder disease than the Small Pox, but has the advantage of not being contagious, and is an effectual remedy against the Small Pox."" - p 168

While the Cow Pox is reputed to be able to go 1st, from one cow to another & then, 2nd, from a cow to a milker to another cow it's somehow "not contagious". Additionally, it's an "infinitely milder disease than the Small Pox" &, yet, people still die from being vaccinated w/ it. Now, doesn't that reassure you? Isn't it better to die from a mild disease that's not contagious than from a vicious one that is?

& just how money-hungry was Jenner's promotion of vaccination?

"The following were the discoveries alleged :—

"Firstly. That Cow Pox was inoculable from cow to man.

"Secondly. That persons so inoculated were for life perfectly secure from Small Pox.

"Jenner added that he had not made a secret of his discoveries, that the progress of Small Pox had already been checked, and that he had been put to so much expense and anxiety : therefore he prayed for remuneration." - p 173

"Admiral Berkeley moved for a grant of £10,000 whch was duly seconded by Sir

Henry Mildmay and carried by a majority of three." - p 175

£10,000 in 1802 is the equivalent of £1,073,495.67 in 2022. That's nothing to sneeze at (unless sneezing is somehow profitable). Then again, "today's prices are 108.35 times higher than average prices since 1802, according to the Office for National Statistics composite price index. A pound today only buys 0.923% of what it could buy back then." (https://www.in2013dollars.com/uk/inflation/1802?amount=10000)

""When the Committee of the House of Commons recommended Dr. Jenner to the munificence of Parliament, it was for a discovery in practice which was never to prove fatal ; which was to excite no new humours, or disorders of the constitution ; and which was to be, not only a perfect security against the Small Pox, but would, if universally adopted, prevent its recurrence for ever.["]" - p 221

""But yet further. In cases where Vaccination did not produce fatal consequences, it gave rise to new, and painful disorders. It was followed sometimes by itchy eruptions ; sometimes by singular ulcerations, and sometimes by glandular swellings of a nature wholly distinct from Scrophula, or any other known glandular disease. Here, again, was a failure in the second point stipulated : and finally,

""It was ascertained that even when Vaccination was performed, from what was called the genuine matter, it would not always prove a preservative against the Small Pox : as several patients, who had been pronounced by the most experienced Vaccinators to have passed regularly through the Cow Pox, were nevertheless attacked with the genuine Small Pox.["]" - pp 222-223

"the question of a further grant was put to the House, and £20,000 was agreed to by a majority of thirteen." - p 227

Jenner must've been a very persuasive fellow, capable of glossing over the numerous observations of his detractors in a manner appealing to the aristocrats in power. In retrospect, Crookshank is certainly among the unconvinced.

"In the year 1804, failures of the new inoculation had multiplied to an alarming degree, and even some of his friends began to lose confidence." - p 178

Nonetheless, Jenner's inventiveness always provided an excuse.

"Jenner had constantly to resort to the theory that if Small Pox occurred after Cow Pox, the vaccination could not have been properly performed."

[..]

"["]Never mind ; you will hear enough of Small Pox after Cow Pox. It must be so. Every bungling vaccinist who excites a pustule on the arm, will swear like G. it was correct, without knowing the nicety of distinction which every man ought to know, before he presumes to take up the vaccine lancet."" - p 181

"But even after perfect vaccination, it was well known that after a little time, patients could be infected by inoculation. To meet this, Jenner urged that the inoculation test should be abandoned." - p 182

"Dunning was now ready to assert that the occurrence of Small Pox after Cow Pox, actually strengthened the theory. Even Jenner was puzzled and wrote :—

""Pray indulge me with a line or two very speedily, to put an end to a little perplexity. You tell me that you know Small Pox will sometimes follow Cow Pox, and nevertheless assert that a case of this sort, which has happened under your immediate observation, places vaccination on higher ground than it has yet stood on.

""Do pray explain, as soon as you can, your meaning. . . .[""]" - p 186

"["]In your postscript, why not ask for cases of Small Pox inoculation, as well as cases of Small Pox after vaccination.""- p 187

Here I was, 187pp into this bk, & I was still confused about the distinction, if any, between inoculation & vaccination, wch seemed to be used sometimes interchangably, sometimes not. I eventually decided that inoculation means the use of the disease that's intended to be protected against as the material used for that protection & that vaccination means using a different disease than that meant to be protected against as the material used as protection for that protection. Hence, Small Pox inoculation meant giving the patient Small Pox under conditions more protected than those under wch they'd naturally get the disease & vaccination meant giving the patient Cow Pox (& other experiments were also tried) to hypothetically protect against Small Pox but under milder conditions. One cd say that there were 3 main camps: One that sd that only inoculation worked, a 2nd that sd that vaccination was safer, & a 3rd that sd that they were both dangerous & ineffectual.

"Birch condemned vaccination as an unnatural experiment, unphilosophical, and unsafe." - p 190

""When therefore it was proposed to me, to *introduce a new Disease into the human system*, I hesitated; but on the assurance given to me, that it was still milder than the Inoculated Small Pox, was productive of no ill consequences, and would equally arrest the progress of variolous Infection, I consented that Abraham Howard, the first Child mentioned at my Examination, should be vaccinated.[""]" - p 190

""Two other Caes however were followed by distinct and unequivocal Small Pox after Vaccination, and then it was admitted that the Cow Pox would not arrest the progress of *variolous* Infection ; although it is well known, Inoculation of the Small Pox within a limited period will *supercede* and *subdue* it.[""]" - p 191

"["]I shall continue firm in my opinion I gave to the Committee of the House of Commons, That what has been called the Cow Pox is not a preservative against the Natural Small Pox."" - p 192

"John Birch, Surgeon Extraordinary to his Royal Highness the Prince of Wales, and Surgeon of St. Thomas's Hospital." - footnote, p 189

"Clergymen warmly advocated the practice of vaccination." - p 192

""Soon after this, I heard with great surprise that an application had been made to the late Archbishop of Canterbury, persuading his Grace to direct the Clergy of the Church of England to recommend Vaccination from their pulpits.["]" - p 205

It's interesting to me that in the 19th century the clergy, admittedly pd or instructed to do so as previously mentioned, were pro-vaccination whereas by the 21st century author Paul A. Offit, M.D., a pro-vaccination researcher, presents religious people as against vaccination, with the implied position that they're backwards & anti-science:

"In his "PROLOGUE" Offit states that:

""I get a lot of hate mail.

""Every week people send letters and e-mails calling me "stupid," "callous," an "SOB," or "a prostitute." People ask, "How in the world can you put money before the health of someone's baby?" or "How can you sleep at night?" or "Why did you sell your soul to the devil?" They say "I don't have a conscience," am "directly responsible for the death and damage of hundreds of children," and "have blood on [my] hands." They "pray that the love of Christ will one day flood [my] darkened heart." They warn that my "day of reckoning is coming."" - p xi"

- tENTATIVELY, a cONVENIENCE's review of Autism's False Prophets - Bad Science, Risky Medicine, and the Search for a Cure - http://idioideo.pleintekst.nl/CriticOffits.html

"The practice was still strongly opposed by an influential section of the medical profession" - p 193

"in 1806, Birch published his reasons for objecting to the practice of vaccination. It was by far the most temperate of the arguments against the new practice, and deserves to be quoted in extenso." - p 193

"["]The bitterness of invective, and the unhandsome sneers, with which the partisans of Vaccination have assailed their opponents, as they offer no argument, merit no reply.["]" - p 195

""The Committee, being at last compelled to acknowledge that cases have been brought before them, in which it was incontestibly proved that persons having passed through the Cow Pox in a regular way, had afterwards received the Small Pox, contrive to destroy the effect of the concession, by the following ambiguous expressions.

""It is admitted that *a few* cases have been brought before them, of persons who had *apparently* passed through the Cow Pox in a regular way, etc.["]" - p 196

""They say, ' *In many of the cases* in which Small Pox has occurred after Inoculation ! ' *Many of the cases !* This expression I presume is to contrast with the *few cases of failure* admitted in Vaccination, and the reader is left to infer that cases of failure in Inoculation are of frequent recurrence ; than which inference nothing can be more unfounded, more contrary to truth.["]"

[..]

""But, in the second place, the fact itself has been uniformly denied by the best and most able practioners. They have always maintained that the Small Pox never has been known to recur after Inoculation ; and however the contrary may be assumed by those who have systems of their own to advance, it is considered as one of the invariable laws of nature, that (and if an exception can be proved, I should be justified in saying, *exceptio probat regulam*) a patient can suffer the Small Pox but once.["]" - p 198

"For granting (what never can be granted) that only one-third of the cases adduced were substantiated, there would remain above one hundred and fifty instances of acknowledged failure : and surely these would be sufficient to convince any dispassionate person, that Vaccination is not, and cannot be, a preservative against the Small Pox. What shall we say then, when, in addition to this, it is proved, that several patients have died of the immediate consequences resulting from the puncture of Vaccination ; while on the other hand it never was, or could be with any truth, asserted that similar fatal consequences had in a single instance resulted from the puncture of Small Pox Inoculation ?["]" - p 199

"["]But they forget that the principle evidence they themselves adduced to support their cause before the House of Commons was that of a Clergyman ; they forget too, that several of the Fanatical Preachers among the Sectaries, have been ever since the most zealous and approved champions of their system, both in their preachings, and practice ; together with some Ladies, who have received their instructions from Dr. Jenner himself.["] - pp 199-200

""The assertion of the Committee in the XXth article, that the Diseases which are said to originate from Cow Pox are scrophulous, and cutaneous, and similar to those which arise from Inoculation, is according to my observation quite incorrect. Many of the eruptions are perfectly novel. As far as my experience and my information go, I will venture to affirm they are eruptions of a nature unknown before the introduction of Vaccination ; and peculiar to those who have been Vaccinated.["]" - p 201

""It is not my intention to pursue further the Report of the Jennerian Committee. I have answered whatever applies materially to my argument : to expose all the errors and fallacies it contains, would be a painful task : I should however be unjust to the Public and myself, did I not state, that besides those I have already noticed, there are assertions so unfounded, and expressions so ambiguous, that these alone would have deterred me from subscribing it.

""This in Article XVI. it is said, that by means of Vaccination, the Small Pox has in some populous Cities been wholly exterminated.

""In Article XVIII, that the prejudice raised against Vaccination has been, in great measure, the cause of the death of near 2,000 persons this present year, in London alone.

""In Article III, that the cases published to prove the failure of Vaccination, have been for the most part refuted ; and

""In Article IV, those Medical Men who dissent from the Jennerian Committee, are stated *generally*, as acting perversely and disingenuously ; persisting in bringing forward unfounded, and refuted reports ; and even misrepresentation, after they have been proved to be such.

""Of these Articles I am compelled to say, and am ready to prove, that the three first are absolutely unfounded. Of the last I must declare, that it seems to me conceived in a spirit of illiberality and ungenerous censure, such as I should have a Committee formed of Gentlemen never would have used ; and which certainly no circumstances can justify.["]" - p 202

""The Royal Patronage, the authority of Parliament, would be made use of, beyond what the sanction given warranted : the command of the Army and Navy would be adduced, not merely as the mean of facilitating the experiment, but as proof of the triumph of the cause : and above all, the monopoly of the press, and the freedom of the Post Office would be employed to circulate the assertions of the friends of Vaccination, and to suppress the arguments of their opponents.

""What I foresaw happened : and such was the influence of the Jennerian Society, that many publishers and booksellers refused to print, or sell such works as might be deemed adverse to Vaccine Inoculation : in consequence of which it was hardly possible, at the first moment, to contradict any thing the Society chose to assert.["]" - p 206

If only people living in 2022 were capable of reading these descriptions of the machinations of the pro-vaccine people in 19th century England. Unfortunately, I know of few, or NO, people w/ the requisite curiosity or attn spans. It's my opinion that if such a reader were to exist they'd quickly recognize the same patterns of manipulation in today's world.

To give an example from personal experience: For decades I've frequented a bkstore in Pittsburgh that has an excellent selection, primarily thanks to its owner, who's also the bk-buyer. B/c of legal troubles, the owner has had to leave the store primarily under the control of the manager. The manager's certainly dedicated to small presses.. but they're also a hypochondriac - something that even their dr told them long before the so-called pandemic started. Once the quarantine started, the manager jumped on the fear-wagon big-time, asserting Draconian measures far beyond even the other businesses in the area, all of wch, except for the bkstore, have long since dropped. In 80°F weather, the manager wd wear a long-sleeve sweater, gloves, 2 masks, probably a plexiglass face shield, & wd shelter behind a plexiglass shield between the cashier area & the customer area. Customers were *required* to wear masks, hand-sanitizer was provided, & gloves were also required.

Under the name of Amir-ul Kafirs & fellow HERETICS I wrote a meticulously researched 1,186pp bk entitled <u>Unconscious Suffocation - A Personal Journey through the PANDEMIC PANIC</u> (http://idioideo.pleintekst.nl/Book2020.09PANDEMIC.html) wch the manager was kind enuf to purchase one copy of for the store. At 1st, the bk was displayed in the area for local authors. However, as the manager got more uncomfortable w/ its heretical position re the so-called pandemic they shifted it to a basement paperback rm wch otherwise houses only fiction. Undoubtedly, the manager never read the bk - such an intellectual exercise wd be beyond their usual fare of writing about boyfriends & working in a bkstore in NYC. Nonetheless, they felt compelled to recontextualize the bk as fiction. This is censorship & an excellent example of a lackey of the propaganda machine making sure that no contradictory opinions are easily available.

""It is allowed on all hands, that Cow Pox is generated by some disorder imparted by the milker. Now if that disorder should happen to be the Small Pox, then the Pustule so occasioned, and the matter coming from it, may inoculate Small Pox, and the patient thus inoculated, may be for ever secure from that disease, for in fact he will have received Small Pox Inoculation. But if the disorder generated on the Cow's teats, have for its base, Itch, as I apprehend has sometimes happened, then the patient will be inoculated with a disorder, which, though it may suspend the capacity for Small Pox for a season in the constitution, will ultimately prove no security.["]" - pp 213-214

I have the highest respect for surgeon Birch's throughly articulated rebuttal of the vaccinists. Still, as I think we will continue to see, nowhere is there any universal agreement, even around "that Cow Pox is generated by some disorder imparted by the milker" & that, therefore, there is no THE SCIENCE.

Otherwise, imagine this, if only as a thought experiment: *every case is different &, therefore, generalities are of limited use.*

""But arguments may be fallacious—let us come to facts. Can anyone disprove the following:

""That Vaccination has too often been fatal :

""That Vaccination has introduced new disorders into the human system :

""That Vaccination is not a perfect security against the Small Pox.

""These facts I maintain can never be disproved.["]" - p 224

While I'm at least temporarily sticking to my distinction between inoculation & vaccination, this bk continues to show the confusion of these terms in the 19th century. On p 226 the term "vaccine inoculation" is used 3 times. There's even a time when "vaccine" seems to be used as a different part of speech: ""IV. Thomas Dyson. His arm was perfectly vaccine in all its stages.["]" (p 290).

"The subject which then occupied Jenner's attention was the prohibition of Small Pox inoculation, for Baron says,

""He knew that vaccination would be comparatively powerless while its virulent and contagious antogonist was permitted to walk abroad uncontrolled."" - p 227

"the Government was now called upon to found an establishment, in place of the Royal Jennerian Institution, which had almost collapsed, from want of funds and from bad management." - p 228

For those of you who consider all medical practice to be sensible, I ask you to consider the merits of "dipping" & to then imagine whether there might not be practices in the 21st century that might be similarly open to question.

"Jenner was also much interested in the treatment of hydrophobia. He corresponded with the Rev. Dr. Worthington on this subject :—"

[..]

"["]I once asked a long-experienced professor what length of time he kept his patients under water ? His reply was, 'As to that I can't tell, but I keep them under till they have done kicking, when I bring them up to recover their senses, and get a little breath, and then down with them again, and so on to a third time, observing the same rule, not to take them up till their struggle is over.'

""You see what a shock the vital principle receives from this process. The modus operandi let us not trouble our heads about, if the fact can be established that it deadens the action of the inserted virus. I have wished to see how far it can be supported by analogy, by getting some vaccinated patient dipped within a few days after the insertion of the vaccine lymph. At all events an inquiry so highly important should be taken up, and it cannot be in better hands than yours."" - p 232

Great idea! Poison a patient & then while the poison's taking effect drown them too! That'll fix 'em right up! NOT.

"["]We touched on hydrophobia. He stated an ingenious idea, that of counteracting the effects of one morbid poison with another. What think you of a viper ? Not its broth, but its fang, as soon as the first symptom of disease appears from *canination*. If this should succeed, we must domiciliate vipers as we have leeches. But from this hint I should be disposed to try, under such an event, Vaccination["]" - p 233

These guys are just FULL of good ideas. NOT.

"Jenner had been summoned to London in the first week in June ; for on the 26th of May, the Hon. Robert Grosvenor was seized with a violent attack of Small Pox. In four days, he became delirious, and an eruption appeared on his face ; but the Small Pox was not expected, because he had been vaccinated by Jenner only ten years previously." - p 238

So much for lifetime guarantees, so much for Jenner's blaming failures on bad practitioners.

"The outbreaks of Small Pox in various parts of the country, and the failures of vaccination, led Jenner to send a circular letter, early in 1821, to the profession, endeavoring to arouse attention to those points of vaccination which he considered essential to afford protection. Even the most ardent supporters of vaccination would now only claim that vaccination modified an attack of Small Pox in future, but Jenner's original opinion remained unchanged. Nothing would shake his belief that persons vaccinated were for ever after secure from the infection of Small Pox." - p 248

"This presented a difficulty. Jenner believed that Cow Pox arose from "grease," and that it protected against Small Pox, yet persons direcly infected with "grease" enjoyed no such immunity. Jenner is ready with an explanation. These cases, in his opinion, decisively proved that the grease could not be relied upon *until it had been passed through the cow.*" - pp 254-255

Jenner devleops this idea that animals are responsible for diseases in humans.

""*May we not then reasonably infer that the source of the Small Pox is the matter generated in the diseased foot of the horse, and that accidental circumstances may have again and again arisen, still working new changes upon it, until it has acquired the contagious and malignant form under which we now see it making its devastations among us ? And from a consideration of the change which the infectious matter from the horse has undergone after it has produced a disease on the cow, may we not conceive that many contagious diseases now prevalent among us, may owe their present appearance not to a simple, but a compound origin ?*" - p 260

One thing that seems to escape Jenner's notice is that domesticated horses, possibly the only ones w/ "grease" on their heels, are provided w/ horse-shoes imposed on them by humans. Horseshoes were invented by humans to protect the horse hooves from the harsh labor imposed on them. These metal shoes are generally *nailed* into the hoof. Gee, I wonder if this practice *might* sometimes cause disease in the horse's foot.

While Jenner seems to consistently overlook over-milking of injured cow teats & the possible negative effects of nailing something into the horse's hoof he's quick to provide endless excuses & cautions against observations that might go contrary to his own claims.

"In other words, any one disposed to apply the variolous test after Cow Pox, was cautioned to employ the Suttonian method, and if an eruption followed, it was not to be hastily concluded that genuine Small Pox had resulted." - p 262

There shd be a *Jenner Car*, the seller can guarantee that it's absolutely indestructible & that anyone driving one cannot be injured in any way in an accident. That way, when the driver *does* get into an accident & gets, say, incinerated, the car manufacturer can simply claim that they spontaneously combusted & that that death is either not a death at all or, at the very least, was in no way caused by destruction of the car - esp given that any claims of the latter are obviously made by people inclined to be delusional.

Children continue to be excellent subjects for experimenting on given that they're more or less defenseless &, besides, it's 'for their own good'.

"Jenner's mind was occupied with the opportunity of making a double experiment ; inoculation of one child with humanised horse-grease, and of another child with matter from the cow's teats." - p 268

"For his experimental purposes, Jenner selected a child five years old, John Baker by name, and on March 16th, 1798, he took matter from a pustule on the hand of Thomas Virgoe, one of the servants who had been infected from the mare's heels.

""He became ill on the sixth day, with symptoms similar to those excited by the Cow Pox matter. On the eight day, he was free from indisposition.[""]" - p 269

"It is not until we read Jenner's *Further Observations* that our attention is again drawn to this matter. In a reference to this case, Jenner insists upon the "similarity to the Cow Pox of the general constituional symptoms which followed," and in a footnote we read :—

""The boy unfortunately died of a fever at a parish workhouse, before I had an opportunity of observing what effects would have been produced by the matter of Small Pox."

"The fact, then, of the boy's death was omitted in the first account, and this is the full meaning of the boy being "rendered unfit for inoculation."" - p 271

Are you getting the idea yet?! IMO, Jenner & his fellow travelers were completely criminally insane. They routinely conducted experiments on children that the victims died from & then covered over the deaths w/ lies & euphemistic language. *This tendency continues to this day.* Again, **hubris**.

"It is evident that in the published account of the boy's case, Jenner had suppressed all details of the progress of the vesicle, the ulceration, and the crysipela, as well as the fatal termination of the case, and he inserted instead that "on the eight day he was free from indisposition," but "was rendered unfit for inoculation from having felt the effects of a contagious fever."" - p 273

Jenner continues to obfuscate clear thinking about his failures by inventing new explanations. He resorts to claiming that there are 2 types of Cow Pox.

"I wish to insist upon the gradual assumption of the existence of a *spurious Cow Pox.* The farmers and cow doctors knew nothing of this *spurious Cow Pox.* They distinguished, from other eruptions such as blistered teats, a disease which produced troublesome ulcerations on the cow's teats, and ulcerations on the hands, enlarged glands, and constitutional symptoms in milkers, and this disease they called the Cow Pox. Jenner was alone responsible for assuming the existence of two kinds of Cow Pox, a true and a spurious." - p 278

"The cases are carelessly jumbled together ; important details are often missing ; dates are omitted ; facts unfavorable to the project are suppressed ; and excuses for failures are ingeniously incorporated. All that the *Inquiry* contained was known to dairymaids and farriers, with the exception of the doctrine of spurious Cow Pox, and certain speculative comments." - p 284

Ceely recounts an accident w/ his assistant. Accidents *will happen*, eh?!

""My assistant, Mr. Taylor, to whom I had entrusted the lancet used in opening the variolous vesicle in the first experiment, on the tenth day, while I was engaged in the tedious process of changing points therefrom, punctured the skin of his own hand, between the thumb and forefinger, with the instrument while moist with lymph, a circumstance with which at the time I was unacquainted. On the fourth day afterwards, he directed my attention to a hard, deep red, papular elevation on the spot, stating the cause, and at the same time assuring me he had been vaccinated in infancy" - p 298

"In the month of December 1840, he inoculated a fine young cow, on the teats and on the external labium, with Small Pox virus."

[..]

"There was one well-developed vesicle on the external labium, and the lymph from it was employed by Badcock for "vaccinating" his son." - p 300

[..]

"It is quite a mistake to speak of this operation as *vaccination*. The method was simply a modification of the Suttonian system of Small Pox inoculation in which, in the first remove, the cow was substituted for the human subject." - pp 300-301

[..]

"In 1836, Dr. Martin, of Attleborough, Mass., inoculated the cow's udder with variolous lymph, and by inoculating children from the variolated cow, produced an epidemic of Small Pox with fatal cases."

[..]

"In 1847, variolation of the cow was successfully performed at Berlin, but the products inoculated in the human subject resulted in retro-variolisation, and one of the experimental children died of confluent Small Pox." - p 301

Are you getting the idea yet? These pompous egomaniacs experiment on giving diseases to both non-humans & humans & then people die &, *somehow*, they still apparently maintain their reputation (&, presumably, their considerable income) as healers. Go figger.

""It seems certain that there are, at least, four animals—the horse, the cow, the sheep, and the goat—which are affected with a disorder communicable to man, and capable of securing him from a malignant form of the same disease."

"The disease which Baron was describing was not Cow Pox but cattle-plague, and the totally erroneous views into which he had drifted arose from that initial nosological error committed by Jenner, who branded Cow Pox as *Variolœ Vaccinœ*, or Small Pox of the Cow. That cattle-plague has a close affinity with human Small Pox is perfectly true, but it has no relation or connection whatever with Cow Pox. I shall give a brief history of the disease referred to by Baron, and then I shall pass on to describe the disastrous results which followed the reception of the *variolœ vaccinœ* theory in India." - p 312

So, AGAIN, people looking for & expecting THE SCIENCE, meaning some sort of agreement among drs & scientists about what's what exactly will be sorely confused. What is Small Pox? What is Cow Pox? What is Cattle Plague? What is *Variolœ Vaccinœ*? If one dr thinks that something's one thing & another thinks that it's another &/or if one is right & the other is wrong, what're the potentials for disaster?!

"Ceely, however, like Baron, maintained that cattle-plague was simply malignant Cow Pox ; and he did so principally from the fact that in the accidental transmission of rinderpest to the human subject, a vesicle was produced presenting the appearances, and running the ordinary course, of inoculated Cow Pox. The following is the case as reported by Ceely :—

""On the 3rd December, 1865, Mr. Henry Hancock, veterinary inspector, Uxbridge, was engaged in superintending the autopsy of a bullock, recently dead of cattle plague. His assistant, who was performing the operation, while occupied in removing the skin from the scrotum, accidentally punctured the back of Mr. Hancock's hand with the point of the knife. The puncture, being slight, was disregarded at the time, but was washed as soon as practicable, and thought of no more. On the 8th, five days afterwards, a small, slightly elevated, hard pimple was felt and seen on the site of the puncture." - pp 314-315

I remember that in the early days of the so-called COVID-19 pandemic there was much fear-mongering being broadcast about the danger of getting sick from contact w/ animals. I witnessed a woman go berserk b/c a man's dog in a park ran close to her when it was fetching a thrown ball. The berserker ran towards the man, an old man wearing a mask, & was barely prevented from beating him by someone she was w/. The emphasis on animals as carriers seemed to die off. Out of curiosity I just went to the CDC website to see what they have to say about human-animal covid transmission.

"Animals infected with SARS-CoV-2 have been documented around the world. Most of these animals became infected after contact with people with COVID-19, including owners, caretakers, or others who were in close contact. We don't yet know all of the animals that can get infected. Animals reported infected worldwide include

- Companion animals, including pet cats, dogs, hamsters, and ferrets.
- Animals in zoos and sanctuaries, including several types of big cats (e.g., lions, tigers, snow leopards), otters, non-human primates, a binturong, a coatimundi, a fishing cat, hyenas, hippopotamuses, and manatees.
- Mink on mink farms.
- Wildlife, including white-tailed deer, mule deer, a black-tailed marmoset, a giant anteater, and wild mink near mink farms."

- https://www.cdc.gov/coronavirus/2019-ncov/daily-life-coping/animals.html

So, you see, now the official story is that humans are a threat to the animals, not vice-versa. I find that to be more believable.

& what about the milk from diseased cows in England?

"and it must be inferred from Dr. Jenner's and other medical writings on the subject that the animal not only continued to secrete milk, but that the milk was used ; while in this country the little that is secreted is never made use of, and perhaps owing to this very circumstance the Guallahs or milkers in India are not affected with Cow Pox, as is the case with this description of persons in Gloucestershire and other counties In England where the disease is most prevalent." - p 320

No doubt, the sanitation conditions in England's dairies have since improved. Still, to those of us born in the mid-20th century, the 19th century wasn't really that long ago. It seems likely that people who will find my position on drs outrageous will resort to acting as if the 19th century were so historically distant that it's almost like the stone age. I, however, think that the next story to be quoted is possible in the 21st century as well.

"On the seventh day, from the commencement of the eruption her mouth and throat became so sore that she was unable to take the breast or any other food : it was necessary to try to support her by a nourishing injection, notwithstanding which she sank on the 20th. The above report, it is hardly necessary to say, is given with great pain ; but I feel that it is right to do it, and to warn my brethren of the danger that sometimes occurs after taking the virus from the cow in this climate. Mhata in the cow of this country is decidedly a much more serious disease than the vaccine diseases in the animal in Europe." - p 324

That was an account of the course of a vaccination made by a Mr. Furnell, in Assam, in 1834. The young girl who died was his own daughter. His attribution of the death to the supposedly greater seriousness of the disease in India seems to me to be his way of making excuses for his having taken his own healthy child & injected a disease into her wch killed her. Sorry (NOT), but from my own POV Furnell was, at best, a complete idiot.

""From these many other native children were inoculated, and no doubts of the genuineness of the lymph were excited until two English children were punctured from one of them, and it was then found that Small Pox supervened in both of these cases, and this was more than suspected to have happened in many of the native children, who had generally dispersed a few days after the operation, and were not afterwards heard of. One of the English children unhappily died."" - p 326

It's all in the interest of science, right? No doubt the vaccinators learned something important from the poor victim's death. NOT.

"From all these independent observations, if we accept them as correct, there would seem to be no doubt that cattle-plague virus inoculated in the human subject will produce a vesicle with the physical characters of the vaccine vesicle, and succeeded occasionally by an eruption which appears to have the characters of the eruption of cattle-plague. That cattle-plague is not infectious to man in the ordinary sense affords no proof that the disease may not be cultivated in the human subject by inoculation.

"But these occurences had to be explained away, for such circumstances were incompatible with the Small Pox theory of Cow Pox. We have only to turn again to Dr. Seaton's *Handbook of Vaccination* to find that more ingenious explanations were forthcoming." - pp 326-327

"["]the fact, that Small Pox of sheep might be substituted for Cow Pox ; but as he had then made only a very few experiments, with a view of ascertaining if it were efficacious or harmful when transmitted to man, he undertook to continue his researches, and then to publish the results. However, as he has recently informed me, he has not been able to do so in consequence of a long and severe illness from which he has been suffering ; it is this which has retarded the publication of these valuable observations" - p 330

Yes, why stop w/ spreading disease between cows & human children when you can throw sheep into the mix?

"[']I only succeeded in meeting with it in the State of Naples, at Capua. Passing through it in 1804, I saw a peasant who was driving a flock of seven sheep to the butcher's ; as I was obliged to stop in this town, I endeavored to profit by the opportunity and to gain information on the subject.

""Having noticed the miserable and dejected appearance of these sheep I stopped ; and after putting various questions to the peasant, and examining the nature and character of the eruption and of the symptoms which accompanied it, I felt sure that the malady was the true Small Pox of sheep. The peasant told me that the malady was common in the neighborhood, that fifty-four sheep had already been slaughtered, and that they would continue this method if the malady should develop in others" - p 330

Yes, you understand aright, the diseased sheep are being butchered for resale as meat before they lose their 'value' altogether.

"[']Dr. Legni. I informed him of my design, and my desire to make experiments with the matter obtained from the sheep, at Capua ; he kindly seconded my project. He procured me six children, who were all inoculated with the matter, which was still fluid ; I also inoculated two other infants with true vaccine, in order to institute a comparison." - p 331

'Hi, I'm a fellow psychopath & I need children to give a disease to, I've never tried this particular experiment before, can you get me some kids?' 'Why, of course!, beloved colleague.' It's practically straight out of de Sade. & the vaccinator in question proceeds to spread disease everywhere he goes.

""Proceeding to Lucca, I used the same *virus* to inoculate various people, and continued to vaccinate also in other places, always renewing the matter which had been originally taken from Sheep Pox, its course being always very regular, and its effect constant, as if it had been derived from a genuine Cow Pox."" - pp 333-334

Keep in mind that this was an *experiment*, he'd never tried this Sheep Pox before, he had no assurance that he knew what wd happen to his victims & he didn't even necessarily stick around to find out. Is that *irresponsible* or what?!

The term "virus" doesn't occur that often in this bk so I started wondering when it originated:

"virus (n.)

"late 14c., "poisonous substance" (a sense now archaic), from Latin *virus* "poison, sap of plants, slimy liquid, a potent juice," from Proto-Italic **weis-o-(s-)* "poison," which is probably from a PIE root **ueis-,* perhaps originally meaning "to melt away, to flow," used of foul or malodorous fluids, but with specialization in some languages to "poisonous fluid" (source also of Sanskrit *visam* "venom, poison," *visah* "poisonous;" Avestan *vish-* "poison;" Latin *viscum* "sticky substance, birdlime;" Greek *ios* "poison," *ixos* "mistletoe, birdlime;" Old Church Slavonic *višnja* "cherry;" Old Irish *fi* "poison;" Welsh *gwy* "poison").

"The meaning "agent that causes infectious disease" emerged by 1790s gradually out of the earlier use in reference to venereal disease (by 1728); the modern scientific use dates to the 1880s. The computer sense is from 1972."

- https://www.etymonline.com/word/virus

Keeping the above etymology in mind, it seems to me that "virus" in this bk is used mainly in the older sense of "agent that causes infectious disease" even though the bk dates from 1889 during the time of its "modern scientific use". Here's an example:

""4. On inspection of the cow which you inoculated at several points with the same matter, I found on the udder a single vesicle, from which I took matter, which was yellowish in colour and not limpid, and used it to inoculate two other boys ; the first had two vesicles on each arm, and on the second I found only one, on the left arm ; in other respects the *virus* contained in the two vesicles was exactly similar to that of true Cow Pox."

""AULLA, *January 29th,* 1807." - footnote, p 334

Now, here's the beginning of a contemporary definition:

"A **virus** is a submicroscopic infectious agent that replicates only inside the living cells of an organism. Viruses infect all life forms, from animals and plants to microorganisms, including bacteria and archaea. Since Dmitri Ivanovsky's 1892 article describing a non-bacterial pathogen infecting tobacco plants and the discovery of the tobacco mosaic virus by Martinus Beijerinck in 1898, more than 9,000 virus species have been described in detail of the millions of types of viruses in the environment. Viruses are found in almost every ecosystem on Earth and are the most numerous type of biological entity. The study of viruses is known as virology, a subspeciality of microbiology.

"When infected, a host cell is often forced to rapidly produce thousands of copies of the original virus. When not inside an infected cell or in the process of infecting a cell, viruses exist in the form of independent particles, or **virions**, consisting of (i) the genetic material, i.e., long molecules of DNA or RNA that encode the structure of the proteins by which the virus acts; (ii) a protein coat, the *capsid*, which surrounds and protects the genetic material; and in some cases (iii) an outside envelope of lipids. The shapes of these virus particles range from simple helical and icosahedral forms to more complex structures. Most virus species have virions too small to be seen with an optical microscope and are one-hundredth the size of most bacteria."

- https://en.wikipedia.org/wiki/Virus

According to that entry, the "modern", non-computer, definition didn't start until 1892 & is, therefore, from after the time of the writing of this bk. The concept of the virus interests me for various reasons. For one, I wonder what the next development of its definition will be. It's come a long way from a plant poison to a self-replicating computer program. For another, ever since the beginning of the so-called pandemic I've been researching the use of the "virus" as *the ultimate scary thing* in pop culture. Here's an excerpt from an email sent to the HERETICS group by me on June 5, 2022:

"I watched the 3rd movie in the "Maze Runner" series tonight. They were made from 2014-2018. A virus results from a solar flare (I think, I didn't watch the 1st 2 movies). It infects most of the people on Earth & turns them into violent homicidal zombies. A few people are immune. They're tormented in order to try to scientifically figure out a way to exploit them to produce a vaccine. Eventually, the main character's blood proves to be the answer.

"Anyway, you can see where this is going. It's mind-boggling how many movies there are that promote the scariest possible version of what viruses can & will do to humanity - they're all doomsday scenarios. In the end, vaccines save the day. Even if this preponderance of such melodramatic nonsense isn't conspiratorial it still supports the idea that *vaccines are the hope for saving humanity*. Needless to say, I think that's complete nonsense. If viruses even exist at all I don't think they're that big of a threat."

Viruses, in the many SciFi movies I've watched, are almost always depicted as causing a state of homicidal zombieism. I don't know of anything medically described as a virus in non-fiction to have that effect. Here's another short email to the HERETICS from me from March 21, 2022:

"I watched "World War Z" last night, starring Brad Pitt. Generally, I think Pitt's a good actor but this one was generic. It was made in 2013. It's about a virally-spread zombieism. The zombies are very fast, they spread the condition by biting, & they're practically indestructible except for head shots. They can take over an entire city w/in hrs. Once they bite someone that person becomes a zombie too w/in 12 seconds. Then they're very spastic but they can jump from high places & land on metal & just get right back up again & continue their spree.

They don't need to eat & they're basically immortal. Unfortunately, there doesn't seem to be much else to recommend their condition. At any rate, as yet-another-pop-movie-promoting-extreme-fear-of-viruses this doesn't have much going for it. At the beginning, a virologist is explaining the 'Spanish Flu', a few people are show[n] wearing face masks, nature is referred to as a "serial killer"."

Note that in none of the pop culture that I've so far encountered are viruses anything but extremely destructive, there's no virus, e.g., that increases the intelligence of the person infected by it or makes them gentler (or, if there is, I've either missed it or forgotten about it). *SO*, what I'm wondering is: will there ever come a time in the evolution of the definition of "virus" where it includes something like: 'an agent that can move from body to body that's capable of changing how certain functions in the body work, either negatively or positively'? I'm not saying that viruses actually have any positive function for anything other than the viruses itself, I'm just wondering if all this fear-based perception of them is skewing unbiased investigation.

Lest you think that it's only humans getting killed, here's a report from a veterinary surgeon in Gloucester named Clayton about Cow Pox, as published by a Mr. Cooke in 1799:

"["]he cannot explain :—that this disease has not a regular process of commencing and terminating without a remedy, because, if not attended to, it would end in a mortification of the teats, and probably death of the animal["]" - pp 340-341

& this is the disease Jenner picked to make a Small Pox vaccine from as something milder than Small Pox itself, a disease that's in this description depicted as fatal to the cow. An early observer of Cow Pox was a "Mr. Lawrence, author of *A Philosophical and Practical Treatise on Horses and on the Moral Duties of Man towards the Brute Creation*." (p 343) In one of the few instances of Crookshank talking about himself he refers, 1st, to Lawrence:

"Lawrence was almost a century before his time. Cow Pox was not again brought forward in this light until 1887-88, when I reported the "filth and nastiness" at a Wiltshire Farm, and advocated the advisability of placing this disease under the Contagious Diseases (Animals) Act." - p 344

Once again, it strikes me as important when human responsibility is brought to the fore. Vaccinators seem to be wearing blinders to the torment domesticated animals were put to, to the low ethical standards of the uses to wch both humans & animals were put to in their experiments, & to the conditions of "filth and nastiness" on farms that represented low standards of sanitation in the use of animals as food producers. I, at least, wonder: how much were the vaccinators guilty of spreading disease rather than ameliorating it?

"["]In some animals, under some circumstances, this state continues little altered till the third or fourth week, rendering the process of milking painful to the animal, and difficult and dangerous to the milker."" - p 349

Once again, it doesn't seem to occur to these people to simply stop milking the cow when it's got an outbreak on its teats - thusly allowing the cow to recover & reducing the likelihood of the cow trying to kick the milker. Despite Jenner's emphasis on Cow Pox as something to be used as a safe vaccine against Small Pox, he nonetheless describes cases of Cow Pox infection that seem severe:

""Mrs. H had **sores** upon her hands, which were communicated to her nose, which became inflamed and very much swollen.

""Sarah Wynne has Cow Pox in such a violent degree that she was confined to her bed, and unable to do any work for ten days.["]" - p 352

Pearson, also, describes such cases:

""Thomas Edinburgh was so lame from the eruption of Cow Pox on the palm of his hand, as to necessitate his being for some time in hospital. For three days he suffered from pain in the armpits, which were swollen and sore to the touch. He described the disease as uncommonly painful, and of long continuance.["]" - pp 352-353

""Annie Francis had pustules on her hands from milking cows. These pustules soon became scabs, which, falling off, discovered ulerating and very painful sores, which were still in healing. Some milk from one of the diseased cows, having squirted on the cheek of her sister and on the breast of her mistress, produced, on those parts of both persons, pustules and sores similar to her own, on her hands."" - p 353

& yet this same milk was presumably being produced for sale?

If this Cow Pox used for vaccination of Small Pox actually worked in producing an ameliorated disease then, of course, its proponents wd have a case for advocating it. However, even Jenner, the lead advocate, witnessed instances where the vaccination sickness was by no means mild.

"Severe symptoms are not limited to milkers casually infected from the cow. Occasionally, intentional inoculation of fresh virus from the cow reproduces the disease without mitigation." - p 364

"Estlin's lymph was employed on sixty-eight children by Messrs. Michell and Prankard, of Langport in Somersetshire, and the results which they reported to Estlin were :—

"In 52 the disease was regular.
    1 Severe erysipelas.
    4 Erythmatous eruptions of a violent character.
    2 Highly inflamed ulcerated arms.

1 No effect after twice vaccinating.
8 Result unknown ; supposed to have been favorable.

"In one of the patients, two months old, erythema appeared on the back, and gradually extended to the feet. The child had much dyspnœa, with croupy cough, and died on the 21st. Mr. Estlin's correspondent wrote :—

""I do not attribute its death to vaccination, nor does the mother wholly, as she lost an infant previously with a similar affection of the air passages, but her neighbors set it down to vaccination entirely."" - pp 368-369

The matter-of-fact way in wch these experiments on children are described by the perpetrators chooses to ignore the trauma that the children must've experienced in the process.

"Mr. Loy, being curious to ascertain whether this disease could be communicated by inoculation, took a quantity of matter from the pustules of this patient and inserted it into the arm of his brother, with the following results :—

""In a few days, some degree of inflammation appeared, and on the eighth day, a vesicle formed ; my patient had now some slight feverish symptoms, which continued a day or two.

""This disease had exactly the appearances of the genuine Cow Pox, and I intended to have tried the effect of the Small Pox virus, had not the fears of the boy's parents prevented me."" - p 381

It's nice to know that the parents intervened on behalf of their children every rare once in a while. Unfortunately, these experimenters never seemed to run out of subjects.

"Dr. John Loy then inoculated a child direct from a horse suffering from grease.

""On the third day, a small degree of inflammation surrounded the wound. On the fourth, the inoculated place was much elevated, and a vesicle, of a purple colour, was formed on the fifth day : on the sixth and seventh, the vesicle increased, and the inflammation extended, and became of a deeper colour ; on the same day, a chilliness came on, attended with nausea and some vomiting. These were soon succeeded by increased heat, pain in the head, and a frequency of breathing ; the pulse was very frequent, and the tongue was covered with a white crust. When in bed, the child was much disposed to sweat.[""]" - p 382

Wow! It sure was a good thing for medical science that Dr. Loy conducted this experiment! NOT.

"With regard to the application of the variolous test, all Loy's experiments were deprived of any value. No conclusions could be drawn when the inoculation was performed at or near the height of the disease, which had been produced by insertion of the virus of the grease."

[..]

"["]This discovery is the more curious and interesting as it places in a new point of view the traditionary account handed down to us by the Arabian physicians that the Small Pox was originally derived from the camel.["]" - p 384

Might as well throw the camel in there. I'm just glad no-one to my knowledge decided to take a disease from a camel & give it to a sheep & go from there to a goat, then to a horse, & next to a cow so that this new, improved!, disease cd be given to a child.. for their own good, of course. NOT.

"In Paris, according to Baron, equination was practised in 1812.

""A coachman who had not had Small Pox, and who dressed a horse affected with the *grease*, had a crop of pustules on his hands, which resembled the vaccine. Two children were inoculated from these pustules, and the genuine vaccine was excited in both : from this stock many successive inoculations were effected, all possessing the proper character. A similar series of inoculations took place from another infant who was infected from one of the scabs taken from the pustules on the hand of the coachman."" - pp 389-390

Lest you've managed to make it this far thru my review & somehow gotten the impression that in the chaos of all these experiments that opinions were unchanging & uniform, consider this:

"In an appendix to the second volume of Jenner's *Biography*, published in 1837, Baron made the following remark :—

""I take this opportunity of expressing my regret that I have employed the word *grease* in alluding to the disease in the horse. *L'ariolœ Equinœ* is the proper designation. It has no necessary connexion with the grease, though the disorders frequently co-exist. This circumstance at first misled Dr. Jenner, and it has caused much misapprehension and confusion."

"In 1840, Ceely remarked that there were farmers and others who had good reason for believing in the origin of Cow Pox from the equine vesicle, which he regarded as *eczema impetiginodes*." - pp 393-394

One of my main take-aways from reading this seemingly thorough look at the history of vaccination is that it's quite possible that it's the very people who're historicized as the heroes of disease prevention that're actually responsible for the spreading of disease in ways it wdn't ordinarily have w/o foolish & delusional human intervention.

"A student named Amyot dressed a horse on which an operation had been performed. The leg which had been operated on (right hind leg) became the seat of a very confluent eruption of Horse Pox"

[..]

"Amyot had a wound on the dorsal aspect of the first interphalangeal joint of the little finger of his right hand ; in spite of this, he continued to dress the horse entrusted to his care. The sore on his finger was the seat of an accidental inoculation with the virus which flowed in such great abundance from the horse's leg. The wound was made on the 3rd of August, and the next day it was swollen and rather painful. On the 5th, Amyot suffered from *malaise* and great weakness, on the 6th, 7th, and 8th, vesicles appeared successively on the fingers of his left hand, on his forehead, on a level with the root of his nose, and between the two eyebrows." - p 398

Now, of course, such sanitary precautions as *rubber gloves* are long-since basic - but what are we missing NOW that'll seem to be solved w/ something else basic in 150 yrs?

""On the 20th of May, 1880, a heifer six and a half months old, in excellent condition, belonging to M. Givelet, was inoculated with this vaccine. A great number of punctures were made around the vulva, and between the thighs, and on the right side of the udder, and on the teats. Several students revaccinated themselves, and on two of them vesicles formed. The inoculation of the heifer took perfectly, so that on the 26th of May each puncture was transformed into a flattened discoid vesicle, umbilicated in the centre, of a yellow-grey colour with an inflammatory areola, presenting, in one word, all the characters of the vesicles of Cow Pox. With the liquid contained in these vesicles I successfully vaccinated several children, and revaccinated some students, some of whom showed vaccinal vesicles. On the 26th of May, Dr. Salamon vaccinated children who had very fine vesicles.["]" - p 408

1st, they made a heifer sick; 2nd, they took the disease from the heifer & gave it to children & students. If you believe that this is somehow healthy then you overlook what, to me, is the obvious sickness of it.

"["]The time has come to apply the data acquired by science respecting the etiology of *dourine* or *maladie du coït*. I cannot do better here than reproduce a passage from a clinical lecture of Professor St. Cyr of the School of Lyons.["]"

[..]

"" ' That the true cause of the *maladie du coït*, when we know how to look for it, will be found in the *importation of a foreign stallion.*'

"" ' That *dourine* has only one known cause, *contagion* ; that all the others to which it has been thought possible to attribute it, are more than problematical" - p 410

[..]

""And M. St Cyr adds '*dourine* is not an autochthonous malady born from local influences, but is, on the contrary, an exotic illness which has been imported, and the origin of which it will always be possible to trace if the practitioner exercises in his etiological investigation the attention and the perspicacity which it requires.' [""]" - p 411

Once again, I feel that vaccinators, & humans in general, are left out of the explanation of contagion's origins. It was *humans* that were infecting some of these animals & it was *humans* who were moving the animals around outside of their normal migratory patterns.

"[""]I examined these animals, I found on only two of them—the asses *Aramis* and *Mexico*—the characteristic eruption of Horse Pox. However, in the first place, Mare No. 3, which had been served by the ass *Mistigry*, on which I did not observe any eruption of Horse Pox, exhibited a splendid vaccinal eruption which had been developed on the under surface of the tail, where I collected crusts, inoculation from which proved, as we have seen, an excellent source of vaccine[""]" - p 412

""Must we in this connection, admit with M. Lafosse, that the infectious agent does not exist before the coitus, that it is formed during the accomplishment of the act of copulation, doubtless at the expense of the male or female secretions, perhaps of both, and under the nervous influx or force which is accumulated in the genital organs by the friction of copulation?

""But I do not see on what principle or on what scientific ground this theory rests, and therefore it is quite useless to pause longer over it.[""]" - pp 412-413

It's odd, in my reviewers's note to self I wrote "Maybe there's something to this" but rereading it I don't know what I was thinking or even if I understand what's written. Does it mean that copulation creates the infectious agent? I can imagine that the friction might lacerate the skin & that biological components of both animals might combine to make something harmful but I'm not quite sure that that's what's being gotten at here.

"[""]the Veterinary School of Toulouse where, in conjunction with M. Cadeac, teacher of the Clinique, I vaccinated two fine Dutch heifers vigorous and in good health, one aged fourteen months, the other seven months. Five days later there were as many vesicles as there had been punctures, and Drs. Armieux, Jougla, Caubet, an Parant, invited by the Director of the Veterinary School to visit the vaccinated heifers, proved the perfect genuineness of the vaccinal eruptions, which had been produced on the perinæum and on the teats.

""This eruption has been the starting point of cultures of vaccine on heifers and calves up to the end of last May, and the vaccine thus kept up, has been used for vaccination of about fifteen hundred persons."" - p 418

How many pro-vaccination people understand that the practice originates w/ drs deliberately infecting domestic animals w/ disease & then taking the disease from those animals & transferring it to humans? If you accept the basic premise as valid then it might seem ok, if you don't, as I don't, it might seem insane & dangerous & causing an increase & strengthening of disease instead of its subjugation.

"From Vienna, lymph was conveyed by Dr. Peschier to Geneva ; but there the new method received a temporary check, for all the persons who were vaccinated, afterwards contracted Small Pox, some by infection, and others by inoculation. These untoward results were explained as the result of "spurious vaccination," and the practice was not therefore abandoned." - p 420

There's always an excuse & a cover story but the bottom line is that Dr. Peschier caused a Small Pox epidemic in Geneva.

"In India, the new method was opposed by the natives, but their objections were overcome by an ingenious device.

""In order to overcome their prejudices, the late Mr. Ellsped Madras, who was well versed in Sanskrit literature, actually composed a short poem in that language on the subject of vaccination. This poem was inscribed on old paper, and said to have been *found*, that the impression of antiquity might assist the effect intended to be produced on the minds of the Brahmins, while tracing the preventative to their sacred cows."" - p 423

In other words, they used lies & trickery. Keep in mind that it was the people of India who had enuf sense to not use the contaminated milk - unlike the dairies of England. Keep also in mind that this was not long before the Opium Wars & that such an imposition of vaccination can be seen as an indicator of a similar lack of ethics.

"The **Opium Wars** (simplified Chinese: 鸦片战争; traditional Chinese: 鴉片戰爭) were two wars waged between China and European powers in the mid-19th century. The First Opium War, was fought in 1839–1842 between China and Great Britain. Opium was illegal in China but a profitable trade good for Britain which initiated war to keep the trade flowing. The Second Opium War was fought between the China and Britain and France, 1856–1860. In each war, the European force's modern military technology led to easy victory over the Chinese forces, with the consequence that its government was compelled to sign unequal treaties to grant favourable tariffs, trade concessions, reparations and territory to the Europeans." - https://en.wikipedia.org/wiki/Opium_Wars

"The principle embodied in the practice of Small Pox inoculation was, the widespread belief that in certain diseases, a mild attack would, as a rule, ward off, or modify, a second attack. Now in the case of Cow Pox there was this initial difficulty, that it was a disease totally distinct from Small Pox. As was soon pointed out, Cow Pox and Small Pox are radically dissimilar. That Jenner foresaw this difficulty, and endeavored to meet it by the invention of the term *variolœ vaccinœ*, or *Small Pox* of the cow, is not at all unlikely ; but whether there was motive or not, the designation of Cow Pox as *variolœ vaccinœ* had, without doubt, a very great effect in rendering the new inoculation acceptable on the Continent." - p 424

"Cow Pox inoculation was introduced into America on the strength of one doubtul experiment, and as on the Continent, under the impression that it was variolæ vaccinæ or Small Pox of the Cow.

"Thus were the scientists in Europe and America deceived. They were led to believe that this English disease was commonly known as Cow Small Pox, whereas it was Jenner who first named it Cow Small Pox. It was really known in England as "the Pox among Cows," or the "Cow Pox."" - p 429

Observant readers will note the variant spellings: *variolœ vaccinœ* & variolæ vaccinæ. I'm trying to stay faithful to the original texts: "œ" & "æ" are both used - although it's possible that in the italicized version the "œ" is really "æ" but doesn't look that way.

"The new inoculation was shortly afterwards tested on a large scale. A Dr. S. obtained lymph from a sailor, who had arrived at Marblehead from London, and was supposed to be suffering from Cow Pox, but in reality had Small Pox. Dr. S. began to use it, and produced an epidemic of Small Pox. Previous to this accident Dr. D. had inoculated about forty persons from the arm of Dr. Waterhouses's son, and all who had been vaccinated took the Small Pox, either casually or by inoculation, one excepted." - p 427

My ongoing point is that the practice of vaccination was established thru trickery, double-talk, lies, etc, & motivated by a desire for fame & wealth. As such, as far as I can tell, nothing substantial has changed to this day & there's still a strong possibility that the medical industry has had a hand, very disguised by faked history, in creating large-scale health problems.

Near the end of History and Pathology of Vaccination - Volume 1, Crookshank recapitulates his criticisms of Jenner who many people consider to be the main culprit in connection w/ establishing vaccination's undeserved good reputation.

"In the first place, the statement which has been recently made, that Jenner believed that Cow Pox was derived from human Small Pox, and hence the term *variolœ vaccinœ* was justifiable, is entirely without foundation. The facts of the case are that Jenner believed that the Cow Pox was derived from the diseased heels of the horse ; he also believed that Small Pox and some other diseases arose from the same source. When the boy Phipps was inoculated with Cow Pox, Jenner was struck with the similarity to some cases of inoculated Small Pox, and he felt convinced that, at least, Cow Pox and Small Pox were derived from the same source. The idea that Cow Pox arose through the agency of milkers suffering from human Small Pox never occurred to Jenner." - p 431

At any rate, vaccination's promise of effectiveness has been disproved over & over again. In this day & age that just means that more & more booster shots are encouraged. Imagine buying a car that you expect to work & then being told that you have to buy a new motor for it once a yr, then once every 6 mnths, then once every 3 mnths in order for it to actually work - only to not have it work no matter how many new motors you buy for it.

"["]At length, about nine years ago, all doubt from my mind was removed, in consequence of my having had ocular and very distinct evidence of **perfect vaccination having failed to produce the promised security.**""

"Dr. Monro not only made his own observations, but he corresponded with other members of the profession. Mr. Cooper informed him that "cases of Small Pox after Cow Pox are now daily occurrences."

"The statements made by Dr. Alexander Ramsay, of Dundee, were still more striking."

[..]

"["]to invalidate the evidence of **Small Pox in its perfect form having succeeded to vaccination in its perfect form.**"" - p 436

"Dr. Smith of Dunse, 2nd June, 1818. Dr. Smith wrote :—

""I had, indeed, seen several cases of Small Pox supervening upon vaccination, which I meantioned at the time to Dr. Farquharson ; but as he seemed to think lightly of them, I judged it prudent to take no further notice of the circumstance. Even now, though I have seen a multitude of cases in which Small Pox has, in every possible shape, taken place after vaccination, I feel myself placed in the painful situation of bringing forth many facts to which gentlemen of the first eminence in the profession will probably give little or no credit. . . ." - p 444

"In fact an alteration in the quality of the lymph had now become one of the stock apologetics for Cow Pox failures, and the profession was still persuaded to believe in "that most precious boon of Jenner to a suffering world." According to Badcock, similar experiences were met with abroad. Out of 547,646 vaccinated, 11,773 were attacked with Small Pox. 1,294 became disfigured or infirm, and 1,379 died in consequence of the disease." - p 450

"Creighton has pointed out how closely the inoculated syphilis runs parallel with the natural Cow Pox" - p 462

Of course, who cd forget the infamous Tuskagee experiment:

"The Tuskegee experiment began in 1932, at a time when there was no known treatment for syphilis, a contagious venereal disease. After being recruited by the promise of free medical care, 600 African American men in Macon County, Alabama were enrolled in the project, which aimed to study the full progression of the disease.

"The participants were primarily sharecroppers, and many had never before visited a doctor. Doctors from the U.S. Public Health Service (PHS), which was running the study, informed the participants—399 men with latent syphilis and a control group of 201 others who were free of the disease—they were being treated for bad blood, a term commonly used in the area at the time to refer to a variety of ailments.

"The men were monitored by health workers but only given placebos such as aspirin and mineral supplements, despite the fact that penicillin became the recommended treatment for syphilis in 1947, some 15 years into the study. PHS researchers convinced local physicians in Macon County not to treat the participants, and instead research was done at the Tuskegee Institute. (Now called Tuskegee University, the school was founded in 1881 with Booker T. Washington at its first teacher.)

"In order to track the disease's full progression, researchers provided no effective care as the men died, went blind or insane or experienced other severe health problems due to their untreated syphilis.

"In the mid-1960s, a PHS venereal disease investigator in San Francisco named Peter Buxton found out about the Tuskegee study and expressed his concerns to his superiors that it was unethical. In response, PHS officials formed a committee to review the study but ultimately opted to continue it—with the goal of tracking the participants until all had died, autopsies were performed and the project data could be analyzed.

"Buxton then leaked the story to a reporter friend, who passed it on to a fellow reporter, Jean Heller of the Associated Press. Heller broke the story in July 1972, prompting public outrage and forcing the study to finally shut down.

"By that time, 28 participants had perished from syphilis, 100 more had passed away from related complications, at least 40 spouses had been diagnosed with it and the disease had been passed to 19 children at birth."

- https://www.history.com/news/the-infamous-40-year-tuskegee-study

That 'study' went on for 40 yrs & the drs were from the Public Health Service. They, too, were ostensibly working for the public good, just as vaccinators continue to claim to do.

"As a result of an investigation into the history and especially the pathology, of "vaccination," I feel convinced that the profession has been misled by Jenner,

Baron, the Reports of the National Vaccine Establishment, and by a want of knowledge concerning the nature of Cow Pox, Horse Pox, and other sources of "vaccine lymph." Though in this country, vaccine lymph is generally taken to mean the virus of Cow Pox, yet the pathology of this disease, and its nature and affinities, have not been made the subject of practical study for nearly half a century. We have submitted instead to purely theoretical teaching, and have been led to regard *vaccination* as inoculation of the human subject with the virus of *a benign disease of the cow*, whereas the viruses in use have been derived from several distinct and severe diseases in different animals." - p 463

"Variolation, though a dangerous practice, can at least lay claim to be based upon scientific grounds, viz., the prevention or modification of a disease by artificially inducing a mild attack of that disease. Jenner's substitution of Cow Pox inoculation was a purely empirical treatment based upon folklore, and involved a totally different pathological principle—the protection from one disease by the artificial induction of a totally distinct disease—a principle which was not, and has not been since, supported by either clinical experience or pathological experiments."

[..]

"Inoculation of Cow Pox does not have the least effect in affording immunity from the analogous disease in man, syphilis, and neither do Cow Pox, Horse Pox, Sheep Pox, Cattle Plague, or any other radically dissimilar disease, exercise any specific protective power against Human Small Pox. Inoculation of Cow Pox, Horse Pox, and Cattle Plague have totally failed to exterminate Small Pox" - p 464

Crookshank is ultimately in favor of what he calls "Variolation" & in quarantine. I'm not particularly in favor of either but might under very limited circumstances find quarantine reasonable. Nonetheless, I found this bk to collate together a remarkable body of compelling evidence against much of what the medical industry forces upon the public.

"There can be no doubt that ere long a system of COMPULSORY NOTIFICATION and ISOLATION will replace vaccination. Indeed, I maintain that where isolation and vaccination have been carried out in the face of an epidemic, it is isolation which has been instrumental in staying the outbreak, though vaccinating has received the credit." - p 465

"It is more probable that when, by means of notification and isolation, Small Pox is kept under control, vaccination will disappear from practice, and will retain only an historical interest." - p 466

Unfortunately, despite all of Crookshank's perspicacity & scholarliness, he was unprepared to anticipate an era in wch all forces wd be brought to bear against a PLANDEMIC in order to NOT control the disease but to control, enslave, & further profit from the public.

www.ingramcontent.com/pod-product-compliance
Lightning Source LLC
Chambersburg PA
CBHW052107020426
42335CB00021B/2671